# Liquid Chromatography in Analysis of Bioactive Compounds for Pharmaceuticals, Cosmetics, and Functional Food Interest

# Liquid Chromatography in Analysis of Bioactive Compounds for Pharmaceuticals, Cosmetics, and Functional Food Interest

Editors

**Jan Oszmianski**
**Sabina Lachowicz-Wiśniewska**

MDPI

Basel • Beijing • Wuhan • Barcelona • Belgrade • Novi Sad • Cluj • Manchester

*Editors*

Jan Oszmianski
Department of Fruit, Vegetables and Nutraceuticals Technology
Wroclaw University of Environmental and Life Sciences
Wrocław, Poland

Sabina Lachowicz-Wiśniewska
Demartment of Health Science
Calisia University
Kalisz, Poland

*Editorial Office*

MDPI
St. Alban-Anlage 66
4052 Basel, Switzerland

This is a reprint of articles from the Special Issue published online in the open access journal *Pharmaceuticals* (ISSN 1424-8247) (available at: https://www.mdpi.com/journal/pharmaceuticals/special_issues/LC_Nature_Products).

For citation purposes, cite each article independently as indicated on the article page online and as indicated below:

Lastname, A.A.; Lastname, B.B. Article Title. *Journal Name* **Year**, *Volume Number*, Page Range.

**ISBN 978-3-0365-9229-9 (Hbk)**
**ISBN 978-3-0365-9228-2 (PDF)**
**doi.org/10.3390/books978-3-0365-9228-2**

# Contents

# About the Editors

**Jan Oszmianski**

Jan Oszmianski, Professor Dr., Retired Professor, completed his Ph.D. degree at Agricultural Academy August Cieszkowski in Poznan in the area of enzymatic transformations of phenolic compounds in model systems and fruit extracts and subsequently carried out postdoctoral research at the Wroclaw University of Environmental and Life Sciences in the area of polyphenolic compounds and functional food. His main research interest is in the phenolic compounds, isolation, and qualitative and quantitative determination of polyphenols and chromatography techniques such as UV-VIS and HPLC-MS-MS. He is a member of the Committee of Food Sciences and Nutrition of the Polish Academy of Sciences.

**Sabina Lachowicz-Wiśniewska**

Sabina Lachowicz-Wiśniewska, Associate Professor, received her Doctor Habilitation (DSc) in Health Science at the Institute of Rural Health in the area of natural products and dietetics. She completed her PhD degree at the Wroclaw University of Environmental and Life Sciences under the direction of Professor Jan Oszmianski in the area of natural product technology and chemistry. Nowadays, she carries out research at Calisia University in the Kalisz Microbiota Research Team in the area of prebiotic functional food and the bioavailability of their bioactive compounds for the prevention of oxidative stress and inflammation. Her past research is in the general area of functional food and pharmaceutical and medicinal plant chemistry, with over 74 peer-reviewed scientific papers covering research fields such as (i) the production of innovative functional food designed to have health-promoting properties; (ii) the bioavailability and digestibility of bioactive compounds in the simulated digestive system by the in vitro method; (iii) the determination of antioxidant, anti-diabetic, anti-obesity, and anti-inflammatory potential; and (iv) the identification and assessment of the health-promoting properties of bioactive compounds from plant materials based on chromatographic techniques. Research collaborations have been established within the Dekaban Foundation with Prof. Anubhav Pratap Singh (Faculty of Land and Food Systems (LFS), University of British Columbia), Prof. Antonio J. Melendez-Martin, the Food Colour and Quality Laboratory, Universidad de Sevilla, and also with many European research centers.

# Preface

Plants produce a wide range of phytochemicals, known as non-nutrition compounds, in order to confirm their identity, and these are used for the production of natural pharmaceuticals, cosmetics, and functional food, among other things, due to their pro-health properties. The use of modern chromatographic techniques allows for accurate quantitative and qualitative identification of the above-mentioned phytochemicals and their natural products. Liquid chromatography is one of the most efficient and robust specific techniques due to the merits of convenience and strong separation ability, as well as a wide range of material applications for identification. Liquid chromatography is widely used to analyze plants, nutraceuticals, pharmaceuticals, and cosmetics and in natural product quality control or quantitative determination of bioactive compounds. The most commonly used for the identification of different plant materials and pharmaceuticals are ultra- and high-performance liquid chromatography with UV-VIS, fluorescence, and diode array, which is equipped with mass spectrometry or tandem mass spectrometry detection methods. Therefore, for this Special Issue, we published studies concerning the latest scientific news, insights, and advances in the field of innovation of liquid chromatography for the analysis of bioactive compounds for pharmaceuticals, cosmetics, and functional food interests. The information presented certainly garners considerable interest among many readers from different disciplines and research fields.

**Jan Oszmianski and Sabina Lachowicz-Wiśniewska**
*Editors*

*pharmaceuticals*

*Article*

# Optimisation of Solid-Phase Extraction and LC-MS/MS Analysis of Six Breast Cancer Drugs in Patient Plasma Samples

**Lu Turković [1], Dragana Mutavdžić Pavlović [2], Zvonimir Mlinarić [1], Anamarija Skenderović [3], Tajana Silovski [4,5] and Miranda Sertić [1,*]**

1 Department of Pharmaceutical Analysis, Faculty of Pharmacy and Biochemistry, University of Zagreb, Ante Kovacica 1, 10000 Zagreb, Croatia; zvonimir.mlinaric@pharma.unizg.hr (Z.M.)
2 Department of Analytical Chemistry, Faculty of Chemical Engineering and Technology, University of Zagreb, Marulicev trg 20, 10000 Zagreb, Croatia; dmutavdz@fkit.hr
3 GxR&D Analytics Zagreb, Global R&D, Teva Pharmaceuticals, Prilaz Baruna Filipovica 25, 10000 Zagreb, Croatia; anamarija.skenderovic@pliva.com
4 Department of Oncology, University Hospital Centre Zagreb, Kispaticeva 12, 10000 Zagreb, Croatia; tsilovsk@kbc-zagreb.hr
5 School of Medicine, University of Zagreb, Salata 3, 10000 Zagreb, Croatia
* Correspondence: miranda.sertic@pharma.unizg.hr

**Abstract:** In the development of bioanalytical LC-MS methods for the determination of drugs in plasma samples in a clinical setting, adequate sample preparation is of utmost importance. The main goals are to achieve the selective extraction of the analytes of interest and attain thorough matrix removal while retaining acceptable ecological properties, cost-effectiveness, and high throughput. Solid-phase extraction (SPE) offers a versatile range of options, from the selection of an appropriate sorbent to the optimisation of the washing and elution conditions. In this work, the first SPE method for the simultaneous extraction of six anticancer drugs used in novel therapeutic combinations for advanced breast cancer treatment—palbociclib, ribociclib, abemaciclib, anastrozole, letrozole, and fulvestrant—was developed. The following sorbent chemistries were tested: octylsilyl (C8), octadecylsilyl (C18), hydrophilic–lipophilic balance (HLB), mixed-mode cation-exchange (MCX and X-C), and mixed-mode weak cation-exchange (WCX), with different corresponding elution solvents. The samples were analysed using LC-MS/MS, with a phenyl column (150 × 4.6 mm, 2.5 μm). The best extraction recoveries (≥92.3%) of all analytes were obtained with the C8 phase, using methanol as the elution solvent. The optimised method was validated in the clinically relevant ranges, showing adequate precision (inter-day RSD ≤ 14.3%) and accuracy (inter-day bias −12.7–13.5%). Finally, its applicability was successfully proven by the analysis of samples from breast cancer patients.

**Keywords:** solid-phase extraction; breast cancer; CDK4/6 inhibitors; therapeutic drug monitoring; liquid chromatography; mass spectrometry; bioanalysis; patient plasma

**Citation:** Turković, L.; Mutavdžić Pavlović, D.; Mlinarić, Z.; Skenderović, A.; Silovski, T.; Sertić, M. Optimisation of Solid-Phase Extraction and LC-MS/MS Analysis of Six Breast Cancer Drugs in Patient Plasma Samples. *Pharmaceuticals* **2023**, *16*, 1445. https://doi.org/10.3390/ph16101445

Academic Editors: Jan Oszmianski and Sabina Lachowicz-Wiśniewska

Received: 18 September 2023
Revised: 7 October 2023
Accepted: 9 October 2023
Published: 12 October 2023



## 1. Introduction

Palbociclib (PAL), ribociclib (RIB), and abemaciclib (ABE) are newly registered anticancer drugs, inhibitors of the cyclin D-dependent kinases 4 and 6 (CDKi). They are prescribed for the treatment of hormone receptor-positive and human epidermal growth factor 2-negative (HR+, HER2−) breast cancer, in synergistic combinations with endocrine therapy (ET): aromatase inhibitors anastrozole (ANA) or letrozole (LET), or an oestrogen receptor antagonist fulvestrant (FUL). It has been shown that, of all HR+, HER2− patients in recent years (2018–2022), about 70–80% were prescribed with CDKi + ET, while chemotherapy was used in about 15% and ET alone in about 10% of the cases [1]. Therapeutic drug monitoring (TDM) of anticancer drugs has great potential in reducing adverse events and improving treatment outcomes for individual patients [2]. The development of an efficient, cost-effective, and ecologically favourable analytical method for the determination of drugs in biological fluids is the first prerequisite for TDM. Simultaneous determination of several drugs with the same analytical conditions would offer better sample turnaround and higher laboratory efficiency, which leads to a faster and more convenient delivery of the results, aiding the subsequent medical decision-making process, as well as having a favourable ecological impact [3].

The six drugs of interest have a wide array of physical-chemical properties, with the CDKi being weak bases, aromatase inhibitors being neutral and relatively hydrophilic, and FUL being highly lipophilic [4,5]. Thus, simultaneous extraction of all six drugs using the same extraction conditions may be challenging.

Solid-phase extraction (SPE) is a popular option for the preparation of various types of samples, including human plasma—the most commonly used sample in TDM. Its selectivity, purification efficacy, versatility, and ease of automation are among the most prominent advantages that make it suitable for complex bioanalytical sample preparation [6].

The most common commercially available SPE sorbents are either silica- or polymer-based, with different functional groups and retention mechanisms, such as reversed-phase sorbents similar to chromatographic stationary phases—C18, C8, phenyl, etc.; polymeric hydrophilic–lipophilic balance (HLB) sorbents, comprised of polar pyrrolidone and non-polar divinylbenzene; and ion-exchange sorbents containing ionisable groups, such as the sulfonic and carboxylic groups for cation exchange, or amines for anion exchange [7–9]. To achieve the simultaneous extraction of both ionisable and non-ionisable analytes with several interaction mechanisms, mixed-mode sorbents containing ion-exchange groups together with the HLB backbone have been introduced. Apart from the choice of sorbent structure, different elution conditions can further fine-tune the SPE performance, while various washing conditions can enhance sample purification and matrix removal [10].

Thus far, several bioanalytical methods for the determination of ABE, PAL, RIB, ANA, LET, and FUL in human plasma samples have been reported. Most of the methods were developed only for selected combinations of the analytes, with three works focused on all six drugs of interest, applying protein precipitation (PPT) and dispersive liquid–liquid microextraction (DLLME) for sample preparation [5,11,12]. For specific analyte combinations, the most common sample preparation procedure applied was PPT with an organic solvent [13–20], liquid–liquid extraction (LLE) [21,22], and SPE [14,23–25]. Of the SPE methods applied for some of the analytes of interest, an Oasis HLB sorbent in combination with methanol (MeOH) as the elution solvent was found suitable for the CDKi [24,25], as well as a Phenomenex C18 sorbent eluted with acetonitrile (ACN) following PPT [14]. Aromatase inhibitors were, on the other hand, successfully extracted using a mixed-mode weak cation-exchange sorbent, Strata X-C, eluted with 5% $NH_4OH$ in MeOH [23]. The structural formulae, as well as important physical-chemical descriptors of the six analytes, are shown in Table 1 [26,27]. Functional groups with the potential for certain intermolecular interactions, which may be important for the retention of different sorbents, are tentatively highlighted. Additional van der Waals forces stemming from dipoles and induced dipoles have not been specifically highlighted but are assumed to be present as well.

To the best of our knowledge, there is no reported protocol for SPE of FUL or a combination of the CDKi with the aromatase inhibitors or FUL from any type of sample thus far. Therefore, the aim of this work was to achieve simultaneous extraction and cleanup with satisfactory extraction yields of all six analytes of interest on a single SPE sorbent.

**Table 1.** Physical-chemical properties and structural formulae of the drugs of interest with highlighted pKa values (corresponding to the proton denoted in red), and functional groups for potential intermolecular interactions: π-π interactions (purple), hydrophobic interactions (green), and hydrogen bonds (blue).

| **Analyte** | **RIB** | **PAL** |
|---|---|---|
| **Structure, interactions, pKa** | 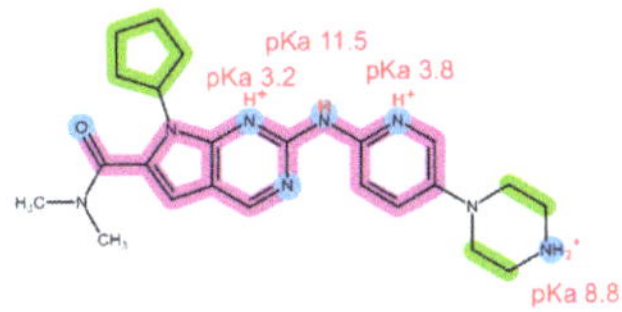  | 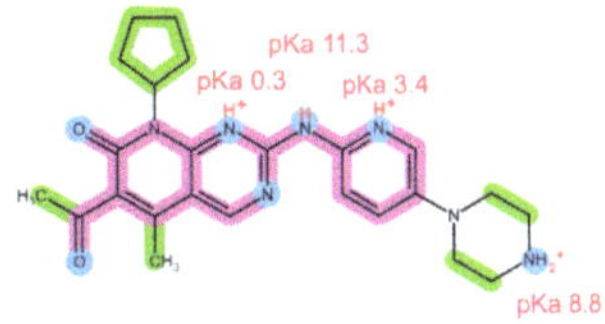  |
| **Emp. formula** | $C_{23}H_{30}N_8O$ | $C_{24}H_{29}N_7O_2$ |
| **Molar mass** | 434.5 g/mol | 447.5 g/mol |
| **logP** | 2.04 | 2.39 |
| **HB donors** | 2 | 2 |
| **HB acceptors** | 5 | 6 |
| **TPSA** | 91.21 Å$^2$ | 105.04 Å$^2$ |
| **Rotatable bonds** | 5 | 5 |
| **Analyte** | **ABE** | **FUL** |
| **Structure, interactions, pKa** | 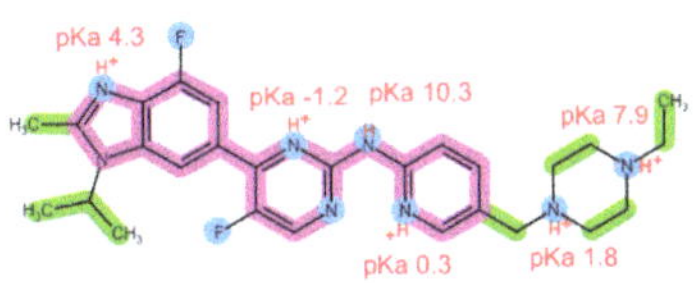  | 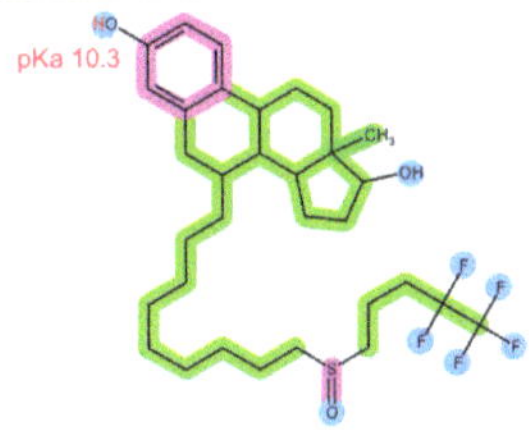  |
| **Emp. formula** | $C_{27}H_{32}F_2N_8$ | $C_{32}H_{47}F_5O_3S$ |
| **Molar mass** | 506.6 g/mol | 606.8 g/mol |
| **logP** | 4.04 | 8.06 |
| **HB donors** | 1 | 2 |
| **HB acceptors** | 8 | 8 |
| **TPSA** | 75.00 Å$^2$ | 76.74 Å$^2$ |
| **Rotatable bonds** | 7 | 14 |
| **Analyte** | **ANA** | **LET** |
| **Structure, interactions, pKa** | 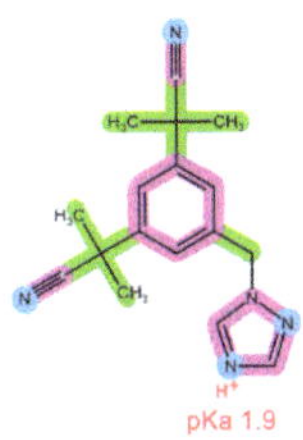  | 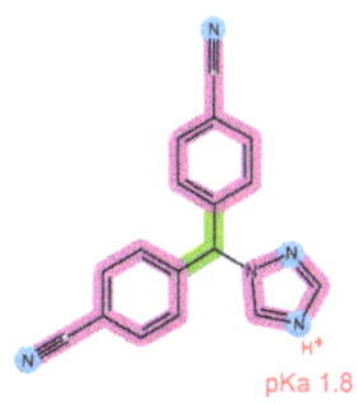  |
| **Emp. formula** | $C_{17}H_{19}N_5$ | $C_{17}H_{11}N_5$ |
| **Molar mass** | 293.4 g/mol | 285.3 g/mol |
| **logP** | 2.35 | 2.32 |
| **HB donors** | 0 | 0 |
| **HB acceptors** | 4 | 4 |
| **TPSA** | 78.29 Å$^2$ | 78.29 Å$^2$ |
| **Rotatable bonds** | 4 | 3 |

Abbreviations: Emp.: empirical; HB: hydrogen bond; TPSA: total polar surface area.

## 2. Results and Discussion

### *2.1. Optimisation of the SPE Procedure*

Several different SPE sorbents were tested: reversed-phase: octylsilyl (C8, Waters Sep-Pak Vac), octadecylsilyl (C18, Waters Sep-Pak Vac), and hydrophilic–lipophilic balance (HLB, Waters Oasis); ion-exchange sorbents: mixed-mode cation-exchange (MCX, Waters Oasis, and X-C, Phenomenex Strata), and weak cation-exchange (WCX, Waters Oasis).

Reversed-phase sorbents were selected as the most widely applicable sorbents suitable for various types of compounds. Cation-exchange and mixed-mode cation-exchange sorbents were tested due to the basic nature of some of the analytes. Some of these sorbents have already been applied in the literature, but only for the extraction of up to two of the mutually similar drugs of interest.

In this phase, all samples were analysed using high-performance liquid chromatography equipped with a diode array and fluorescence detection (HPLC-DAD-FLD). This technique was found suitable for the preliminary experiments due to its simplicity, relatively low cost of operation, and the ability to observe the chromatographic profile of the leftover plasma interferences, thus enabling an easier and broader evaluation of the sample cleanup efficiency.

The extraction recoveries were calculated as the ratio of peak areas in spiked plasma samples and standard solutions of the same concentration and are shown in figures as the median and range. Two spiked samples and one blank plasma sample were prepared for each tested condition. Blank sample chromatograms were inspected for matrix interferences at the retention times of the analytes.

#### 2.1.1. Reversed-Phase Sorbents

The structures of the tested reversed-phase and HLB sorbents, with the tentatively highlighted groups responsible for potential intermolecular interactions, are shown in Table 2 [7].

**Table 2.** Structural formulae of the tested reversed-phase sorbents with highlighted functional groups and pKa for potential intermolecular interactions: π-π interactions (purple), hydrophobic interactions (green), hydrogen bonds (blue), and dipole-based interactions (yellow).

| Sorbent | Sep-Pak Vac C18 | Sep-Pak Vac C8 | Oasis HLB |
|---|---|---|---|
| Structure | 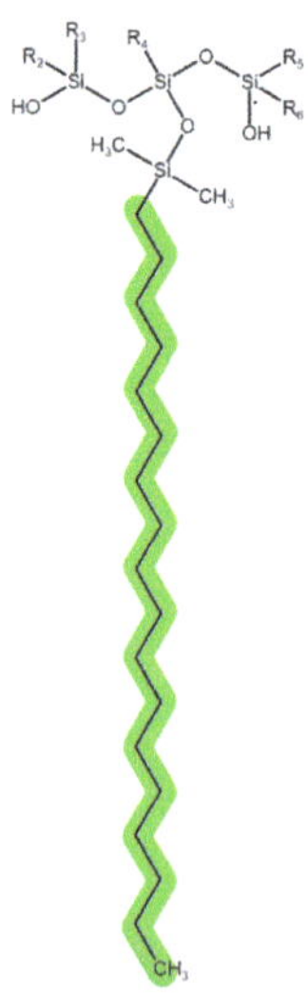 | 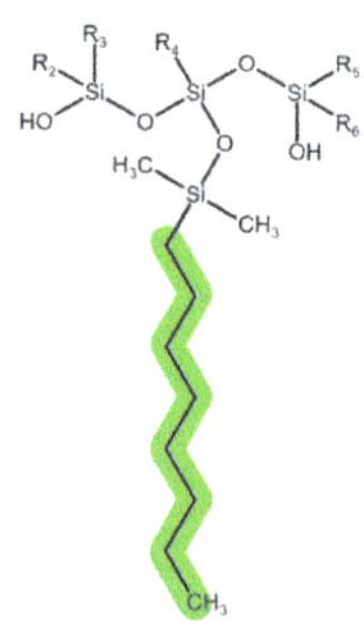 | 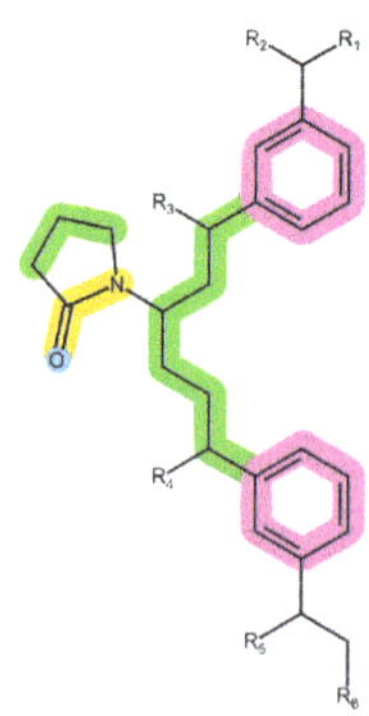 |

For the experiments, plasma samples were diluted with Milli-Q water in the volume ratio of 4:5. The sorbent-conditioning step included loading with MeOH and Milli-Q water, as per the manufacturer's instructions [7]. After the application of the sample, a washing step with Milli-Q water and 5% MeOH in water was performed in order to remove the unretained matrix interferences, such as plasma proteins. Different solvents were tested for analyte elution: MeOH and ACN, as organic solvents with different elution strengths, as well as acidified and alkalised MeOH, to test the effect of the pH. Varying the volumes of MeOH was also assessed for the C8 sorbent. All the conditions are summarised in Table 3.

**Table 3.** The SPE procedures used for the reversed-phase sorbents.

| **Experiment** | **I.** | **II.** |
|---|---|---|
| **SPE cartridges** | Oasis HLB 60 mg/3 mL<br>Sep-Pak Vac C18 200 mg/3 mL<br>Sep-Pak Vac C8 200 mg/3 mL | Oasis HLB 60 mg/3 mL<br>Sep-Pak Vac C8 500 mg/3 mL |
| **1. Conditioning** | 2 mL MeOH<br>2 mL $H_2O$ | 2 mL MeOH<br>2 mL $H_2O$ |
| **2. Sample addition (400 μL diluted plasma sample)** | | |
| **3. Washing** | 1 mL $H_2O$<br>1 mL 5% MeOH | 1 mL $H_2O$<br>1 mL 5% MeOH |
| **4. Elution** | 2 × 750 μL MeOH<br>or 2 × 750 μL ACN<br>or 2 × 750 μL 5% $NH_4OH$ in MeOH<br>or 2 × 750 μL 2% HCOOH in MeOH | **HLB:**<br>750 μL MeOH and 750 μL 2% HCOOH in MeOH<br>or 750 μL 2% HCOOH in MeOH and 750 μL MeOH<br>**C8:**<br>2 × 250 μL MeOH<br>or 2 × 500 μL MeOH<br>or 2 × 750 μL MeOH |

The results are presented in Figure 1. As seen in Figure 1a, the highest recoveries on the highly hydrophobic C18 sorbent were obtained with MeOH and HCOOH in MeOH as elution solvents. This is also in accordance with our previous chromatographic experience: while using mobile phases without pH modifiers on a C18 column, increased retention of the weakly basic CDKi was observed, which was drastically reduced by the addition of HCOOH and their subsequent protonation. When comparing the SPE elution efficacy of ACN and MeOH, it can be concluded that the formation of hydrogen bonds with MeOH, a protic solvent, greatly increased analyte solubility, whereas the same effect could not be achieved with ACN, an aprotic solvent. Finally, the addition of ammonia to MeOH provided similar recoveries of the CDKi compared to MeOH alone since the degree of ionisation and, thus, the affinity of the CDKi for the sorbent was relatively similar, as seen from their pKa values (Table 1). A chromatogram of a blank plasma sample prepared using the C18 sorbent eluted with HCOOH in MeOH is shown in Supplementary Figure S1.

On the C8 sorbent, significantly higher recoveries of all the analytes for all tested conditions except for ACN were observed (Figure 1b). It is possible that FUL was too strongly retained in the structure of the C18 sorbent, whereas C8 offered a less lipophilic environment, which also proved favourable for the other analytes. The extracts obtained with MeOH as the eluent were both rich in the analytes of interest and free from the interfering matrix components (Supplementary Figure S2).

Increasing volumes of MeOH were also tested on the C8 sorbent to assess if greater recoveries could be obtained. As discernible from Figure 1c, lower MeOH volumes were already sufficient for ANA and LET, while for the CDKi and FUL, maximal volumes were necessary to achieve the highest recoveries and precision. This indicates relatively stronger interactions of the CDKi and FUL with the sorbent. In the case of the CDKi, this is probably related to their low degree of ionisation at neutral pH, which causes their lower distribution into the elution solvent. Increasing the volume of the solvent increases the possibility of hydrogen bond formation, shifting this balance. For FUL, a strong interaction with the sorbent may be based on its steric properties—the long alkyl chain easily becomes trapped between the C8 chains of the sorbent; however, a large enough volume of the elution solvent is able to sufficiently solubilise the molecules. ANA and LET, on the other hand, are smaller compounds, less abundant in saturated alkyl groups, and are, therefore, likely to be more easily eluted from the sorbent, even with lower volumes.

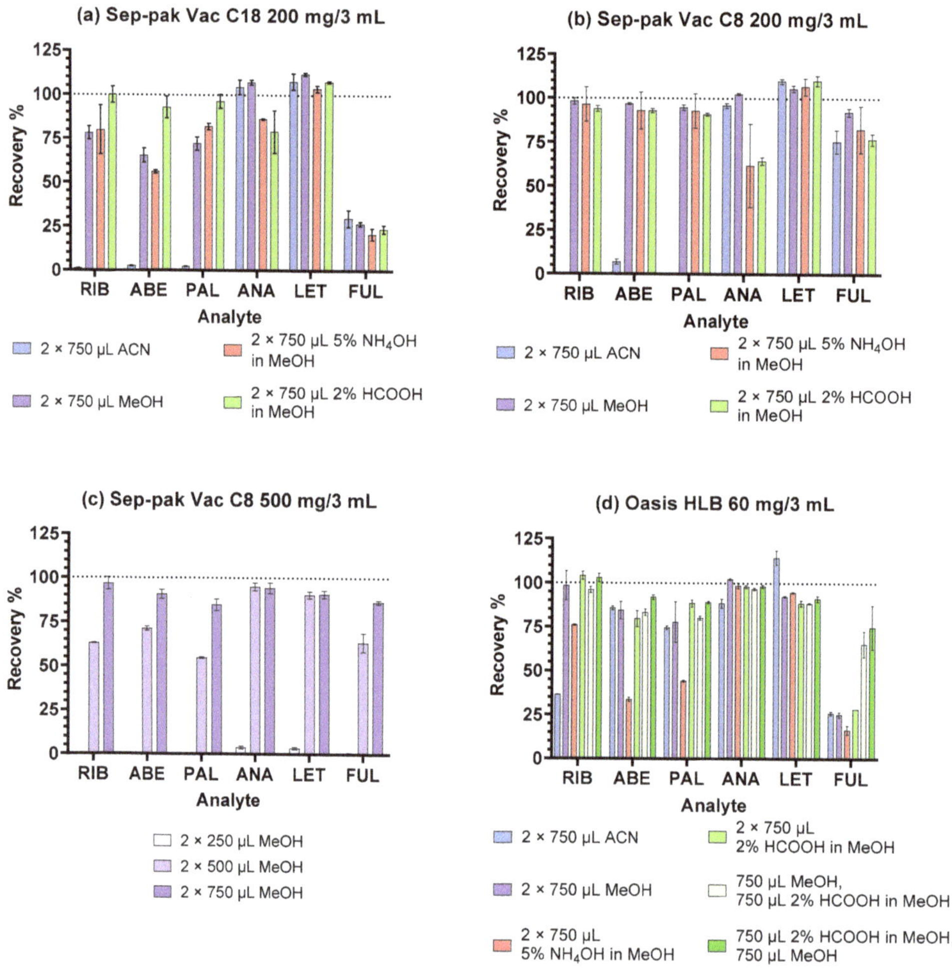


**Figure 1.** Extraction recoveries obtained with different eluents on reversed-phase sorbents.

In contrast to the silica-based C8 and C18 sorbents, Oasis HLB is a polymeric sorbent comprised of vinylpyrrolidone and divinylbenzene groups that enable both hydrophilic, dipole–dipole interactions and hydrophobic, especially $\pi$-$\pi$ interactions, with the compounds from the sample [9]. It is to be expected that the analytes with more polar or aromatic groups will be more strongly retained. The results are shown in Figure 1d. Similar to the C8 and C18 sorbents, the CDKi are more easily solubilised and eluted when in their ionised forms or in a protic solvent. Consequently, their diminished recoveries were observed with ACN and $NH_4OH$ in MeOH, whereas solvent combinations including HCOOH in MeOH and/or only MeOH, proved favourable, which is also supported by the literature [24,25]. ANA and LET are eluted with similar efficiency, regardless of the elution solvent composition, since they are neutral in the whole studied pH range; therefore, their solubilisation is not dependent on their ionisation. For FUL, retention may be based on steric positioning, allowing for the $\pi$-$\pi$, hydrophobic, and dipole–dipole interactions to take place. Although HCOOH provided improved recoveries, especially for FUL, a significant amount of matrix interferences was visible in the chromatograms (Supplementary Figure S3); therefore, an SPE sorbent yielding a more selective extraction may be more adequate.

2.1.2. Ion-Exchange Sorbents

As shown in Table 1, the CDKi are weak, basic compounds that predominantly exist in the cationic form below pH $\approx$ 9 [4]; therefore, different mixed-mode cation-exchange sorbents may be of interest. These sorbents are usually based on a polymeric backbone similar to HLB, thus offering additional retention mechanisms for ion exchange, which can benefit the non-ionised analytes, such as ANA, LET, and FUL. What is more, ANA and LET were successfully extracted using Strata X-C in the previously reported literature [23]. The structures of the ion-exchange sorbents used in this work are shown in Table 4 [7,28].

**Table 4.** Structural formulae of the tested ion-exchange sorbents with highlighted functional groups and pKa for potential intermolecular interactions: $\pi$-$\pi$ interactions (purple), hydrophobic interactions (green), hydrogen bonds (blue), electrostatic interactions (red), and dipole-based interactions (yellow).

| Sorbent | Oasis MCX | Oasis WCX | Strata X-C |
|---|---|---|---|
| Structure | 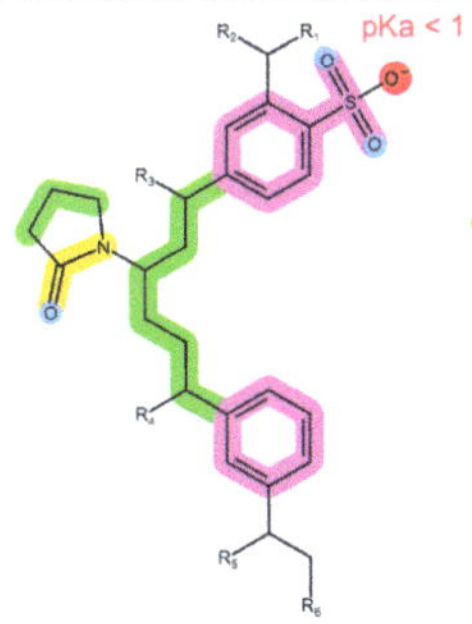  | 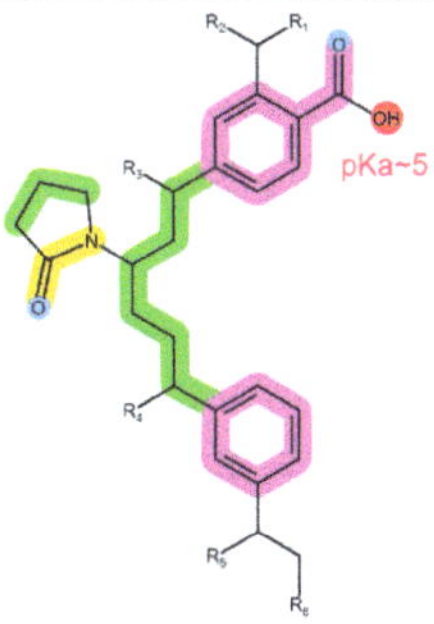  | 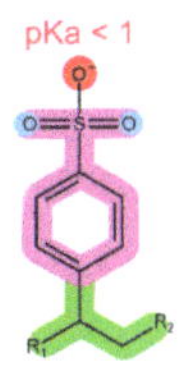  |

All tested conditions are listed in Table 5. To achieve analyte protonation, plasma samples were diluted with 2% $H_3PO_4$ and Milli-Q water at a ratio of 5:1:5 (experiments IV and V in Table 5) or with 100 mM sodium acetate (Na-Ac), pH 5.6, at a ratio of 4:5 (experiment VI) [29]. The washing step included flushing the sorbent with aqueous acid solutions, while MeOH with different pH modifiers was tested as the elution solvent.

The results are shown in Figure 2. Oasis MCX is a sorbent comprised of the same polymeric base as HLB, however, with the addition of sulfonic groups (pKa < 1) [8]. Thus, retention mechanisms of HLB are available together with the cation-exchanging properties. The retained cationic analytes can be eluted using an alkaline solution, shifting them to their non-ionic form [7]. The results, shown in Figure 2a, are in accordance with the expectations: no or very low recoveries of the CDKi and aromatase inhibitors were observed when applying neutral or acidic eluents, while high recoveries were obtained with $NH_4OH$ in MeOH. It is probable that the CDKi mostly entered electrostatic interactions with the sulfonic groups, while aromatase inhibitors and FUL also interacted with the divinylbenzene groups via the $\pi$-$\pi$ and hydrophobic interactions. Chromatograms of the samples eluted with $NH_4OH$ in MeOH (Supplementary Figure S4) revealed some severe matrix interferences and low selectivity (also evident from the seemingly elevated recovery of ANA), which, along with the low recoveries of FUL, warrant the choice of different extraction conditions.

**Table 5.** SPE procedures used for the ion-exchange sorbents.

| Experiment | IV. | V. | VI. |
|---|---|---|---|
| **SPE cartridges** | Oasis MCX 30 mg/1 mL<br>Oasis WCX 60 mg/3 mL | Oasis WCX 60 mg/3 mL<br>Strata-X-C 60 mg/3 mL | Oasis WCX 60 mg/3 mL<br>Strata-X-C 60 mg/3 mL |
| **1. Conditioning** | 2 mL MeOH<br>2 mL 0.2% $H_3PO_4$ | 2 mL MeOH<br>2 mL 0.2% $H_3PO_4$ | 1 mL MeOH<br>1 mL 100 mM Na-Ac (pH 5.6) |
| **2. Sample addition (400 µL diluted plasma sample)** | | | |
| **3. Washing** | 3 mL 0.2% $H_3PO_4$<br>2 mL 0.1 M HCl | 3 mL 0.2% $H_3PO_4$<br>2 mL 0.1 M HCl | 1 mL Na-Ac (pH 5.6)<br>1 mL MeOH:100 mM Na-Ac (pH 5.6) = 2:8 |
| **4. Elution** | 2 × 750 µL MeOH<br>or 2 × 750 µL 5% $NH_4OH$ in MeOH<br>or 2 × 750 µL 2% HCOOH in MeOH | 750 µL MeOH and 750 µL 5% $NH_4OH$ in MeOH | 2 × 750 µL 5% $NH_4OH$ in MeOH |

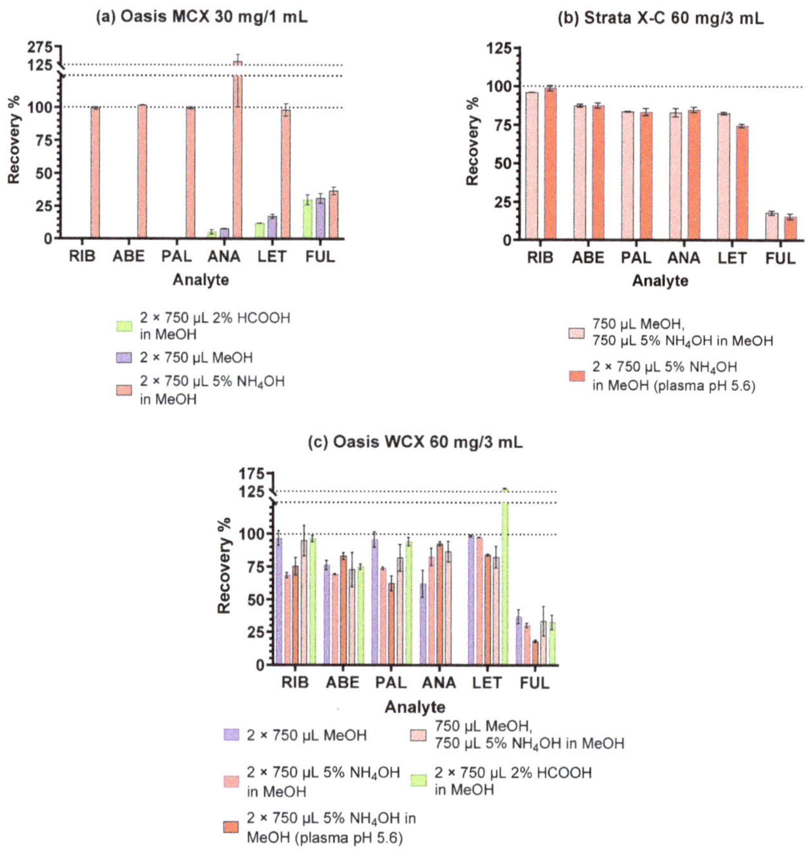


**Figure 2.** Extraction recoveries obtained with different eluents on ion-exchange sorbents.

Strata X-C is another polymeric, mixed-mode cation-exchange sorbent for the extraction of weak bases, similar to the Oasis MCX; however, it was comprised only of divinylbenzene-sulfonic groups without the vinylpyrrolidone [8,30]. It is visible in Figure 2b that elution with $NH_4OH$ in MeOH or a two-step combination of MeOH and $NH_4OH$ in MeOH yielded practically equal results for each analyte and was slightly lower than with the Oasis MCX. It is not clear from the manufacturer data whether the degree of divinylbenzene sulfonation is higher than in the MCX, but it could, together with the lack of the vinylpyrrolidone groups, account for the observed lower recoveries. It can be assumed that ANA and LET could still be easily retained via interactions with divinylbenzene when unprotonated, due to their small molecular masses and significant number of $\pi$ bonds, whereas the large and hydrophobic FUL could not. Judging by the structure of this sorbent (Table 4), a steric challenge to the binding of FUL may also be possible.

Oasis WCX is a polymer of vinylpyrrolidone and divinylbenzene, modified with carboxylic groups (pKa $\approx$ 5). It is meant for the extraction of stronger bases from the sample, assuming they remain protonated at higher pH values of the alkaline washing solvent. Ion exchange occurs when eluting with an acidic eluent and the protons from the solvent switch place with the analytes [7]. However, since none of the analytes of interest are strong bases, as shown in Table 1, the strong alkaline washing step proposed by the manufacturer was avoided in order to preserve their retention. The results obtained with this sorbent (Figure 2c) indicate the least precision of all the tested sorbents as well as the poorest cleanup. Variable matrix interferences coeluting with the analytes of interest were observed at all the tested conditions. Vague similarities with the results obtained with HLB can be observed—which is not surprising—since the same retention mechanism was most likely present for the analytes of interest. MeOH and HCOOH in MeOH provided slightly better overall recoveries than the $NH_4OH$ combinations; however, they showed a significant coelution of interferences with ANA and LET (evident from Supplementary Figures S5 and S6 and reflected in their recoveries). Since the full potential of the WCX sorbent could not be realised for these analytes, other sorbents may be more suitable.

To determine the best extraction conditions for all six analytes of interest, the most favourable conditions for each sorbent were mutually compared, as summarised in Figure 3.

As can be seen, the extraction recoveries of FUL were the poorest in most of the tested conditions, except for the C8 eluted with MeOH and the HLB eluted with a combination of MeOH and HCOOH in MeOH. For the other analytes, superior or similar extraction efficacies were achieved with the C8 eluted with MeOH and also with narrower error bars than in some other cases. The C18 sorbent eluted with HCOOH in MeOH and the X-C eluted with a combination of MeOH and $NH_4OH$ in MeOH also showed potential for the extraction of the combinations of only CDKi and aromatase inhibitors. However, SPE with a Sep-Pak Vac C8 column, 200 mg/3 mL, eluted with 2 × 750 µL MeOH, proved optimal for all six of the drugs of interest. This method was therefore transferred to LC-MS/MS, as described in Section 3.5., validated, and applied to real patient plasma samples.

### 2.2. Method Validation

The following validation parameters were assessed for the newly developed SPE-LC-MS/MS method: linearity, calibration range, accuracy, precision, selectivity, carry-over, and matrix effects. The stability of the samples at the same storage and working conditions as used in this work has already been confirmed [12].

#### 2.2.1. Linearity and Calibration range

Fresh calibration curves were prepared each time analyte quantitation was planned. All calibration curves were weighted by $1/x^2$. At least 75% (six) of the calibration standard levels met the criteria that the accuracies of the back-calculated concentrations were within ±15% of the nominal values. The calibration results are summarised in Table 6. The extracted ion chromatograms (EIC) of all the analytes at the LLOQ concentration level are shown in Supplementary Figure S7.

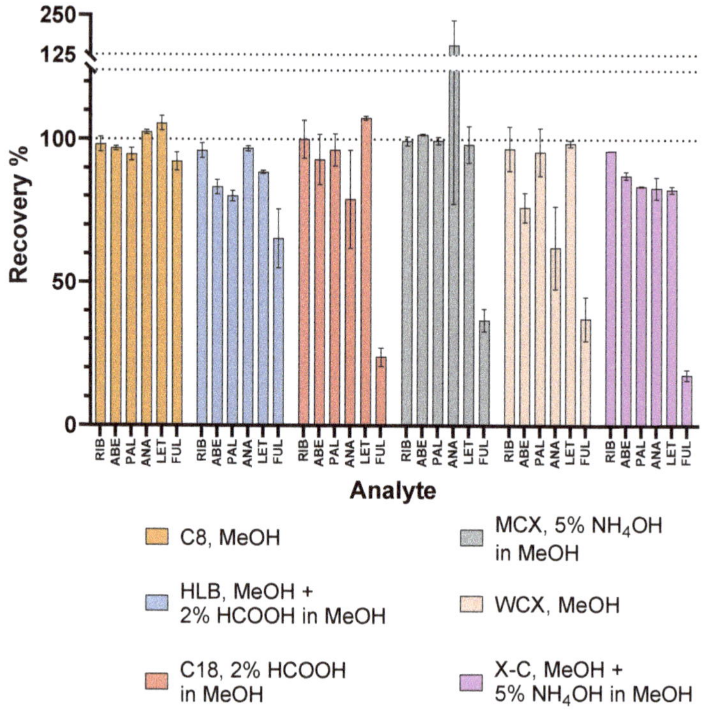


**Figure 3.** Cross-comparison of the best extraction recoveries with all the tested SPE sorbents.

**Table 6.** Calibration results.

| Analyte | RIB | ABE | PAL | ANA | LET | FUL |
|---|---|---|---|---|---|---|
| **Range (ng/mL)** | 700–3500 | 80–400 | 40–200 | 20–100 | 40–200 | 10–50 |
| **Slope** | 1103 | 10,846 | 11,875 | 68,230 | 5066 | 3013 |
| **Intercept** | 620,726 | 613,810 | −6744 | −28,605 | 4075 | −17,390 |
| **R** | 0.9953 | 0.9933 | 0.9970 | 0.9983 | 0.9948 | 0.9972 |
| **N(points)** | 7 | 7 | 7 | 7 | 7 | 7 |
| **Max. %bias *** | 5.77 | −7.85 | 4.62 | −5.01 | 8.06 | −6.60 |

* The accuracy of the back-calculated concentrations of each calibration standard.

### 2.2.2. Accuracy and Precision

Accuracy and precision were assessed at three quality control (QC) concentration levels across the linear range: lower limit of quantitation (LLOQ), low, and high, on 10 samples per concentration level within one day and a total of 15 samples per concentration level between days. They were found to be acceptable at all concentration levels (the bias was within ±15%, and the RSD was below 15%). The results are shown in Table 7. All QCs were quantitated according to a fresh calibration curve.

### 2.2.3. Selectivity and Carry-Over

Selectivity was assessed during the method development by reviewing chromatograms of blank plasma samples. The blank chromatogram obtained with the optimised C8-SPE method (Supplementary Figure S2) showed no significant interferences at the retention times of the analytes.

The presence of carry-over was checked by injecting blank samples after the highest calibration concentrations. A slight carry-over of RIB and PAL was observed. An additional blank injection was introduced for the needle and column cleanup, and the needle was washed for 100 s with 50% MeOH after each injection. Thus, the carry-over of RIB and PAL was reduced to less than 3% of the LLOQ (Supplementary Figure S8).

**Table 7.** Results of accuracy and precision tests.

| Analyte | Concentration (ng/mL) | Intra-Day, n = 10 | | Inter-Day, n = 15 | |
|---|---|---|---|---|---|
| | | Bias% | RSD% | Bias% | RSD% |
| **RIB** | 700 | 0.6 | 7.1 | 1.3 | 6.6 |
| | 1120 | 9.5 | 1.8 | 10.4 | 2.9 |
| | 2800 | −3.2 | 1.8 | −2.8 | 2.0 |
| **ABE** | 80 | −12.1 | 12.2 | −7.7 | 11.3 |
| | 128 | −1.3 | 3.9 | 4.2 | 6.1 |
| | 320 | −2.3 | 3.1 | −3.6 | 3.0 |
| **PAL** | 40 | 7.0 | 9.1 | 4.7 | 8.2 |
| | 64 | 10.0 | 4.7 | 8.1 | 4.3 |
| | 160 | 3.1 | 3.0 | −0.5 | 4.9 |
| **ANA** | 20 | −6.0 | 9.6 | −2.0 | 8.8 |
| | 32 | −4.3 | 6.6 | −0.4 | 7.3 |
| | 80 | −5.8 | 4.4 | 1.1 | 8.6 |
| **LET** | 40 | 7.3 | 9.2 | 3.2 | 9.0 |
| | 64 | 6.0 | 8.6 | 4.6 | 9.1 |
| | 160 | −3.9 | 12.4 | 1.4 | 13.4 |
| **FUL** | 10 | 8.2 | 14.2 | 13.5 | 14.3 |
| | 16 | 8.7 | 14.8 | 2.0 | 14.3 |
| | 40 | −11.5 | 6.6 | −12.7 | 12.9 |

### 2.2.4. Matrix Effects

The matrix effects were tested at two QC concentration levels (low and high) in triplicate on plasma samples from six different individuals, including a haemolysed and lipemic plasma, and calculated according to Equation (1):

$$\text{Matrix effect } (\%) = \frac{\text{Signal in the post-extraction spiked sample}}{\text{Signal in the standard solution}} * 100\% - 100\% \quad (1)$$

The results are depicted in Figure 4 and Supplementary Table S1. Differences in the matrix effects between the low and high QC concentrations were negligible. It is discernible that stronger ion suppression was present in the cases of LET and FUL. All analytes exhibited ion suppression in lipemic plasma, with some ion enhancement otherwise present for the CDKi. The variability between different sources of plasma was most pronounced for FUL (peak area RSD 25.21%) and the least for RIB (peak area RSD 5.16%). Evidently, although simultaneous extraction of all the analytes, as well as the cleanup of most of the UV-absorbing interferences, was achieved, plasma lipid removal was not as thorough.

It should be noted that extensive signal instability was also observed in the case of FUL, which sometimes manifested in a nonlinear response. This phenomenon is attributed to the drug's high lipophilicity, low ionisation efficiency in the positive ESI mode, and high susceptibility to ionisation effects from any leftover plasma interferences accumulated at the LC-MS interface. Regular, thorough rinsing of the spray shield with 50% i-propanol alleviated the issue to a certain extent; however, the introduction of an isotopically labelled analogue as the internal standard is highly recommended in the case of routine applications in clinical practice. In this work, the main goal was to achieve successful simultaneous extraction of the analytes from plasma samples using SPE and identify potential pitfalls for the method's application. Further addressing the MS issues and routine application of the method are outside the scope of this paper but will be the subject of our future efforts.

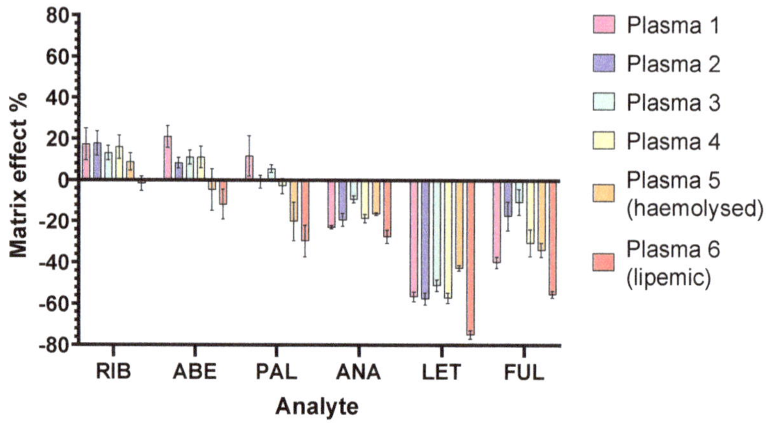


**Figure 4.** Matrix effects: mean and SD of all the collected data from six different lots; n = 6 samples per plasma lot.

### 2.3. *Analysis of Patient Samples*

To prove the applicability of the method on real patient samples, plasma from four breast cancer patients treated with CDKi + ET combinations was prepared using the developed procedure and quantitated according to a fresh calibration curve. Each of the analytes was present in at least one of the patients. The results were as follows: patient 1: RIB 981.7 ng/mL, ANA 41.7 ng/mL; patient 2: ABE 276.4 ng/mL, LET 69.8 ng/mL; patient 3: PAL 74.2 ng/mL, FUL 21.0 ng/mL; patient 4: PAL 119.3 ng/mL, FUL 14.2 ng/mL. All determined concentrations are within the method's validated linear ranges as well as within the expected concentration ranges of these drugs in patient plasma [31–36], which proves the feasibility of the proposed extraction method. Chromatograms of the patient samples are shown in Supplementary Figure S9.

### 2.4. *Comparison with Previously Published Methods*

Key parameters of the proposed method and the relevant previously published works are shown in Table 8. The procedure developed in this work is the first reported SPE method for the simultaneous extraction of all six breast cancer drugs of interest from plasma samples. In regard to the SPE methods for any of these analytes [14,23–25], novel SPE conditions were optimised. Plasma samples were diluted with Milli-Q water and applied onto a C8 SPE column after sorbent conditioning with MeOH and water. The sorbent was washed with water and 5% MeOH, thus removing the plasma proteins, and the analytes were eluted with two portions of MeOH. The obtained extraction recoveries for all the analytes were higher than in any of the previously published works. The method's linear ranges are clinically relevant, according to the expected patient plasma concentrations, and were found suitable in the conducted patient samples' analysis. Adequate precision and accuracy of the method in the upper and lower determination ranges have been confirmed.

In relation to other methods for these six breast cancer drugs, where samples are prepared using less selective PPT or labour-intensive DLLME [5,11,12], the proposed method explores a novel sample preparation approach. SPE is simple and selective when fully developed, and this work offers insight into the potential advantages as well as drawbacks of certain commercially available sorbents in the described application. The newly developed and validated C8-SPE method shows excellent extraction recoveries and sample cleanup, offering a novel, automatable approach to clinical application for the drugs of interest. What is more, since different interaction mechanisms are explored in depth in this article, the method development procedure may also be helpful for other structurally related compounds. The silica-based C8 sorbent itself is proven a versatile and efficient tool, retaining its applicability even beside the new-generation polymeric sorbents.

**Table 8.** Comparison of the developed SPE-LC-MS/MS method with previously published methods.

| Reference | Analytes | Analytical Technique | Sample Preparation Technique | Mean Extraction Recovery | Linear Range |
|---|---|---|---|---|---|
| Beer et al., 2010. [23] | ANA, LET, tamoxifen | LC-MS/MS | SPE w/Strata X-C (200 mg/3 mL) | 92% ANA, 89% LET | 5–200 ng/mL ANA, 10–300 ng/mL LET |
| Chavan et al., 2018. [14] | PAL | LC-MS/MS | PPT w/ACN, evaporation, SPE w/Phenomenex C18 | n.a. | n.a. |
| Nalanda et al., 2018. [24] | PAL | LC-DAD | SPE w/Oasis HLB (30 mg/1 mL) | 72.1% | 100–3000 ng/mL |
| Leenhardt et al., 2021. [25] | PAL, RIB | LC-MS/MS | SPE w/Oasis HLB (30 mg/1 mL) | n.a. | 3.9–129 ng/mL |
| Sato et al., 2021. [11] | PAL, RIB, ABE, ANA, LET, FUL | LC-MS/MS | PPT w/ACN:MeOH 9:1 | n.a. | 1–75 ng/mL ANA and FUL, 5–375 ng/mL PAL, 10–750 ng/mL ABE and LET, 100–7500 ng/mL RIB |
| Turković et al., 2022. [12] | PAL, RIB, ABE, ANA, LET, FUL | LC-MS/MS | PPT w/ACN | >85% | 1–200 ng/mL ANA, 2.5–500 ng/mL LET, 3.1–500 ng/mL PAL, 5–1000 ng/mL FUL, 15–3000 ng/mL ABE, 25–5000 ng/mL RIB |
| Turković et al., 2023. [5] | PAL, RIB, ABE, ANA, LET, FUL | LC-DAD-FLD | PPT w/ACN, DLLME | 81.7–95.6% | 2.50–60.0 μg/mL ANA, 0.04–1.00 μg/mL LET, 0.08–1.92 μg/mL PAL, 0.50–12.0 μg/mL FUL, 0.11–2.61 μg/mL ABE, 0.25–5.95 μg/mL RIB |
| This work | PAL, RIB, ABE, ANA, LET, FUL | LC-MS/MS | SPE w/Sep-Pak Vac C8 (200 mg/3 mL) | 92.3–105.5% | 20–100 ng/mL ANA, 40–200 ng/mL LET and PAL, 10–50 ng/mL FUL, 80–400 ng/mL ABE, 700–3500 ng/mL RIB |

w/: with; n.a.: not available.

## 3. Materials and Methods

### *3.1. Chemicals and Reagents*

MeOH for the HPLC and MS, as well as ACN for the HPLC, were purchased from J.T. Baker (Phillipsburg, NJ, USA); HCOOH for the LC-MS was purchased from Supelco (St. Louis, MO, USA); $H_3PO_4$ (85%) was purchased from T.T.T. (Sveta Nedelja, Croatia); $NH_4OH$ (25%) and Na-Ac were from Alkaloid (Skopje, North Macedonia); and HCl (37%) was from VWR (Radnor, PA, USA). Ultrapure water was obtained using a Merck Millipore Milli-Q IQ 7015 system (Darmstadt, Germany). Analytical grade standards (purity > 97%) of PAL, RIB, and ABE were from Toronto Research Chemicals (Toronto, Canada); ANA and

LET were from Tokyo Chemical Industry (Tokyo, Japan); and FUL was from MilliporeSigma (Burlington, MA, USA).

### 3.2. Preparation of the Standard Solutions

The primary stock solutions of RIB, ABE, ANA, LET, and FUL were prepared to the concentration of 1 mg/mL in MeOH. The primary stock solution of PAL was prepared to the concentration of 225 μg/mL in $H_2O$:ACN 50:50 *v*/*v*. These solutions were mixed and diluted in MeOH to obtain working solutions of the appropriate concentrations. The concentrations of the calibrants used in the method validation are shown in Table 9. All the solutions were kept refrigerated at 4 °C and were stable for at least three months.

**Table 9.** Concentration levels of the calibrant plasma samples (ng/mL).

| Calibrant | RIB | ABE | PAL | ANA | LET | FUL |
|---|---|---|---|---|---|---|
| 1 | 700 | 80 | 40 | 20 | 40 | 10 |
| 2 | 1120 | 128 | 64 | 32 | 64 | 16 |
| 3 | 1400 | 160 | 80 | 40 | 80 | 20 |
| 4 | 1750 | 200 | 100 | 50 | 100 | 25 |
| 5 | 2100 | 240 | 120 | 60 | 120 | 30 |
| 6 | 2520 | 288 | 144 | 72 | 144 | 36 |
| 7 | 2800 | 320 | 160 | 80 | 160 | 40 |
| 8 | 3500 | 400 | 200 | 100 | 200 | 50 |

### 3.3. Plasma Sample Pretreatment

Blood from the patients treated with the drugs of interest, as well as drug-free blood from healthy volunteers, was collected in vials with the $K_2$-EDTA anticoagulant. After centrifugation at 1500× *g* for 10 min, the supernatant was stored short-term at −18 °C and long-term at −80 °C. The plasma was thawed at room temperature for 30 min before any experiments were conducted. Research approvals were obtained from the Ethics Committee of the University of Zagreb Faculty of Pharmacy and Biochemistry (approval 251-62-03-19-30) and the Ethics Committee of University Hospital Centre Zagreb (approval 02/21-JG).

### 3.4. Plasma Sample Preparation

The tested extraction phases, Oasis MCX 30 mg/1 mL, Oasis HLB 60 mg/3 mL, Oasis WCX 60 mg/3 mL, Sep-Pak Vac C18 200 mg/3 mL, Sep-Pak Vac C8 200 mg/3 mL, and Sep-Pak Vac C8 500 mg/3 mL, were obtained from Waters (Milford, MA, USA), while the Strata-X-C 60 mg/3 mL was from Phenomenex (Torrance, CA, USA).

Before commencing with the SPE procedure, blank plasma samples were spiked with a working analyte standard solution at the ratio of plasma:standard 9:1 *v*/*v* and the samples were diluted with an appropriate solvent. The extractions were carried out using a vacuum manifold (Supelco, Bellefonte, PA, USA) equipped with a vacuum pump. A known volume of the sample eluate (80% of the nominal eluate volume) was evaporated to dryness on a vacuum concentrator (Eppendorf, Hamburg, Germany) and reconstituted in 80 μL of 65% MeOH. Once prepared, the samples were kept on the autosampler at 10 °C for no longer than 10 h.

### 3.5. Chromatographic Conditions

A Waters XBridge BEH phenyl column, with the dimensions of 150 × 4.6 mm and particle size 2.5 μm, with a corresponding VanGuard phenyl guard column (Waters, Milford, MA, USA), was used as the stationary phase, thermostated at 35 °C. Ultrapure water (phase A) and MeOH (phase B), both containing 0.1% HCOOH, were used as the mobile phase in the gradient elution. The initial experiments were performed on an Agilent 1100 HPLC equipped with DAD and FLD (Agilent Technologies, Santa Clara, CA, USA). The flow rate was set to 0.8 mL/min, and 10 μL of the sample was injected. The applied mobile phase gradient is shown in Table 10.

**Table 10.** Mobile phase gradient applied on the Agilent 1100 HPLC.

| Time (min) | Mobile Phase B (%) |
|---|---|
| 0.0 | 30.0 |
| 5.5 | 85.0 |
| 9.5 | 90.0 |
| 10.0 | 100.0 |
| 16.5 | 100.0 |
| 17.0 | 30.0 |
| 30.0 | 30.0 |

The method was then transferred to an Agilent 1290 Infinity II UHPLC coupled to a 6470 triple-quadrupole mass spectrometer (QQQ-MS) equipped with an Agilent Jet Stream electrospray (AJS-ESI) source (Agilent Technologies, Santa Clara, CA, USA) to achieve the necessary detection sensitivity. With the same stationary and mobile phases, the flow rate was adjusted to 0.6 mL/min using the gradient elution described in Table 11. An adjusted gradient composition was necessary due to the smaller dwell volume of the UHPLC system. The column temperature was 35 °C, and 10 µL of the sample was injected into the system.

**Table 11.** Mobile phase gradient applied on the Agilent 1290 UHPLC.

| Time (min) | Mobile Phase B (%) |
|---|---|
| 0.0 | 5.0 |
| 10.5 | 85.0 |
| 14.5 | 90.0 |
| 15.0 | 100.0 |
| 20.0 | 100.0 |
| 20.5 | 30.0 |
| 26.0 | 30.0 |

### *3.6. Detection Conditions*

In the preliminary studies, the analytes were detected using DAD on 360 nm for PAL, RIB, and ABE, together with FLD at a 212 nm excitation (Ex) and a 310 nm emission (Em) for ANA, LET, and FUL, as previously reported [5]. The AJS-ESI-MS conditions were as follows: gas temperature 300 °C, gas flow 13 L/min, nebulizer pressure 35 psi, sheath gas temperature 350 °C, sheath gas flow 12 L/min, positive ESI mode with capillary voltage 3500 V, nozzle voltage 500 V, and cell accelerator voltage 5 V. Analyte-specific parameters of the monitored mass transitions are shown in Table 12, while the representative chromatograms, obtained using both LC methods, are shown in Figure 5. Exemplary mass spectra of all the analytes are provided in Supplementary Figure S10.

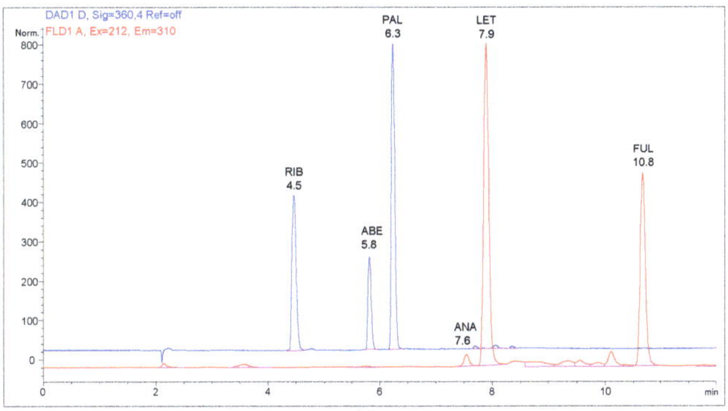


(a)

**Figure 5.** *Cont.*

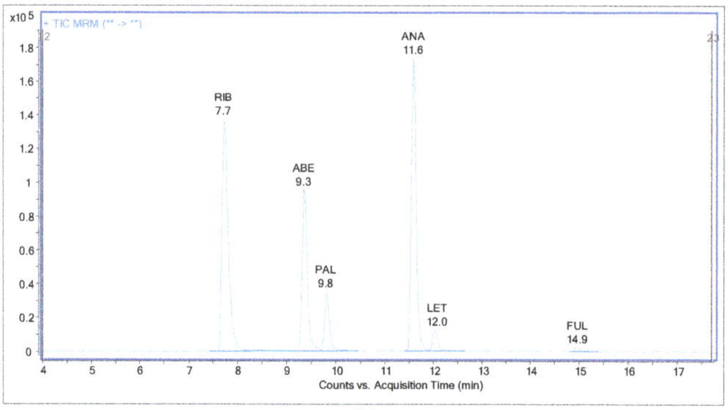


(**b**)

**Figure 5.** Representative chromatograms of a spiked plasma sample prepared using the optimal C8-SPE sample preparation method: (**a**) HPLC-DAD-FLD: DAD 360 nm (blue), FLD Ex 212, Em 310 nm (red), concentration of all analytes 10 µg/mL; (**b**) UHPLC-MS/MS: total ion chromatogram (TIC) at the LLOQ concentration level.

**Table 12.** Optimised MS/MS transitions for each analyte.

| Analyte | Precursor Ion (*m*/*z*) | Product Ion (*m*/*z*) | CE (V) | Dwell Time (ms) | Fragmentor (V) |
|---|---|---|---|---|---|
| **RIB** | 435.2 | 322.1<br>252.1 | 25<br>30 | 200 | 200 |
| **ABE** | 507.2 | 393.1<br>245.0 | 25<br>75 | 60 | 140 |
| **PAL** | 448.0 | 380.3<br>362.0 | 30<br>45 | 60 | 180 |
| **ANA** | 294.1 | 225.4<br>115.2 | 25<br>70 | 60 | 140 |
| **LET** | 217.0 | 190.3<br>164.1 | 25<br>50 | 60 | 120 |
| **FUL** | 607.4 | 589.0<br>467.0 | 15<br>29 | 200 | 160 |

CE: collision energy.

### *3.7. Data Collection and Analysis*

Data were collected using the Agilent Chemstation 10 software (Santa Clara, CA, USA) for the Agilent 1100 HPLC, and the Agilent MassHunter Workstation 10.0 software was used for the data acquisition and qualitative analysis (Santa Clara, CA, USA) for the Agilent 1290 UHPLC. Data analyses were performed using Microsoft 365 Excel (Redmond, WA, USA) and GraphPad Prism 8 (Boston, MA, USA).

## 4. Conclusions

In recent years, the importance of TDM of anticancer drugs has been an emerging topic. An increasing number of bioanalytical methods are being reported in search of fulfilling the first precondition for TDM testing: the development of a reliable procedure for the quantitative determination of drugs-candidates in biological matrices suitable for the busy clinical setting. SPE is currently one of the most common sample preparation techniques due to its versatility. It is simple and fast once optimised; however, its implementation requires extensive method development. In this work, different extraction conditions were tested (HLB, reversed-phase C8 and C18, and ion-exchange sorbents WCX, MCX, and

X-C, combined with various elution solvents) for the extraction of six breast cancer drugs, palbociclib, ribociclib, abemaciclib, anastrozole, letrozole, and fulvestrant, from human plasma samples.

Of all the tested conditions, the Sep-Pak Vac C8 (200 mg/3 mL) sorbent eluted with MeOH shows the best overall extraction yields and cleanup for all six drugs of interest, while the Sep-Pak Vac C18 (200 mg/3 mL) eluted with HCOOH in MeOH and the Strata X-C (60 mg/3 mL) eluted with a combination of MeOH and $NH_4OH$ in MeOH show potential for combinations of only CDKi and aromatase inhibitors. These findings, backed by the detailed analysis of the potential interaction mechanisms, may also be a useful starting point for the development of SPE methods for other structurally similar drugs of interest.

The proposed C8-SPE-LC-MS/MS method is simple and fast, ecologically acceptable, precise, accurate, and sensitive, with adequate, clinically relevant linear ranges. Potential pitfalls in the method development and its application are acknowledged, such as the high levels of interference, regardless of the acceptable extraction yields in the case of the MCX and WCX sorbents, or the observed matrix effects with the C8 sorbent for some of the analytes. This can be avoided in routine clinical settings with regular and thorough ion source rinsing and by using isotopically labelled internal standards. Carry-over was observed during the method development and was resolved with additional column rinsing. Nevertheless, the method's feasibility is established by its application on four samples from patients treated with the drugs of interest. To the best of our knowledge, this is the first SPE method for the simultaneous extraction of ABE, RIB, PAL, ANA, LET, and FUL from plasma samples, demonstrating satisfactory sensitivity, selectivity, and precision, together with exceedingly high recoveries ($\geq$ 92.3%).

**Supplementary Materials:** The following supporting information can be downloaded at: https://www.mdpi.com/article/10.3390/ph16101445/s1, Figure S1: Chromatogram of a blank plasma sample prepared using the Sep-Pak Vac C18 sorbent (200 mg/3 mL) eluted with HCOOH in MeOH. Detection wavelengths: DAD 360 nm (blue), FLD Ex 212, Em 310 nm (red); Figure S2: Chromatogram of a blank plasma sample prepared using the Sep-Pak Vac C8 sorbent (200 mg/3 mL) eluted with MeOH. Detection wavelengths: DAD 360 nm (blue) FLD Ex 212, Em 310 nm (red); Figure S3: Chromatogram of a blank plasma sample prepared using the Oasis HLB sorbent (60 mg/3 mL) eluted with HCOOH in MeOH. Detection wavelengths: DAD 360 nm (blue) FLD Ex 212, Em 310 nm (red); Figure S4: Chromatogram of a blank plasma sample prepared using the Oasis MCX sorbent (30 mg/1 mL) eluted with 5% $NH_4OH$ in MeOH. Detection wavelengths: DAD 360 nm (blue) FLD Ex 212, Em 310 nm (red); Figure S5: Chromatogram of a blank plasma sample prepared using the Oasis WCX sorbent (60 mg/3 mL) eluted with MeOH. Detection wavelengths: DAD 360 nm (blue) FLD Ex 212, Em 310 nm (red); Figure S6: Chromatogram of a blank plasma sample prepared using the Oasis WCX sorbent (60 mg/3 mL) eluted with HCOOH in MeOH. Detection wavelengths: DAD 360 nm (blue) FLD Ex 212, Em 310 nm (red); Figure S7: Extracted ion chromatograms of the quantifier ion transition for the analytes at the LLOQ concentration level: a) RIB, b) ABE, c) PAL, d) ANA, e) LET, and f) FUL; Figure S8: Overlaid extracted ion chromatograms of the quantifying ion transitions for all analytes at the LLOQ concentration levels and in a blank sample injected after the ULOQ concentration level. Brown: RIB; dark blue: ABE; dark red: PAL; light green: ANA; light blue: LET; dark green: FUL; Figure S9: MS chromatograms of the samples from a) patient 1 (RIB, ANA); b) patient 2 (ABE, LET); c) patient 3 (PAL, FUL); and d) patient 4 (PAL, FUL); Figure S10: Exemplary MS spectra of the analytes: a) RIB (precursor ion $m/z$ 435.2; CE 40 V, fragmentor 200 V), b) ABE (precursor ion $m/z$ 507.2; CE 30 V, fragmentor 140 V), c) PAL (precursor ion $m/z$ 448.0; CE 40 V, fragmentor 180 V), d) ANA (precursor ion $m/z$ 294.1; CE 30 V, fragmentor 140 V), e) LET (precursor ion $m/z$ 217.0; CE 50 V, fragmentor 120 V), and f) FUL (precursor ion $m/z$ 607.4; CE 30 V, fragmentor 160 V); Table S1: Overall matrix effects results—average value of matrix effects and peak area RSD between all tested samples for each analyte.

**Author Contributions:** Conceptualization, M.S. and D.M.P.; methodology, D.M.P., L.T. and A.S.; validation, L.T., Z.M., A.S. and T.S.; formal analysis, A.S. and L.T.; investigation, D.M.P., L.T. and Z.M.; resources, M.S. and T.S.; data curation, L.T. and D.M.P.; writing—original draft preparation, L.T. and D.M.P.; writing—review and editing, Z.M. and M.S.; visualization, L.T.; supervision, M.S. and D.M.P.; project administration, M.S.; funding acquisition, M.S. All authors have read and agreed to the published version of the manuscript.

**Funding:** This research was funded by the Croatian Science Foundation, grant number UIP-2019-04-8461 and DOK-2021-02-4995, as well as the European Regional Development Fund, project FarmInova, grant number KK.01.1.1.02.0021. The APC was funded by the Croatian Science Foundation, grant number UIP-2019-04-8461.

**Institutional Review Board Statement:** The study was conducted according to the guidelines of the Declaration of Helsinki and approved by the Ethics Committee of the University of Zagreb Faculty of Pharmacy (approval number 251-62-03-19-30, date of approval, 23 April 2019) and by the Ethics Committee of University Hospital Centre Zagreb (approval number 02/21-JG, date of approval, 20 August 2019).

**Informed Consent Statement:** Informed consent was obtained from all subjects involved in the study.

**Data Availability Statement:** Data are contained within the article and Supplementary Material.

**Acknowledgments:** We are grateful to Kristina Javorić, Juraj Vuić, Sandro Makarić, Vladimir Radić, and Biljana Nigović for their help and support.

**Conflicts of Interest:** T.S. received speaker honoraria from Elly Lili, Novartis, and Pfizer, a conference reimbursement from Novartis, and is a co-investigator in a Novartis clinical study. The other authors declare no conflicts of interest that might be relevant to the content of this manuscript. The funders had no role in the design of the study, in the collection, analyses, or interpretation of data, in the writing of the manuscript, or in the decision to publish the results.

## References

1. Engler, T.; Fasching, P.A.; Lüftner, D.; Hartkopf, A.D.; Müller, V.; Kolberg, H.C.; Hadji, P.; Tesch, H.; Häberle, L.; Ettl, J.; et al. Implementation of CDK4/6 Inhibitors and its Influence on the Treatment Landscape of Advanced Breast Cancer Patients—Data from the Real-World Registry PRAEGNANT. *Geburtshilfe Frauenheilkd.* **2022**, *82*, 1055–1067. [CrossRef]
2. Mueller-Schoell, A.; Groenland, S.L.; Scherf-Clavel, O.; van Dyk, M.; Huisinga, W.; Michelet, R.; Jaehde, U.; Steeghs, N.; Huitema, A.D.R.; Kloft, C. Therapeutic drug monitoring of oral targeted antineoplastic drugs. *Eur. J. Clin. Pharmacol.* **2021**, *77*, 441–464. [CrossRef] [PubMed]
3. Pena-Pereira, F.; Wojnowski, W.; Tobiszewski, M. AGREE—Analytical GREEnness Metric Approach and Software. *Anal. Chem.* **2020**, *92*, 10076–10082. [CrossRef] [PubMed]
4. Mlinarić, Z.; Turković, L.; Begović, I.; Nigović, B.; Sertić, M. Rapid Capillary Electrophoresis Method for Simultaneous Determination of Abemaciclib, Ribociclib, and Palbociclib in Pharmaceutical Dosage Forms: A Green Approach. *Molecules* **2022**, *27*, 7603. [CrossRef] [PubMed]
5. Turković, L.; Koraj, N.; Mlinarić, Z.; Silovski, T.; Crnković, S.; Sertić, M. Optimisation of dispersive liquid-liquid microextraction for plasma sample preparation in bioanalysis of CDK4/6 inhibitors in therapeutic combinations for breast cancer treatment. *Heliyon* **2023**, *9*, e18880. [CrossRef]
6. Sentellas, S.; Saurina, J.; Núñez, O. Solid-phase extraction in bioanalytical applications. In *Solid-Phase Extraction*; Poole, C.F., Ed.; Elsevier: Amsterdam, The Netherlands, 2020; pp. 673–698. ISBN 9780128169063.
7. Waters Oasis Sample Extraction Products—Taking the Complexity Out of SPE Method Development. Available online: https://www.waters.com/webassets/cms/library/docs/720005685en.pdf (accessed on 25 August 2023).
8. Fontanals, N.; Marcé, R.M.; Borrull, F.; Cormack, P.A.G. Mixed-mode ion-exchange polymeric sorbents: Dual-phase materials that improve selectivity and capacity. *TrAC Trends Anal. Chem.* **2010**, *29*, 765–779. [CrossRef]
9. Dias, N.C.; Poole, C.F. Mechanistic study of the sorption properties of Oasis® HLB and its use in solid-phase extraction. *Chromatographia* **2002**, *56*, 269–275. [CrossRef]
10. Mutavdžić Pavlović, D.; Babić, S.; Horvat, A.J.M.; Kaštelan-Macan, M. Sample preparation in analysis of pharmaceuticals. *TrAC Trends Anal. Chem.* **2007**, *26*, 1062–1075. [CrossRef]
11. Sato, Y.; Shigeta, K.; Hirasawa, T.; Sato, T.; Ogura, J.; Maekawa, M.; Ebata, A.; Hamanaka, Y.; Tada, H.; Ishida, T.; et al. Establishment of an analytical method for simultaneous quantitation of CDK4/6 inhibitors, aromatase inhibitors, and an estrogen receptor antagonist in human plasma using LC-ESI-MS/MS. *J. Chromatogr. B Anal. Technol. Biomed. Life Sci.* **2021**, *1173*, 122655. [CrossRef]

12. Turković, L.; Bočkor, L.; Ekpenyong, O.; Silovski, T.; Lovrić, M.; Crnković, S.; Nigović, B.; Sertić, M. Development and Validation of a Novel LC-MS/MS Method for the Simultaneous Determination of Abemaciclib, Palbociclib, Ribociclib, Anastrozole, Letrozole, and Fulvestrant in Plasma Samples: A Prerequisite for Personalized Breast Cancer Treatment. *Pharmaceuticals* **2022**, *15*, 614. [CrossRef]
13. Gumustas, M.; Sengel-Turk, C.T.; Hascicek, C.; Ozkan, S.A. Optimization of a validated stability-indicating RP-LC method for the determination of fulvestrant from polymeric based nanoparticle systems, drugs and biological samples. *Biomed. Chromatogr.* **2014**, *28*, 1409–1417. [CrossRef]
14. Chavan, B.B.; Tiwari, S.; Shankar, G.; Nimbalkar, R.D.; Garg, P.; Srinivas, R.; Talluri, M.V.N.K. In vitro and in vivo metabolic investigation of the Palbociclib by UHPLC-Q-TOF/MS/MS and in silico toxicity studies of its metabolites. *J. Pharm. Biomed. Anal.* **2018**, *157*, 59–74. [CrossRef] [PubMed]
15. Martínez-Chávez, A.; Rosing, H.; Hillebrand, M.; Tibben, M.; Schinkel, A.H.; Beijnen, J.H. Development and validation of a bioanalytical method for the quantification of the CDK4/6 inhibitors abemaciclib, palbociclib, and ribociclib in human and mouse matrices using liquid chromatography-tandem mass spectrometry. *Anal. Bioanal. Chem.* **2019**, *411*, 5331–5345. [CrossRef]
16. Soledad Poetto, A.; Posocco, B.; Zanchetta, M.; Gagno, S.; Orleni, M.; Canil, G.; Alberti, M.; Puglisi, F.; Toffoli, G. A new LC-MS/MS method for the simultaneous quantification of abemaciclib, its main active metabolites M2 and M20, and letrozole for therapeutic drug monitoring. *J. Chromatogr. B Anal. Technol. Biomed. Life Sci.* **2022**, *1207*, 123403. [CrossRef] [PubMed]
17. Posocco, B.; Buzzo, M.; Poetto, A.S.; Orleni, M.; Gagno, S.; Zanchetta, M.; Iacuzzi, V.; Guardascione, M.; Puglisi, F.; Basile, D.; et al. Simultaneous quantification of palbociclib, ribociclib and letrozole in human plasma by a new LC-MS/MS method for clinical application. *PLoS ONE* **2020**, *15*, e0228822. [CrossRef]
18. Habler, K.; Kalla, A.S.; Rychlik, M.; Vogeser, M.; Teupser, D. Therapeutic drug monitoring in breast cancer therapy—LC-MS/MS method for quantification of the CDK4/6 inhibitors abemaciclib, palbociclib, ribociclib, and major metabolites abemaciclib M20 and M2 in human serum. *J. Pharm. Biomed. Anal.* **2023**, *225*, 115211. [CrossRef] [PubMed]
19. Margaryan, T.; Elliott, M.; Sanai, N.; Tovmasyan, A. Simultaneous determination of LY3214996, abemaciclib, and M2 and M20 metabolites in human plasma, cerebrospinal fluid, and brain tumor by LC-MS/MS. *J. Pharm. Anal.* **2022**, *12*, 601–609. [CrossRef]
20. Burke, S.M.; Kamal, M.K.; Goey, A.K. Development and Validation of a Quantitative LC-MS/MS Method for CDK4/6 Inhibitors Palbociclib, Ribociclib, Abemaciclib, and Abemaciclib-M2 in Human Plasma. *Ther. Drug Monit.* **2023**, *45*, 327–336. [CrossRef]
21. Alegete, P.; Kancherla, P.; Albaseer, S.S.; Boodida, S. A validated liquid chromatography–tandem mass spectrometric (LC-MS/MS) method for the estimation of fulvestrant in human plasma. *Orient. J. Chem.* **2017**, *33*, 1146–1155. [CrossRef]
22. Jolibois, J.; Schmitt, A.; Royer, B. A simple and fast LC-MS/MS method for the routine measurement of cabozantinib, olaparib, palbociclib, pazopanib, sorafenib, sunitinib and its main active metabolite in human plasma. *J. Chromatogr. B Anal. Technol. Biomed. Life Sci.* **2019**, *1132*, 121844. [CrossRef]
23. Beer, B.; Schubert, B.; Oberguggenberger, A.; Meraner, V.; Hubalek, M.; Oberacher, H. Development and validation of a liquid chromatography-tandem mass spectrometry method for the simultaneous quantification of tamoxifen, anastrozole, and letrozole in human plasma and its application to a clinical study. *Anal. Bioanal. Chem.* **2010**, *398*, 1791–1800. [CrossRef]
24. Nalanda, R.B.; Srinivasa Rao, A.; Gowri Sankar, D. Determination of palbociclib in human plasma using high performance liquid chromatography—Ultraviolet detection. *Int. J. Pharm. Sci. Res.* **2018**, *9*, 3883–3888. [CrossRef]
25. Leenhardt, F.; Gracia, M.; Perrin, C.; Muracciole-Bich, C.; Marion, B.; Roques, C.; Alexandre, M.; Firmin, N.; Pouderoux, S.; Mbatchi, L.; et al. Liquid chromatography–tandem mass spectrometric assay for the quantification of CDK4/6 inhibitors in human plasma in a clinical context of drug-drug interaction. *J. Pharm. Biomed. Anal.* **2020**, *188*, 113438. [CrossRef]
26. Daina, A.; Michielin, O.; Zoete, V. SwissADME: A free web tool to evaluate pharmacokinetics, drug-likeness and medicinal chemistry friendliness of small molecules. *Sci. Rep.* **2017**, *7*, 42717. [CrossRef] [PubMed]
27. ChemAxon Chemicalize. Available online: www.chemicalize.com (accessed on 1 October 2023).
28. Phenomenex Strata X-C SPE Products. Available online: https://www.phenomenex.com/products/strata-x-solid-phase-extraction-products/strata-x-c (accessed on 1 October 2023).
29. Phenomenex SPE Method Development Tool. Available online: https://www.phenomenex.com/Tools/SPEMethodDevelopment (accessed on 17 May 2022).
30. Mutavdžić Pavlović, D.; Babić, S.; Dolar, D.; Ašperger, D.; Košutić, K.; Horvat, A.J.M.; Kaštelan-Macan, M. Development and optimization of the SPE procedure for determination of pharmaceuticals in water samples by HPLC-diode array detection. *J. Sep. Sci.* **2010**, *33*, 258–267. [CrossRef]
31. Samant, T.S.; Dhuria, S.; Lu, Y.; Laisney, M.; Yang, S.; Grandeury, A.; Mueller-Zsigmondy, M.; Umehara, K.; Huth, F.; Miller, M.; et al. Ribociclib Bioavailability Is Not Affected by Gastric pH Changes or Food Intake: In Silico and Clinical Evaluations. *Clin. Pharmacol. Ther.* **2018**, *104*, 374–383. [CrossRef] [PubMed]
32. Fujiwara, Y.; Tamura, K.; Kondo, S.; Tanabe, Y.; Iwasa, S.; Shimomura, A.; Kitano, S.; Ogasawara, K.; Turner, P.K.; Mori, J.; et al. Phase 1 study of abemaciclib, an inhibitor of CDK 4 and 6, as a single agent for Japanese patients with advanced cancer. *Cancer Chemother. Pharmacol.* **2016**, *78*, 281–288. [CrossRef] [PubMed]
33. Tamura, K.; Mukai, H.; Naito, Y.; Yonemori, K.; Kodaira, M.; Tanabe, Y.; Yamamoto, N.; Osera, S.; Sasaki, M.; Mori, Y.; et al. Phase I study of palbociclib, a cyclin-dependent kinase 4/6 inhibitor, in Japanese patients. *Cancer Sci.* **2016**, *107*, 755–763. [CrossRef] [PubMed]

34. Dowsett, M.; Cuzick, J.; Howell, A.; Jackson, I. Pharmacokinetics of anastrozole and tamoxifen alone, and in combination, during adjuvant endocrine therapy for early breast cancer in postmenopausal women: A sub-protocol of the "Arimidex$^{TM}$ and tamoxifen alone or in combination" (ATAC) trial. *Br. J. Cancer* **2001**, *85*, 317–324. [CrossRef]
35. Desta, Z.; Kreutz, Y.; Nguyen, A.; Li, L.; Skaar, T.; Kamdem, L.; Henry, N.; Hayes, D.; Storniolo, A.; Stearns, V.; et al. Plasma Letrozole Concentrations in Postmenopausal Women with Breast Cancer Are Associated with CYP2A6 Genetic Variants, Body Mass Index, and Age. *Clin. Pharmacol. Ther.* **2011**, *90*, 693–700. [CrossRef]
36. European Medicines Agency (EMA). EPAR Product Information: Faslodex. Available online: https://www.ema.europa.eu/en/documents/product-information/faslodex-epar-product-information_en.pdf (accessed on 15 February 2022).

*pharmaceuticals*

MDPI

*Article*

# Study of Mono and Di-*O*-caffeoylquinic Acid Isomers in *Acmella oleracea* Extracts by HPLC-MS/MS and Application of Linear Equation of Deconvolution Analysis Algorithm for Their Characterization

**Maria Bellumori *, Marco Pallecchi, Beatrice Zonfrillo, Luigi Lucio, Marta Menicatti, Marzia Innocenti, Nadia Mulinacci and Gianluca Bartolucci ***

NEUROFARBA Department, Section of Pharmaceutical and Nutraceutical Sciences, University of Florence, Via U. Schiff 6, Sesto Fiorentino, 50019 Firenze, Italy; marco.pallecchi@unifi.it (M.P.); beatrice.zonfrillo@unifi.it (B.Z.); luigi.lucio@unifi.it (L.L.); marta.menicatti@unifi.it (M.M.); marzia.innocenti@unifi.it (M.I.); nadia.mulinacci@unifi.it (N.M.)

* Correspondence: maria.bellumori@unifi.it (M.B.); gianluca.bartolucci@unifi.it (G.B.)

**Abstract:** Chlorogenic acids, the esters of caffeic and quinic acids, are the main phenolic acids detected in *Acmella oleracea* extracts and have gained increasing interest in recent years due to their important biological activities. Given their structural similarity and instability, the correct analysis and identification of these compounds in plants is challenging. This study aimed to propose a simple and rapid determination of the *A. oleracea* caffeoylquinic isomers, applying an HPLC-MS/MS method supported by a mathematical algorithm (Linear Equation of Deconvolution Analysis (LEDA)). The three mono- and the three di-caffeoylquinic acids in roots of *Acmella* plants were studied by an ion trap MS analyzer. A separation by a conventional chromatographic method was firstly performed and an MS/MS characterization by energetic dimension of collision-induced dissociation mechanism was carried out. The analyses were then replicated using a short HPLC column and a fast elution gradient (ten minutes). Each acquired MS/MS data were processed by LEDA algorithm which allowed to assign a relative abundance in the reference ion signal to each isomer present. Quantitative results showed no significant differences between the two chromatographic systems proposed, proving that the use of LEDA algorithm allowed the distinction of the six isomers in a quarter of the time.

**Keywords:** chlorogenic acids; tandem mass spectrometry; ion trap MS analyzer; isomers recognition; ERMS

**Citation:** Bellumori, M.; Pallecchi, M.; Zonfrillo, B.; Lucio, L.; Menicatti, M.; Innocenti, M.; Mulinacci, N.; Bartolucci, G. Study of Mono and Di-O-caffeoylquinic Acid Isomers in *Acmella oleracea* Extracts by HPLC-MS/MS and Application of Linear Equation of Deconvolution Analysis Algorithm for Their Characterization. *Pharmaceuticals* **2023**, *16*, 1375. https://doi.org/10.3390/ph16101375

Academic Editors: Jan Oszmianski and Sabina Lachowicz-Wiśniewska

Received: 24 August 2023
Revised: 20 September 2023
Accepted: 26 September 2023
Published: 28 September 2023



## 1. Introduction

One of the most distinguished members of the genus *Acmella* is *Acmella oleracea* (L.) R.K. Jansen, an annual herb native to Brazil that is occurring throughout tropical and subtropical regions around the world. It has been widely cultivated for horticultural, medicinal, insecticidal, and culinary purposes, and recently, multidisciplinary studies on this herb have been promoted, with an increasing number of commercial products appearing on the market as personal care, health care, and culinary products [1].

Different extracts of *A. oleracea* have been reported to hold numerous important biological activities, including a local anesthetic property, which is the main reason why *A. oleracea* has been used since ancient times to relieve toothache. In addition, anti-inflammatory, analgesic, cytotoxic, antioxidant, antimicrobial, anthelmintic, antiwrinkling, aphrodisiac, and insecticidal/acaricidal properties are reported [1–7].

The richness and variety of secondary metabolites are mainly responsible for *A. oleracea* bioactivities. Nascimento et al. (2020) and Bellumori et al. (2022) [8,9] characterized many bioctive compounds, including mainly phenolic acids, glycosylated flavonoids, alkamides, and fatty acids, some of which specific to certain anatomical parts of the plant or cultivation system [10,11]. Of particular interest are the chlorogenic acids, a class of esters between caffeic and quinic acids which represent the main phenolic acids detected in this plant. These isomer compounds of *Acmella*, also found in many other plant species, are biologically important and have gained increasing interest in recent years [12–14]. However, due to their structural similarity and instability, and the fact that the number of commercially available standards is limited to just a few, the correct analysis and identification of these compounds in plants is a challenge. The most widespread analytical approach to solve this problem involves the chromatographic separation of the component of sample extract and their characterization by ultraviolet-visible (UV-Vis), mass spectrometry (MS), or tandem mass spectrometry (MS/MS) spectra [15–17]. Nevertheless, the application of this strategy often requires the set up of ad hoc sample preparation and the evaluation of different chromatographic columns or mobile phase composition to obtain adequate analytes separation. All these procedures are usually molecule-specific and rarely can be extended to other compounds. From this point of view, the application of MS detection can be a promising technology as it combines both the "universal" revelation features and compounds distinction proprieties, separating the analyte-specific ion component from the total acquired signal. Unfortunately, the MS-specific revelation capability find a strong limitation in the isomers analysis, when their characteristics lead to a common ion signals. However, numerous strategies have been proposed to solve the problem and some of these have involved the study of MS/MS features allowing the characterization and quantification of isomers and/or isobars in mixtures via a standardized approach, applicable to different compounds [18–28]. Our proposal aims to exploit the separation characteristics of mass spectrometry, reversing the guarantees of specificity of the analytical determination from the chromatographic system to the mass detector. We addressed this topic by introducing and developing Linear Equation of Deconvolution Analysis (LEDA), a mathematical tool that allows the recognition of isomers by processing MS/MS data without their chromatographic separation [29–32]. In this study an HPLC-MS/MS method supported by LEDA was applied in determination of the three caffeoylquinic acids (CQAs) and three di-caffeoylquinic acids (diCQAs) in *A. oleracea* (Figure 1).

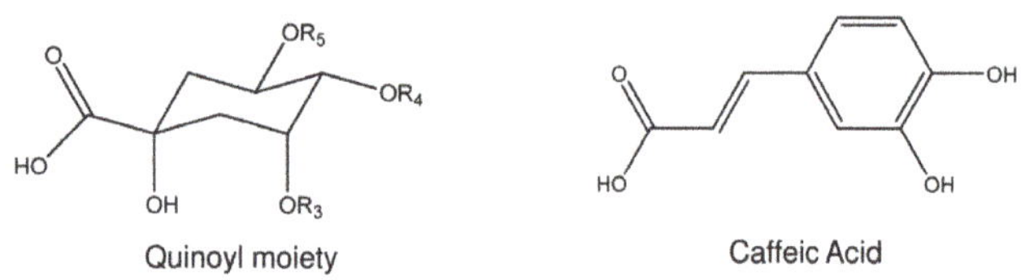


| Common Name | Abbreviation | R3 | R4 | R5 | MW (Da) |
|---|---|---|---|---|---|
| 3-*O*-Caffeoylquinic acid | 3-CQA | Caffeoyl- | H | H | 354 |
| 4-*O*-Caffeoylquinic acid | 4-CQA | H | Caffeoyl- | H | 354 |
| 5-*O*-Caffeoylquinic acid | 5-CQA | H | H | Caffeoyl- | 354 |
| 3,4-di-*O*-Caffeoylquinic acid | 3,4-diCQA | Caffeoyl- | Caffeoyl- | H | 516 |
| 3,5-di-*O*-Caffeoylquinic acid | 3,5-diCQA | Caffeoyl- | H | Caffeoyl- | 516 |
| 4,5-di-*O*-Caffeoylquinic acid | 4,5-diCQA | H | Caffeoyl- | Caffeoyl- | 516 |

**Figure 1.** General structures, common names, abbreviations, and molecular weights of the studied analytes.

The introduced HPLC approach significantly simplifies the operating parameters, using a short column and a fast elution gradient. Following these arrangements, the chromatographic column was used only to avoid or limit the interference towards the analytes' ionization process by the sample matrix (matrix effects), allowing the extension to recognition of other compounds present in the sample only modifying the MS/MS conditions [31,32].

The aim was to propose a simple and rapid determination and differentiation of the *A. oleracea* caffeoylquinic acid isomers (Figure 1), resolving the MS/MS spectra and assigning the correct signal to each isomer, even given chromatographically unresolved peaks. A single chromatographic set up was used, tuned to minimize the run time without requiring high efficiency or resolution between analytes, exploiting the distinguishing capabilities of MS/MS experiments.

## 2. Results and Discussion

Recognition of isomers by applying the LEDA approach involves using a multistage MS system and collecting as much information as possible on the MS/MS behavior of each analyte. Regarding the MS system, in this study an ion trap (IT) MS analyzer was used that shows a peculiarity compared with other multistage instruments: the MS/MS experiments are performed in the same site, using time to modify the conditions applied on studied ions [33]. Then, a sequence of time-dependent steps was carried out to perform the selection of precursor ion, its fragmentation by collision-induced dissociation (CID) mechanism and detection of formed product ions (Pis) [34,35]. Thus, the sum of these time-steps leads to a longer MS/MS acquisition cycle in the IT compared with other multistage instruments, reducing the collection frequency of sample data collection and affecting its chromatographic profile. However, IT instruments show some advantages in MS/MS analysis, such as the sensitivity of the Pis scan acquisition (complete MS/MS spectrum), the different management of excitation energy, and the reiteration on a selected product ion of another tandem MS experiment ($MS^n$) [36]. Concerning the collection of MS/MS data for each studied compounds, unfortunately, a complete panel of pure standards was not available in our lab. Therefore, the required information was gathered from the analysis of *Acmella* working solution (Acmella WS) through a conventional chromatographic separation of each caffeoylquinic-isomer and following its MS/MS characterization. Taking this information into account, we planned a series of studies to acquire the MS/MS data of each analyte, investigate the IT features, and define the most suitable MS/MS parameters for the isomers recognition using the LEDA approach.

The information collected for each caffeoylquinic-isomer present in Acmella WS was as follows:

- Separation by a conventional chromatographic method;
- MS/MS characterization by energetic dimension of CID mechanism;
- Interpretation of obtained MS/MS spectra (CID study);
- Application of the LEDA approach in Acmella WS;
- Assessment of the LEDA quantitative performances.

### 2.1. *Conventional Chromatographic Separation of Caffeoylquinic-Isomers in Acmella WS*

The complete separation of each studied analyte was necessary as their pure standards were not available in our lab. Hence, the separative capabilities of HPLC were exploited to "purify" the analyte signals, allowing a study to evaluate their individual MS/MS behavior.

This purpose was achieved by using a conventional HPLC-MS approach which involved the use of a lengthy column, a slow elution gradient program and a long run-time (Chromatography system 1 or ChromSys 1, see Section 3.4). Considering only the peaks over 1% of relative abundance, the obtained chromatographic profiles showed two groups of isomers: the CQAs (3-, 5- and 4-CQA), characterized by $[M\text{-}H]^-$ ion species at 353 $m/z$ and the diCQAs (3,4-, 3,5- and 4,5-diCQA) that exhibit the $[M\text{-}H]^-$ species at 515 $m/z$. The identification of the 5-CQA was achieved by correlation between the retention times and

MS/MS spectra obtained with the pure standard, while the preliminary attribution of the other CQAs and diCQAs compounds were carried out by comparing the MS/MS spectra of unknown peaks with those published by M.N. Clifford et al. [37,38]. The proposed separation was found to be suitable for distinguishing each group of isomers, allowing the use of an appropriate MS/MS event, distributed over two different time segments. Figure 2 shows an example of the chromatographic profile of the HPLC-MS analysis, carried out with ChromSys 1, of the negative ions acquired in $m/z$ range between 150 and 750 of the Acmella WS.

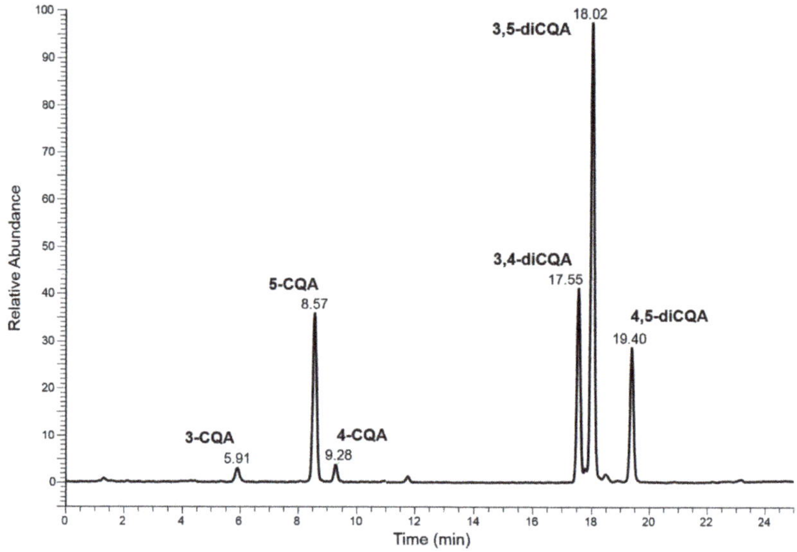


**Figure 2.** Chromatographic profile of the HPLC-MS analysis (ChromSys 1) of the negative ions in $m/z$ range between 150 and 750 of the Acmella WS.

In the Supplementary Materials (Table S1) the calculated peak parameters (i.e., retention times, peaks width, efficiency, etc.) for each analyte have been reported. The resolution value ($\geq$1.5) calculated for each contiguous pair of analytes assured the "purity" of every isomer peak, avoiding mutual interferences in following MS/MS characterization.

### 2.2. MS/MS Characterization by Energetic Dimension of CID Mechanism

In the IT MS/MS experiment, the precursor ion is energized by applying an excitation amplitude (ExA) for an established time (excitation time or ExT) in order to raise the number of collisions with the helium gas, normally present inside the Its. The combination of ExA and ExT values allows the activation of the fragmentation channels of the precursor ion and the formation of the Pis. As reported above, pure standards for each component studied were not available; therefore, in order to obtain information on their CID behavior, a series of HPLC-MS/MS analyses were performed on $[M-H]^-$ ion species of every caffeoylquinic-isomer of Acmella WS. Each analysis was carried out with the same chromatographic separation, using ChromSys 1, but different precursor ion energization. In a previous work, it was verified that the energization of the precursor ion during the CID process reached an equilibrium for the ExT values between 50 and 100 milliseconds (ms) [39]. Therefore, due to the long time for each HPLC-MS/MS analysis, the ExA values were varied in the range of 0 to 50 arbitrary units (a.u.), while the ExT value was kept constant at 50 ms. In this way, an energy resolved mass spectrometry experiment (ERMS) on each caffeoylquinic-isomer present in Acmella WS was simulated and the MS/MS data collected were used to evaluate the energetic dimension of CID process. This evaluation was performed by plotting the survival yield curves of precursor ion (SY), the Pis formation (PiF) and the Pis yield (PiY), as detailed in Section 3.3). A typical graph used to describe the CID energetic study was reported in Figure 3, while the graphs of the other isomeric compounds can be found in the Supplementary Materials (Figures S1–S5).

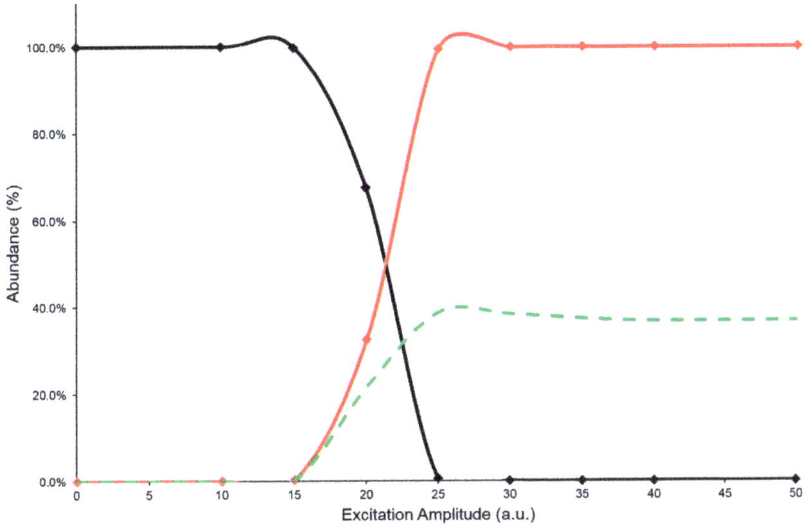


**Figure 3.** SY (black line), PiF (red line), and PiY (green dashed line) curves of the 5-CQA isomer at ExT 50 ms.

The PiF curve (red line) represents the ratio between the sum of the intensities of Pis and the total abundance of detected ions (see Equation (2)). The PiF trend should describes the efficiency of CID process but the abundance ratios, used to build it, are not fully representative of the mechanism. Indeed, these ratio values neglect the signal loss due to the precursor ion ejection, during the excitation process, or caused by undetected product ions, because their $m/z$ are below the low mass cut-off of the IT MS/MS experiment. As a result, the calculated PiF values often suffer from overestimation, depicting an incorrect CID efficiency. Therefore, to evaluate a reliable value of fragmentation yield from the CID process, the sum of the abundances of Pis was referred to the averaged intensity of the precursor ion before its fragmentation (PiY, see Equation (3)). By plotting the calculated values of PiY vs. ExA applied, a curve (green dashed line) is obtained that should best describe the efficiency of the CID process under the tested conditions. In this graph, the trend of the PiY curve increases achieving the highest value (PiYmax); the corresponding ExA value (ExAmax) represents the energy amplitude to be applied in the MS/MS process to reach the greatest abundance of Pis. Furthermore, observing the profiles of the SY plots, it is worth noting that the SY curve (black line) crosses the PiF curve (red line) at 50% abundance value: the related ExA value is characteristic for each compound subjected to the CID process and represents the ExA required to fragment 50% of the precursor ion ($SY_{ExA50}$) [20,21]. In Table 1 are reported the data of SYExA50, PiYmax, and ExAmax obtained for the studied isomers.

The collected data described the energy requirement during the IT-CID process to fragment the precursor ion of the studied analytes. Most of the analytes showed an $ExA_{max}$ of 25 a.u., with the exception of 3,4-diCQA, which required 30 a.u. to achieve the highest Pis formation. Therefore, in order to maintain the same CID condition for the two class of compounds (mono- and di-CQAs), the MS/MS time-event was set with an ExA value of 25 a.u. for the mono CQAs isomers, while an ExA value of 30 a.u. was used for the diCQAs isomers (see Section 3.5). It is worth noting that, for all the studied compounds, the $PiY_{max}$ values were estimated to be around 50% (between 39% and 56%) compared with the abundance of the precursor ion, confirming a loss of ion signal during the MS/MS process. This decrease in signal was unavoidable and can be attributed to both the MS/MS mechanism used and the characteristics of the analyte.

**Table 1.** The characteristic parameters of CID process such as efficiency ($PiY_{max}$), $SY_{ExA50}$ values, and related applied ExA ($ExA_{max}$) calculated from ERMS experiments carried out on caffeoylquinic-isomers of Acmella WS. A.u., arbitrary units.

| Compound | $SY_{ExA50}$ (a.u.) | $ExA_{max}$ (a.u.) | $PiY_{max}$ (%) |
|---|---|---|---|
| 3-CQA | 21.0 | 25 | 56 |
| 5-CQA | 20.5 | 25 | 53 |
| 4-CQA | 19.9 | 25 | 39 |
| 3,4-diCQA | 24.4 | 30 | 41 |
| 3,5-diCQA | 18.1 | 25 | 39 |
| 4,5-diCQA | 22.1 | 25 | 43 |

## 2.3. Collision-Induced Dissociation Study

The CQAs and diCQAs isomers studied (Figure 1), showed MS spectra in negative ions electrospray ionization (ESI) with only an abundant cluster signal representative of the deprotonated molecules ($[M-H]^-$ species), with the same $m/z$ value within the isomeric set (Supplementary Materials, Figures S6 and S7). Therefore, an ERMS study (see Section 3.4) was performed on all these isomers to enhance the differences between their fragmentation pathways and/or to find any specific Pis, which would allow them to be distinguished. The ERMS data of each analyte were processed and the collision breakdown curves were plotted; the obtained graphs were reported in Figure 4 and in the Supplementary Materials (Figures S8–S12).

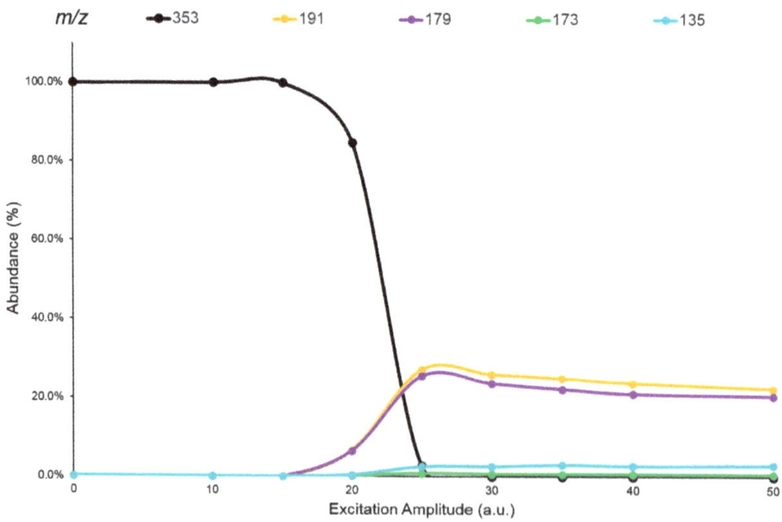


**Figure 4.** The breakdown curves of 3-CQA isomer obtained by elaboration of ERMS data. The legend reports each monitored ion ($m/z$) and its representative color in the graph.

These graphs showed, after activation of their fragmentation channels, a constant Pis composition and the related formation curves followed the same trend achieving a highest value at the same ExA. The pattern of these breakdown curves demonstrates that the fragmentation channels of precursor ion in IT are simultaneously activated at a defined ExA. Hence, after complete dissociation of the precursor ion, the IT MS/MS spectra are the same despite the ExA increase; only PiY changed, reaching its maximum value at ExAmax. Concerning the fragmentation pathway of the analytes, M.N. Clifford et al. proposed an extensive study on the fragmentation behavior of CQAs and diCQAs. The CQAs isomers showed mainly three common negative ions: the quinic acid residue ion at 191 $m/z$, the caffeic acid ion at 179 $m/z$ and a dehydrated quinic acid ion at 173 $m/z$ [37]. The relative abundance of these ions differed between the considered isomers and can be useful for their distinction. Conversely, the diCQAs showed the formation of a favorite Pi at 353 $m/z$ (relative abundance $\geq 25\%$), representative of the loss of a caffeoyl residue from the $[M\text{-}H]^-$ species [38]. Other fragment ions were detected at low intensity (relative abundance $\leq 5\%$) which, together with the 353 $m/z$ ion, can be useful for the isomer recognition. Indeed, all these ions showed different abundances depending on the origin precursor isomer. Therefore, the use of an appropriate 'mathematical device' suitable to distinguish the common signals by processing the MS/MS data and assigning the correct abundance to the isomer(s) present in the sample would be extremely useful.

### 2.4. HPLC-MS/MS Supported by LEDA Approach

The LEDA post-processing mathematical device allows compounds to be distinguished in the analysis of complex samples by the elaboration of the acquired MS/MS data. The case of isomer recognitions is complicated by their similar characteristics such as the same elemental composition, molecular weight, and often common fragment ions. The novelty of the LEDA approach is based on the exploration of the "energetic dimension" MS/MS experiment to enhance possible intrinsic differences between the isomer compounds and allowing their recognition in a complex sample. This algorithm is based on the rationale that the MS/MS spectrum is represented as the sum of the contribution of each isomer present in the sample. As shown in the breakdown curves of Figure 4, each isomer produces the same panel of Pis with a different formation yield. To increase the reliability of these data and improve the significance value of differences on Pi yield, the abundance ratio of each Pi to a reference ion (Ri) was calculated. For this purpose, the Ri signal was chosen among the MS/MS signals having the same abundances, under given MS/MS conditions, among the studied isomers. Unfortunately, none of the Pis detected in MS/MS spectra of analytes had the required Ri features. However, in the analysis of isomers by the ESI source, it is reasonable to consider a similar behavior in their ionization process and then in the MS abundance signals (e.g., $[M\text{-}H]^-$ species). Based on this hypothesis, the signal of the precursor ion was selected as Ri at the maximum intensity in the breakdown curve, without any fragmentation process occurring. In this way, the MS/MS experiment must be split into two events characterized by two different energetic levels (low and high) which allow to acquire both the unfragmented precursor ion (Ri) and its Pis spectrum. Also, by playing with "energetic dimension" of the MS/MS experiment, the acquisition of related ions (e.g., precursor and Pis) is permitted, monitoring the distinctive features between the isomers emphasized in the CID study. In this way, the ratio between the abundances of each Pi acquired and Ri represented the yield of formation of each Pi at selected ExA. In Figure 5 the comparison between the MS/MS spectra of 4,5-diCQA isomer obtained at low and high energetic level of MS/MS experiment is shown, while the corresponded MS/MS spectra related to other isomers were reported in Supplementary Materials (Figures S13–S17).

In an IT application, the use of the precursor ion signal as Ri requires inserting a low collision energy MS/MS event to enable acquisition of the unfragmented precursor ion (MS/MS event 1). Knowing the MS/MS behavior of each isomer, the LEDA tool elaborates each MS/MS ratio signal, distinguishes the possible components, and assigns the correct abundance to the isomers present. Profiting by LEDA features, it was possible to introduce a "standardized" methodological approach by simplifying the HPLC parameters, without worrying about the separation of the isomers [39]. To verify the proposed approach, we replicated the analysis of Acmella WS with a HPLC-MS/MS system assembled with a short column and using a fast elution gradient (see Section 3.5, Chromatography system 2 or ChromSys 2). The tuning of the elution parameters was carried out optimizing only the separation among CQAs and diCQAs isomers groups. This distinction was necessary to allow the setting up of the two MS/MS time segments, each dedicated to monitoring a set of isomers. Within each time segment, two MS/MS events are introduced to alternatively acquire the Ri and Pis signals (see Section 3.5). The proposed HPLC method, as expected, achieved only the partial components separation of the two major group of peaks, representative of the CQAs and diCQAs isomers. However, this result was reached in just ten minutes of chromatographic run without renouncing the solvents gradient elution, a desirable feature to avoid cross-contamination, due to unknown components, between the sample. To solve the partial coelution of the peaks and allow the distinction of the studied isomers, each acquired MS/MS data point was processed by the LEDA algorithm ('scan-by-scan' method) and a relative abundance in the Ri signal was assigned to each isomer present. Indeed, the abundance of Ri signal represents the sum of abundances of the precursor ions of all the isomers present in the elaborated data point. Hence, after the LEDA computing, a relative amount of Ri signal was allocated to each isomer present, with a consequent deconvolution of unresolved chromatographic peak (reconstructed chromatographic profile). Observing a typical reconstructed chromatographic profile, the performance of LEDA can be appreciated in the separation of co-eluted peak of the Ri signal that splits its abundance between the isomers, generating a reconstructed chromatographic trace of each recognized isomer. The comparison between a typical chromatogram obtained by analysis of Acmella WS with ChromSys 2 and its LEDA reconstructed chromatographic profiles are reported in Figure 6.

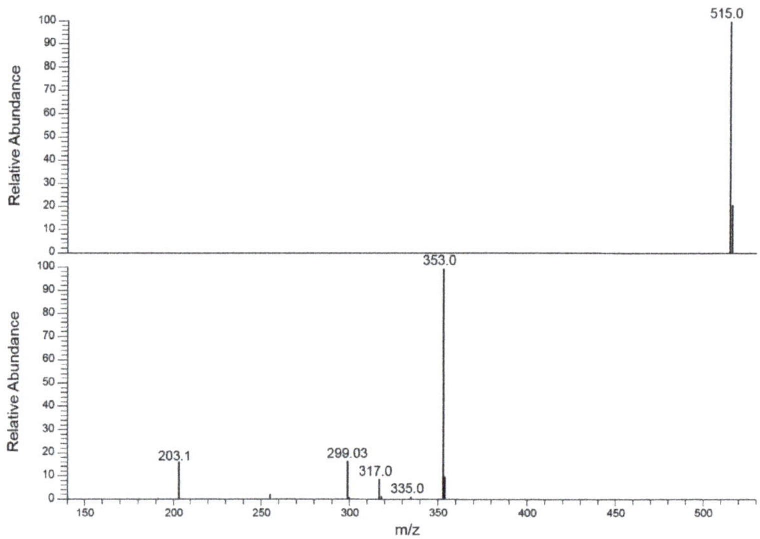


**Figure 5.** Comparison of the MS/MS spectra obtained at low ExA (MS/MS event 1, (**above**)), used to monitor the Ri signal, and at high ExA (MS/MS event 2, (**below**)), to acquire the Pis signals of 4,5-diCQA.

It should be noted that in the chromatogram reconstructed with LEDA, all the isomers studied (e.g., 5-CQA and 4-CQA or 3,4-diCQA and 3,5-diCQA) can be distinguished even under conditions of coelution or different relative abundance. Naturally, the reliability of LEDA in the signal assignment must be verified, by comparing the quantitative data obtained from ChromSys 2 with those obtained from conventional HPLC-MS/MS analysis (ChromSys 1).

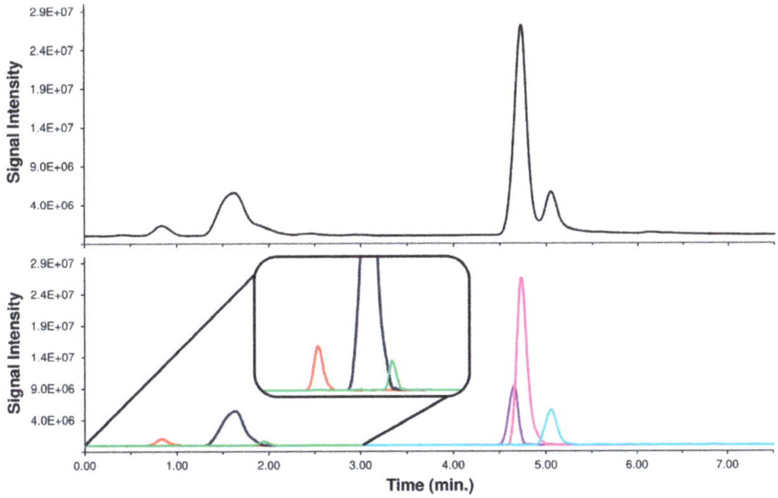


**Figure 6.** Chromatographic profile of the Ri signals (black lines) of Acmella WS with ChromSys 2 (**above**) and LEDA reconstructed chromatographic profiles of the 3-CQA (red line), 5-CQA (blue line), 4-CQA (green line), 3,4-diCQA (purple line), 3,5-diCQA (pink line), and 4,5-diCQA (light blue line) (**below**). A tenfold zoom of the abundance (inset) was reported to highlight the CGAs detected at low amount.

### 2.5. Evaluation of Quali-Quantitative Performance of LEDA Approach

The lack of pure analyte standards did not allow us to prepare spiked samples, with known concentration and composition, to evaluate the analytical performance (accuracy and precision) of the LEDA approach. Therefore, the same sample of Acmella WS were independently analyzed by the two proposed chromatographic systems and the results of quali-quantitative analyses of CQAs and diCQAs were compared. The determination of the CQAs and diCQAs in the conventional HPLC-MS/MS analysis (ChromSys 1) was performed by integration of every peak area on the Ri signal corresponding to the separated analytes (see Figure 2). Each integrated area was converted in quantitative data through the external standard method by referring to a five levels calibration curve built analyzing standard solutions of chlorogenic acid (5-CQA) with the same HPLC-MS/MS method. Also, for the 5-CQA, the Ri signal (unfragmented precursor ion) was used for the integration of peak areas. Indeed, as previously reported, these analytes have a different behavior during the CID process, with formation of Pis at different yields. Thus, using the unfragmented precursor ion (Ri) as the quantitative signal appeared to be the best compromise as reference for a quantitative evaluation. The analysis performed with chromatographic system 2 did not allow the quantitative estimation of the analytes, as the poor separation of the isomers prevented their recognition. Therefore, it was necessary to integrate all the MS/MS signals selected during the LEDA setup, calculate each Pi vs. Ri ratio, and process them with the proposed "mathematical device". Also in this case, the Ri area was distributed among the isomers present in chromatographic peak, respecting their relative abundances (peak purity). The Ri assigned area of each recognized isomer was converted into quantitative data by referring to the same 5-CQA calibration curve previously used. The Acmella WS was analyzed six times for each proposed chromatographic system to allow evaluation of the precision of quantitative approaches by calculating the deviation standard (SD) of the results. Table 2 showed the quantitative data obtained by analyzing the Acmella WS with the conventional HPLC separation (ChromSys 1) and the new proposed LEDA approach (ChromSys 2).

**Table 2.** Comparison between the quantitative data obtained by analyzing the Acmella WS with ChromSys 1 and 2.

| Isomers | ChromSys 1 Acmella WS ± SD (mg L$^{-1}$) | ChromSys 1 Acmella ES ± SD (mg L$^{-1}$) | ChromSys 2 Acmella WS ± SD (mg L$^{-1}$) | ChromSys 2 Acmella ES ± SD (mg L$^{-1}$) |
|---|---|---|---|---|
| 3-CQA | 0.6 ± 0.1 | 12 ± 2 | 0.8 ± 0.1 | 15 ± 2 |
| 5-CQA | 6.0 ± 0.5 | 120 ± 10 | 5.5 ± 0.2 | 110 ± 4 |
| 4-CQA | 0.4 ± 0.1 | 8 ± 2 | 0.4 ± 0.1 | 8 ± 2 |
| 3,4-diCQA | 6.5 ± 0.5 | 130 ± 10 | 7.0 ± 0.5 | 140 ± 10 |
| 3,5-diCQA | 16.5 ± 1.0 | 330 ± 20 | 18.0 ± 0.5 | 360 ± 10 |
| 4,5-diCQA | 4.0 ± 0.5 | 80 ± 10 | 4.5 ± 0.5 | 90 ± 10 |

Quantitative results showed no significant differences between the ChromSys used, demonstrating that both the analytical approaches lead to the same results, albeit ChromSys 2 in a quarter of the time (Supplementary Materials, Figure S17). Particularly interesting were the discriminating properties of the LEDA algorithm in the determination of 4-CQA, estimated at a low concentration level in the Acmella WS. In fact, the 4-CQA in the ChromSys 2 was coeluted with 5-CQA, present at a ten times higher concentration in the sample; despite this quantitative significant difference, the LEDA effectively processed the MS/MS data distinguishing the correct amount of the 4-CQA signal from the peak tail of the 5-CQA (Figure 6, inset).

## 3. Materials and Methods

### 3.1. Chemicals and Instruments

Ultrapure water (resistivity 18 MΩ cm) was obtained by the Milli-Q-system Millipore (Milan, Italy). Acetonitrile, ethanol, *n*-hexane (Chromasolv), formic acid, ammonium formate (MS grade), and chlorogenic acid or 5-CQA (purity ≥ 95%) were purchased from Merck (Milan, Italy). The HPLC-MS/MS analysis was carried out by using a Thermo LCQ Deca XP plus ion trap (Waltham, MA, USA) equipped with Surveyor liquid chromatography system and an electrospray ion source (ESI). Raw data were collected and processed by Excalibur version 1.4 software.

### 3.2. Standard and Calibration Solutions

Stock solution of 5-CQA was prepared in acetonitrile at 1.0 mg mL$^{-1}$ and stored at 4 °C. The 5-CQA working solutions were freshly prepared by diluting stock solutions up to a concentration of 100 mg L$^{-1}$ and 10 mg L$^{-1}$ (working solutions 1 and 2, respectively) in mixture of ultrapure water:acetonitrile 50:50 (*v*/*v*). The quantitative data of each analyte of Acmella WS were calculated by an external standard method, referring to the 5-CQA calibration curve. This curve was built analyzing five level calibration solutions prepared by adding a proper volume of working solution (1 or 2) and diluted up to 1 mL with ultrapure water:acetonitrile 50:50 (*v*/*v*) solution. Final concentrations of 5-CQA calibration solutions were: 1.0, 2.5, 5.0, 10.0, and 20.0 mg L$^{-1}$, respectively. Each calibration solution was analyzed six times by the ChromSys 1 proposed method.

### 3.3. Preparation of Phenolic Extracts

The powdered roots obtained from in vitro seedling plants from regenerating lines derived by organogenesis of Acmella oleracea were extracted as reported in Bellumori et al. [9]. Briefly, 0.5 g were extracted twice with 10 mL of ethanol 80% (*v*/*v*) at 60 °C for 10 min under sonication and centrifuged at 5000 g for 10 min. The two supernatants were collected and defatted with n-hexane (1:2 *v*/*v*); the hydroalcoholic extract was then dried under vacuum at 35 °C, re-dissolved with exactly 5 mL of the same extractive mixture and stored a −20 °C (Acmella extract sample or Acmella ES). To perform the MS and MS/MS study, a diluted solution of the Acmella extract sample was daily prepared, collecting

50 µL and diluting up to 1 mL by ultrapure water:acetonitrile 50:50 (*v*/*v*) mixture (Acmella working solution or Acmella WS).

### 3.4. MS and ERMS Experiments

In all the reported MS and MS/MS analysis in this manuscript, the ESI source operates in negative ion mode by using the following setting: 5 kV source voltage, 45 arbitrary units (a.u.) sheath gas, 5 a.u. auxiliary gas, the capillary voltage was 25 V while its temperature was set at 280 °C, and 40 V of the tube lens. The isomers studied are structurally characterized by a free carboxylic function in position 1 of the quinoyl-moiety and therefore their dissociation as carboxylate ions in ESI source is favored. The MS analyzes, used to characterize the Acmella WS, were carried out in ions scan mode by monitoring the $m/z$ range from 150 to 750 with 50 ms of maximum ion trap fill time. The MS/MS study, scheduled to study the energetics of the fragmentation of $[M\text{-}H]^-$ species of each analyte and build its breakdown curves [39], were carried out increasing the ExA (named Normalized Collision Energy in the control panel of the instrument) stepwise in the range 0–50 a.u. To obtain the information about the fragmentation and CID behavior of each analyte, a series of HPLC-MS/MS analysis of the same Acmella WS was performed by using a common HPLC elution (ChromSys 1) with different MS/MS applied energies in CID process. The entire MS/MS analyzes batch represents the ERMS experiment and it allows the description of the energetic dimension of CID process. The other common parameters used for all the ERMS experiment were: 3 $m/z$ of precursor isolation width, 50 ms of ExT, and 0.25 of q value. The obtained ERMS data were used to build the graphs that describe the energetic dimension of CID process (SY curves of precursor ion, the Pis formation, and the Pis yield). The SY curve describes the energetics of degradation of precursor ion according to grow of the ExA. Each SY value at defined ExA is calculated as follows:

$$SY(\%) = \frac{\text{Precursor ion Abundance}}{\text{Precursor ion Abundance} + \sum_1^n \text{Pis Abundances}} \times 100 \quad (1)$$

Likewise, the Pis formation (PiF) curve describes the energetics of formation of all Pis with the increment of applied ExA. Each PiF value at defined ExA is calculated as follows:

$$PiF(\%) = \frac{\sum_1^n \text{Pis Abundances}}{\text{Precursor ion Abundance} + \sum_1^n \text{Pis Abundances}} \times 100 \quad (2)$$

The Pis yield (PiY) is calculated to depict the efficiency of CID process and is estimated by ratio between the sum of abundances of the Pis at defined ExA vs. the average intensity of precursor ion before its decay. Normally, the higher abundance of precursor ion is observed at low ExA values, before the activation of fragmentation channels. The average of these abundances represents the quantity of precursor ion signal (Precursor ion max) available for CID process, while the sum of Pis abundances is the signal remained. Then, the yield of CID process should be described as follows:

$$PiY(\%) = \frac{\sum_1^n \text{Pis Abundances}}{\text{Precursor ion max}} \times 100 \quad (3)$$

Finally, the CID graphs were plotted to describe the dependence of fragmentation/formation processes (breakdown curves) from the energy provided to the precursor ion. The breakdown curves were plotted by reporting the ratio between the intensity peak values of each ion signal in the MS/MS spectra versus the precursor ion max at the ExA applied. In this way, the ratio of the abundances represents the yield of formation of each Pi respect to the precursor ion to the applied ExA.

### 3.5. HPLC-MS/MS Methods

In this manuscript two different chromatographic set ups were employed to achieve the planned aims. The first system, abbreviated as ChromSys 1, was used to separate each analyte in Acmella WS and was coupled with both MS and MS/MS detection mode. ChromSys 1 is distinguished by the use of a column Agilent Poroshell 120 EC-C18 2.1 × 150 mm, 2.7 mm particle size (Santa Clara, CA, USA), eluting with a gradient mobile phase. The program of elution gradient was set up as follows: initial at 95% solvent A, then decreased to 75% in 20.00 min, and to 10% in 3.25 min, kept at 10% for 9.75 min, returned to initial conditions in 0.01 min and maintained for 9.99 min to a total run time of 45.00 min. A second chromatographic system (ChromSys 2) was used to analyze the sample with the LEDA approach. The column used by this system was an Agilent Poroshell 120 EC-C18 2.1 × 30 mm, 2.7 mm particle size (Santa Clara, CA, USA), also employing a mobile phase elution gradient. This program of elution gradient was set up as follows: initial at 95% solvent A, then decreased to 70% in 5.00 min and to 10% in 0.10 min respectively, kept at 10% for 2.40 min, returned to initial conditions in 0.01 min and maintained for 2.49 min to a total run time of 10.00 min. The common parameters used by both chromatographic systems were: the mobile phase was maintained at constant flow of 0.25 mL $min^{-1}$, column temperature at 40 °C and the injection volume was 5 µL. Finally, the used solvents were 10 mM formic acid in ultrapure water (solvent A) and 10 mM formic acid in acetonitrile (solvent B).

The elaboration of data from the ERMS experiment allowed the definition of the CID conditions used to compile the MS/MS methods. Indeed, each used chromatographic system was coupled by a specific MS/MS method that assured the correct application of CID parameters for the detection of the studied analytes. These conditions can be divided into common parameters, such as isolation width 3 $m/z$, 0.25 as q value, and ExT of 50 ms, or MS/MS event specifics, listed and reported in Table 3.

**Table 3.** Time segments and MS/MS parameters used for acquisition of the CQAs and diCQAs isomers by using ChromSys 1 or 2.

| | Time Segment | Time Segment (min) | Precursor Ion ($m/z$) | MS/MS Event | Pis Scan Range ($m/z$) | ExA (a.u.) |
|---|---|---|---|---|---|---|
| ChromSys 1 | | | | | | |
| CQAs | 1 | 0.0–11.0 | 353 | Ri | 300–365 | 15 |
| | | | | Pis | 95–250 | 25 |
| diCQAs | 2 | 11.0–25.0 | 515 | Ri | 450–530 | 10 |
| | | | | Pis | 140–365 | 30 |
| ChromSys 2 | | | | | | |
| CQAs | 1 | 0.0–3.0 | 353 | Ri | 300–365 | 15 |
| | | | | Pis | 95–250 | 25 |
| diCQAs | 2 | 3.0–7.5 | 515 | Ri | 450–530 | 10 |
| | | | | Pis | 140–365 | 30 |

The MS/MS methods are both characterized by two acquisition time segments, each one split into two MS/MS events. The acquisition time segments allow the correct application of MS/MS events to CQAs or diCQAs isomers and their time range value depends on the by the chromatography system used. While the MS/MS events defined the acquisition of the ion signals that, alternately applied, allow the recording in ions scan mode of the Ri and Pis signals of the studied isomers. Both the HPLC-MS/MS methods were applied to analyze the Acmella WS, repeating the examination six times for each system to compare the quantitative results between the conventional and LEDA approaches.

### 3.6. The LEDA Alghorithm

The MS/MS spectrum of mixture of isomer or isobar compounds, after complete fragmentation of precursor ion, is represented by the sum of contribution of each component present in the mixture. As described above, the isomer compounds often produce, during MS/MS experiments, common Pis but with different yield of formation. To enhance the compound-dependent Pi yield differences and managing reliable data, the relative abundances of selected Pis with respect to Ri were calculated (see Section 3.6). Therefore, knowing the characteristic abundance ratios (Pi/Ri) of pure isomer, a deconvolution of acquired MS/MS signals is possible based on a series of linear regression equation as follows:

$$\left(\frac{\mathrm{Pi}}{\mathrm{Ri}}\right)_{\mathrm{m}} = \sum_{\mathrm{x}=1}^{\mathrm{n}} \left(\frac{\mathrm{Pi}}{\mathrm{Ri}}\right)_{\mathrm{x}} * [\%]_{\mathrm{x}} \tag{4}$$

- $(\mathrm{Pi}/\mathrm{Ri})_{\mathrm{m}}$ is the abundance ratio between the product ion (Pi) vs. reference ion (Ri) measured (m) in the sample;
- $(\mathrm{Pi}/\mathrm{Ri})_{\mathrm{x}}$ is the characteristic abundance ratio between the Pi vs. Ri of pure isomer;
- $[\%]_{\mathrm{x}}$ is the concentration (%) of the isomer in the sample.

Theoretically, considering a simple binary mixture of isomers (A–B), a single Equation (4) related to only a product ion ratio (Pi/Ri) could be sufficient. Indeed, by assuming that only the pair of isomers constitutes the MS/MS signal, the concentration of B is calculated as B% = (1 − A)%. However, in this case the possible contribution of signals from unknown isomers (or any co-eluting compound having the same precursor and product ions) is neglected. Therefore, the MS/MS signal processing "mathematical device" for recognizing mixtures of n isomers (LEDA algorithm), is preferable to be a matrix with at least n linear Equation (4). Naturally, to increase the specificity and reliability of isomers speciation, an overdetermined system of linear equations can be assembled; in this case the LEDA matrix was composed by a number > n of linear Equation (4). This is the case of IT that operates in MS/MS acquiring the Pis in a defined range of *m*/*z* (product ions scan); then, all Pis signals included in the range are recorded and can be used to set up the LEDA matrix. In detail, the proposed application leads to the determination of two isomers groups (CQAs and diCQA), acquired with distinct CID conditions and different precursor and product ions monitored, then two LEDA matrices were setup to elaborate the MS/MS data referring to each group of isomers. Considering only the Pi/Ri ratios ≥ 2%, two overdetermined system of linear equations can be assembled to solve the CQAs and diCQA isomers sets (supplementary materials, Equations (S1) and (S2)). A general equation to solve each overdetermined LEDA matrix was reported in supplementary materials (Equation (S3)). The characteristic abundance ratios were calculated by data obtained from ERMS experiment described above. The ratios between Pi vs. Ri selected in the MS/MS methods were calculated and the resulting values were reported in Table 4.

**Table 4.** Characteristic ion abundance ratios (Pi/Ri) ± standard deviation (SD) calculated by MS/MS data from 100 ng mL$^{-1}$ solution of each pure isomer by ChromSys 1 described in Section 3.5.

| Isomers Group | Ratio Pi/Ri (*m*/*z*) | 3-CQA Ratio Value ± SD | 5-CQA Ratio Value ± SD | 4-CQA Ratio Value ± SD |
|---|---|---|---|---|
| CQAs | 191/353 | 0.23 ± 0.01 | 0.30 ± 0.01 | 0.02 ± 0.01 |
| | 179/353 | 0.20 ± 0.01 | 0.02 ± 0.01 | 0.20 ± 0.01 |
| | 173/353 | 0.01 ± 0.01 | 0.01 ± 0.01 | 0.17 ± 0.01 |
| | 135/353 | 0.02 ± 0.01 | 0.01 ± 0.01 | 0.01 ±0.01 |
| | **Ratio Pi/Ri (*m*/*z*)** | **3,4-diCQA Ratio Value ± SD** | **3,5-diCQA Ratio Value ± SD** | **4,5-diCQA Ratio Value ± SD** |
| diCQAs | 353/515 | 0.28 ± 0.01 | 0.37 ± 0.01 | 0.26 ± 0.01 |
| | 335/515 | 0.02 ± 0.01 | 0.01 ± 0.01 | 0.01 ± 0.01 |
| | 317/515 | 0.01 ± 0.01 | 0.01 ± 0.01 | 0.02 ± 0.01 |
| | 299/515 | 0.02 ± 0.01 | 0.01 ± 0.01 | 0.04 ± 0.01 |
| | 203/515 | 0.02 ± 0.01 | 0.01 ± 0.01 | 0.04 ± 0.01 |

The deconvolution was performed by applying the algorithm either to the area abundances, obtained from the integrated peak intensities of each Pi signal, or to individual MS/MS data point of the chromatographic sample profile. In the first case, the LEDA provides the relative amounts (%) of each known component present in the sample. In the second approach, each MS/MS signal is deconvoluted 'scan-by-scan' and assigned to the present isomers, allowing a graphical separation of the processed chromatographic profiles. All calculations for the deconvolution of MS/MS data are processed using an Excel™ macro.

### 3.7. Calibration Curve of HPLC-MS/MS Methods

The quantitative data on CQAs and diCQAs isomers, present in *Acmella* sample, were calculated by an external standard method referring on 5-CQA standard solutions. The HPLC-MS/MS analyzes with ChromSys 1 of the calibration solutions of 5-CQA were used to build its calibration curve, obtained by plotting the Ri peak area versus the nominal concentration of each calibration solution. A linear regression analysis was applied to obtain the best fitting function between the calibration points (Supplementary Materials, Figure S19). In order to obtain reliable LOD and LOQ values, the standard error (SE) of response and slope approach was employed [40]. The estimated SEs of responses of each analyte were obtained by the SE of y-intercepts (Y-SE) of regression lines elaboration of data obtained from the HPLC-MS/MS analysis of calibration solutions [31]. The results of 5-CQA calibration curve obtained for MS/MS Ri quantitation signal, defined as linear regressions parameters (slope and y-intercept), the determination coefficient ($R^2$), and the estimated LOD and LOQ values for each analyte are reported in Supplementary Materials (Table S2). The Ri signal should be remembered that represents the precursor ion of the analyte not yet fragmented in the IT-MS/MS experiment; hence, its intensity depends by the molecule characteristics (e.g., weight, functional groups, ionization proprieties, etc.). Since the analytes are isomers or homologous derivatives, it is reasonable that their ionization yield inside the ESI source are similar and, consequently, their response factor (ratio between Ri abundance versus concentration). Then, this 5-CQA calibration curve represents the best compromise as quantitative reference for the determination of CQAs and diCQAs in *Acmella* sample.

## 4. Conclusions

This study proposed a simple and rapid determination of the *A. oleracea* caffeoylquinic isomers, applying an HPLC-MS/MS method supported by LEDA algorithm. The three mono- and the three di-caffeoylquinic acids in roots of *Acmella* plants were characterized by LEDA which allowed to assign a relative abundance in the Ri signal to each isomer present. In fact, although ChromSys2 allowed to obtain only a partial separation of the isomers which did not permit their quantitative estimation, the processing of the MS/MS data with LEDA led to the deconvolution of the unresolved chromatographic peaks. This result was achieved in just ten minutes of chromatographic run, without renouncing the solvent elution gradient and allowing the distinction of the six isomers in a quarter of the time compared with conventional chromatographic methods.

The obtained results demonstrate the effectiveness of the LEDA algorithm in recognizing and separating the different isomers present in complex samples without the need of good chromatographic resolution.

**Supplementary Materials:** The following supporting information can be downloaded at: https://www.mdpi.com/article/10.3390/ph16101375/s1, Table S1. Chromatographic parameters for each analyte obtained with the proposed HPLC-MS analysis; Table S2. Results of the calibration curve obtained for Ri signal of chlorogenic acid (or 5-CQA), defined as linear regressions parameters (slope and y-intercept), the determination coefficient ($R^2$) and the estimated LOD and LOQ values; Figure S1. Precursor SY curve (black line), PiF (red line) and PiY (green dashed line) of the 3-CQA isomer at ExT 50 ms; Figure S2. Precursor SY curve (black line), PiF (red line) and PiY (green dashed line) of the 4-CQA isomer at ExT 50 ms; Figure S3. Precursor SY curve (black line), PiF (red line) and PiY (green dashed line) of the

3,4-diCQA isomer at ExT 50 ms; Figure S4. Precursor SY curve (black line), PiF (red line) and PiY (green dashed line) of the 3,5-diCQA isomer at ExT 50 ms; Figure S5. Four precursor SY curve (black line), PiF (red line) and PiY (green dashed line) of the 4,5-diCQA isomer at ExT 50 ms; Figure S6. ESI negative ions spectra of 3-CQA (upper), 5-CQA (middle) and 4-CQA (bottom); Figure S7. ESI negative ions spectra of 3,4-diCQA (upper), 3,5-diCQA (middle) and 4,5-diCQA (bottom); Figure S8. The breakdown curves of 5-CQA isomer obtained by elaboration of ERMS data; Figure S9. The breakdown curves of 4-CQA isomer obtained by elaboration of ERMS data; Figure S10. The breakdown curves of 3,4-diCQA isomer obtained by elaboration of ERMS data; Figure S11. The breakdown curves of 3,5-diCQA isomer obtained by elaboration of ERMS data; Figure S12. The breakdown curves of 4,5-diCQA isomer obtained by elaboration of ERMS data; Figure S13. Comparison of the MS/MS spectra obtained at low ExA (MS/MS event 1, above), used to monitor the Ri signal, and at high ExA (MS/MS event 2, below), to acquire the Pis signals of 3-CQA; Figure S14. Comparison of the MS/MS spectra obtained at low ExA (MS/MS event 1, above), used to monitor the Ri signal, and at high ExA (MS/MS event 2, below), to acquire the Pis signals of 5-CQA; Figure S15. Comparison of the MS/MS spectra obtained at low ExA (MS/MS event 1, above), used to monitor the Ri signal, and at high ExA (MS/MS event 2, below), to acquire the Pis signals of 4-CQA; Figure S16. Comparison of the MS/MS spectra obtained at low ExA (MS/MS event 1, above), used to monitor the Ri signal, and at high ExA (MS/MS event 2, below), to acquire the Pis signals of 3,4-diCQA; Figure S17. Comparison of the MS/MS spectra obtained at low ExA (MS/MS event 1, above), used to monitor the Ri signal, and at high ExA (MS/MS event 2, below), to acquire the Pis signals of 3,5-diCQA; Figure S18. Comparison between the HLPC-MS/MS analysis of Acmella WS with ChromSys 1 (above) and ChromSys 2 (bottom); Figure S19. The calibration curve of 5-CQA; LEDA algorithm, Equations (S1)–(S3).

**Author Contributions:** Conceptualization, M.B., M.I., N.M. and G.B.; Data curation, M.B. and G.B.; Formal analysis, M.P., B.Z., L.L. and M.M.; Investigation, M.B., M.P., B.Z., L.L. and M.M.; Methodology, M.P. and G.B.; Writing—original draft, M.B., M.I., N.M. and G.B. All authors have read and agreed to the published version of the manuscript.

**Funding:** This research received no external funding.

**Data Availability Statement:** Data is contained within the article.

**Conflicts of Interest:** The authors declare no conflict of interest.

## References


1. Uthpala, T.G.G.; Navaratne, S.B. *Acmella oleracea* plant; identification, applications and use as an emerging food source—Review. *Food Rev. Int.* **2020**, *37*, 399–414. [CrossRef]
2. Pinheiro, M.S.S.; Moysés, D.A.; Galucio, N.C.R.; Santos, W.O.; Pina, J.R.S.; Oliveira, L.C.; Silva, S.Y.S.; Silva, S.C.; Frazão, N.F.; Marinho, P.S.B.; et al. Cytotoxic and molecular evaluation of spilanthol obtained from *Acmella oleracea* (L.) R. K. Jansen (jambu) in human gastric cancer cells. *Nat. Prod. Res.* **2023**, 1–6. [CrossRef] [PubMed]
3. Lobina, C.; Sau, R.; Fara, F.; Maccioni, P.; Carai, M.A.M.; Colombo, G. Analgesic properties of a food grade lecithin delivery system of *Zingiber officinale* and *Acmella oleracea* standardized extracts in rats. *Nat. Prod. Res.* **2021**, *35*, 3078–3082. [CrossRef] [PubMed]
4. Yien, R.M.K.; Gomes, A.C.C.; Fiorot, R.G.; Miranda, A.L.P.; Neves, G.A.; Andrade, B.S.A.; Costa, F.C.; Tributino, J.L.M.; Simas, N.K. Alkylamides from *Acmella oleracea*: Antinociceptive effect and molecular docking with cannabinoid and TRPV1 receptors. *Nat. Prod. Res.* **2023**, *37*, 3136–3144. [CrossRef] [PubMed]
5. Lalthanpuii, P.B.; Lalchhandama, K. Chemical composition and broad-spectrum anthelmintic activity of a cultivar of toothache plant, *Acmella oleracea*, from Mizoram, India. *Pharm. Biol.* **2020**, *58*, 393–399. [CrossRef]
6. Rondanelli, M.; Riva, A.; Allegrini, P.; Faliva, M.A.; Naso, M.; Peroni, G.; Nichetti, M.; Gasparri, C.; Spadaccini, D.; Iannello, G.; et al. The Use of a New Food-Grade Lecithin Formulation of Highly Standardized Ginger (*Zingiber officinale*) and *Acmella oleracea* Extracts for the Treatment of Pain and Inflammation in a Group of Subjects with Moderate Knee Osteoarthritis. *J. Pain Res.* **2020**, *13*, 761–770. [CrossRef]
7. Spinozzi, E.; Ferrati, M.; Baldassarri, C.; Cappellacci, L.; Marmugi, M.; Caselli, A.; Benelli, G.; Maggi, F.; Petrelli, R. A review of the chemistry and biological activities of *Acmella oleracea* ("jambù", Asteraceae), with a view to the development of bioinsecticides and acaricides. *Plants* **2022**, *11*, 2721. [CrossRef]
8. Nascimento, L.E.S.; Arriola, N.D.A.; da Silva, L.A.L.; Faqueti, L.G.; Sandjo, L.P.; de Araújo, C.E.S.; Biavatti, M.W.; Barcelos-Oliveira, J.L.; Dias de Mello Castanho Amboni, R. Phytochemical profile of different anatomical parts of jambu (*Acmella oleracea* (L.) R.K. Jansen): A comparison between hydroponic and conventional cultivation using PCA and cluster analysis. *Food Chem.* **2020**, *332*, 127393. [CrossRef]

9. Bellumori, M.; Zonfrillo, B.; Maggini, V.; Bogani, P.; Gallo, E.; Firenzuoli, F.; Mulinacci, N.; Innocenti, M. *Acmella oleracea* (L.) R.K. Jansen: Alkylamides and phenolic compounds in aerial parts and roots of in vitro seedlings. *J. Pharm. Biomed. Anal.* **2022**, *220*, 11499. [CrossRef]
10. Li, A.N.; Li, S.; Zhang, Y.J.; Xu, X.-R.; Chen, Y.M.; Li, H.B. Resources and Biological Activities of Natural Polyphenols. *Nutrients* **2014**, *6*, 6020–6047. [CrossRef]
11. Huang, W.Y.; Yi-Zhong Cai, Y.Z.; Zhang, Y. Natural Phenolic Compounds From Medicinal Herbs and Dietary Plants: Potential Use for Cancer Prevention. *Nutr. Cancer* **2009**, *62*, 1–20. [CrossRef] [PubMed]
12. Alcázar Magaña, A.; Kamimura, N.; Soumyanath, A.; Stevens, J.F.; Maier, C.S. Caffeoylquinic acids: Chemistry, biosynthesis, occurrence, analytical challenges, and bioactivity. *Plant J.* **2021**, *107*, 1299–1319. [CrossRef] [PubMed]
13. Lu, H.; Tian, Z.; Cui, Y.; Liu, Z.; Ma, X. Chlorogenic acid: A comprehensive review of the dietary sources, processing effects, bioavailability, beneficial properties, mechanisms of action, and future directions. *Compr. Rev. Food Sci. Food Saf.* **2020**, *19*, 3130–3158. [CrossRef] [PubMed]
14. Wianowska, D.; Gil, M. Recent advances in extraction and analysis procedures of natural chlorogenic acids. *Phytochem. Rev.* **2019**, *18*, 73–302. [CrossRef]
15. Zhang, Y.; Cai, P.; Cheng, G.; Zhang, Y. A Brief Review of Phenolic Compounds Identified from Plants: Their Extraction, Analysis, and Biological Activity. *Nat. Prod. Commun.* **2022**, *17*. [CrossRef]
16. Khoddami, A.; Wilkes, M.A.; Roberts, T.H. Techniques for Analysis of Plant Phenolic Compounds. *Molecules* **2013**, *18*, 2328. [CrossRef]
17. Wianowska, D.; Typek, R.; Dawidowicz, A.L. Chlorogenic Acid Stability in Pressurized Liquid Extraction Conditions. *J. AOAC Int.* **2015**, *98*, 415–421. [CrossRef]
18. Crotti, S.; Menicatti, M.; Pallecchi, M.; Bartolucci, G. Tandem mass spectrometry approaches for recognition of isomeric compounds mixtures. *Mass Spectrom. Rev.* **2021**, *42*, e21757. [CrossRef]
19. Kurimoto, A.; Daikoku, S.; Mutsuga, S.; Kanie, O. Analysis of energy-resolved mass spectra at MSn in a Pursuit to characterize structural isomers of oligosaccharides. *Anal. Chem.* **2006**, *78*, 3461–3466. [CrossRef]
20. Kertesz, T.M.; Hall, L.H.; Hill, D.W.; Grant, D.F. CE50: Quantifying collision induced dissociation energy for small molecule characterization and identification. *J. Am. Soc. Mass Spec.* **2009**, *20*, 1759–1767. [CrossRef]
21. Memboeuf, A.; Jullien, L.; Lartia, R.; Brasme, B.; Gimbert, Y. Tandem mass spectrometric analysis of a mixture of isobars using the survival yield technique. *J. Am. Soc. Mass Spec.* **2011**, *22*, 1744–1752. [CrossRef]
22. Dit Fouque, J.D.; Maroto, A.; Memboeuf, A. Purification and Quantification of an Isomeric Compound in a Mixture by Collisional Excitation in Multistage Mass Spectrometry Experiments. *Anal. Chem.* **2016**, *88*, 10821–10825. [CrossRef] [PubMed]
23. Dit Fouque, J.D.; Lartia, R.; Maroto, A.; Memboeuf, A. Quantification of intramolecular click chemistry modified synthetic peptide isomers in mixtures using tandem mass spectrometry and the survival yield technique. *Anal. Bioanal. Chem.* **2018**, *410*, 5765–5777. [CrossRef] [PubMed]
24. Maroto, A.; Dit Fouque, D.J.; Memboeuf, A. Ion trap MS using high trapping gas pressure enables unequivocal structural analysis of three isobaric compounds in a mixture by using energy-resolved mass spectrometry and the survival yield technique. *J Mass Spectrom.* **2020**, *55*, e4478. [CrossRef]
25. Dit Fouque, J.D.; Maroto, A.; Memboeuf, A. Structural analysis of a compound despite the presence of an isobaric interference by using in-source Collision Induced Dissociation and tandem mass spectrometry. *J. Mass Spectrom.* **2021**, *56*, e4698. [CrossRef] [PubMed]
26. Campbell, M.T.; Chen, D.; Glish, G.L. Identifying the D-Pentoses Using Water Adduction to Lithium Cationized Molecule. *J. Am. Soc. Mass Spectrom.* **2017**, *28*, 1420–1424. [CrossRef] [PubMed]
27. Campbell, M.T.; Chen, D.; Wallbillich, N.J.; Glish, G.L. Distinguishing Biologically Relevant Hexoses by Water Adduction to the Lithium-Cationized Molecule. *Anal. Chem.* **2017**, *89*, 10504–10510. [CrossRef]
28. Campbell, M.T.; Chen, D.; Glish, G.L. Distinguishing Linkage Position and Anomeric Configuration of Glucose-Glucose Disaccharides by Water Adduction to Lithiated Molecules. *Anal. Chem.* **2018**, *90*, 2048–2054. [CrossRef]
29. Menicatti, M.; Guandalini, L.; Dei, S.; Floriddia, E.; Teodori, E.; Traldi, P.; Bartolucci, G. The power of energy resolved tandem mass spectrometry experiments for resolution of isomers: The case of drug plasma stability investigation of multidrug resistance inhibitors. *Rapid Commun. Mass Spectrom.* **2016**, *30*, 423–432. [CrossRef]
30. Menicatti, M.; Pallecchi, M.; Bua, S.; Vullo, D.; Di Cesare Mannelli, L.; Ghelardini, C.; Carta, F.; Supuran, C.T.; Bartolucci, G. Resolution of co-eluting isomers of anti-inflammatory drugs conjugated to carbonic anhydrase inhibitors from plasma in liquid chromatography by energy-resolved tandem mass spectrometry. *J. Enzyme Inhib. Med. Chem.* **2018**, *33*, 671–679. [CrossRef]
31. Menicatti, M.; Pallecchi, M.; Ricciutelli, M.; Galarini, R.; Moretti, S.; Sagratini, G.; Vittori, S.; Lucarini, S.; Caprioli, G.; Bartolucci, G. Determination of coeluted isomers in wine samples by application of MS/MS deconvolution analysis. *J. Mass Spectrom.* **2020**, *55*, e4607. [CrossRef] [PubMed]
32. Pallecchi, M.; Braconi, L.; Menicatti, M.; Giachetti, S.; Dei, S.; Teodori, E.; Bartolucci, G. Simultaneous Degradation Study of Isomers in Human Plasma by HPLC-MS/MS and Application of LEDA Algorithm for Their Characterization. *Int. J. Mol. Sci.* **2022**, *23*, 13139. [CrossRef] [PubMed]
33. March, R.E.; Todd, J.F. *Quadrupole Ion Trap Mass Spectrometry*, 2nd ed.; John Wiley & Sons: Hoboken, NJ, USA, 2005; pp. 2–416.
34. Cooks, R.G.; Glish, G.L.; McLuckey, S.A.; Kaiser, R.E. Ion Trap Mass Spectrometry. *Chem. Eng. News* **1991**, *69*, 26–41. [CrossRef]

35. Johnson, J.V.; Yost, R.A.; Kelley, P.E.; Bradford, D.C. Tandem-in-Space and Tandem-in-Time Mass Spectrometry: Triple Quadrupoles and Quadrupole Ion Traps. *Anal. Chem.* **1990**, *62*, 2162–2172. [CrossRef]
36. Jonscher, K.R.; Yates, J.R. The Quadrupole Ion Trap Mass Spectrometer—A Small Solution to a Big Challenge. *Anal. Biochem.* **1997**, *244*, 1–15. [CrossRef]
37. Clifford, M.N.; Johnston, K.L.; Knight, S.; Kuhnert, N. Hierarchical Scheme for LC-$MS^n$ Identification of Chlorogenic Acids. *J. Agric. Food Chem.* **2003**, *51*, 2900–2911. [CrossRef]
38. Clifford, M.N.; Knight, S.; Kuhnert, N. Discriminating between the six isomers of dicaffeoylquinic acid by LC-$MS^n$. *J Agric Food Chem.* **2005**, *53*, 3821–3832. [CrossRef]
39. Pallecchi, M.; Lucio, L.; Braconi, L.; Menicatti, M.; Dei, S.; Teodori, E.; Bartolucci, G. Isomers Recognition in HPLC-MS/MS Analysis of Human Plasma Samples by Using an Ion Trap Supported by a Linear Equations-Based Algorithm. *Int. J. Mol. Sci.* **2023**, *24*, 11155. [CrossRef]
40. ICH Q2B, Validation of Analytical Procedure: Methodology, International Conference on Harmonisation of Technical Requirements for Registration of Pharmaceuticals for Human Use. 1996. Available online: https://database.ich.org/sites/default/files/Q2%28R1%29%20Guideline.pdf (accessed on 27 March 2023).

pharmaceuticals

Article

# Unraveling the Radioprotective Mechanisms of UV-Resistant *Bacillus subtilis* ASM-1 Extracted Compounds through Molecular Docking

**Asim Ur Rahman [1,†], Aftab Ali [1], Faisal Ahmad [2,†], Sajjad Ahmad [3,4,5], Metab Alharbi [6], Abdullah F. Alasmari [6], Amna Fayyaz [7], Qurrat ul ain Rana [1,8], Samiullah Khan [1], Fariha Hasan [1], Malik Badshah [1] and Aamer Ali Shah [1,*]**

[1] Department of Microbiology, Quaid-i-Azam University, Islamabad 45320, Pakistan; arahman@bs.qau.edu.pk (A.U.R.); aftabkhan21995@gmail.com (A.A.); qurana@lbl.gov (Q.u.a.R.); samikhan@qau.edu.pk (S.K.); farihahasan@yahoo.com (F.H.); malikbadshah@qau.edu.pk (M.B.)
[2] National Center for Bioinformatics, Quaid-i-Azam University, Islamabad 45320, Pakistan; faisalahmad@bs.qau.edu.pk
[3] Department of Health and Biological Sciences, Abasyn University, Peshawar 25000, Pakistan; sajjad.ahmad@abasyn.edu.pk
[4] Gilbert and Rose-Marie Chagoury School of Medicine, Lebanese American University, Beirut P.O. Box 36, Lebanon
[5] Department of Natural Sciences, Lebanese American University, Beirut P.O. Box 36, Lebanon
[6] Department of Pharmacology and Toxicology, College of Pharmacy, King Saud University, P.O. Box 2455, Riyadh 11451, Saudi Arabia; mesalharbi@ksu.edu.sa (M.A.); afalasmari@ksu.edu.sa (A.F.A.)
[7] Department of Environmental Sciences, Quaid-i-Azam University, Islamabad 45320, Pakistan; amnafayyaz505@gmail.com
[8] Joint Genome Institute, Lawrence Berkely National Laboratory, Berkley, CA 94720, USA
* Correspondence: alishah@qau.edu.pk; Tel.: +92-51-9064-3116
† These authors contributed equally to this work.

**Citation:** Rahman, A.U.; Ali, A.; Ahmad, F.; Ahmad, S.; Alharbi, M.; Alasmari, A.F.; Fayyaz, A.; Rana, Q.u.a.; Khan, S.; Hasan, F.; et al. Unraveling the Radioprotective Mechanisms of UV-Resistant *Bacillus subtilis* ASM-1 Extracted Compounds through Molecular Docking. *Pharmaceuticals* **2023**, *16*, 1139. https://doi.org/10.3390/ph16081139

Academic Editors: Jan Oszmianski and Sabina Lachowicz-Wiśniewska

Received: 1 July 2023
Revised: 26 July 2023
Accepted: 1 August 2023
Published: 11 August 2023



**Abstract:** Radioresistant microorganisms possess inimitable capabilities enabling them to thrive under extreme radiation. However, the existence of radiosensitive microorganisms inhabiting such an inhospitable environment is still a mystery. The current study examines the potential of radioresistant microorganisms to protect radiosensitive microorganisms in harsh environments. *Bacillus subtilis* strain ASM-1 was isolated from the Thal desert in Pakistan and evaluated for antioxidative and radioprotective potential after being exposed to UV radiation. The strain exhibited 54.91% survivability under UVB radiation ($5.424 \times 10^3$ J/m$^2$ for 8 min) and 50.94% to mitomycin-C (4 μg/mL). Extracellular fractions collected from ASM-1 extracts showed significant antioxidant potential, and chemical profiling revealed a pool of bioactive compounds, including pyrrolopyrazines, amides, alcoholics, and phenolics. The E-2 fraction showed the maximum antioxidant potential via DPPH assay (75%), and $H_2O_2$ scavenging assay (68%). A combination of ASM-1 supernatant with E-2 fraction (50 μL in a ratio of 2:1) provided substantial protection to radiosensitive cell types, *Bacillus altitudinis* ASM-9 (MT722073) and *E. coli* (ATCC 10536), under UVB radiation. Docking studies reveal that the compound supported by literature against the target proteins have strong binding affinities which further inferred its medical uses in health care treatment. This is followed by molecular dynamic simulations where it was observed among trajectories that there were no significant changes in major secondary structure elements, despite the presence of naturally flexible loops. This behavior can be interpreted as a strategy to enhance intermolecular conformational stability as the simulation progresses. Thus, our study concludes that *Bacillus subtilis* ASM-1 protects radiosensitive strains from radiation-induced injuries via biofilm formation and secretion of antioxidative and radioprotective compounds in the environment.

**Keywords:** antioxidants; radioresistant microorganisms; radioprotection; biofilm; co-operative growth; oxidative stress

## 1. Introduction

Microorganisms that survive in extreme environmental conditions, like high desiccation and prolonged exposure to radiation, exhibit unique survival strategies which enable them to grow optimally in such life-threatening conditions [1]. Radiation triggers DNA damage and induces oxidative injury to vital biomolecules through the radiolysis of water [2,3]. In response to these detrimental effects of radiation, radioresistant microorganisms developed advanced genome repair mechanisms accompanied by various radiation-responsive secondary metabolic products such as extremolytes, including photoprotective pigments, that can absorb radiation [4,5]. In addition to these, the enzymatic and non-enzymatic antioxidant systems also assist in the survival of these microorganisms by neutralizing and scavenging reactive radicals and other unstable compounds which are produced due to radiation exposure [6].

Microbial bio-pigments that have an array of targeted biological properties are now being evaluated by researchers worldwide against numerous human diseases in the form of skin care agents, immunosuppressants, antioxidants, and anti-tumor agents, and are, therefore, considered potential candidates for the pharmaceutical and cosmetic industries. So far, the extremolytes utilized as radioprotectants in skin care utilities are sioxanthin, astaxanthin, shinorine, scytonemine, pannarin, melanin, biopterin, bacterioruberin, and microsporine-like amino acids [7]. In a nutshell, radioresistant extremophiles act as a reservoir for novel bioactive compounds that are not only important for their structural and biochemical diversity but also have extensive biotechnological applications.

Besides this, there are a vast number of scientific reports available on the metabolites-aided survival capabilities of radioresistant microorganisms inhabiting such an inhospitable environment. However, in the process of sampling, isolation, and screening, you will also obtain a few radiosensitive microorganisms, which raises a question of concern: how are radiosensitive microorganisms able to survive in such hostile environments? In one of the previous reports, Igor et al. studied the co-operative growth between radioresistant and radiosensitive microbial cells inhabiting the same ecological settings, using a genetic approach targeting only catalase enzymes (by creating mutants) [8]. However, the actual survival mechanism of such radiosensitive microorganisms is still unclear and poses a new era of research, investigating the symbiotic relationships between microorganisms living in extreme environments.

In this connection, the present study evaluates the in vitro biological activities and characterization of bioactive metabolites from a radioresistant bacterium, *Bacillus subtilis* ASM-1 isolated from Thal desert of Pakistan. Furthermore, the radioprotective properties of the extracted bioactive compounds isolated from strain ASM-1 were also assessed using a new approach, i.e., protecting radiosensitive microbial strains *Bacillus altitudinis* ASM-9 and *E. coli* (ATCC 10536) from radiation-induced damage.

## 2. Results

### 2.1. Isolation and Screening for Radioresistant Bacteria

A total of 33 bacterial isolates recovered from Chashma (TMC) and the Makarwal region (ASM) were subjected to UV radiation to determine their behavior. Among 33 isolates, 18 isolates from Chashma and 8 from the Makarwal region were found to be resistant to UV radiation in primary screening for 5 min. In secondary screening (radiation dose for more than 5 min), the UVB dosage to all resistant isolates was gradually increased by increasing exposure time from 3 to 20 min. Strain ASM-1 was selected based on resistance to a maximum energy dose, as shown in Table 1.

**Table 1.** Primary and secondary screening for radioresistant isolates with radiation exposure time and energy dose.

| Exposure Time | Radiant Exposure in (J/m$^2$)<br>He = Ee × t (Sec) | UV Resistant Isolates |
|---|---|---|
| 3 min | $2.034 \times 10^3$ J/m$^2$ | * TMC 1 to TMC 18<br>** ASM 1 to ASM 8 |
| 5 min | $3.390 \times 10^3$ J/m$^2$ | TMC 2, TMC 7, TMC 6, TMC 9, TMC 11, TMC 12, TMC15, ASM 2, ASM 1, ASM 3, ASM 5, TMM 6, ASM 8 |
| 8 min | $5.424 \times 10^3$ J/m$^2$ | TMC 2, TMC 7, TMC 6, TMC 9, TMC 12, TMC15, ASM 1, ASM 5, ASM 6, ASM 8 |
| 10 min | $6.780 \times 10^3$ J/m$^2$ | ASM-1, ASM-6, ASM-8 |
| 12 min | $8.136 \times 10^3$ J/m$^2$ | ASM-1 |

* Chashma sample denoted by TMC; ** Makarwal sample denoted by ASM.

### 2.2. Characterization of Strain ASM-1

Strain ASM-1 was observed to be Gram-positive, growing in dry, opaque, circular colonies with dark red colored pigmentation on TGY agar plates. The strain was positive for amylase, catalase, superoxide dismutase, peroxidase, cellulase, and lipase enzymes, while negative for protease enzyme. A maximum similarity of 100% was shown against *Bacillus* sp. when a 16S rRNA gene sequence of our potential strains was subjected to BLAST analysis. This was identified as *Bacillus subtilis* strain ASM-1. A phylogenetic tree was constructed using the Neighbor-Joining method to deduce the evolutionary relationship of ASM-1 with other strains already reported in NCBI GenBank (Figure 1). The nucleotide sequence was submitted to NCBI GenBank with accession number (OK559666).

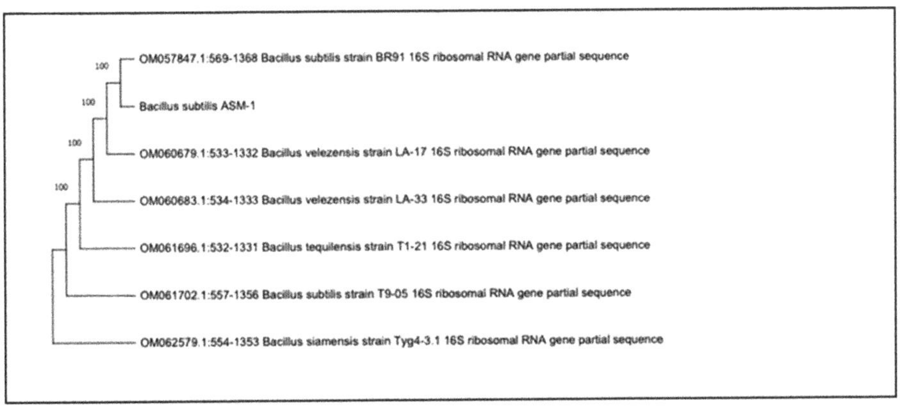


**Figure 1.** The phylogenetic tree was constructed for ASM-1 using the Neighbor-Joining approach using Molecular Evolutionary Genetics Analysis (MEGA-X) software with a bootstrap value (1000 replicates) of 100 to *Bacillus subtilis* strain ASM-1(OK559666).

### 2.3. Survival Rate of Bacillus subtilis Strain ASM-1 under UVB Radiation and Mitomycin-C

*B. subtilis* strain ASM-1 exhibited 54.91% survivability against a control strain *E. coli* (ATCC 10536) that showed 20.23% under the same dose of UVB radiation, i.e., $5.424 \times 10^3$ J/m$^2$ (8 min). Likewise, the control strain showed no growth as compared to ASM-1 under UVB dose of $8.136 \times 10^3$ J/m$^2$ (12 min), as shown in Figure 2A. Similarly, ASM-1 showed 50.90% survivability at 4 µg/mL under Mitomycin-C exposure as compared to a control *E. coli* strain which showed only 25.63%, respectively (Figure 2B).

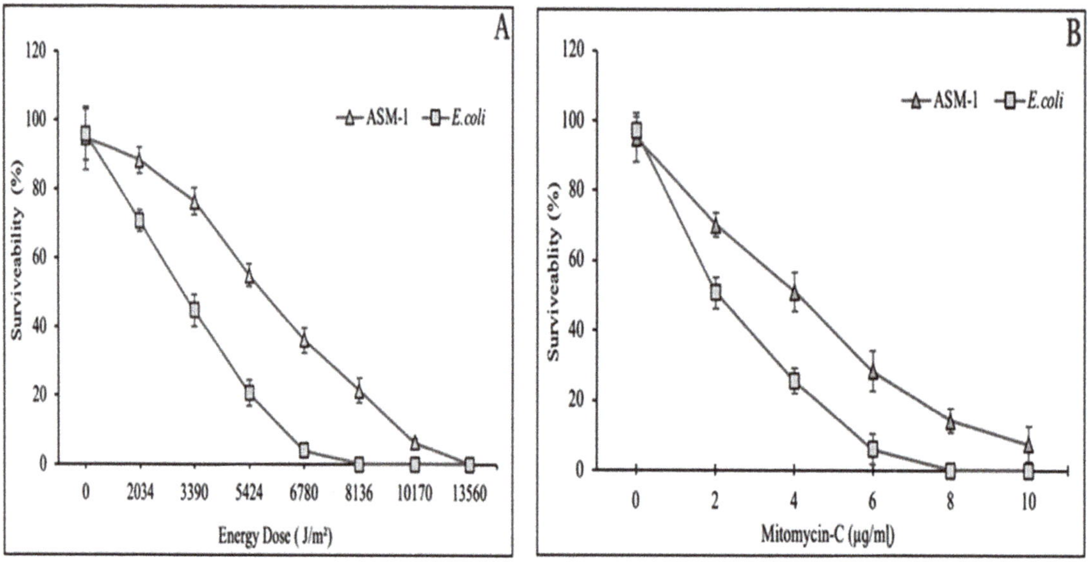


**Figure 2.** Survivability rate determination for *Bacillus subtilis* strain ASM-1 under UVB radiation (**A**), and Mitomycin-C (**B**). The data is calculated and expressed from the average of three replicates ± SDs.

### 2.4. Purification and Selection of the Extracellular Crude Extracts

The extracellular crude extracts (AS-2) resulted in four fractions in their corresponding solvents. All fractions were primarily evaluated for their antioxidant capability via DPPH assay. Fractions E-2 and D-2 possessed maximum antioxidant activity, while the remaining fractions W-2 and M-2 exhibited insignificant activities. The fractions E-2 and D-2 were suspended in their respective solvents and stored at 4 °C for bioassays and analytical purposes.

### 2.5. In Vitro Evaluation for the Antioxidant Potential of the Selected Fractions

#### 2.5.1. DPPH Assay

Results demonstrated that scavenging activity of the fractions E-2 and D-2 was increased with their increasing concentrations. The E-2 fraction accounted for 75.55% of DPPH free radical scavenging potential as compared to ascorbic acid (ASA) as a positive control, which showed 80.43% activity at a concentration of 250 µg/mL, while at this concentration, the D-2 fraction exhibited the lowest scavenging potential of 60.37% (Figure 3A).

#### 2.5.2. Hydrogen Peroxide Scavenging Assay

Hydrogen peroxide scavenging was also observed in a concentration-dependent manner. However, the E-2 fraction showed 68.53% scavenging potential in comparison to the D-2 fraction (63.37%) against a positive control ASA that provided 66.84% scavenging at a concentration of 250 µg/mL (Figure 3B).

#### 2.5.3. Cytotoxic and Antibacterial Potential

The cytotoxic effect on brine shrimp larvae was perceived only for the D-2 fraction as compared to the E-2 fraction. An approximately 61% killing effect was detected for D-2 as compared to 74% for Mitomycin-C at the maximum concentration of 250 µg/mL. All fractions exhibited insignificant antibacterial activities, except for the D-2 fraction that possessed a slight antibacterial effect against a clinical strain of *Staphylococcus aureus* with

ZIB of (7 mm). However, no fraction was found to have antibacterial activity against ASM-9 and *E. coli* (test strains).

2.5.4. Biofilm Detection Assay

The in vitro evaluation for biofilm formation ability was tested for both isolates, i.e., ASM-1, and ASM-9 at three-time intervals (24, 48, and 72 h). Both isolates showed biofilm formation potential in a diminishing manner. ASM-1 showed maximum biofilm potential as compared to ASM-9 after 24 h (Figure 3C).

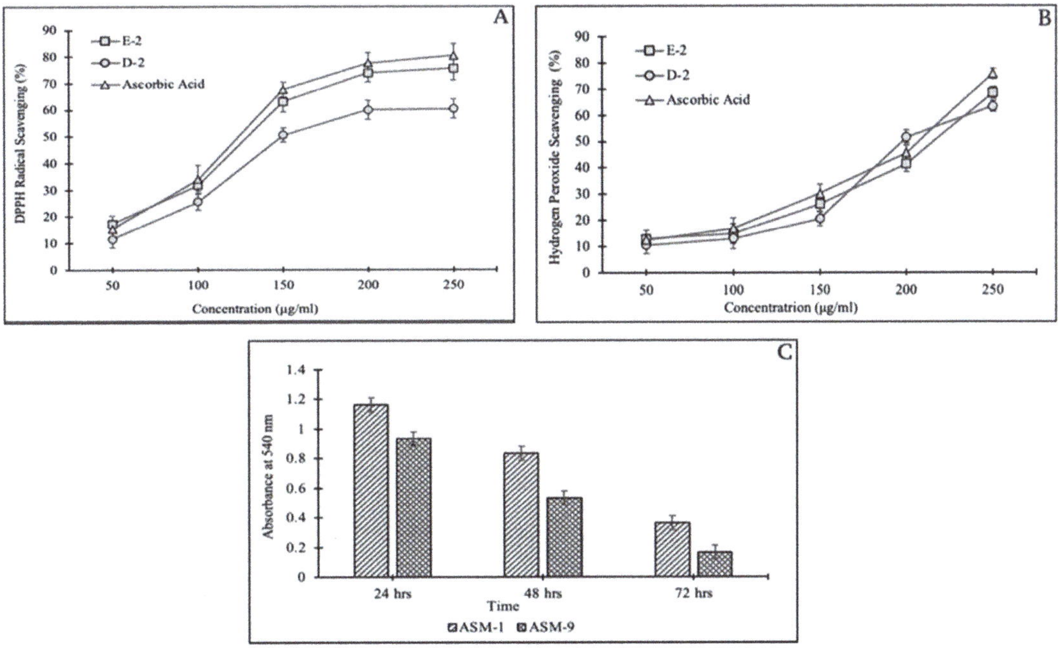


**Figure 3.** The selected fractions obtained from ASM-1 extracts were evaluated for their antioxidant potential via DPPH free-radical scavenging assay (**A**), hydrogen peroxide assay (**B**). Similarly, the biofilm formation potential for both isolates (ASM-1, and ASM-9) are depicted in the bar graph (**C**). The data expressed here was calculated as a mean value obtained from the triplicates showing ± SDs.

2.5.5. GC–MS Analysis of the Selected Fractions

The selected ASM-1 fractions (E-2 and D-2) which exhibited maximum antioxidant activities were subjected to their chemical profiling through GC/MS analysis. The GC/MS spectra unveiled a pool of volatile compounds that were carefully identified based on their spectral masses upon comparing them to the online National Institute of Standards and Technology (NIST) databases. In the present study, a pool of chemical compounds from different classes in all perceived fractions was Pyrrolopyrazine, Alkane, Phenol, Dicarboxylic acid butyl ester, Dicarboxylic acid octyl ester, Fatty acid, Amide, β-carboline alkaloid and Alcohol. Dicarboxylic acid butyl ester was the only compound detected at RT 21.83 in fraction D-2. The detailed information of the chemical compounds including their names, retention time, molecular weight, molecular formula, and class are listed in Table 2.

Table 2. Major constituents detected in the GC–MS spectra for the selected fractions of ASM-1 extracts.

| Fractions | Retention Time | Content (%) | Chemical Compounds | Molecular Formula | Molecular Weight | Class |
|---|---|---|---|---|---|---|
| E-2 | 3.083 | 1.03 | Pyrrolo [1,2-a] pyrazine -1,4-dione, hexahydro-3-(2-methyl propyl) | $C_{11}H_{18}N_2O_2$ | 210 | Pyrrolopyrazine |
| | 4.146 | 0.14 | 9H-Pyrido [3,4-b] indole | $C_{11}H_8N_2$ | 168 | β-carboline alkaloid |
| | 4.123 | 0.56 | Pyrrolo [1,2a] pyrazine-1,4-dione, hexahydro -3-(phenylmethyl) | $C_7H_{10}N_2O_2$ | 154 | Pyrrolopyrazine |
| | 7.521 | 0.77 | n-Nonadecanol-1 | $C_{19}H_{40}O$ | 284 | Alcohol |
| | 11.614 | 0.21 | Dodecane | $C_{12}H_{26}$ | 170 | Alkane |
| | 15.291 | 1.91 | Phenol,2,2′-methylenebis [6-(1,1-dimethylethyl)-4-methyl | $C_{23}H_{32}O_2$ | 340 | Phenol |
| D-2 | 3.412 | 1.09 | Benzamide | $C_7H_7NO$ | 121 | Amide |
| | 4.243 | 0.28 | 1,2-benzenedicarboxylic acid, monobutyl ester | $C_{12}H_{14}O_4$ | 222 | Dicarboxylic acid Butyl ester |
| | 5.461 | 0.32 | 1,2-Benzenedicarboxylic acid, diisooctyl ester | $C_{24}H_{38}O_4$ | 390 | Dicarboxylic acid Octyl ester |
| | 11.065 | 0.37 | (S)-6-Hydroxyheptanoic acid | $C_7H_{14}O_2$ | 130 | Fatty acid |
| | 11.534 | 0.04 | 1-Octanol | C8H18O | 130 | Alcohol |

### 2.6. Radioprotective Activity of ASM-1 Extracts on Selected Radiosensitive Strains

*Bacillus subtilis* strain ASM-1 extracts had been assessed for their potential to facilitate radioprotective effect on selected radiosensitive bacterial strains ASM-9 and *E. coli* under different doses of UVB radiation. For this purpose, the E-2 fraction was selected against the D-2 fraction because of its significant antioxidant potential and having no cytotoxic and antibacterial capacity. The survivability of test strains was reduced under UVB radiation in a dose-dependent manner. With no extracts added, *E. coli* (ATCC 10536) exhibited 19.69% survivability, while strain ASM-9 showed 0.33% survivability under an energy dose of $3.390 \times 10^3$ J/m$^2$ (5 min) (Figure 4A). Their survivability was significantly enhanced from (19.69% to 44.92%) for *E. coli*, and for ASM-9 survivability was enhanced from (0.33% to 21.62%) after the addition of the mixture (E-2 fraction and supernatant) under an acute UVB dose of $3.390 \times 10^3$ J/m$^2$ for 5 min (Figure 4B).

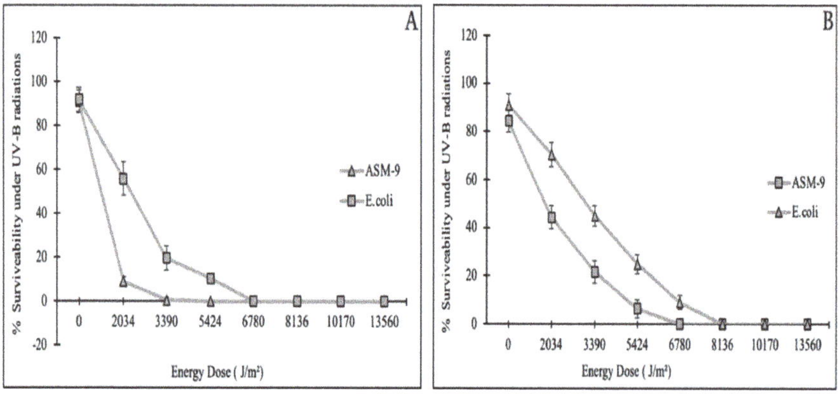


**Figure 4.** Growth and survivability rate of radiosensitive strains (*E. coli* and ASM-9) under UVB radiation with no ASM-1 extracts (**A**). The second figure shows survivability of radiosensitive strains after the addition of ASM-1 extracts (**B**). The data are expressed as mean values of the triplicates ± SDs.

## 2.7. Ligand Preparations and Molecular Docking

Molecular docking of the compounds was carried out using the PyRx interface of AutoDock Vina and GOLD software [9]. For the docking process, the following parameters were employed: an exhaustive value of 300, the binding site was defined as the residues of the N-terminal domain involved in the binding pocket, and a maximum of 10 poses were allowed. The PyRx tool generated binding affinity values as negative values (where a more negative value indicates a stronger binding affinity). The docking results of the top inhibitors can be found in Figures 5 and 6. To validate the docking protocol used, extracted compounds were docked into the N-terminal structure of (PDB ID = 5F1A and 5ZKP). The binding mode as the one observed in crystallization studies for the known inhibitors was obtained. The docking reproducibility results obtained using AutoDock Vina and GOLD are presented in Figures 5 and 6.

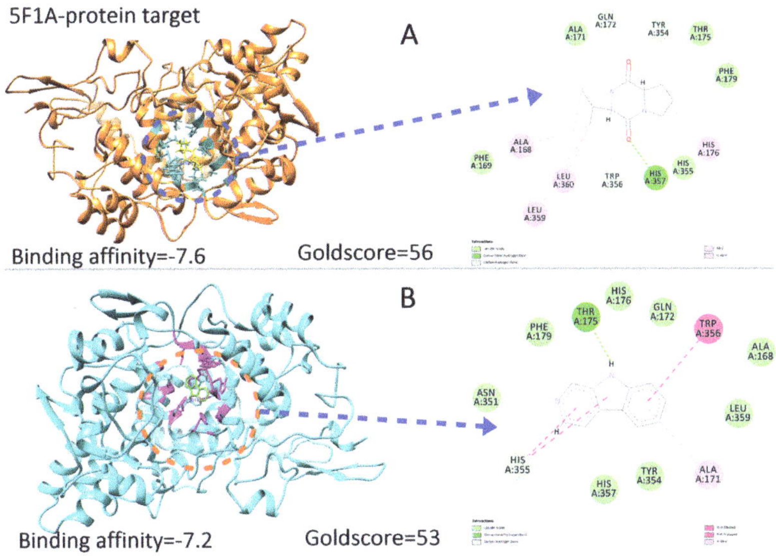


**Figure 5.** Depicting the protein target with binding interactions of both compounds against the target 5F1A with binding affinities and Goldscore. (**A**,**B**) Presenting the active cavity of the protein where both ligand molecules are attached as shown in blue and orange circle on left whereas 2D image of both (**A**,**B**) shows the binding residues of the active pocket especially hydrogen bonding.

Protein–ligand docking is an important technique for accurately predicting the orientation of a ligand within its target protein [10]. Compounds that exhibited strong binding affinity and correct binding poses were selected. The binding affinity represents the total torsional energy, internal energy, and intermolecular energy subtracted from the unbound energy system. A high binding affinity indicates a stable protein–ligand complex. Interestingly, the compounds docked in similar conformations to both the full-length 5F1A structure and the N-terminal structure alone. As the C-terminal region did not significantly contribute to the binding of compounds at the N-terminal, this study focused on the N-terminal structure alone. To determine the binding poses, interactions, and binding affinity of both the inhibitors, they were docked into the binding pocket of 5F1A and 5ZKP using the PyRx tool and GOLD software. Each inhibitor generated 10 poses, and those with strong binding affinity were chosen (Figures 5 and 6). The docking results revealed a range of binding affinities from −7.6 to −7.1 kcal/mol with Goldscore of 56 and 53 against 5F1A, whereas 53 to 51 against 5ZKP for compounds **1** and **2**. The analysis showed that compound **1** and compound **2** shared a motif that bound to the same residues (Thr175, His176, Gln172, Trp356, Ala168, Leu359, Ala171, Tyr354, His357, His355, Asn351, Phe155). Moreover, a hydrogen bond interaction with the active site residues (His357 and Thr175) was observed for the shared motif in compound **1** and compound **2**, whereas against protein target 5ZKP, the shared active residues involved during binding interaction are Tyr17, Tyr72, Leu400, Trp68, Ala88, Gly89, Phe92, Phe155, His396, Gln397, Arg9, Glu156, Cys154 and Tyr71, while Tyr17, Tyr72 and Glu are involved in hydrogen bond formations in both compounds **1** and **2** in Figures 5 and 6.

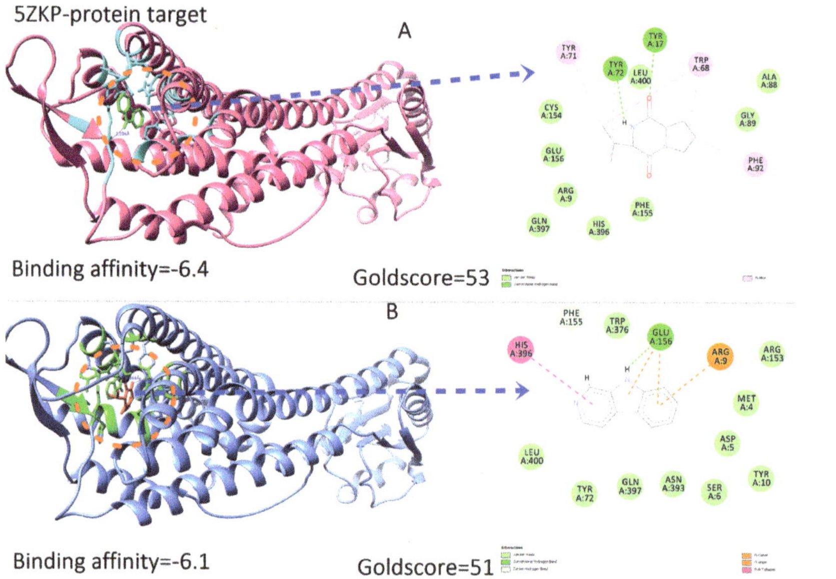


**Figure 6.** Depicting the protein target with binding interactions of both compounds against the target 5ZKP with binding affinities and Goldscore. Both (**A**,**B**) show the binding cavity where both inhibitors are docked to specific active residues as shown in orange circle on left side. Whereas the 2D image on right side presenting the hydrogen bonds, van der waals and other interactions holding the ligand inside the pocket.

## *2.8. Molecular Dynamics Simulation Analysis*

To validate the docking, the findings and validate intermolecular stable conformation between the enzymes and compounds, molecular dynamics simulation analysis was conducted. The simulation was carried out for 20 ns. The simulation trajectories were used root mean square deviation (RMSD) analysis to shed light on enzyme carbon alpha atoms deviations during simulation time. As can be seen in Figure 7, all four docked complexes revealed stable dynamics with no drastic local or global changes observed. The 5f1A-a-56 system was noticed as the most stable with a mean RMSD value of 0.51 Å. The mean RMSD of 5faA-2-53, 5zkp2-51 and 5zkp1-51 was 0.89 Å, 1.32 Å, and 1.33 Å, respectively (Figure 7). Upon trajectory visualizations, no major secondary structure elements changes were witnessed despite naturally flexible loops, which may be explained as an approach to provide more intermolecular conformational stability as the simulation time proceeds.

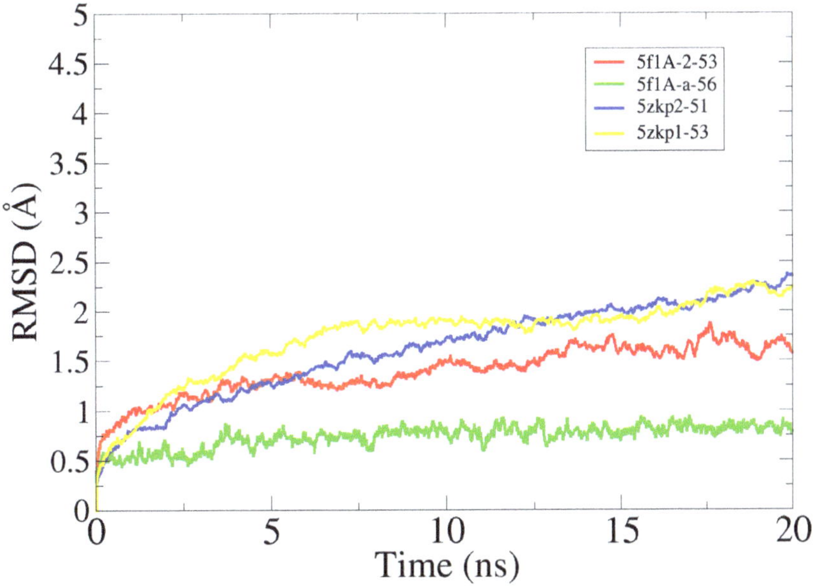


**Figure 7.** RMSD analysis of simulation complexes based on carbon alpha atoms.

## 3. Discussion

Actinomycetes, followed by Bacillus, are considered a reservoir for their numerous vital metabolites including antibacterials, antifungals, antiparasitics, antioxidants, growth-promoting substances, immune modifiers, immunosuppressants, and enzyme inhibitors [11]. These natural products have bioactive properties that have been extracted from microorganisms for decades to serve as potential candidates for a variety of industrial sectors [12]. As a result, there is always a need for novel and highly potent metabolites that can be obtained from previously untapped extremophilic microbial sources [1,13]. Similarly, understanding the effects of acute and chronic radiation on radiosensitive and radioresistant microbial cell types is a potentially important new tool for understanding the symbiotic relationships among microorganisms inhabiting such hostile environments [7].

*Bacillus subtilis* strain ASM-1, a radioresistant bacterium, was isolated from an unexplored environment in Pakistan's Thal desert. *Bacillus* sp. are thought to be one of the most common inhabitants in such hostile environments due to their endospore and biofilm formation abilities, as well as the secretion of several vital metabolites. Similarly, the isolate's production of enzymes, such as catalases and peroxidases, demonstrated the strain's enzymatic antioxidant potential to counterbalance the oxidative stress produced intracellularly and extracellularly by multiple oxidants [11].

The survival efficiency of the ASM-1 strain was investigated by subjecting it to various physical and chemical stressors, i.e., UVB radiation and Mitomycin-C. UVB radiation can cause oxidative stress or impose the formation of thiamine dimers, which are difficult to transcribe and repair, ultimately resulting in cell death. Mitomycin-C, on the other hand, is an anticancer agent that inhibits abnormal cell proliferation. Studies revealed that ASM-1 could withstand the damaging effects of these stressors up to a certain level. This is most likely due to the fact that radioresistant microorganisms evolved multiple defense mechanisms, such as the production of UV-absorbing compounds, enzymatic and non-enzymatic antioxidant systems, and an efficient DNA repair system, allowing them to survive in such harsh environments [14]. The ASM-1 strain was chosen for further investigation due to its elevated resistance to both stressors.

Radiation-induced oxidative stress is harmful to all life forms, so compounds that counteract this oxidative stress are recognized as radioprotective agents. These compounds' primary methods of radioprotection include cellular oxygen tension decrease, free radical scavenging, and hydrogen transfer [15]. This study used two antioxidant assays to assess the potency of ASM-1 crude extracts as an antioxidative and radioprotective agent against radiation-induced oxidative stress. The assays chosen were solely based on the extracts' potential future implications (investigating their potential role to aid in the survival of radiosensitive test strains under radiation-induced oxidative stress). ASM-1 extracts demonstrated significant antioxidant potential in a dose-dependent manner by scavenging DPPH free radicals and hydrogen peroxide species [16,17]. The use of multiple bioassays revealed that ASM-1 extracts had significant bioactivity against multiple reactive oxygen species (ROS). These findings imply that ASM-1 extracts may be useful in preventing ROS-mediated damage.

Both fractions were promoted for further bioassays, i.e., cytotoxic, antibacterial, and biofilm formation potential, to achieve the study's intended goal. As the E-2 fraction had no significant antibacterial and cytotoxic potential compared to the D-2 fraction, which had little cytotoxic and antibacterial potential, only the E-2 fraction was preferred for the radioprotective experiment. Similarly, the biofilm formation potential of both strains was observed (ASM-1 and ASM-9). However, ASM-1 demonstrated significantly greater biofilm formation ability than ASM-9 after 24 h, which then declined after 72 h of incubation [18].

Similarly, ASM-1 fractions, particularly E-2 and D-2, were chemically characterized by the employment of GC–MS analysis. The GC–MS analysis for the ASM-1 fractions revealed several chemical compounds including alcohols, phenols, pyrrolopyrazine, alkaloids, amides, and different types of acids. Some of the chemical constituents of the GC–MS spectra have been previously reported to have antioxidative, cytoprotective, and radioprotective capacities. In the present study, the E-2 fraction comprised of two pyrrolopyrazine compounds, Pyrrolo [1,2-a] pyrazine -1,4-dione, hexahydro-3-(2- methyl propyl) and Pyrrolo [1,2a] pyrazine-1,4-dione, hexahydro -3-(phenylmethyl), that are well-known compounds to have antioxidative and radioprotective potential, as previously described [19]. Similarly, 9H-Pyrido [3,4-b] indole is a tricyclic indole β-carboline alkaloid that inhibits radiation-induced oxidation transformation reactions [20]. Dodecane is another long-chain alkane detected in GC–MS spectra to have a significant antioxidant potential [21]. Likewise, n-Nonadecanol-1 is long-chain alcohol and 1, 2-benzene dicarboxylic acid that possesses antimicrobial and cytotoxic properties, respectively [22]. Likewise, the D-2 fractions were comprised of 1-Octanol, previously detected in *Bacillus subtilis* CF-3 extracts to have bioactive potential, While several benzoic acids in the extracts have already been reported in *Bacillus* sp. to have biological activities [23], benzamide was the only compound possessing anticancer and other biological activities. The chemical constituents detected in the GC/MS spectra of ASM-1 fractions suggested the antioxidant potential of *Bacillus subtilis* strain ASM-1. Therefore, this study provides a shred of evidence on the antioxidative and radioprotective role of the metabolites derived from *Bacillus subtilis* ASM-1.

Several tests were performed to determine the potential of ASM-1 extracts against oxidative stress, bacterial competitors, and cytotoxicity. These findings assumed the E-2 fraction's safe and intended effect in the radioprotective experiment. In a radioprotective experiment, administering only the E-2 fraction did not result in a significant change. Therefore, ASM-1 supernatant was also considered because it contains antioxidant enzymes, UV-absorbing compounds, and exopolysaccharides (a prerequisite for biofilm development). ASM-1 supernatant from a 24-hour-old culture was collected for this purpose because significant biofilm formation was detected at this stage. As a result, a recipe (Supernatant and E-2 fraction in 2:1) was developed after several trials to obtain enhanced survivability. ASM-1 was found to have significant radioprotective activity, meaning that the extracellular extracts protected radiosensitive *E. coli* and ASM-9 from the harmful effects of UVB rays far beyond the standard dose (5 min). The reason for this improved survivability could be that the E-2 fraction provided the necessary components (antioxidants), whereas the supernatant provided antioxidative enzymes, UV-absorbing compounds, and exopolysaccharides. Furthermore, after the irradiation dose, well-defined clumps of ASM-9 cells were detected on TGY agar plates, which could be due to the exopolysaccharides that glued the ASM-9 cells for enhanced survival. However, the precise mechanism is unknown and requires further exploration. The findings were like those of Igor et al., who used a molecular model to investigate the radioprotective and antioxidative potential of radioresistant *Dinococcus radiodurans* on radiosensitive *E. coli*, specifically silencing the catalase enzyme gene [8]. This experimental setup, however, was modified to investigate the radioprotective effect of radioresistant ASM-1 extracts on radiosensitive strains in the environment. Furthermore, some of the compounds obtained were searched out against the UV protections where two compounds were identified as reported to be used against UV radiation in sunscreen. These compounds includes Pyrrolo [1,2-a]pyrazine-1,4-dione, hexahydro-3-(2-methylpropyl) and 9H-Pyrido [3,4-b]indole. Both compounds were docked against the reported protein targets which is responsible for involved in UV-induced photodamage, such as serotonin-receptor subtype 5-HT2A, platelet-activating factor receptor (5ZKP), and Aspirin and other nonsteroidal anti-inflammatory drugs target the cyclooxygenase enzymes (COX-1 and COX-2) to block the formation of prostaglandins (5F1A). Molecular docking inferred potential binding affinities of the hit compounds against both the protein targets which open a way towards further investigations in future. Further these docking results have been validated via molecular dynamic simulations which emphasize the confirmational behavior in a real time environment. Interestingly, it has been deduced that no confirmational changes were recorded after 20 ns of simulation time interval against all the complex systems, despite some loop changes and the ligand remaining in contact with active residues of the active pocket.

These results shed light on the symbiotic relationships between radioresistant and radiosensitive microbial cells that co-exist in the same environmental niche exposed to chronic radiation. The radioresistant microorganisms' indefinite defense capabilities might aid the survival means for radiosensitive strains by secreting numerous potent metabolites, and by forming a protective barn (biofilm) for the same and different microbial species in such an unrelenting environment with extreme oxidative stress caused by radiation, as per the first-time study in this regard [24]. In a symbiotic relationship, radioresistant microbial species could allow radiosensitive microbial cells (effective degraders) to detoxify radioactive waste sites. Similarly, metabolites derived from radioresistant microorganisms can protect humans and other life forms from radiation-induced toxicity, particularly carcinogenesis.

## 4. Materials and Methods

### 4.1. Isolation and Screening of Radioresistant Bacteria

Soil samples were aseptically collected from the surface and subsurface (2 cm depth) using standard microbiological protocol from two different locations., i.e., the Chashma and Makarwal regions of the Thal desert, Pakistan. The samples were serially diluted in double distilled water, followed by inoculation on Tryptone Glucose Yeast (TGY) medium by spread plate method. The composition of TGY medium was (g/L): tryptone 10, yeast extract 5; glucose 1. The TGY plates were exposed to UV radiation for 3 and 5 min in the UV chamber (119 cm × 69 cm × 52 cm) supplied with a 20 W, 280 nm UV light before incubation. The test sample's UV fluence rate was determined using the following equation [25].

$$He = Ee \times t \quad (1)$$

where He is the energy that reaches a surface area due to irradiance (Ee) maintained for the time duration (t). The UV fluence rate is defined as the energy of radiation that reaches a surface area in a definite period (energy/area/time) measured in $J/m^2$. Exposure duration to the UV fluence rate was used to compute the total UV dosage. Subsequently, plates were wrapped in aluminum foil and incubated at 37 °C for 72 h. For further validation and to evaluate their survival curve, the bacterial isolates from radioactively contaminated plates were subcultured.

### 4.2. Identification of Radioresistant Bacterial Strain ASM-1

Strain ASM-1 isolated from the Makarwal region was selected based on its maximum survival rate under UVB exposure as well as its maximum antioxidant potential. The ASM-1 strain was identified morphologically as well as biochemically as per previously described approaches [26]. The molecular characterization was performed by employing 16S rRNA gene sequencing.

A DNA extraction kit (QIAGEN, Hilden Germany) was used to extract the DNA, and two universal primers were used to amplify the 16S rRNA gene sequence [27F′: AGAGTTTGATCMTGGCTCAG, 1492R′: TACGGYTACCTTGTTACGACTT] in PCR reactions. The Macrogen Service Center sequenced the amplified product (Geunchun-gu, Seoul, South Korea) and analyzed by a Nucleotide sequence alignment tool (nBLAST) available online in the National Center for Biotechnology Information (NCBI) database to ascertain the best matching genus. The ASM-1 strain, along with other homologs obtained from NCBI database, were subjected to computational analysis using Molecular Evolutionary Genetics Analysis (MEGA-X) to find out their phylogenetic relationships [27–29]. To identify the ASM-1 strain and analyzing the diversity of UV-resistant extremophiles, a neighbor-joining tree was built. To receive an accession number, the acquired sequence was uploaded to the NCBI GenBank.

### 4.3. Survival Curve of Strain ASM-1 at UVB and Oxidative Stress

After being exposed to various UVB radiation dosages, the ASM-1 strain's capacity to survive was assessed using the procedure as described earlier by (Mattimore and Battista 1996) and the survival curve was plotted. Cells of strain ASM-1 were serially diluted (1:1000), using phosphate-buffered saline (PBS), and spread on tryptone glucose yeast (TGY) agar plates. The plates were exposed to various doses of 280 nm UV radiation and then incubated at 37 °C for 72 h. The rate of survival was calculated by dividing the total number of colonies on radioactive plates by the total number of colonies on non-radioactive plates [30]. Oxidative stress and Mitomycin-C tolerance were determined by diluting an overnight grown culture of the ASM-1 strain with sterile normal saline up to an optical density (OD600) 0.5, then treated with different molar concentrations (0–10 mmol $L^{-1}$) of hydrogen peroxide ($H_2O_2$) for 30 min and mitomycin C (2–10 µg/mL) for 20 min, before inoculation on TGY plates. The number of colonies formed from the treated and untreated samples were compared after the plates were incubated at 37 °C for 72 h to evaluate the survival rate [25,31]. Each experiment was performed in triplicate.

### 4.4. Extraction of Extracellular Bioactive Compounds from Bacillus subtilis ASM-1 Strain

A total of 2.5 L of the overnight grown culture of strain ASM-1 in TGY broth was centrifuged at 10,000× *g* for 10 min at 4 °C and supernatant was collected. The extracellular metabolites were harvested from cell-free supernatant through liquid phase solvent extraction using an equal volume of ethyl acetate. The extracellular crude extracts (AS-2) were obtained in dried form using rotary evaporator.

### 4.5. Partial Purification of Extracellular Bioactive Compounds from B. subtilis ASM-1

The extracellular crude extract AS-2 from strain ASM-1 was subjected to purification through solid-phase extraction using HC-C18 SPE manual cartridge column (4.6 × 100 mm, 62 Å average pore size, 58 µm average particle size) with an average flow rate of 1.0 mL/min. The cartridge was loaded with 480 mg of crude extract (AS-2) and eluted using a series of solvents from polar to non-polar [water (W)—methanol (M)—ethyl acetate (E)—dichloromethane (D)], resulting in 4 different fractions in their respective solvents (Figure 8).

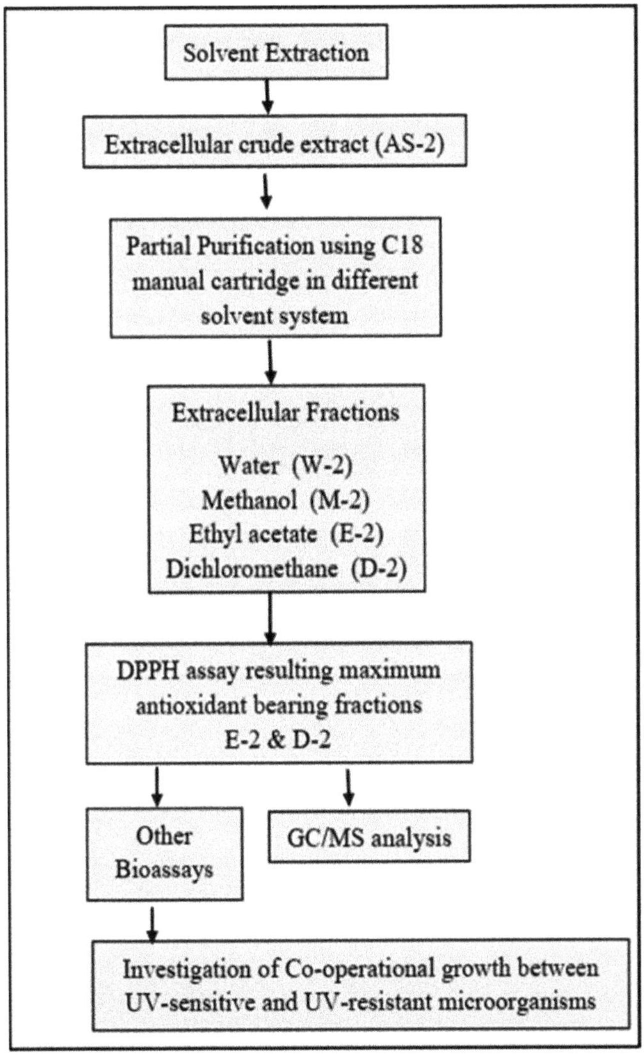


**Figure 8.** A schematic representation of the stepwise extraction, fractionation, and evaluation process for the extracellular crude extracts obtained from strain ASM-1.

### 4.6. *In Vitro Bioassays of Purified Fractions from B. subtilis ASM-1*

#### 4.6.1. DPPH Radical Scavenging Assay

The antioxidant activity of selected purified fractions E-2 and D-2 was determined by 2,2-diphenyl-1-picrylhydrazyl (DPPH) radical scavenging assay [31]. A total of 100 µL of fractions was mixed in different concentrations (50, 100, 150, 200, and 250 µg/mL) with 100 µL of 0.1 mM of DPPH solution in microtiter plate and incubated at 37 °C for 30 min. Ascorbic acid was used as positive control while DPPH solution (200 µL) without a sample as negative control. After incubation in dark, the absorbance was read at 517 nm using an ELISA plate reader. The experiment was run in triplicates and the percent scavenging was calculated using Equation (2):

$$\% \text{ Scavenging} = (\text{Abs of control} - \text{Abs of sample})/(\text{Abs of control}) \times 100 \quad (2)$$

4.6.2. Hydrogen Peroxide Scavenging Assay

A total of 120 μL of purified fractions (E-2 and D-2) in different concentrations (100–250 μg/mL) was taken in a microtiter plate and then 80 μL of the hydrogen peroxide solution (40 mM) in 100 mM phosphate buffer (pH 7.4) was added into it [32]. The plates were incubated at 37 °C for 15 min and absorbance was measured at 230 nm using ELISA plate reader. A phosphate-buffered solution as a negative and ascorbic acid as a positive control was run separately. The experiment was run in triplicate and the scavenging percent was calculated using Equation (2).

4.6.3. Cytotoxic and Antibacterial Potential of ASM-1 Fractions

The brine shrimp lethality assay was carried out to anticipate the cytotoxic potential of the extracts, with Mitomycin-C used as a standard by the method previously described in [33]. The antibacterial potential of the test extracts was investigated using a well diffusion assay against a range of different bacterial species, i.e., *Bacillus altitudinis* ASM-9, *E. coli* (ATCC 10536), *Pseudomonas aeruginosa*, *Listeria monocytogenes*, *Klebsiella pneumoniae*, *Staphylococcus aureus*, and *Staphylococcus epidermidis* (clinical strains).

4.6.4. Biofilm Detection through Microtiter Plate Method

Both *Bacillus subtilis* ASM-1 and *Bacillus altitudinis* ASM-9 strains were evaluated for their biofilm formation ability through the tissue culture plate method as described by Christensen et al. [34]. A total of 200 μL of freshly grown cultures of ASM-1 and ASM-9 in LB broth with 1% glucose were inoculated in sterile 96 well plate and incubated at 37 °C for 72 h. Negative control wells contained uninoculated sterile broth. After incubation, the plates were washed and stained with 125 μL of 1% (*w*/*v*) crystal violet for 15 min. Absorbance of the remaining dye was measured at λ540 nm using an ELISA Microplate reader.

4.6.5. Gas Chromatography–Mass Spectrometry Analysis for ASM-1 Fractions

The chemical constituents in most promising fractions (E-2 and D-2) with maximum antioxidant activity were analyzed by gas chromatography–mass spectrometry (GC–MS) (Agilent, Model number 6890 N) [35]. Agilent JW Scientific DB-5 MS capillary column (dimensions 30 m × 0.25 mm × 0.25 μm) fused with 5% phenyl methyl polysiloxane; Helium was used as carrier gas at a flow rate of 1 mL/min. The column temperature was initially set at 70 °C for 2 min followed by an increase of 3 °C/min to attain 250 °C for the total sample run time of 30 min. The ionization voltage for MS was operated at 70 eV with an ionization current of 0–315 μA. The instrument was set to detect compounds in a mass range of 50–1000 a.m.u. The chemical compounds were identified after comparing the mass spectral data with the National Institute of Standards and Technology (NIST) Ver. 02 library.

### *4.7. In Vitro Analysis of ASM-1 Extracts in Protecting Radiosensitive Microorganisms under UVB Radiation*

The radioprotective property of extract from ASM-1 was analyzed against radiosensitive microorganisms under UV radiation stress. Two radiosensitive bacterial strains ASM-9 and *E. coli* (ATCC 10536) were spread on TGY agar plates (90 mm) and followed by spreading of 100 μL of liquid mixture containing E-2 fraction (1 mg/mL) and supernatant (24-hour-old culture) of a radioresistant ASM-1 strain in a ratio of (2:1). The plates were kept in a UVB chamber for UVB doses in exponential form for different time intervals. Thereafter, the plates were wrapped in aluminum foils and incubated for 3 days at 37 °C. A control with radiosensitive strains without spreading ASM-1 extracts was run in parallel as a control. The results were recorded after 3 days using the following formula and the data is expressed as a mean value obtained from triplicates as shown in Figure 9.

$$\%Survival = \frac{No\ of\ colonies\ on\ UV\ irradiated\ plates}{No\ of\ colonies\ on\ control\ plates} \times 100 \quad (3)$$

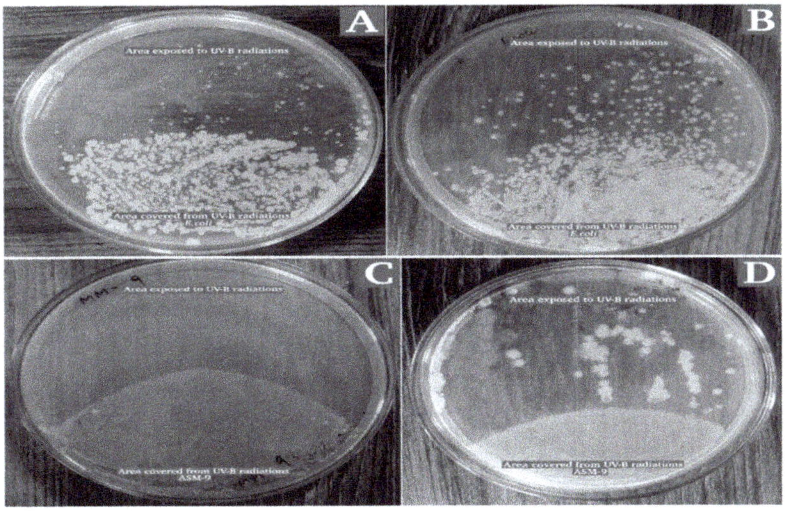


**Figure 9.** Growth and survivability rate of radiosensitive strains (*E. coli* and ASM-9) under UVB radiation with no ASM-1 extracts (**A**). The second figure shows survivability of radiosensitive strains after the addition of ASM-1 extracts (**B**). The data are expressed as mean values of the triplicates ± SDs. (**C**) Depicts the survival of ASM 9 under UV radiation without the addition of ASM 1 extracts. Whereas, (**D**) shows the survival of ASM 9 under UV radiation after the addition of ASM 1 extracts.

### 4.8. Ligand Preparation

The compounds were studied based on literature search where two compounds were obtained having role in inhibiting UV radiation among the compounds obtained via GC/MS analysis. These compounds are Pyrrolo [1,2-a] pyrazine -1,4-dione, hexahydro-3-(2- methyl propyl and 9H-Pyrido [3,4-b] indole. Ligand preprocessing, which included protonation, ionization, and the addition of explicit counter ions, hydrogen atoms, or atomic partial charges, was performed using Discovery Studio (DS) and UCSF Chimera. Energy minimization of small molecules was carried out using the AMBER ff14SB forcefield. The refined dataset obtained from these steps was then utilized for further computational experiments. To facilitate the study, ChemDraw (Mendelsohn, 2004), a chemical drawing tool, was used to draw the 2D structures of the compounds and convert them into 3D structures. Subsequently, ligand preprocessing was performed on these compounds.

### 4.9. Molecular Docking

Molecular docking of the compounds was carried out using the PyRx interface [36] and GOLD software [9]. For the docking process, the following parameters were employed: an exhaustiveness value of 10, the binding site was defined as the residues of the N-terminal domain involved in the binding pocket, and a maximum of 10 poses were allowed. The PyRx tool generated binding affinity values in the negative (where a more negative value indicates a stronger binding affinity). The obtained inhibitors demonstrated a high binding affinity were selected for further analysis. The docking results of the top inhibitors can be found in Figures 5 and 6. To validate the docking protocol used, an inhibitor was docked into the N-terminal structure of (PDB ID = 51FA and 5KZP) using 10 iterations. The same binding mode as the one observed in crystallization studies for the known inhibitors was obtained. The docking reproducibility results obtained using AutoDock Vina are presented in Figures 5 and 6.

### 4.10. Molecular Dynamics Simulations

The docked solutions and inhibitor were subjected to a 20 ns molecular dynamics (MD) simulation using AMBER16 [37]. The inhibitor was optimized with the AMBER (GAFF) force field [38], while the targets protein parameters were generated using the ff14SB force field [39]. To integrate the complex into a TIP3P water box, a padding distance of 12 between target proteins and the box borders was set. Sodium ions were added to neutralize the system. The system was heated to 300 K (NVT) using Langevin dynamics [40] for 20 ps to maintain a constant temperature [41]. A time step of 2 fs was used, and a 5 kcal/mol-$Å^2$ restriction was applied to carbon alpha atoms. The system was relaxed for 100 milliseconds during equilibration. To maintain system pressure, a 50 ps NPT ensemble was employed. Finally, a 20 ns production run was conducted at a 2 fs rate. The trajectories generated were analyzed for structural parameters using the AMBER CPPTRAJ program [42]. The hydrogen bonds formed between protein targets and the inhibitor throughout the trajectories were visualized.

## 5. Conclusions

In summary, this study demonstrates the isolation of a radioresistant bacterial strain from the Thal desert, Pakistan. The strain was identified as *Bacillus subtilis* strain ASM-1 (OK559666) and was able to tolerate the detrimental effects of UVB radiation and Mitomycin-C up to a significant level. The extracellular ethyl acetate fraction as compared to other fractions possessed considerable antioxidant potential evaluated via DPPH and $H_2O_2$ scavenging assays. The chemical profile of the partially purified fractions concluded from GC–MS spectra was revealed to have phenolic compounds, pyrrolopyrazine, and other important constituents having significant antioxidant and radioprotective potential. Consequently, ASM-1 unveiled its hidden potential of having a radioprotective role to aid in the survival quest of radiosensitive microorganisms under UVB radiation. Thus, the present study concluded the antioxidative, and radioprotective potential of *Bacillus subtilis* strain ASM-1 extracts, which could play a crucial role in the bioremediation studies of radioactive wastes via growth cooperation with the efficient radiosensitive microbial degraders.

**Author Contributions:** Conceptualization of the idea is presented by A.A.S.; methodology section performed by A.U.R. and A.A.; Data analysis software usage performed by A.U.R., A.F.A. and F.A.; validation of the data performed by A.U.R. and A.A.S.; formal analysis performed by S.K. and Q.u.a.R.; investigation performed by M.B.; resources, S.A.; data curation, A.F.A. and M.A.; writing—original draft preparation, A.U.R., S.K., F.H. and M.B.; writing—review and editing, S.A., M.A. and A.F.; visualization, A.F.A.; funding acquisition, M.A. and A.F.A.; supervision, A.A.S. All authors have read and agreed to the published version of the manuscript.

**Funding:** This research was funded by Quaid-i-Azam University, Islamabad, Pakistan. The authors are thankful to the Researchers Supporting Project number (RSP2023R335), King Saud University, Riyadh, Saudi Arabia.

**Institutional Review Board Statement:** Not applicable.

**Informed Consent Statement:** Not applicable.

**Data Availability Statement:** Data is contained within the article.

**Acknowledgments:** The authors are thankful to the Researchers Supporting Project number (RSP2023R335), King Saud University, Riyadh, Saudi Arabia.

**Conflicts of Interest:** The authors declare no conflict of interest.

## References

1. Giddings, L.-A.; Newman, D. *Bioactive Compounds from Terrestrial Extremophiles*; Springer International Publishing: Berlin/Heidelberg, Germany, 2015; ISBN 978-3-319-13259-4.
2. Nocker, A.; Shah, M.; Dannenmann, B.; Schulze-Osthoff, K.; Wingender, J.; Probst, A.J. Assessment of UV-C-Induced Water Disinfection by Differential PCR-Based Quantification of Bacterial DNA Damage. *J. Microbiol. Methods* **2018**, *149*, 89–95. [CrossRef]

3. Sakudo, A.; Toyokawa, Y.; Nakamura, T.; Yagyu, Y.; Imanishi, Y. Nitrogen Gas Plasma Treatment of Bacterial Spores Induces Oxidative Stress that Damages the Genomic DNA. *Mol. Med. Rep.* **2017**, *15*, 396–402. [CrossRef] [PubMed]
4. Yu, L.Z.-H.; Luo, X.-S.; Liu, M.; Huang, Q. Diversity of Ionizing Radiation-Resistant Bacteria Obtained from the Taklimakan Desert. *J. Basic Microbiol.* **2015**, *55*, 135–140. [CrossRef]
5. Becker, J.; Wittmann, C. Microbial Production of Extremolytes—High-Value Active Ingredients for Nutrition, Health Care, and Well-Being. *Curr. Opin. Biotechnol.* **2020**, *65*, 118–128. [CrossRef]
6. Latifi, A.; Ruiz, M.; Zhang, C.-C. Oxidative Stress in Cyanobacteria. *FEMS Microbiol. Rev.* **2009**, *33*, 258–278. [CrossRef] [PubMed]
7. Chaudhri, A.; Nadeem, M.; Rahman, A.; Alam, T.; Sajjad, W.; Hasan, F.; Badshah, M.; Khan, S.; Rehman, F.; Shah, A. Antioxidative and Radioprotective Properties of Glycosylated Flavonoid, Xanthorhamnin from Radio-Resistant *Bacterium Bacillus* Indicus Strain TMC-6. *Curr. Microbiol.* **2020**, *77*, 1245–1253. [CrossRef]
8. Shuryak, I.; Matrosova, V.Y.; Gaidamakova, E.K.; Tkavc, R.; Grichenko, O.; Klimenkova, P.; Volpe, R.P.; Daly, M.J. Microbial Cells Can Cooperate to Resist High-Level Chronic Ionizing Radiation. *PLoS ONE* **2017**, *12*, e0189261. [CrossRef]
9. Verdonk, M.L.; Cole, J.C.; Hartshorn, M.J.; Murray, C.W.; Taylor, R.D. Improved Protein–Ligand Docking Using GOLD. *Proteins Struct. Funct. Bioinform.* **2003**, *52*, 609–623. [CrossRef]
10. Ahmad, S.; Abbasi, H.W.; Shahid, S.; Gul, S.; Abbasi, S.W. Molecular Docking, Simulation and MM-PBSA Studies of Nigella Sativa Compounds: A Computational Quest to Identify Potential Natural Antiviral for COVID-19 Treatment. *J. Biomol. Struct. Dyn.* **2020**, *39*, 4225–4233. [CrossRef]
11. Riesenman, P.J.; Nicholson, W.L. Role of the Spore Coat Layers in *Bacillus Subtilis* Spore Resistance to Hydrogen Peroxide, Artificial UV-C, UV-B, and Solar UV Radiation. *Appl. Environ. Microbiol.* **2000**, *66*, 620–626. [CrossRef] [PubMed]
12. Paillié-Jiménez, M.E.; Stincone, P.; Brandelli, A. Natural Pigments of Microbial Origin. *Front. Sustain. Food Syst.* **2020**, *4*, 590439. [CrossRef]
13. Nayak, T.; Sengupta, I.; Dhal, P.K. A New Era of Radiation Resistance Bacteria in Bioremediation and Production of Bioactive Compounds with Therapeutic Potential and Other Aspects: An in-Perspective Review. *J. Environ. Radioact.* **2021**, *237*, 106696. [CrossRef] [PubMed]
14. Laroussi, M.; Leipold, F. Evaluation of the Roles of Reactive Species, Heat, and UV Radiation in the Inactivation of Bacterial Cells by Air Plasmas at Atmospheric Pressure. *Int. J. Mass Spectrom.* **2004**, *233*, 81–86. [CrossRef]
15. Célariès, B.; Rima, G.; Court, L.; Lion, C.; Laval, J.D. Radioprotective Activity and Synthesis of Siladithioacetals and Germadithioacetals Derived from N-Substituted Naphthylethylimidazoline. *Met.-Based Drugs* **2001**, *8*, 199–210. [CrossRef]
16. Brand-Williams, W.; Cuvelier, M.E.; Berset, C. Use of a Free Radical Method to Evaluate Antioxidant Activity. *LWT Food Sci. Technol.* **1995**, *28*, 25–30. [CrossRef]
17. Kadaikunnan, S.; Rejiniemon, T.S.; Khaled, J.M.; Alharbi, N.S.; Mothana, R. In-Vitro Antibacterial, Antifungal, Antioxidant and Functional Properties of *Bacillus amyloliquefaciens*. *Ann. Clin. Microbiol. Antimicrob.* **2015**, *14*, 9. [CrossRef]
18. Lemon, K.P.; Earl, A.M.; Vlamakis, H.C.; Aguilar, C.; Kolter, R. Biofilm Development with an Emphasis on *Bacillus subtilis*. *Curr. Top. Microbiol. Immunol.* **2008**, *322*, 1–16. [CrossRef] [PubMed]
19. Ghorab, M.M.; Ragab, F.A.; Alqasoumi, S.I.; Alafeefy, A.M.; Aboulmagd, S.A. Synthesis of Some New Pyrazolo[3,4-d]Pyrimidine Derivatives of Expected Anticancer and Radioprotective Activity. *Eur. J. Med. Chem.* **2010**, *45*, 171–178. [CrossRef] [PubMed]
20. Sverdlov, R.L.; Brinkevich, S.D.; Shadyro, O.I. Effects of Tryptophan Derivatives and β-Carboline Alkaloids on Radiation- and Peroxide-Induced Transformations of Ethanol. *Radiat. Phys. Chem.* **2014**, *98*, 77–85. [CrossRef]
21. Liu, M.-H.; Zhang, Q.; Zhang, Y.-H.; Lu, X.-Y.; Fu, W.-M.; He, J.-Y. Chemical Analysis of Dietary Constituents in Rosa Roxburghii and Rosa Sterilis Fruits. *Molecules* **2016**, *21*, 1204. [CrossRef]
22. Elamin, M.M.; Abdelrahim, N.A.; Elhag, D.E.A.; Joseph, M.R.P.; Hamid, M.E. Bioactive Pyrrole-Pyrazine Derivative from a Novel *Bacillus* Species and Review of the Literature. *Afr. J. Pharm. Pharmacol.* **2021**, *15*, 138–151. [CrossRef]
23. Gao, H.; Li, P.; Xu, X.; Zeng, Q.; Guan, W. Research on Volatile Organic Compounds From *Bacillus subtilis* CF-3: Biocontrol Effects on Fruit Fungal Pathogens and Dynamic Changes During Fermentation. *Front. Microbiol.* **2018**, *9*, 456. [CrossRef]
24. Sajjad, W.; Ahmad, S.; Aziz, I.; Azam, S.S.; Hasan, F.; Shah, A.A. Antiproliferative, Antioxidant and Binding Mechanism Analysis of Prodigiosin from Newly Isolated Radio-Resistant Streptomyces Sp. Strain WMA-LM31. *Mol. Biol. Rep.* **2018**, *45*, 1787–1798. [CrossRef] [PubMed]
25. Li, Q.; Chen, X.; Jiang, Y.; Jiang, C.; Li, Q.; Chen, X.; Jiang, Y.; Jiang, C. Cultural, Physiological, and Biochemical Identification of Actinobacteria. In *Actinobacteria-Basics and Biotechnological Applications*; IntechOpen: London, UK, 2016; ISBN 978-953-51-2248-7.
26. Wright, M.H.; Adelskov, J.; Greene, A.C. Bacterial DNA Extraction Using Individual Enzymes and Phenol/Chloroform Separation. *J. Microbiol. Biol. Educ.* **2017**, *18*, 10–1128. [CrossRef]
27. Weisburg, W.G.; Barns, S.M.; Pelletier, D.A.; Lane, D.J. 16S Ribosomal DNA Amplification for Phylogenetic Study. *J. Bacteriol.* **1991**, *173*, 697–703. [CrossRef]
28. Kumar, S.; Stecher, G.; Li, M.; Knyaz, C.; Tamura, K. MEGA X: Molecular Evolutionary Genetics Analysis across Computing Platforms. *Mol. Biol. Evol.* **2018**, *35*, 1547–1549. [CrossRef]
29. Mattimore, V.; Battista, J.R. Radioresistance of Deinococcus Radiodurans: Functions Necessary to Survive Ionizing Radiation Are Also Necessary to Survive Prolonged Desiccation. *J. Bacteriol.* **1996**, *178*, 633–637. [CrossRef] [PubMed]
30. Kedare, S.B.; Singh, R.P. Genesis and Development of DPPH Method of Antioxidant Assay. *J. Food Sci. Technol.* **2011**, *48*, 412–422. [CrossRef]

31. Elmastaş, M.; Gülçin, İ.; Işildak, Ö.; Küfrevioğlu, Ö.İ.; İbaoğlu, K.; Aboul-Enein, H.Y. Radical Scavenging Activity and Antioxidant Capacity of Bay Leaf Extracts. *J. Iran. Chem. Soc.* **2006**, *3*, 258–266. [CrossRef]
32. Zhang, Y.; Mu, J.; Han, J.; Gu, X. An Improved Brine Shrimp Larvae Lethality Microwell Test Method. *Toxicol. Mech. Methods* **2012**, *22*, 23–30. [CrossRef]
33. Christensen, G.D.; Simpson, W.A.; Younger, J.J.; Baddour, L.M.; Barrett, F.F.; Melton, D.M.; Beachey, E.H. Adherence of Coagulase-Negative Staphylococci to Plastic Tissue Culture Plates: A Quantitative Model for the Adherence of Staphylococci to Medical Devices. *J. Clin. Microbiol.* **1985**, *22*, 996–1006. [CrossRef] [PubMed]
34. Tahir, H.A.S.; Gu, Q.; Wu, H.; Raza, W.; Hanif, A.; Wu, L.; Colman, M.V.; Gao, X. Plant Growth Promotion by Volatile Organic Compounds Produced by *Bacillus subtilis* SYST2. *Front. Microbiol.* **2017**, *8*, 171. [CrossRef]
35. Dallakyan, S.; Olson, A.J. Small-Molecule Library Screening by Docking with PyRx. In *Chemical Biology*; Springer: Berlin/Heidelberg, Germany, 2015; pp. 243–250.
36. Morris, G.M.; Huey, R.; Lindstrom, W.; Sanner, M.F.; Belew, R.K.; Goodsell, D.S.; Olson, A.J. AutoDock4 and AutoDockTools4: Automated Docking with Selective Receptor Flexibility. *J. Comput. Chem.* **2009**, *30*, 2785–2791. [CrossRef]
37. Case, D.A.; Betz, R.M.; Cerutti, D.S.; Cheatham, T.; Darden, T.; Duke, R.E.; Giese, T.J.; Gohlke, H.; Götz, A.W.; Homeyer, N.; et al. *Amber 16, University of California, San Francisco*; University of California: San Francisco, CA, USA, 2016. [CrossRef]
38. Sprenger, K.G.; Jaeger, V.W.; Pfaendtner, J. The General AMBER Force Field (GAFF) Can Accurately Predict Thermodynamic and Transport Properties of Many Ionic Liquids. *J. Phys. Chem. B* **2015**, *119*, 5882–5895. [CrossRef] [PubMed]
39. Case, D.A.; Babin, V.; Berryman, J.T.; Betz, R.M.; Cai, Q.; Cerutti, D.S.; Cheatham III, T.E.; Darden, T.A.; Duke, R.E.; Gohlke, H.; et al. The FF14SB Force Field. *Amber* **2014**, *14*, 29–31.
40. Izaguirre, J.A.; Catarello, D.P.; Wozniak, J.M.; Skeel, R.D. Langevin Stabilization of Molecular Dynamics. *J. Chem. Phys.* **2001**, *114*, 2090–2098. [CrossRef]
41. Price, D.J.; Brooks, C.L. A Modified TIP3P Water Potential for Simulation with Ewald Summation. *J. Chem. Phys.* **2004**, *121*, 10096–10103. [CrossRef]
42. Roe, D.R.; Cheatham, T.E. PTRAJ and CPPTRAJ: Software for Processing and Analysis of Molecular Dynamics Trajectory Data. *J. Chem. Theory Comput.* **2013**, *9*, 3084–3095. [CrossRef]

pharmaceuticals

MDPI

Article

# Bioanalytical LC-QTOF/MS Method for a *N*-phenylpiperazine Derivate (LQFM05): An Anxiolytic- and Antidepressant-like Prototype Drug Applied to Pharmacokinetic and Biodistribution Studies

**Ana Cláudia M. Ramos [1], Kênnia R. Rezende [1,*], Carolina M. Teixeira [1], Aline R. Fernandes [2], Heloa Santos [2], Rúbia Darc Machado [1], Ricardo Menegatti [3], Boniek G. Vaz [2] and Andréa R. Chaves [2,*]**

1 Laboratory of Biopharmacy and Pharmacokinetics (BioPK), Faculty of Pharmacy, Federal University of Goiás, Goiânia 74605-170, GO, Brazil
2 Laboratory of Chromatography and Mass Spectrometry (LaCEM), Chemistry Institute, Federal University of Goiás, Goiânia 74690-900, GO, Brazil; boniek@ufg.br (B.G.V.)
3 Laboratory of Pharmaceutical Medicinal Chemistry (LQFM), Faculty of Pharmacy, Federal University of Goiás, Goiânia 74605-170, GO, Brazil
* Correspondence: kennia@ufg.br (K.R.R.); andrea_chaves@ufg.br (A.R.C.); Tel.: +55-(62)3209-6470 (K.R.R.); +55-(62)3521-1127 (A.R.C.)

**Citation:** Ramos, A.C.M.; Rezende, K.R.; Teixeira, C.M.; Fernandes, A.R.; Santos, H.; Machado, R.D.; Menegatti, R.; Vaz, B.G.; Chaves, A.R. Bioanalytical LC-QTOF/MS Method for a *N*-phenylpiperazine Derivate (LQFM05): An Anxiolytic- and Antidepressant-like Prototype Drug Applied to Pharmacokinetic and Biodistribution Studies. *Pharmaceuticals* **2023**, *16*, 930. https://doi.org/10.3390/ph16070930

Academic Editors: Jan Oszmianski and Sabina Lachowicz-Wiśniewska

Received: 5 October 2022
Revised: 8 November 2022
Accepted: 11 November 2022
Published: 26 June 2023



**Abstract:** The LQFM05 is a prototype drug designed for treatment of psychiatric disorders, such as schizophrenia, exhibiting anxiolytic- and antidepressant-like (12 or 24 μmol/kg) effects in classical behavioral tests. In order to evaluate its pharmacokinetic properties, a liquid chromatography method coupled to a quadrupole time of flight mass spectrometry system (LC-QTOF/MS) was developed and fully validated for LQFM05 analysis in rat plasma and tissue samples (brain, heart, liver, and kidneys). Liquid–liquid extraction, solid phase extraction and protein precipitation were assessed as clean-up procedures for biological samples and analyte enrichment. Plasma and tissue samples underwent protein precipitation as a preliminary step, using acetonitrile. Linearity was fully demonstrated for the dynamic range (10.0 to 900.0 ng/mL), with $r^2$ values higher than 0.99 ($RSD_{slope} \leq 2\%$, $F_{cal} < F_{tab}$, $C_{cal} < C_{tab}$). Biodistribution studies in rats revealed high brain tissue concentrations (12.4 μg/g), suggesting elevated drug affinity to the main therapeutic target tissue, showing a blood partition coefficient of 1.9. Kidneys also showed great exposure and tissue affinity, suggesting a potential extrahepatic clearance. Likewise, all examined tissues exhibited satisfactory LQFMF05 distribution. The mass fragmentation spectrum indicated the presence of its main metabolite, LQFM235, yielded by high hepatic hydroxylation route, an equally bioactive derivative. Lastly, the developed LC-QTOF/MS method was shown to be sensitive (LOQ = 10 ng/mL), precise and accurate for LQFM05 determination in tissue homogenates and plasma samples.

**Keywords:** antipsychotic drug LQFM05; sample pretreatment; tissue distribution; bioanalytical methods; LC-QTOF/MS

## 1. Introduction

Schizophrenia is a severe chronic psychiatric disorder and one of 25 leading causes of disability worldwide [1,2]. Clozapine is the most used drug for schizophrenia disorder control, and is therefore constantly related to hematological and metabolic disorders as a side effect for some clinical patients [3], limiting its prescription and use. To fill this gap, 1-(4-methoxyphenyl)−4-((1-phenyl-1H-pyrazol-4-yl)methyl)piperazine (LQFM05) was investigated as a promising prototype antipsychotic *N*-phenylpiperazine drug, showing anxiolytic- and antidepressant-like effects in classical behavioral tests (12 or 24 μmol/kg LQFM005), putatively through activation of 5-HT1A receptors [4].

Preclinical pharmacokinetics studies are key aspects of the new drug development phase as a way of identifying an adequate bioavailability profile, tissue distribution and excretion rates. Furthermore, it can support data on dosage schedules, pharmacological information, and possible toxicity [5–7]. In this sense, a reliable and accurate analytical method is required.

Several methods for quantification of phenylpiperazine antipsychotics and atypical antidepressants have been reported, often using liquid chromatography (LC) with ultraviolet (UV), diode array (DAD), fluorescence (FLD) and mass spectrometry detector (MS) [8–11].

Moreover, a sample pretreatment procedure also seems like a crucial step for analytical method development. Techniques such as protein precipitation, liquid–liquid extraction (LLE), solid phase extraction (SPE), solid phase microextraction (SPME) and hollow fiber-based liquid phase microextraction (HF-LPME) have been used to extract and/or pre-concentrate drugs in the biological fluid before analytical system analysis [9,12–15].

Mass spectrometry methods have several applications in pharmacokinetic studies due to the better selectivity of the higher resolution mass spectrometers, such as the quadrupole time-of-flight mass spectrometry system (QTOF/MS). The higher resolution, sensitivity and selectivity of these systems, especially when coupled to a liquid chromatography system (LC-QTOF/MS), allow for techniques with reduced time consumption and sample pretreatment costs.

Thus, here, we present a bioanalytical LC-QTOF/MS method applied to the pharmacokinetics and biodistribution studies of a new potential antipsychotic drug with anxiolytic and antidepressant-like effect (LQFM05) using a straightforward sample pretreatment procedure. Pharmacokinetics and drug tissue/plasma ratio (Kp) to the brain, heart, liver and kidneys were also evaluated.

## 2. Results and Discussion

### 2.1. Method Development and Sample Preparation Strategies

Carbamazepine, dexamethasone, olanzapine, ethinylestradiol, fluoxetine macohydrochloride and diazepam were evaluated as internal standards (ISs) because some of their physicochemical properties, such as log P and parent fragmentation in the MS/MS spectrometer, were similar to LQFM05 [16,17]. Accordingly, diazepam was selected for MS ionization condition, fragmentation profile, matrix endogenous compounds interference, as well as chromatographic profile. The LQFM05 and diazepam MS/MS obtained spectrum are shown in Figure 1. The most abundant ion (157.08 $m/z$) was selected as the LQFQ05 quantifier.

For sample clean-up and analyte enrichment procedures, solid phase extraction (SPE), liquid–liquid (LLE) and protein precipitation extraction (PPE) were evaluated. Some SPE drawbacks may include multiple steps and slow filtration. Moreover, the observed analyte run time elution was inadequate (~20 min) as well as its selectivity. Instead, LLE was shown to be a faster and less costly sample preparation procedure than SPE. Experiments were performed based on the protocol established by Kumar and Ramanathan [10], although considerable residual effects were observed. Eventually, PPE was also evaluated. Regardless of some often-reported disadvantages, such as low selectivity and ion signal suppression, herein data from PPE procedure showed better sample clean-up, with no MS signal interferences. In summary, the efficiency of the evaluated techniques is expressed by a slightly higher determination coefficient ($r^2$) for PPE (0.9985) compared to LLE ($r^2$ = 0.9872), as exemplified by the liver extraction procedure in Figure 2.

### 2.2. Analytical Validation

The total ion chromatograms (TICs) of blank plasma, brain, heart, liver and kidney blank samples and heart compared to matrix spiked with IS (250 ng/mL) and LQFM05 (10 ng/mL) are shown in Figure 3. LQFM05 and IS retention time (RT) were 5.6 min and 6.3 min, respectively. Endogenous or exogenous interferences were not seen for LQFM05 and IS peaks in the matrices studied.

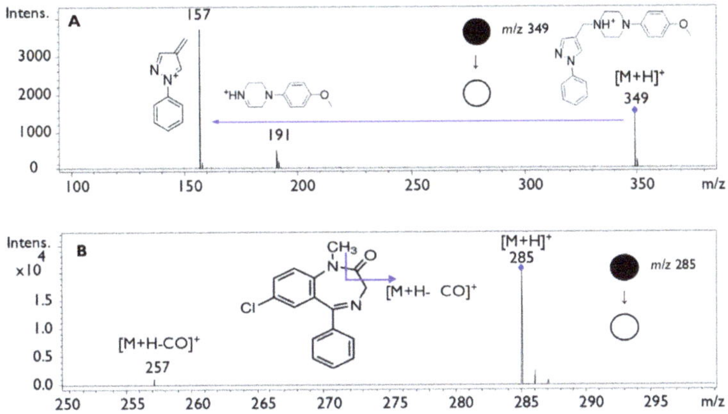


**Figure 1.** MS/MS spectrum of (**A**): LQFM05 at 30 ng/mL (retention time at 6.3 min), $[M + H]^+$ 349 $m/z$, and (**B**): diazepam 250 ng/mL (retention time at 5.6 min), $[M + H]^+$ 285 $m/z$.

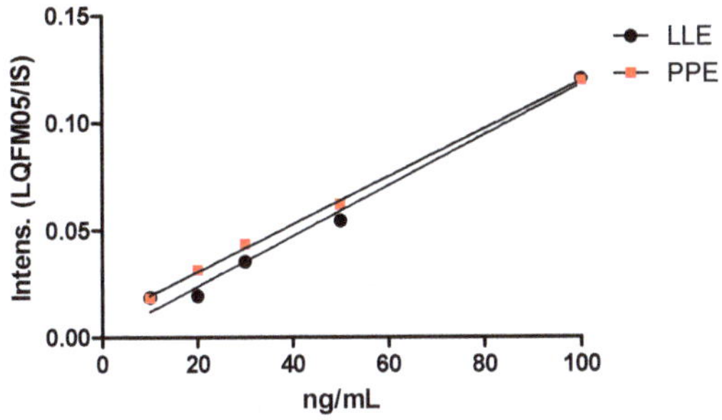


**Figure 2.** LLE and PPE calibration curves (10.0–100.0 ng/mL, $n$ = 5) in liver homogenate.

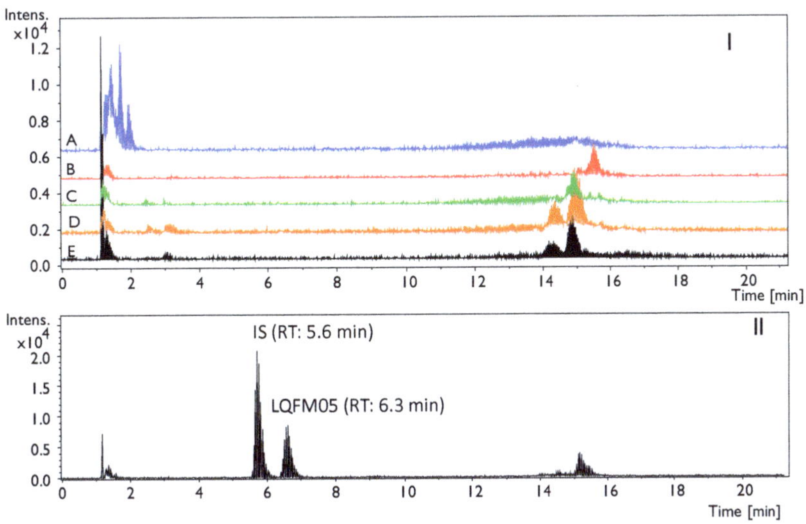


**Figure 3.** Total ion chromatogram (TIC) of: (**I**) reference blank samples (A: plasma; B: brain; C: liver; D: kidneys and E: heart); (**II**) heart homogenates (IS: 250 ng/mL and LQFM05: 250 ng/mL).

Analytical parameters assessed during the LC-MS method validation process (linearity, precision, accuracy, etc.) aim to provide evidence of confidence for the applied sample preparation and analysis procedure. Methods of calibration are especially meaningful for assaying studies. Linearity in the LQFM05 concentration range of 10.0 to 900.0 ng/mL was primarily assessed by means of an $r^2 > 0.99$. The lower limit of quantification (LLOQ) concentration was at 10.0 ng/mL based on the signal/noise higher than 10 with accuracy and precision. A summary of these values can be found in Table 1.

**Table 1.** Calibration curve data for LQFM05 (range of 10.0 to 900.0 ng/mL) at all studied matrices.

| Parameter | Plasma | Brain | Heart | Liver | Kidneys |
|---|---|---|---|---|---|
| **Slope** [a] (±SD, *n* = 4) | $1.761 \times 10^{-3} \pm 1.587 \times 10^{-5}$ | $9.882 \times 10^{-4} \pm 8.168 \times 10^{-6}$ | $1.239 \times 10^{-3} \pm 1.985 \times 10^{-5}$ | $1.692 \times 10^{-3} \pm 7.928 \times 10^{-6}$ | $1.127 \times 10^{-3} \pm 1.655 \times 10^{-5}$ |
| **Intercept** [a] (±SD, *n* = 4) | $1.081 \times 10^{-3} \pm 7.043 \times 10^{-3}$ | $0.905 \times 10^{-2} \pm 0.362 \times 10^{-2}$ | $-0.7043 \times 10^{-3} \pm 0.8809 \times 10^{-2}$ | $7.30 \times 10^{-3} \pm 3.518 \times 10^{-3}$ | $8.292 \times 10^{-3} \pm 7.345 \times 10^{-3}$ |
| $r^2$ [b] | 0.9969 | 0.9974 | 0.9903 | 0.9992 | 0.9919 |
| $RSD_{slope}$ (%) [c] | 0.901 | 0.827 | 1.602 | 0.469 | 1.469 |
| $F_{calculated}$ [d] | 1.351 | 1.257 | 0.285 | 1.062 | 0.641 |
| $C_{calculated}$ [e] | 0.201 | 0.226 | 0.204 | 0.073 | 0.191 |

[a] Peak intensity of the analyte/IS vs. concentration. [b] Correlation coefficient. [c] Relative standard deviation of the slope ($RSD_{slope} \leq 2\%$). [d] Fisher ratio; $F_{cal} < F_{tab}$ (linear relationship) = calculated and tabulated value ($\alpha = 0.05$), respectively [18]. [e] Cochran test; $C_{cal} < C_{tab}$ (homoscedastic residuals) = calculated and tabulated value ($\alpha = 0.05$), respectively [19].

In general, the relative standard deviation of the slope (RSDslope) used to evaluate linearity and the significance of the angular coefficient is an indication of the experimental dispersion data around the regression line [20]. The RSDslope values in all matrices are in the range of 0.47 to 1.6% (Table 1). According to the obtained RSDslope values, the regressions can be considered linear (≤2%) [20].

Additionally, the Cochran hypothesis test was performed in order to evaluate the homogeneity of variance of the residues from the (y) axis. In other words, if the calculated (Ccal) value at a 5% level of significance (P = 0.05) found for (J = 4 and I = 10) was lower than 0.373 (Ctabulated). As the calculated value (Ccal) was lower than Ctabulated in all the matrices, there is homoscedasticity of responses [19,20]. Hence, once the hypothesis was accepted, the least squares method was adequate for estimating the best regression line that passes through the points obtained experimentally from the calibration curve.

Fisher-Snedecor's F-Ratio was calculated for comparison between calibration data variability. At the 95% confidence level, F-tabulated (0.05, 8, 30) for 10 concentrations levels (I) and two different days were 2.27 [18–20]. Accordingly, Fcal was lower than Ftab in all matrices, and the obtained results describe a linear calibration. See Table 1.

Graphical representation of the residues between observed and predicted Y (response) values is also a very important procedure for detection of outliers, points of influence, lack of adjustment or unequal variation. However, Jurado et al. [21] adds the differentiation of the use of studentized residuals by the residual standard deviation ($S_{y/x}$) and the leverage effect of each point on the regression line. Studentized residues (Figure A1) distributed around zero showed no tendency and corresponded to the acceptance criteria (±2) in the range of analyzed concentrations. Thus, the regression model used was considered adequate to this aim.

The intra- and inter-assay precision and accuracy were estimated at four concentration levels, each concentration in replicates (*n* = 6 and *n* = 12, respectively). All results are shown in Table 2.

**Table 2.** Precision and accuracy of LQFM05 in rat plasma and tissue homogenates (brain, heart, liver and kidneys) by LC-QTOF/MS.

| Matrix | $C_{spiked}$ (ng/mL) | Intra (*n* = 6) | | | Inter (*n* = 12) | | |
|---|---|---|---|---|---|---|---|
| | | $C_M$ Mean ± SD (ng/mL) | RSD% | RE% | $C_M$ Mean ± SD (ng/mL) | RSD% | RE% |
| Plasma | 10 | 9.66 ± 0.51 | 5.3 | −3.4 | 9.98 ± 0.72 | 7.2 | −0.2 |
| | 30 | 31.00 ± 1.37 | 4.4 | 3.3 | 30.71 ± 1.57 | 5.1 | 2.4 |
| | 450 | 455.58 ± 9.50 | 2.1 | 1.2 | 453.78 ± 8.97 | 2.0 | 0.8 |
| | 750 | 754.80 ± 10.30 | 1.4 | 0.6 | 754.24 ± 10.06 | 1.3 | 0.6 |
| Brain | 10 | 9.44 ± 0.56 | 6.0 | −0.2 | 9.76 ± 0.79 | 8.1 | −2.4 |
| | 30 | 29.31 ± 1.47 | 5.0 | −0.1 | 29.36 ± 1.39 | 4.7 | −2.1 |
| | 450 | 432.76 ± 3.20 | 0.7 | −0.2 | 441.38 ± 9.34 | 2.1 | −1.9 |
| | 750 | 727.68 ± 7.53 | 1.0 | −4.2 | 743.67 ± 18.95 | 2.5 | −0.8 |
| Heart | 10 | 9.85 ± 0.30 | 3.4 | −1.5 | 9.92 ± 0.25 | 2.6 | −0.8 |
| | 30 | 30.06 ± 0.12 | 0.4 | 0.2 | 30.22 ± 0.42 | 1.4 | 0.7 |
| | 450 | 451.04 ± 8.07 | 1.8 | 0.2 | 449.65 ± 6.04 | 1.3 | −0.1 |
| | 750 | 750.11 ± 11.61 | 1.5 | 0.0 | 751.43 ± 9.65 | 1.3 | 0.2 |
| Liver | 10 | 10.22 ± 0.37 | 3.6 | 2.2 | 10.30 ± 0.35 | 3.4 | 3.0 |
| | 30 | 30.16 ± 0.86 | 2.9 | 0.5 | 30.36 ± 0.76 | 2.5 | 1.2 |
| | 450 | 449.78 ± 7.87 | 1.8 | 0.1 | 449.19 ± 6.16 | 1.4 | −0.2 |
| | 750 | 751.24 ± 8.68 | 1.2 | 0.2 | 749.57 ± 8.92 | 1.2 | −0.1 |
| Kidneys | 10 | 10.59 ± 0.71 | 6.7 | 5.9 | 10.49 ± 0.71 | 6.7 | 4.9 |
| | 30 | 29.84 ± 0.77 | 2.6 | −0.5 | 29.94 ± 0.75 | 2.5 | −0.2 |
| | 450 | 452.49 ± 10.95 | 2.4 | 0.6 | 453.60 ± 8.57 | 1.9 | 0.8 |
| | 750 | 756.41 ± 10.11 | 1.3 | 0.9 | 752.52 ± 8.93 | 1.2 | 0.3 |

$C_{spiked}$: spiked concentration; CM: measured concentration; mean determination; SD: standard de-viation; RSD%: relative standard deviation percentage; RE%: relative error.

Recovery of a bioanalytical method measures efficiency of the extraction procedure. Regarding drug recovery rates from all investigated tissues, the lowest one was found for the brain (around 45%), probably as a result of the high drug affinity to brain tissue. Additionally, biotransformation by enzymatic route can also bias to lower recovery rates. However, is worth mentioning that high analytical precision was found at all drug concentration levels (RSD% $\leq$ 10.9%), providing reliable recovery values. See Table 3.

The IS matrix effects showed values in the range of 0.98 (30 ng/mL: heart and liver) to 1.09 (30 ng/mL: plasma, and RSD% $\leq$ 8.4% (Table 3) in agreement to worldwide acceptance criteria [22]. Thus, results imply no significant drug ionization suppression or enhancement in plasma or other homogenates [23]. All RSD% values were lower than 15%.

Adsorption of the analyte or IS into the injector, chromatographic column, LC connections, ionization interface or other section of the chromatographic system can result in a late peak appearance of them into the next following run [24]. Potential interfering peaks at the analyte retention time should not exceed 20% of the obtained value when setting the LLOQ. In the present developed method, no carryover was observed at retention time of both IS (5.6 min) and LQFM05 (6.3 min) after injecting the highest calibrator concentration (900.0 ng/mL).

The stability was assessed at three different conditions (FTC: stability after freezing and thawing cycles, SST: short time stability and PPS: post-processing stability). The obtained data showed acceptable accuracy and precision values below 15% (Figure 4) [22].

### 2.3. Tissue Distribution Study

Tissue distribution of LQFM05 was investigated in male Wistar rats at 0.08, 0.17, 0.25, 1, 3, 6 and 12 h after *intravenous* (*i.v*). dosing (10 mg/Kg) by collecting tissue samples from the brain, heart, liver, kidneys and blood. The concentration-time profile of LQFM05 in various tissues is shown in Figure 5.

**Table 3.** Recovery and matrix effect data expressed by standard deviation (SD) and precision for all evaluated matrices ($n$ = 6 for QC and $n$ = 18 for IS) by LC-QTOF/MS.

| Matrix | | $C_{spiked}$ (ng/mL) | Recovery | | Matrix Effect (NMF) | |
|---|---|---|---|---|---|---|
| | | | Mean ± SD (%) | RSD (%) | Mean ± SD | RSD (%) |
| Plasma | LQFM05 | 30 | 55.53 ± 4.63 | 8.3 | 1.02 ± 0.09 | 8.8 |
| | | 450 | 56.82 ± 5.0 | 8.7 | 1.04 ± 0.06 | 5.8 |
| | | 750 | 56.22 ± 3.70 | 6.6 | 1.03 ± 0.03 | 2.9 |
| | IS | 250 | 79.50 ± 4.93 | 6.2 | 1.09 ± 0.04 | 4.1 |
| Brain | LQFM05 | 30 | 45.84 ± 3.82 | 8.3 | 1.09 ± 0.04 | 4.1 |
| | | 450 | 47.54 ± 2.03 | 4.3 | 1.05 ± 0.04 | 3.6 |
| | | 750 | 43.26 ± 1.62 | 3.7 | 1.06 ± 0.03 | 2.7 |
| | IS | 250 | 94.38 ± 4.50 | 4.8 | 0.99 ± 0.07 | 7.3 |
| Heart | LQFM05 | 30 | 68.94 ± 3.87 | 5.6 | 1.04 ± 0.04 | 3.5 |
| | | 450 | 63.47 ± 1.84 | 2.9 | 1.00 ± 0.05 | 4.5 |
| | | 750 | 65.28 ± 3.05 | 4.7 | 0.98 ± 0.04 | 4.4 |
| | IS | 250 | 99.61 ± 5.89 | 5.9 | 1.02 ± 0.05 | 5.1 |
| Liver | LQFM05 | 30 | 71.82 ± 4.79 | 6.7 | 0.99 ± 0.04 | 3.7 |
| | | 450 | 59.13 ± 2.92 | 4.9 | 0.98 ± 0.06 | 6.3 |
| | | 750 | 70.75 ± 5.03 | 7.1 | 1.00 ± 0.04 | 3.7 |
| | IS | 250 | 82.38 ± 7.04 | 8.5 | 1.00 ± 0.03 | 3.0 |
| Kidneys | LQFM05 | 30 | 63.77 ± 6.97 | 10.9 | 1.05 ± 0.04 | 3.6 |
| | | 450 | 55.40 ± 2.53 | 4.6 | 1.06 ± 0.03 | 2.7 |
| | | 750 | 66.50 ± 2.38 | 3.6 | 0.99 ± 0.07 | 7.3 |
| | IS | 250 | 98.31 ± 3.70 | 3.8 | 1.04 ± 0.04 | 3.5 |

IS: Diazepam; $C_{spiked}$: spiked concentration; NMF: normalized matrix factor; mean determination; SD: standard deviation; RSD%: relative standard deviation percentage; QC: quality control samples.

Non-compartmental analysis (NCA) was employed without assumption of any previous PK model, as it could be used in all cases, being the main advantage of the method [25]. All tissues investigated (Table 4) showed decreasing tissue concentrations from 0.08 to 12 h, except for liver tissue ($t_{max}$ at 0.17 h). The highest tissue concentration (Figure 5) was obtained in the brain tissue (12,357.0 ng/g), indicating higher tissue affinity as desired to any psychoactive drug candidates.

The highest exposure to LQFM05 was found in the kidneys (14,595.2 h*ng/g) followed by the heart (10,919.6 h*ng/g) and brain (10,460.6 h*ng/g). The lowest exposure was observed in the liver (8235.3 h*ng/g) in a similar manner as that found for the LASSBio-579 [26].

The liver and brain half-lives were higher (3.6 and 2.5 h, respectively) compared to other tissue values around to 2.2 h. The largest mean residence time (3.1 h) was observed in the liver. Penetration coefficient values (Kp) were larger for the kidneys (2.7), heart (2.0) and brain (1.9). Thus, according to Kp data, LQFM05 was widely distributed to tissues (Kp > 1). Similar results were found for phenylpiperazine analogs, such as LASSBio-579 and 581, except for adipose tissue, which presented a value of 2.24 for penetration of LASSBio-579 [8,11].

Although LQFM05 penetration was much larger when compared to LASSBio-579 (log P 3.48) and LASSBio-581 (log P 2.91), it can be related to its higher lipophilicity (log P 3.68), favoring drug passage through biological phospholipid bilayer membranes. Other works also reported Kp > 1 as a tissue penetration pharmacokinetic parameter. In order to compare different formulations of quetiapine, an atypical antipsychotic, Carreno et al. [27] reported values ranging from 1.5 for the brain to 3.6 for the liver, according to the formulation administrated; Laxman et al. [28] also explores the tissue affinity of a new inhibitor, with Kp values ranging from 0.03 in the brain to 36.33 in the small intestine, where it showed remarkable accumulation.

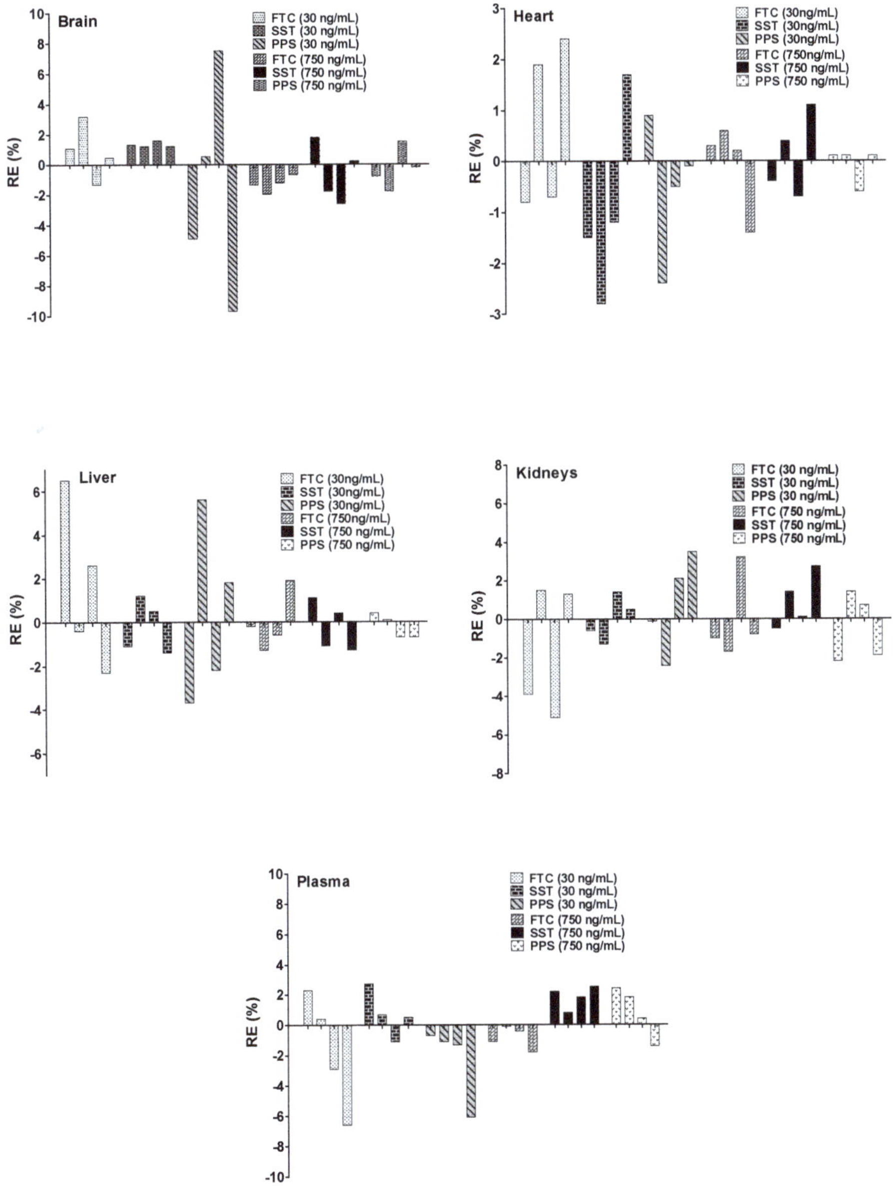


**Figure 4.** LQFM05 and IS stability in rat plasma and tissue homogenates (brain, heart, liver, kidneys) under different storage conditions ($n = 4$) by LC-QTOF/MS compared to fresh samples. FTC: stability under freezing and thaw cycles; SST: stability after short time analysis; PPS: postprocessing analysis; RE%: relative error.

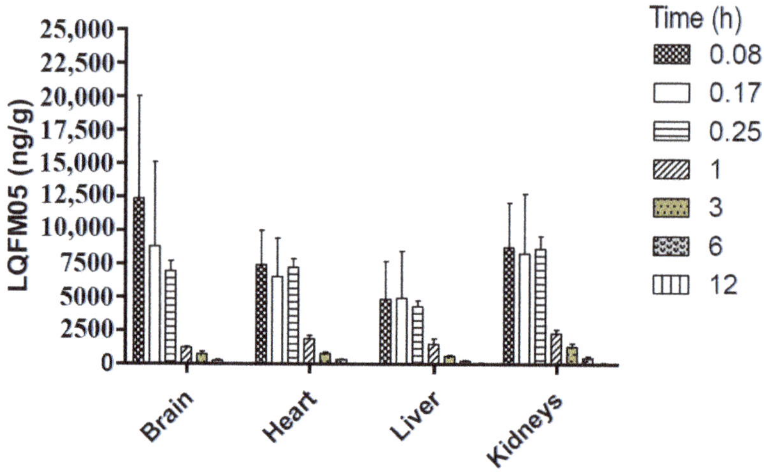


**Figure 5.** LQFM05 concentration-time profile in four tissue homogenates after an *i.v.* administration (10 mg/Kg) to rats. The final amount (ng/g) was calculated after considering the appropriate dilution factor.

**Table 4.** Tissue pharmacokinetic parameters after *i.v.* administration of LQFM05 (10 mg/Kg) to male Wistar rats (mean, $n = 3$) out of 21 rats randomly distributed into seven groups. Data obtained from Phoenix WinNonlin® 8.1.

| Matrix | $AUC_{(0-\infty)}$ (h*ng/g) | $t_{max}$ (h) | $C_{max}$ (ng/g) | $K_\beta$ ($h^{-1}$) | $t_{1/2}\beta$ (h) | MRT (h) | Kp [b] |
|---|---|---|---|---|---|---|---|
| Plasma | 5366.7 [a] | - | - | 0.31 | 2.3 | 1.7 | - |
| Brain | 10,460.6 | 0.08 | 12,357.0 | 0.28 | 2.5 | 2.0 | 1.9 |
| Heart | 10,919.6 | 0.08 | 7430.9 | 0.33 | 2.1 | 2.1 | 2.0 |
| Liver | 8235.3 | 0.17 | 4922.9 | 0.19 | 3.6 | 3.1 | 1.5 |
| Kidneys | 14,595.2 | 0.08 | 8768.0 | 0.30 | 2.3 | 2.4 | 2.7 |

[a]: h*ng/mL. [b] Kp: tissue/plasma ratio = $AUC_{(0-\infty)\ tissue}/AUC_{(0-\infty)\ plasma}$. $AUC_{(0-\infty)}$: area under the matrix concentration-time curve from time zero to infinity; $t_{max}$: time to peak concentration; $C_{max}$: peak concentration; $K_\beta$: elimination rate constant; $t_{1/2}\beta$: terminal half-life; MRT: mean residence time.

In this sense, supplementary LC analysis of rat tissue samples also detected the 4-(4-((4-(4-methoxyphenyl)piperazin-1-yl)methyl)−1H-pyrazol-1-yl)phenol (LQFM235) [4], a bioactive metabolite of the LQFM05 at 3.3 min (Figure 6).

As a metabolite, it is assumed to be a less hydrophobic compound than LQFM05 and more easily eliminated from the body. Although the metabolite quantification was not the main purpose of the present study, the LQFM235 concentration was tentatively estimated in all tissues from *in vivo* rat samples (Figure 6) as it also showed anxiolytic-like properties, as previously demonstrated [4]. Both LC-MS drug analyses were run under the same experimental conditions, meaning that the same sample preparation and instrumental conditions were used (Section 3.4.2), except for the MRM transition ions of LQFM235 (Figure 6). The putative average area of the chromatographic peak was a direct comparison to a LQFM235 (500 ng/mL, $n = 3$) peak area obtained from an original sample (not fully validated method).

From the biodistribution study, the metabolite could be detected and tentatively quantified in all tissues (Figure 7), with the exception of brain and heart matrices after 6 and 12 h post LQFM05 administration. The liver had the highest LQFM235 concentration at all sample times as compared to other evaluated tissues.

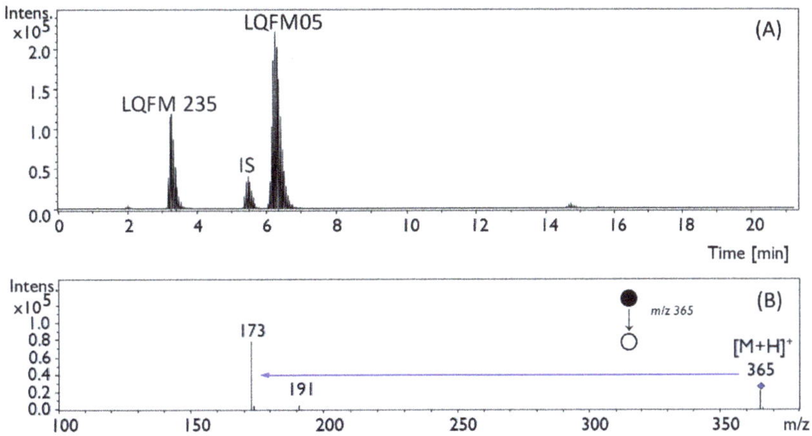


**Figure 6.** (**A**) Total ion chromatogram (TIC) of liver homogenate sample at 0.17 h; (**B**) fragmentation spectrum of LQFM235 (retention time: 3.3 min) by LC-QTOF/MS. MS spectra characterization in [4].

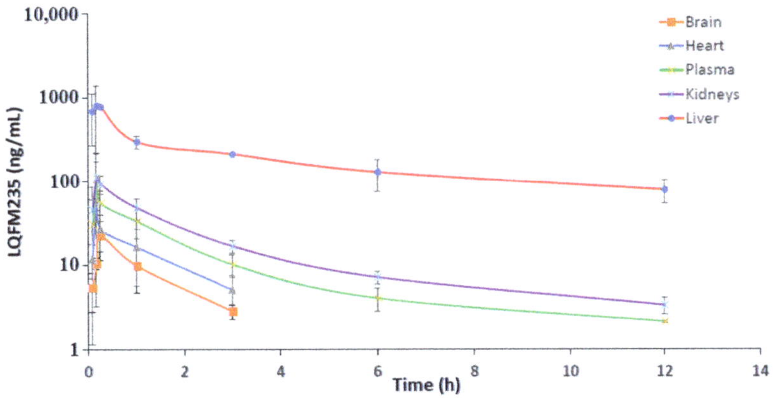


**Figure 7.** Putative mean tissue concentration (±tandard deviation)—time curves of metabolite LQFM235 after *intravenous* administration of LQFM05 (10 mg/kg).

## 3. Materials and Methods

### 3.1. Chemicals and Materials

Diazepam (IS), purity ≥99%, was purchased from the National Institute for Quality Control in Health (Rio de Janeiro, Brazil). Acetonitrile (ACN) and methanol (MeOH) of LC-grade were obtained from Merck (Darmstadt, Germany). Ammonium acetate and hydroxypropyl-β-cyclodextrin were purchased from Sigma-Aldrich (St. Louis, MO, USA). Ultrapure water (≥18 M Ω cm) was obtained from an Elga Purelab Option-Q water purification system (Elga, Purelab Option-Q, EUA).

LQFM05 was synthesized at Laboratório de Química Farmacêutica Medicinal (LQFM, Universidade Federal de Goiás, Goiânia, Brazil). Purity (>98% ± 0.95 *w*/*w*; *n* = 2) was determined by $^1$H spectra, as described elsewhere [4].

### 3.2. Animals

Male Wistar rats weighing 300 ± 50 g (10–12 weeks of age) were provided from the University Animal House (UFG, Goiânia, Brazil). The experimental protocol (022/17; 22 August 2017) was approved by the Ethics Committee on the Use of Animals of Universidade Federal de Goiás

(Goiânia, Brazil). One week before experiments, animals were acclimatized under controlled conditions of temperature (24 ± 2 °C), humidity (50 ± 10%) and 12 h light-dark cycle, with free access to water and standard laboratory rodent food.

### 3.3. *LQFM05 Dosage Form Preparation*

LQFM05 (*ca.* 600.0 mg) was dissolved in 50 mL of 30% (*w/w*) hydroxypropyl-β-cyclodextrin (HP-β-CD) aqueous solution and shaken in a borosilicate flask (1500 rpm, 24 h at 21 ± 1 °C). The sample was filtered and sterilized (Nylon filter; 0.22 μm) to obtain a clear LQFM05 drug solution of 11.87 ± 0.44 mg/mL ($n$ = 6) prior to *in vivo* analysis. The concentration of LQFM05 was determined by the validated LC-QTOF/MS method over the linear concentration range (0.0001–0.08 mg/mL).

#### 3.3.1. Drug Administration to Pharmacokinetic Studies

A single dose of LQFQM05 solution was administered intravenously (IV, 10 mg/kg) and orally in two different doses (20 and 40 mg/kg, max vol. 0.4 mL/100 g body weight) to groups of four healthy male Wistar rats: G1 (10 mg/mL, $n$ = 4) received an intravenous bolus dose (ca. 30 sec bolus injection; ca. 0.1 mL/100 mg BW) via catheterization of the carotid artery and, G2 (20 mg/Kg, $n$ = 4; *ca.* 0.2 mL/100 mg BW) and G3 (40 mg/kg, $n$ = 4; ca. 0.4 mL/100 g BW) received a single oral dose by gavage.

Blood samples (150–200 μL) were withdrawn from the lateral tail vein (max 6% BW) at 0.08; 0.25; 0.50; 0.75; 1.0; 1.5; 2.0; 2.5; 3.0; 4.0; 5.0 and 6.0 h after IV drug administration. After oral dosing, samples were taken at 0.17; 0.33; 0.50; 0.75; 1.0; 1.50; 2.0; 3.0; 4.0; 6.0; 9.0 and 12.0 h. The study design was based on a LQFM05 pilot study and previous work from Tasso et al. (2005) [11] and Conrado et al. (2008) [26].

Pharmacokinetic parameters were estimated by using non-compartmental analysis (NCA) without the assumption of any previous PK model and performed using Phoenix (WinNonlin® version 8.1; Pharsight Corp., Mountain View, CA, USA), plus Microsoft Excel software.

#### 3.3.2. Drug Administration to Tissue Distribution Study

For the LQFM05 tissue distribution study, 21 rats were randomly divided into seven groups. After *i.v.* dosing (10 mg/kg; 28.7 μmol/kg), biological fluids and tissue samples (heart, liver, kidneys, brain and blood) were collected from animals at 0.08, 0.17, 0.25, 1, 3, 6 and 12 h post-dosing time intervals.

The LQFM05 concentration in rat tissues was expressed in ng/g and calculated by Equation (1) [11,26].

$$C_t = (C_s \times Vs)/P \tag{1}$$

where $C_t$ represented the tissue concentration (ng/g), and $C_s$, $V_s$ and P were the concentration (ng/mL), volume (mL) and weight (g) of the tissue samples, respectively.

### 3.4. *Preparation of Samples for LC-QTOF/MS Analysis*

#### 3.4.1. Stock Solutions, Calibration Standards and Quality Control Samples

Aiming to provide evidence of confidence for the applied sample preparation and analysis procedure, LQFM05 and diazepam (IS) stock solutions were prepared in acetonitrile (ACN 1.0 mg/mL). Next, the stock was diluted to yield working solutions in ACN-water (1:1 *v/v*). Calibration curve samples (10.0, 20.0, 25.0, 50.0, 100.0, 250.0, 400.0, 600.0, 750.0 and 900.0 ng/mL) were prepared by spiking blank plasma/tissue samples with equal volumes of standard working solutions aliquots. Quality control (QC) samples (10.0, 30.0, 450.0, 750.0 and 1500.0 ng/mL) were prepared separately from a different drug weighting batch. The IS working solution (250 ng/mL) was prepared from stock standard solution in ACN.

For LC-QTOF/MS analysis, the sample residue was reconstituted with ACN:10 mM ammonium acetate (1:1, *v/v*, 100 μL), vortexed (10 min), centrifuged (12,000 rpm/12 min) and then injected into an LC system (15 μL).

3.4.2. Unkown Rat Plasma and Tissue Samples

Blood was collected in heparinized tubes and immediately centrifuged (1000 rpm/111× *g*, 10 min) to separate the plasma portion. The supernatant was transferred into clean tubes and stored at −80 °C until analysis by LC-QTOF/MS. Tissue samples were gently blotted with absorbent paper to remove the excess blood, weighed, and stored at −80 °C until analysis.

Plasma, calibration standards and QC samples were processed by protein precipitation. Previously, plasma (100 μL) was spiked with IS (100 μL, 250 ng/mL) and deproteinized with ACN (500 μL), vortexed (10 min), centrifuged (10,000 rpm/10 min) and the supernatant was transferred to clean 1.5 mL tubes and vacuum dried (45 °C, 90 min). Tissue samples from biodistribution studies were first homogenized (Ultra-Turrax® model T10 basic, IKA, Staufen, Germany) in 10 mM monobasic potassium phosphate buffer (pH 9.0) at the ratio of 1:3 (organ: buffer, *w*/*v*), except for liver which was 1:2 (*w*/*v*). After, IS solution (100 μL, 250 ng/mL) was added to tissue homogenates (100 μL) and submitted to extraction as previously mentioned. If any of the quantified values were higher than ULOQ, an appropriate sample aliquot of supernatant was additionally diluted using the same biological matrix as an actual sample. The final amount (ng/g) was calculated after considering the dilution factor.

For LC-QTOF/MS analysis, the sample residue was reconstituted with ACN:10 mM ammonium acetate (1:1, *v*/*v*, 100 μL), vortexed (10 min), centrifuged (12,000 rpm/12 min) and then injected into an LC system (15 μL).

### 3.5. *Instrumentation*

LC-QTOF/MS Conditions

Data were acquired on a Shimadzu LC liquid chromatography system (LC-20AD, Shimadzu Corporation, Kyoto, Japan) with two pumps (LC-20AD), an autosampler (SIL-20ACHT) and a UV detector (SPD-20AV) and also coupled to a micrOTOF-Q III mass spectrometer (Brucker Daltonics, Bremen, Germany) equipped with electrospray source (ESI) operated in a positive-ion mode (ESI(+)) using Brucker Data Analysis software. Data acquisition was carried out in the "multiple reaction monitoring" (MRM) mode, and the most abundant ion was selected as the quantifier ion. Recall that TOF detectors simultaneously collect both precursor and product ion information over a mass range without isolating a particular ion. The two most intense product ions of LQFM05 were selected as quantifier (349.21 $m/z \rightarrow$ 157.08 $m/z$) and qualifier ions (349.21 $m/z \rightarrow$ 191 $m/z$). MS parameters of the analyte and IS are listed in Table 5.

**Table 5.** MRM transition in positive ion mode, collision energy and retention time for the determination of LQFM05 and diazepam (IS).

| No. | Compounds | MRM Transition (*m*/*z*) | | CE *(eV) | RT ** (min) |
|---|---|---|---|---|---|
| | | Quantifier | Qualifier | | |
| 1 | Diazepam (IS) | 285.08 → 257.09 | | 20 | 5.6 |
| 2 | LQFM05 | 349.21 → 157.08 | 349.21 → 191 | 15 | 6.3 |

* Collision energy; ** Retention time.

Chromatographic separation was performed using the X-Terra RP18 column (150 × 3.0 mm, 3.5 μm, Waters), at 40 °C. The binary mobile phase of 10 mM of ammonium acetate (A) and ACN (B) with gradient elution: 0–8 min (55% A), 8–9 min (50% A), 9–10 min (45% A), 10–11 min (40% A), 11–12 min (35% A), 12–13 min (30% A), 13–14 min (40% A), 14–16 min (50% A) and 16–21 min (55% A) at a flow of 0.5 mL/min. The auto-sampler was conditioned at 15 °C, and the injection volume was 15 μL.

The Q-TOF settings were as follows: nebulizer nitrogen gas with 220 °C temperature, 5.5 bar pressure, drying gas of 10 L/min, capillary voltage was set to 4.5 kV; end plate offset

500 V, transfer 200 Vpp, quadrupole ion energy 3 eV, collision cell energy of 7 eV e transfer time 72 μS. The micrOTOF_Q III was programmed for scanning low mass (100–380 $m/z$).

The LC-QTOF/MS method was only validated for the quantification of LQFM05 (not for LQFM235) in plasma and tissues according to the Brazilian Health Surveillance Agency [22,29] guidelines: selectivity, linearity, precision and accuracy, recovery, matrix effect, carryover and stability of the analyte and IS were the analytical parameters assessed. For LQFM 235, the Q-TOF settings were kept as displayed on Table 5, although $m/z$ transitions were set at 365 →173.

### 3.6. Analytical Validation

#### 3.6.1. Selectivity

Six different blank plasma batches and animal tissue homogenates (brain, heart, liver and kidneys) were investigated by assessing possible endogenous interferences at the same retention time of analyte and IS.

#### 3.6.2. Linearity

Linearity of the calibration range (10.0, 20.0, 25.0, 50.0, 100.0, 250.0, 400.0, 600.0, 750.0 and 900.0 ng/mL) was assessed by triplicate samples on two different days. Blank samples were spiked with LQFM05 and IS (analytical curve), and zero samples (blank samples spiked only with IS) of the biological matrix were also analyzed.

Calibration curves were analyzed by the linear least squares regression method applied to a graphical plot of the analyte/IS peak intensity ratio (*y axis*) *versus* the theoretical concentration (*x axis*) of the analyte. Linearity was assessed by correlation (r) and determination ($r^2$) coefficients and other additional criteria, such as, graphically, %RSDslope, studentized residual and homoscedasticity (Cochran's hypothesis test and *F*-test values).

#### 3.6.3. Precision and Accuracy

Precision and accuracy were evaluated by spiked QC plasma samples in replicates ($n$ = 6) at four concentration levels in the same day (intra-day repeatability). Intermediate precision and accuracy (inter-day) were also investigated at the same QC levels on two different days ($n$ = 12). Precision was expressed as a percentage of the relative standard deviation (RSD%) of the specified concentrations, while accuracy was determined by the relative error (RE%) of the experimental samples [22].

#### 3.6.4. Recovery of the Extraction Procedure

Relative recovery was determined by comparison of analyte response in the spiked matrix samples before and after extraction procedure at all three QC levels (30.0, 450.0 and 750.0 ng/mL) [22,24].

#### 3.6.5. Carry-Over Effects

Carryover effect was assessed by comparing the chromatographic profile of blank samples' homogenates spiked with LQFM05 (900.0 ng/mL). One blank plasma homogenate was run before spiked sample analysis and repeated twice after the upper concentration limit of quantification (900.0 ng/mL).

#### 3.6.6. Matrix Effect

In order to evaluate the matrix effect, blank samples were spiked with IS (250 ng $mL^{-1}$) and drug aliquots at three concentration levels (30.0, 450.0, 750 ng $mL^{-1}$). Next, the drug/IS peak ratio was compared to drug response in solution in order to calculate the normalized matrix factor (NMF) for each sample, according to Equation (2) [29].

$$\text{NMF} = \frac{\text{LQFM05 response in matrix/IS response in matrix}}{\text{LQFM05 response in solution/IS response in solution}} \tag{2}$$

3.6.7. Stability

The stability of the analyte and IS was determined by the following tests: three freeze-thaw cycles, short-term stability and post-processing assay at two concentration levels (low and high QCs). Short-term stability samples were kept in an auto-sampler (15 °C) and analyzed after 35 h of storage. Post-processing stability samples were injected after being kept in the refrigerator (2–8 °C, 58 h), followed by the auto-sampler storage (15 °C, 24 h).

## 4. Conclusions

In the presented study, an LC-QTOF/MS method for pharmacokinetic and tissue biodistribution of a new prototype drug named LQFM05 was developed and duly validated. Different sample preparation techniques (SPE, LLE and PPE) were evaluated. The PPE method was preferred due to its simplicity, reduced cost, lower preparation time, higher sensitivity and accuracy. The LC-QTOF/MS method was shown to be linear over the therapeutic range, selective and sensitive for LQMF05 tissue biodistribution studies after *i.v.* administration in rats.

The LQMF05 biodistribution study demonstrated fast and wide body distribution to all tissues analyzed. The greater exposure and tissue affinity were seen for kidneys, thus suggesting a potential extrahepatic clearance. Additionally, LQFM05 crosses the blood-brain barrier, reaching the target tissue with a high tissue-blood Kp (1.9) and half-life $t_{1/2\beta}$ of 2.5 h, accounting for a proper pharmacodynamic effect. In addition, brain tissue showed the highest tissue concentration (Cmax: 12,357.0 ng/g), thus evidencing its higher tissue affinity rate.

In conclusion, the developed and validated LQFM05 method was successfully applied to different tissue samples. Accordingly, our results showed that LQFM05 is a promising antipsychotic drug, mainly metabolized in the liver.

**Author Contributions:** Conceptualization, K.R.R., R.M. and A.R.C.; methodology, A.C.M.R., C.M.T., A.R.F., H.S. and R.D.M.; validation, A.C.M.R.; formal analysis, A.C.M.R., C.M.T., A.R.F., H.S., R.D.M., R.M., K.R.R., B.G.V. and A.R.C.; investigation, A.C.M.R., K.R.R. and A.R.C.; resources, K.R.R., R.M., B.G.V. and A.R.C.; data curation, K.R.R. and A.R.C.; writing—review and editing, A.C.M.R., R.D.M., K.R.R. and A.R.C.; supervision, K.R.R., R.M. and A.R.C.; project administration, K.R.R., R.M. and A.R.C.; funding acquisition, K.R.R., R.M., B.G.V. and A.R.C. All authors have read and agreed to the published version of the manuscript.

**Funding:** This research was funded by CAPES, CNPq, FINEP and FAPEG, and the APC fee was provided by the postgraduation program funds (PROAP/UFG/PPGCS).

**Institutional Review Board Statement:** The animal study protocol was approved by the Ethics Committee on Animal Use of the Federal University of Goiás, Brazil (protocol # 022/2017; 22 August 2017).

**Informed Consent Statement:** Not applicable.

**Data Availability Statement:** Not applicable.

**Conflicts of Interest:** The authors declare no conflict of interest. The funders had no role in the design of the study, in the collection, analyses or interpretation of data, in the writing of the manuscript or in the decision to publish the results.

## Appendix A

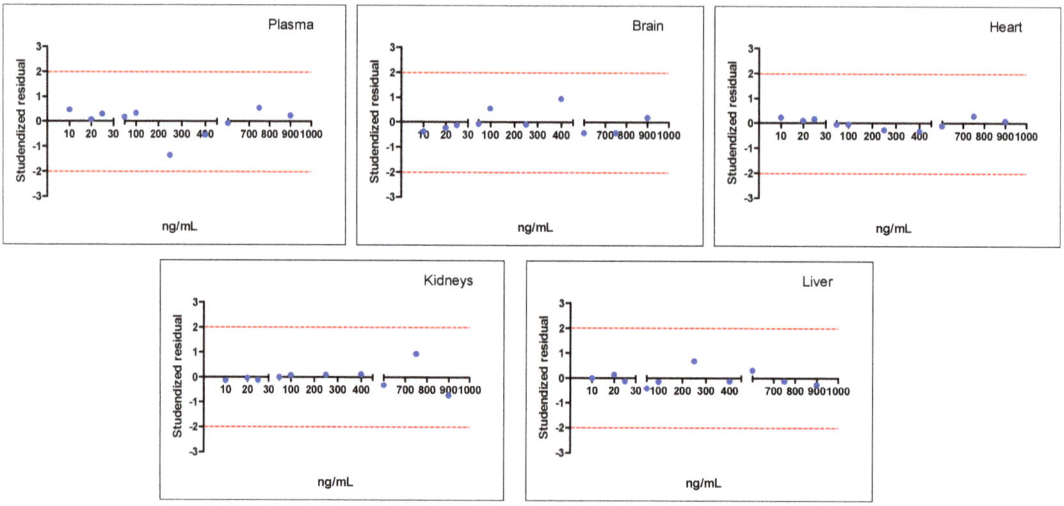


**Figure A1.** Studentized residues of LQFM05 in rat plasma and tissue homogenates (brain, heart, liver and kidneys) obtained from the calibration curve by LC-QTOF/MS.

## References


1. Chen, X.-W.; Sun, Y.-Y.; Fu, L.; Li, J.-Q. Synthesis and Pharmacological Characterization of Novel N -(Trans -4-(2-(4-(Benzo[d]Isothiazol-3-Yl)Piperazin-1-Yl)Ethyl)Cyclohexyl) Amides as Potential Multireceptor Atypical Antipsychotics. *Eur. J. Med. Chem.* **2016**, *123*, 332–353. [CrossRef] [PubMed]
2. Chong, H.Y.; Teoh, S.L.; Wu, D.B.-C.; Kotirum, S.; Chiou, C.-F.; Chaiyakunapruk, N. Global Economic Burden of Schizophrenia: A Systematic Review. *Neuropsychiatr. Dis. Treat.* **2016**, *12*, 357–373. [CrossRef] [PubMed]
3. Mena, C.; Gonzalez-Valderrama, A.; Iruretagoyena, B.; Undurraga, J.; Crossley, N.A. Early Treatment Resistance in a Latin-American Cohort of Patients with Schizophrenia. *Schizophr. Res.* **2018**, *199*, 380–385. [CrossRef] [PubMed]
4. da Silva Moreira, L.K.; Silva, R.R.; da Silva, D.M.; Mendes, M.A.S.; de Brito, A.F.; de Carvalho, F.S.; Sanz, G.; Rodrigues, M.F.; da Silva, A.C.G.; Thomaz, D.V.; et al. Anxiolytic- and Antidepressant-like Effects of New Phenylpiperazine Derivative LQFM005 and Its Hydroxylated Metabolite in Mice. *Behav. Brain Res.* **2022**, *417*, 113582. [CrossRef]
5. Magotra, A.; Sharma, A.; Singh, S.; Ojha, P.K.; Kumar, S.; Bokolia, N.; Wazir, P.; Sharma, S.; Khan, I.A.; Singh, P.P.; et al. Physicochemical, Pharmacokinetic, Efficacy and Toxicity Profiling of a Potential Nitrofuranyl Methyl Piperazine Derivative IIIM-MCD-211 for Oral Tuberculosis Therapy via in-Silico—In-Vitro—In-Vivo Approach. *Pulm. Pharmacol. Ther.* **2018**, *48*, 151–160. [CrossRef]
6. Pires, C.C.; Kaiser, M.; Grünspan, L.D.; Barreto, F.; Innocente, A.; Gnoatto, S.; Laureano, J.V.; Araujo, B.V.; Dalla Costa, T.; Tasso, L. Development and Validation of a Sensitive and Selective LC—MS/MS Method for the Determination of an Antimalarial Drug Candidate in Rat Plasma, and Its Application to a Preclinical Pharmacokinetic Study. *Acta Chromatogr.* **2016**, *28*, 439–453. [CrossRef]
7. Verbanac, D. Predictive Methods as a Powerful Tool in Drug Discovery. *Biochem. Med.* **2010**, *20*, 314–318. [CrossRef]
8. Conrado, D.J.; Palma, E.C.; Fraga, C.A.M.; Barreiro, E.J.; Rates, S.M.K.; Dalla Costa, T. Development and Validation of a LC-MS/MS Method with Electrospray Ionization for Determination of LASSBio-579 in Rat Plasma. *J. Pharm. Biomed. Anal.* **2007**, *43*, 677–682. [CrossRef]
9. de Freitas, D.F.; Porto, C.E.D.; Vieira, E.P.; de Siqueira, M.E.P.B. Three-Phase, Liquid-Phase Microextraction Combined with High Performance Liquid Chromatography-Fluorescence Detection for the Simultaneous Determination of Fluoxetine and Norfluoxetine in Human Plasma. *J. Pharm. Biomed. Anal.* **2010**, *51*, 170–177. [CrossRef]
10. Siva Selva Kumar, M.; Ramanathan, M. Concurrent Determination of Olanzapine, Risperidone and 9-Hydroxyrisperidone in Human Plasma by Ultra Performance Liquid Chromatography with Diode Array Detection Method: Application to Pharmacokinetic Study. *Biomed. Chromatogr.* **2016**, *30*, 263–268. [CrossRef]
11. Tasso, L.; Neves, G.; Menegatti, R.; Fraga, C.A.M.; Barreiro, E.; Eifler-Lima, V.; Rates, S.M.K.; Costa, T.D. Pharmacokinetics and Tissue Distribution of a New Heterocyclic N-Phenylpiperazine Derivative (LASSBio-581) in Rats. *Eur. J. Pharm. Sci.* **2005**, *26*, 194–202. [CrossRef]

12. Haas, S.E.; Brum, L.; de Andrade, C.; Azeredo, F.J.; Pigatto, M.; Silva Torres, B.G.; Guterres, S.S.; Costa, T.D. Highly Sensitive LC-MS/MS Method for the Determination of Clozapine in Rat Plasma: Application to a Preclinical Pharmacokinetic Study. *J. Liq. Chromatogr. Relat. Technol.* **2012**, *35*, 2873–2883. [CrossRef]
13. Marghade, S.; Musmade, P.B.; Moorkoth, S. High-Performance Liquid Chromatographic Assay for Ziprasidone in Plasma Samples: Application to Pharmacokinetic Studies in Rats. *J. Chromatogr. Sci.* **2012**, *50*, 902–908. [CrossRef]
14. Petruczynik, A.; Wróblewski, K.; Szultka-Młyńska, M.; Buszewski, B.; Karakuła-Juchnowicz, H.; Gajewski, J.; Morylowska-Topolska, J.; Waksmundzka-Hajnos, M. Determination of Some Psychotropic Drugs in Serum and Saliva Samples by HPLC-DAD and HPLC MS. *J. Pharm. Biomed. Anal.* **2016**, *127*, 68–80. [CrossRef]
15. Silva, B.J.G.; Lanças, F.M.; Queiroz, M.E.C. Determination of Fluoxetine and Norfluoxetine Enantiomers in Human Plasma by Polypyrrole-Coated Capillary in-Tube Solid-Phase Microextraction Coupled with Liquid Chromatography-Fluorescence Detection. *J. Chromatogr. A* **2009**, *1216*, 8590–8597. [CrossRef]
16. Brazil Agência Nacional de Vigilância Sanitária (ANVISA). *Farmacopeia Brasileira, Monografias: Insumos e Especialidades*, 6th ed.; Agência Nacional de Vigilância Sanitária SIA Trecho 5: Brasília, Brazil, 2019; Volume 2.
17. Chemicalize. Calculate Properties Instantly, Search Chemical Data, and Draw Molecules Online. Available online: www.chemicalize.com (accessed on 30 November 2017).
18. Araujo, P. Key Aspects of Analytical Method Validation and Linearity Evaluation. *J. Chromatogr. B* **2009**, *877*, 2224–2234. [CrossRef]
19. Miller, J.N.; Miller, J.C. *Statistics and Chemometrics for Analytical Chemistry*, 5th ed.; Pearson Education Limited: Edinburgh Gate, London, UK, 2005; ISBN 0 131 29192 0.
20. Raposo, F. Evaluation of Analytical Calibration Based on Least-Squares Linear Regression for Instrumental Techniques: A Tutorial Review. *TrAC Trends Anal. Chem.* **2016**, *77*, 167–185. [CrossRef]
21. Jurado, J.M.; Alcázar, A.; Muñiz-Valencia, R.; Ceballos-Magaña, S.G.; Raposo, F. Some Practical Considerations for Linearity Assessment of Calibration Curves as Function of Concentration Levels According to the Fitness-for-Purpose Approach. *Talanta* **2017**, *172*, 221–229. [CrossRef]
22. Kadian, N.; Raju, K.S.R.; Rashid, M.; Malik, M.Y.; Taneja, I.; Wahajuddin, M. Comparative Assessment of Bioanalytical Method Validation Guidelines for Pharmaceutical Industry. *J. Pharm. Biomed. Anal.* **2016**, *126*, 83–97. [CrossRef]
23. Chambers, E.; Wagrowski-Diehl, D.M.; Lu, Z.; Mazzeo, J.R. Systematic and Comprehensive Strategy for Reducing Matrix Effects in LC/MS/MS Analyses. *J. Chromatogr. B* **2007**, *852*, 22–34. [CrossRef]
24. Moein, M.M.; El Beqqali, A.; Abdel-Rehim, M. Bioanalytical Method Development and Validation: Critical Concepts and Strategies. *J. Chromatogr. B* **2017**, *1043*, 3–11. [CrossRef] [PubMed]
25. Rowland, M.; Tozer, T.N. *Clinical Pharmacokinetics and Pharmacodynamics: Concepts and Applications*, 4th ed.; Lippincott William & Wilkins: Philadelphia, PA, USA, 2011; ISBN 9780781750097.
26. Conrado, D.J.; Verli, H.; Neves, G.; Fraga, C.A.M.; Barreiro, E.J.; Rates, S.M.K.; Costa, T.D. Pharmacokinetic Evaluation of LASSBio-579: An N -Phenylpiperazine Antipsychotic Prototype. *J. Pharm. Pharmacol.* **2008**, *60*, 699–707. [CrossRef] [PubMed]
27. Carreño, F.; Paese, K.; Silva, C.M.; Guterres, S.S.; Dalla Costa, T. Pre-Clinical Investigation of the Modulation of Quetiapine Plasma Pharmacokinetics and Tissues Biodistribution by Lipid-Core Nanocapsules. *J. Pharm. Biomed. Anal.* **2016**, *119*, 152–158. [CrossRef] [PubMed]
28. Laxman, T.S.; Puttrevu, S.K.; Pradhan, R.; Mishra, A.; Verma, S.; Chhonker, Y.S.; Srivastava, S.; Singh, S.P.; Sashidhara, K.V.; Bhatta, R.S. Pharmacokinetics, Metabolism, Bioavailability, Tissue Distribution and Excretion Studies of 16α-Hydroxycleroda-3, 13(14) Z -Dien-15, 16-Olide—A Novel HMG-CoA Reductase Inhibitor. *Naunyn. Schmiedebergs. Arch. Pharmacol.* **2018**, *391*, 965–973. [CrossRef]
29. Brazil Ministério da Saúde; ANVISA. RESOLUÇÃO RDC No 166, DE 24 DE JULHO DE 2017. Available online: https://www.in.gov.br/materia/-/asset_publisher/Kujrw0TZC2Mb/content/id/19194581/do1-2017-07-25-resolucao-rdc-n-166-de-24-de-julho-de-2017-19194412 (accessed on 27 September 2022).

pharmaceuticals

MDPI

Article

# A Novel UHPLC-MS/MS Method for the Quantification of Seven Opioids in Different Human Tissues

Alessandra Manca [1,†], Amedeo De Nicolò [1,*,†], Elisa Delia De Vivo [1], Micol Ferrara [2], Sharon Oh [3], Sahar Khalili [3], Niamh Higgins [3], Robert G. Deiss [3], Stefano Bonora [2], Jessica Cusato [1], Alice Palermiti [1], Jacopo Mula [1,4], Sara Gianella [3,‡] and Antonio D'Avolio [1,4,‡]

1 Laboratory of Clinical Pharmacology and Pharmacogenetics, Department of Medical Sciences, University of Turin, Amedeo di Savoia Hospital, Corso Svizzera, 164, 10149 Turin, Italy; alessandra.manca@unito.it (A.M.)
2 Unit of Infectious Diseases, Department of Medical Sciences, University of Turin, Amedeo di Savoia Hospital, 10149 Turin, Italy
3 San Diego Division of Infectious Diseases and Global Public Health, University of California San Diego, La Jolla, CA 92037, USA
4 CoQua Lab s.r.l., 10149 Turin, Italy
* Correspondence: amedeo.denicolo@unito.it; Tel.: +39-011-4393841; Fax: +39-011-4393996
† These authors equally contributed to this work.
‡ These authors equally contributed to this work.

**Abstract:** Background: Opioids are considered the cornerstone of pain management: they show good efficacy as a first-line therapy for moderate to severe cancer pain. Since pharmacokinetic/pharmacodynamic information about the tissue-specific effect and toxicity of opioids is still scarce, their quantification in post-mortem autoptic specimens could give interesting insights. Methods: We describe an ultra-high-performance liquid chromatography coupled with tandem mass spectrometry method for the simultaneous quantification of methadone, morphine, oxycodone, hydrocodone, oxymorphone, hydromorphone and fentanyl in several tissues: liver, brain, kidney, abdominal adipose tissue, lung and blood plasma. The presented method has been applied on 28 autoptic samples from different organs obtained from four deceased PLWH who used opioids for palliative care during terminal disease. Results: Sample preparation was based on tissue weighing, disruption, sonication with drug extraction medium and a protein precipitation protocol. The extracts were then dried, reconstituted and injected onto the LX50 QSight 220 (Perkin Elmer, Milan, Italy) system. Separation was obtained by a 7 min gradient run at 40 °C with a Kinetex Biphenyl 2.6 μm, 2.1 × 100 mm. Concerning the analyzed samples, higher opioids concentrations were observed in tissues than in plasma. Particularly, O-MOR and O-COD showed higher concentrations in kidney and liver than other tissues (>15–20 times greater) and blood plasma (>100 times greater). Conclusions: Results in terms of linearity, accuracy, precision, recovery and matrix effect fitted the recommendations of FDA and EMA guidelines, and the sensitivity was high enough to allow successful application on human autoptic specimens from an ethically approved clinical study, confirming its eligibility for post-mortem pharmacological/toxicological studies.

**Keywords:** LC-MS; tissue; morphine; fentanyl; opioids

**Citation:** Manca, A.; De Nicolò, A.; De Vivo, E.D.; Ferrara, M.; Oh, S.; Khalili, S.; Higgins, N.; Deiss, R.G.; Bonora, S.; Cusato, J.; et al. A Novel UHPLC-MS/MS Method for the Quantification of Seven Opioids in Different Human Tissues. *Pharmaceuticals* **2023**, *16*, 903. https://doi.org/10.3390/ph16060903

Academic Editor: Jan Oszmianski

Received: 26 May 2023
Revised: 15 June 2023
Accepted: 16 June 2023
Published: 19 June 2023



## 1. Introduction

Opioids have been the most useful drugs for the management of severe pain for more than 200 years [1]. They show good efficacy as a first-line therapy for moderate to severe cancer pain with greater analgesic efficacy than non-steroidal anti-inflammatory drugs.

These drugs act by binding opioid receptors located along the nociceptive pathway [2] name μ (in turn including μ-1 and μ-2 subtypes), κ and δ [3,4]. Several opioids are available for clinical use in the management of chronic pain, with the most common including morphine (MOR), methadone (MTD), hydromorphone (H-MOR), hydrocodone (H-COD), oxycodone (O-COD), oxymorphone (O-MOR), and fentanyl (FENT) [5,6]. MOR is a phenanthrene derivative and the prototypical μ-receptor opiate [3]; it is the first-line treatment of severe pain and cancer [7]. After oral administration, about 40 to 50% of the administered dose reaches the central nervous system. In small amounts, MOR is also metabolized in H-MOR: this last is present in 66% of MOR consumers without excessive drug response [3,8]. FENT is an opioid agonist that is about 80–100 times more potent than morphine, highly lipophilic and highly bound to plasma proteins [3,9]. H-COD is the most used opioid indicated for moderate to severe pain. O-COD has similarities with H-COD and has activity at multiple opiate receptors including the k-receptor. O-COD has high affinity for the μ receptor and is about 10 times more potent than MOR, and it is not affected by CY2D6 or CY3A4 [3]. H-MOR is more potent than MOR (7–10 times more potent), and it has a good solubility in water which allows for concentrated formulations [10].

MTD is a synthetic μ opioid receptor agonist; in addition to its opioid activity, it is also an antagonist of the N-methyl-D aspartate (NMDA) receptor [3].

The management of these drugs is complex due to their effects on the central nervous system but also on other systems, causing respiratory depression, orthostatic hypotension, constipation, urinary retention, nausea and vomiting [3] and, in most critical cases, coma and death [11].

Most opioids are subjected to wide first-pass metabolism in the liver before reaching systemic circulation. Metabolism allows facilitating renal excretion, improving drug hydrophilicity. CYP450 and UDP-glucuronosyltransferases (UGTs) are the two major enzyme systems involved in opioids metabolism: this process results in the production of both inactive and active products [12].

Some opioids are pro-drugs, and they become active after the metabolism process. Others opioids are transformed in more potent drugs after an initial metabolism [13,14].

As an example, COD is a pro-drug that shows pharmacological activity after metabolism to MOR in the liver. About 80% of COD is eliminated by glucuronization through the UGT enzyme. A minor pathway (6–9% of the dose) is represented by N-demethylation to nor-COD and O-demethylation to MOR by the CYP2D6 enzyme [12].

Indeed, H-COD is metabolized by CYP enzymes. More than 50% of total H-COD clearance is mediated by CYP2D6 and CYP3A4, resulting in the formation of H-MOR and norhydrocodone, respectively. H-COD is metabolized with glucuronization, and it is about 10 times more potent and less polar than its parent drug, codeine. H-COD may be a pro-drug, requiring further metabolism to H-MOR, which is an active opioid agonist [12].

Since pharmacokinetic/pharmacodynamic information about the tissue-specific penetration and toxicity of opioids is still scarce, their quantification in post-mortem autoptic specimens could give interesting insights about the tissue distribution of these drugs.

Some opioids are lipophilic and can be stored in body tissues for a long time: thanks to their good solubility in lipids, some opioids are rapidly distributed in tissues. Consequently, they can be gradually released, causing tissue redistribution [15]. In this scenario, investigation about opioid distribution in tissues could be a useful tool to better understand tissue-specific toxicity: in this scenario, the post-mortem toxicological investigation of different specimens is needed. Consequently, thoroughly validated analytical methods capable of a reliable quantification of these drugs in different tissues are required.

Most tissue distribution studies are performed in animals models [16,17], while poor information is available about distribution in humans [18–20]. In addition, the currently available observations in humans are mainly focused on toxicological/forensic qualitative or semi-quantitative applications, serving more than a quantitative purpose in the context of pain management.

Furthermore, a very small variety of matrices are normally evaluated even in the toxicological analysis, such as hair [21], nails [22,23], saliva [24], plasma, blood, urine and, in rare post-mortem analyses, brain or liver [11,25–28].

The evaluation of these molecules in different biological matrices should be carried out with sensitive and specific methodologies, such as liquid chromatography coupled with mass spectrometry (UHPLC-MS/MS) [29,30]. This technique has been also applied in several fields, for example for the quantification of different phytocostituents and metabolites, as shown in the works of Thakur et al. and Perez De Souza et al. [31,32].

For these reasons, in this work, we reported on the development and validation, following EMA and FDA guidelines [33,34], of an ultra-high performance liquid chromatography coupled with tandem mass spectrometry (UHPLC-MS/MS) method for the simultaneous quantification of MOR, O-COD, H-COD, O-MOR, H-MOR, FENT and MTD in several animal tissues (e.g., liver, brain, kidney, heart, lung) and its application in human autoptic samples. Interestingly, these samples were from a unique cohort of terminally sick participants with concomitant HIV infection who agreed to participate in the study, donating their post-mortem tissues, which were withdrawn and snap frozen within 6 h after death. Furthermore, compared to the literature, the present method is able to quantify a larger number of opioids in different specimen types.

## 2. Results

### 2.1. Calibration Curve and Dilution Integrity

During method validation, all drugs showed linear calibration curves. The coefficient of determination ($R^2$) of all calibration curves ranged from 0.996 to 0.999, confirming good fitting to the calibration models. The equations are reported in Table 1.

**Table 1.** Overall evaluation of IS-nME and IS-nREC by comparison of the slopes as suggested by Matuszewsky et al. [35,36]. All deviation data are related to the curve prepared in pure solvent (reference calibration curve); the CV% is referred to the comparison between different tissues. M = calibration slope; k = calibration intercept.

| | Morphine | | | Oxymorfone | | | Hydromorfone | | | Oxycodone | | | Hydrocodone | | | Fentanyl | | | Methadone | | |
|---|---|---|---|---|---|---|---|---|---|---|---|---|---|---|---|---|---|---|---|---|---|
| | m | k | Dev. from Ref. | m | k | Dev. from Ref. | m | k | Dev. from Ref. | m | k | Dev. from Ref. | m | k | Dev. from Ref. | m | k | Dev. from Ref. | m | k | Dev. from Ref. |
| Pure solvent (reference) | 1.030 | 0.007 | n.a. | 5.437 | 0.022 | n.a. | 3.752 | 0.027 | n.a. | 2.544 | 0.026 | n.a. | 3.980 | −0.012 | n.a. | 4.451 | −0.001 | n.a. | 2.236 | 0.002 | n.a. |
| Heart | 1.062 | 0.006 | 3.1% | 4.994 | 0.046 | −8.1% | 3.730 | 0.007 | −0.6% | 2.510 | 0.006 | −1.3% | 4.042 | −0.014 | 1.6% | 4.422 | 0.005 | −0.7% | 2.442 | 0.024 | 9.2% |
| Lung | 1.015 | 0.006 | −1.5% | 4.808 | 0.038 | −11.6% | 3.716 | −0.004 | −1.0% | 2.488 | 0.015 | −2.2% | 4.050 | −0.013 | 1.8% | 4.422 | 0.002 | −0.7% | 2.406 | 0.014 | 7.6% |
| Kidney | 1.011 | 0.007 | −1.8% | 5.26 | −0.011 | −3.3% | 3.712 | 0.003 | −1.1% | 2.525 | 0.013 | −0.7% | 4.006 | 0.007 | 0.7% | 4.420 | 0.003 | −0.7% | 2.35 | 0.017 | 5.1% |
| Liver | 1.022 | −0.001 | −0.8% | 5.62 | 0.042 | 3.4% | 3.820 | 0.001 | 1.8% | 2.678 | −0.009 | 5.3% | 3.982 | 0.016 | 0.1% | 4.453 | −0.006 | 0.0% | 2.449 | 0.017 | 9.5% |
| Intestine | 1.032 | 0.001 | 0.2% | 5.476 | 0.046 | 0.7% | 3.678 | 0.012 | −2.0% | 2.678 | 0.061 | 5.3% | 4.053 | −0.005 | 1.8% | 4.416 | 0.001 | −0.8% | 2.36 | 0.009 | 5.5% |
| Subcut. fat | 1.044 | −0.012 | 1.4% | 5.408 | −0.053 | −0.5% | 3.598 | 0.003 | −4.1% | 2.385 | 0.024 | −6.3% | 3.932 | −0.003 | −1.2% | 4.344 | −0.005 | −2.4% | 2.203 | 0.015 | −1.5% |
| Plasma | 0.961 | 0.012 | −6.7% | 4.851 | 0.021 | −10.8% | 3.891 | −0.031 | 3.7% | 2.451 | 0.001 | −3.7% | 4.038 | −0.022 | 1.5% | 4.32 | 0.005 | −2.9% | 2.361 | 0.016 | 5.6% |
| | Inter-tissue CV 3.11% | | Mean Dev. −0.9% | Inter-tissue CV 6.17% | | Mean Dev. −4.3% | Inter-tissue CV 2.55% | | Mean Dev. −0.5% | Inter-tissue CV 4.37% | | Inter-tissue CV −0.5% | Inter-tissue CV 1.11% | | Mean Dev. 0.9% | Inter-tissue CV 1.10% | | Mean Dev. −1.2% | Inter-tissue CV 3.50% | | Mean Dev. 5.9% |

Samples spiked with concentrations higher than STD 6 were quantified with a mean bias lower than 10% after a 3-fold dilution with extraction solvent, highlighting a good dilution integrity.

### 2.2. Specificity and Selectivity

The chromatographic separation of all the analytes and their IS medium standard point (STD 5) have been depicted in Figure 1: a summary of the RT of opioids is provided in Table 2.

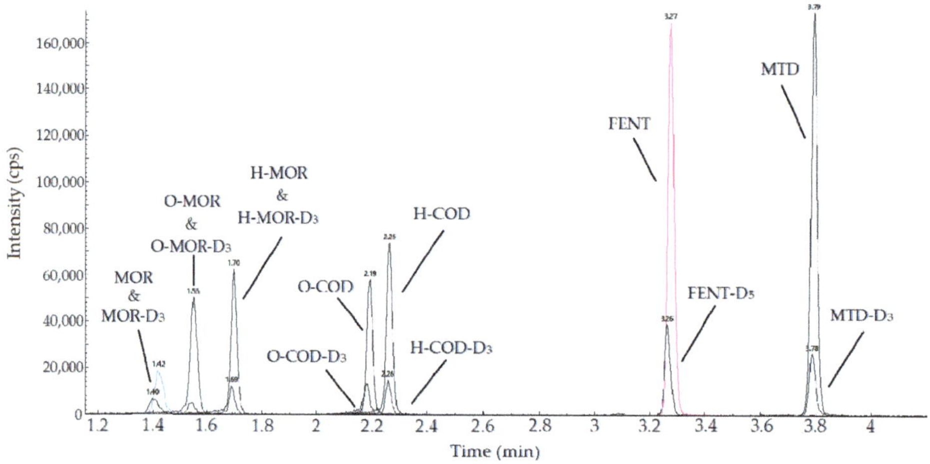


**Figure 1.** Overlaid chromatograms of the target analytes and their internal standards from the analysis of a medium level of the standard curve (STD 5).

**Table 2.** For each drug are reported, in order: retention time (RT), the concentration at the highest standard point of the calibration curve (STD 7 to STD1/LLOQ), dwell times and mass transitions (precursor and product ions), with the corresponding entrance voltages and collision energies. All concentration data are referred to the initial plasma sample.

| DRUGs | RT (min) | Calibration Range (ng) | $[M+H]^+$ (*m*/*z*) | Dwell Time (ms) | Quantification Trace (*m*/*z*) | Entrance Voltage (V) | Collision Energy Second Product Ion Trace (eV) | Qualifier Trace (*m*/*z*) | Entrance Voltage (V) | Collision Energy First Ion Product Trace (eV) |
|---|---|---|---|---|---|---|---|---|---|---|
| **MRPH** | 1.32 | 0.027–20 | 286.10 | 25 | 201.10 | 40 | −32 | 165.10 | 40 | −50 |
| **MRPH-$D_3$** | 1.30 | - | 289.10 | 25 | 201.10 | 40 | −33 | 165.10 | 40 | −50 |
| **O-MRPH** | 1.47 | 0.020–15 | 302.10 | 25 | 227.10 | 30 | −35 | 198.10 | 28 | −60 |
| **O-MRPH -$D_3$** | 1.45 | - | 305.10 | 25 | 230.10 | 30 | −35 | 201.10 | 28 | −60 |
| **H-MRPH** | 1.64 | 0.014–10 | 286.10 | 25 | 185.10 | 46 | −40 | 128.10 | 47 | −79 |
| **H-MRPH -$D_3$** | 1.63 | - | 289.10 | 25 | 185.10 | 46 | −40 | 128.10 | 47 | −79 |
| **O-COD** | 2.15 | 0.020–15 | 316.20 | 25 | 212.10 | 30 | −55 | 241.10 | 30 | −55 |
| **O-COD -$D_3$** | 2.14 | - | 319.20 | 25 | 215.10 | 30 | −35 | 244.10 | 30 | −55 |
| **H-COD** | 2.22 | 0.014–10 | 300.10 | 15 | 241.10 | 40 | −34 | 199.10 | 40 | −39 |
| **H-COD -$D_3$** | 2.22 | - | 303.10 | 25 | 241.10 | 40 | −34 | 199.10 | 40 | −39 |
| **FEN** | 3.24 | 0.007–5 | 337.30 | 25 | 188.20 | 33 | −31 | 105.10 | 34 | −57 |
| **FEN -$D_5$** | 3.23 | - | 342.30 | 25 | 188.20 | 33 | −31 | 105.10 | 34 | −57 |
| **MET** | 3.76 | 0.027–20 | 310.30 | 25 | 105.10 | 20 | −40 | 91.00 | 20 | −64 |
| **MET -$D_3$** | 3.75 | - | 313.30 | 25 | 105.10 | 20 | −40 | 92.10 | 20 | −64 |

The blank tissues samples did not yield any significant "noise" (20% of the signal of the analytes at the LLOQ or 5% of the IS) due to endogenous components at the analytes' RT (Figure 2).

Similarly, the addition of antiretroviral drugs (Abacavir, Amprenavir, Atazanavir, Bictegravir, Cabotegravir, Cobicistat, Darunavir, Doravirine, Dolutegravir, Efavirenz, Elvitegravir, Emtricitabine, Etravirine, Lamivudine, Lopinavir, Maraviroc, Nevirapine, Raltegravir, Rilpivirine, Ritonavir, Tenofovir Disoproxil Fumarate and Tenofovir Alafenamide) did not yield significant interference neither in terms of additional signal nor matrix effect (ME).

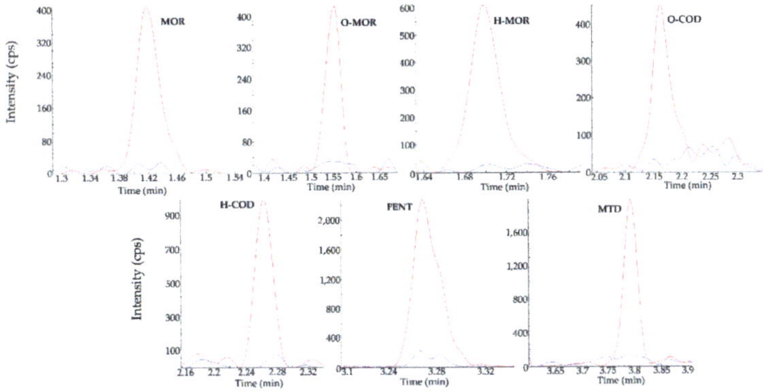


**Figure 2.** Overlaid chromatograms from the injection of the lowest level of the standard curve (STD 1 = LLOQ) and from blank liver sample (STD 0) for each analyte.

### 2.3. Lower Limit of Quantification (LLOQ) and Limit of Detection (LOD)

The lower limits of quantification (LOQ) and of detection (LOD) are reported in Table 3.

**Table 3.** Summary of the overall trueness, intra-day and inter-day precision for each drug in all the tested tissues. Lower limits of quantification (LOQ) and of detection (LOD).

| Analyte | QC Level | Trueness (%) | Precision | |
|---|---|---|---|---|
| | | | Intra-Day (CV%) | Inter-Day (CV%) |
| **Morphine** | H (10 ng) | 101.5 | 6.7 | 5.5 |
| | M (1 ng) | 98.4 | 4.4 | 7.5 |
| | L (0.1 ng) | 97.7 | 12.1 | 3.0 |
| | LLOQ (0.027 ng) | 111.2 | 11.3 | 8.2 |
| | LOD (0.009 ng) | - | - | - |
| **Oxymorphone** | H (7.5 ng) | 103.3 | 4.3 | 2.2 |
| | M (0.75 ng) | 104.6 | 6.5 | 3.0 |
| | L (0.075 ng) | 101.5 | 11.8 | 6.2 |
| | LLOQ (0.020 ng) | 107.8 | 12.1 | 10.2 |
| | LOD (0.007 ng) | - | - | - |
| **Hydromorphone** | H (5 ng) | 104.2 | 4.8 | 5.5 |
| | M (0.5 ng) | 99.8 | 7.9 | 7.1 |
| | L (0.05 ng) | 96.9 | 8.3 | 3.2 |
| | LLOQ (0.014 ng) | 108.2 | 10.8 | 12.1 |
| | LOD (0.005 ng) | - | - | - |
| **Oxycodone** | H (7.5 ng) | 101.1 | 7.8 | 1.5 |
| | M (0.75 ng) | 102.3 | 7.2 | 0.5 |
| | L (0.075 ng) | 96.3 | 10.9 | 8.7 |
| | LLOQ (0.020 ng) | 106.2 | 11.6 | 10.1 |
| | LOD (0.007 ng) | - | - | - |
| **Hydrocodone** | H (5 ng) | 102.2 | 5.6 | 2.7 |
| | M (0.5 ng) | 101.5 | 8.6 | 8.6 |
| | L (0.05 ng) | 97.6 | 12.5 | 6.8 |
| | LLOQ (0.014 ng) | 110.5 | 13.1 | 14.0 |
| | LOD (0.005 ng) | - | - | - |
| **Fentanyl** | H (2.5 ng) | 98.8 | 1.8 | 1.9 |
| | M (0.25 ng) | 96.8 | 2.2 | 2.2 |
| | L (0.025 ng) | 95.9 | 5.2 | 2.8 |
| | LLOQ (0.007 ng) | 98.2 | 9.3 | 9.1 |
| | LOD (0.002 ng) | - | - | - |
| **Methadone** | H (10 ng) | 97.5 | 5.2 | 0.7 |
| | M (1 ng) | 97.0 | 3.7 | 0.9 |
| | L (0.1 ng) | 93.6 | 7.0 | 3.6 |
| | LLOQ (0.027 ng) | 91.8 | 8.9 | 9.6 |
| | LOD (0.009 ng) | - | - | - |

The LLOQ resulted at least equal to the STD1, as required by FDA guidelines.

The overlaid chromatograms for each analyte at the LLOQ and in blank plasma are reported in Figure 2.

### 2.4. Stability

Long-term stability data showed deviation lower than 15% after 4 months at −80 °C. Similarly, short-term stability bench-top and in an autosampler has resulted in CV% in accordance to reference guidelines for method validation. All drugs were not affected by 2 freeze and thaw cycles, as reported in Supplementary Table S1.

### 2.5. Recovery and Matrix Effect

The recovery (REC) data, both in terms of absolute REC and in terms of IS-normalized REC, were consistent and highly reproducible among different matrices and matrix lots for each analyte. Similarly, the matrix effect (ME) results were quite variable among different tissues: nevertheless, the evaluation of IS-normalized ME (IS-nME) showed the very good performance of the chosen IS compounds to correct the variability accounted by ME, which was in accordance with previous reports and with EMA guidelines [33,37]. Data are summarized in Table 4.

**Table 4.** Summary of the evaluation of raw and IS-normalized REC and ME by post-extraction addition technique at 3 different concentrations in all the tested tissues.

| Recovery and Matrix Effect by Post-Extraction Addition | | | | | |
|---|---|---|---|---|---|
| **Analyte** | **QC Level** | **REC (%)** | **IS-nREC (%)** | **EM (%)** | **IS-nEM (%)** |
| **Morphine** | H | 88.8 (18.5) | 91.8 (5.6) | −1.0 (21.9) | 9.9 (5.3) |
| | M | 92.9 (21.9) | 99.0 (4.6) | −13.9 (22.6) | −5.0 (5.9) |
| | L | 91.6 (26.5) | 95.3 (10.0) | −19.7 (23.9) | 4.8 (9.5) |
| **Oxymorphone** | H | 90.2 (11.1) | 101.2 (5.0) | 0.1 (14.9) | 0.3 (5.5) |
| | M | 91.9 (14.9) | 101.4 (3.7) | −12.5 (18.0) | 1.2 (5.9) |
| | L | 93.1 (17.7) | 105.8 (10.3) | −15.5 (16.8) | −0.4 (5.1) |
| **Hydromorphone** | H | 89.9 (13.0) | 101.5 (4.2) | −1.5 (4.1) | −3.1 (2.4) |
| | M | 94.3 (14.2) | 100.4 (3.9) | −16.1 (10.6) | −3.0 (3.6) |
| | L | 88.2 (20.2) | 100.2 (8.2) | −2.9 (14.9) | 8.8 (9.6) |
| **Oxycodone** | H | 102.1 (24.9) | 97.6 (8.4) | −22.5 (24.8) | −0.8 (9.9) |
| | M | 97.7 (25.7) | 100.1 (7.4) | −23.2 (17.4) | 1.5 (5.8) |
| | L | 95.5 (19.9) | 97.9 (10.2) | −23.1 (19.2) | −6.9 (7.7) |
| **Hydrocodone** | H | 108.8 (22.5) | 99.7 (4.6) | −12.1 (25.3) | 1.6 (5.7) |
| | M | 99.6 (17.9) | 101.5 (7.1) | −7.5(15.5) | 2.5 (5.9) |
| | L | 101.5 (19.7) | 103.7 (6.6) | −2.3 (19.5) | 1.0 (5.2) |
| **Fentanyl** | H | 106.3 (32.5) | 99.8 (1.7) | −33.2 (35.0) | −1.1 (1.1) |
| | M | 102.2 (30.9) | 99.3 (2.1) | −33.2 (19.9) | −1.3 (2.2) |
| | L | 95.4 (29.1) | 99.3 (4.8) | −23.8 (28.5) | 0.3 (5.1) |
| **Methadone** | H | 100.8 (30.3) | 100.7 (5.1) | −28.8 (28.2) | −0.8 (4.2) |
| | M | 104.9 (29.7) | 101.8 (3.8) | −36.6 (19.4) | −2.6 (2.9) |
| | L | 94.7 (26.3) | 98.6 (7.2) | −30.3 (27.7) | 6.1 (5.5) |

### 2.6. Carry-Over

Carry-over was investigated by injecting blank plasma extracts after the injection of a sample extract prepared at a twice higher concentration than the highest standard sample (STD 6). Resulted data showed the absence of significant carry-over.

### 2.7. Testing of Participants' Samples

The presented method has been applied on 28 autoptic samples from different organs obtained from 4 deceased PLWH who used opioids for palliative care during terminal disease.

All samples were successfully quantified for each drug. Concentrations are reported in Table 5 and expressed as ng/g. In order to evaluate incurred sample reanalysis, these samples have been re-analyzed, showing acceptable bias in compliance to EMA and FDA guidelines: 8.6% for MOR, 5.7% for O-MOR, 7.9% for H-MOR, 7.01% for O-COD, 5.9% for FENT and 7.7% for MTD. No sample containing H-COD was tested.

**Table 5.** Summary of opioid concentrations in human autoptic samples.

| | | | | Participant 1 | | | |
|---|---|---|---|---|---|---|---|
| | Morphine (ng/g) Mean (CV%) | Oxymorphone (ng/g) Mean (CV%) | Hydromorphone (ng/g) Mean (CV%) | Oxycodone (ng/g) Mean (CV%) | Hydrocodone (ng/g) Mean (CV%) | Fentanyl (ng/g) Mean (CV%) | Methadone (ng/g) Mean (CV%) |
| Heart | n.d. | 33.4 (14.0) | n.d. | 703.4 (18.2) | n.d. | n.d. | n.d. |
| Lung | n.d. | 24.9 (10.7) | n.d. | 1046.4 (6.9) | n.d. | n.d. | n.d. |
| Kidney | n.d. | 380.7 (2.8) | n.d. | 2580.6 (1.2) | n.d. | n.d. | n.d. |
| Liver | n.d. | 317.5 (4.5) | n.d. | 4421.7 (5.3) | n.d. | n.d. | n.d. |
| Left colon | n.d. | 18.9 (0.8) | n.d. | 455.7 (7.4) | n.d. | n.d. | n.d. |
| Abdominal adipose tissue | n.d. | 3.9 (2.6) | n.d. | 108.8 (3.4) | n.d. | n.d. | n.d. |
| Plasma (pre-mortem) | n.d. | 2.0 (4.3) | n.d. | 38.0 (6.7) | n.d. | n.d. | n.d. |
| | | | | Participant 2 | | | |
| | Morphine (ng/g) Mean (CV%) | Oxymorphone (ng/g) Mean (CV%) | Hydromorphone (ng/g) Mean (CV%) | Oxycodone (ng/g) Mean (CV%) | Hydrocodone (ng/g) Mean (CV%) | Fentanyl (ng/g) Mean (CV%) | Methadone (ng/g) Mean (CV%) |
| Heart | 75,753 (5.7) | n.d. | n.d. | n.d. | n.d. | n.d. | n.d. |
| Lung | 13,119 (13.6) | n.d. | n.d. | n.d. | n.d. | n.d. | n.d. |
| Kidney | 11,739 (15.1) | n.d. | n.d. | n.d. | n.d. | n.d. | n.d. |
| Liver | 10,166 (2.8) | n.d. | n.d. | n.d. | n.d. | n.d. | n.d. |
| Left colon | 413,709 (3.5) | n.d. | n.d. | n.d. | n.d. | n.d. | n.d. |
| Abdominal adipose tissue | 510 (4.2) | n.d. | n.d. | n.d. | n.d. | n.d. | n.d. |
| Plasma (post-mortem) | 4013 (7.8) | n.d. | n.d. | n.d. | n.d. | n.d. | n.d. |
| | | | | Participant 3 | | | |
| | Morphine (ng/g) Mean (CV%) | Oxymorphone (ng/g) Mean (CV%) | Hydromorphone (ng/g) Mean (CV%) | Oxycodone (ng/g) Mean (CV%) | Hydrocodone (ng/g) Mean (CV%) | Fentanyl (ng/g) Mean (CV%) | Methadone (ng/g) Mean (CV%) |
| Heart | n.d. | n.d. | n.d. | n.d. | n.d. | 60.5 (2.6) | 569 (1.1) |
| Lung | n.d. | n.d. | n.d. | n.d. | n.d. | 80.7 (0.5) | 1699 (15.9) |
| Kidney | 323.2 (16.3) | n.d. | n.d. | n.d. | n.d. | 57.1 (19.5) | 710 (18.2) |
| Liver | n.d. | n.d. | n.d. | n.d. | n.d. | 96 (3.3) | 853 (1.3) |
| Left colon | n.d. | n.d. | n.d. | n.d. | n.d. | 14.1 (13.4) | 171 (8.0) |
| Abdominal adipose tissue | n.d. | n.d. | n.d. | n.d. | n.d. | 45.3 (1.2) | 239 (4.4) |
| Plasma (post-mortem) | n.d. | n.d. | n.d. | n.d. | n.d. | 10.6 (1.3) | 181 (4.9) |

**Table 5.** *Cont.*

| | Participant 4 | | | | | | |
|---|---|---|---|---|---|---|---|
| | **Morphine (ng/g) Mean (CV%)** | **Oxymorphone (ng/g) Mean (CV%)** | **Hydromorphone (ng/g) Mean (CV%)** | **Oxycodone (ng/g) Mean (CV%)** | **Hydrocodone (ng/g) Mean (CV%)** | **Fentanyl (ng/g) Mean (CV%)** | **Methadone (ng/g) Mean (CV%)** |
| Heart | n.d. | n.d. | 221.6 (11.1) | n.d. | n.d. | n.d. | n.d. |
| Lung | n.d. | n.d. | 221.6 (15.8) | n.d. | n.d. | n.d. | n.d. |
| Kidney | n.d. | n.d. | 134.4 (1.4) | n.d. | n.d. | n.d. | n.d. |
| Liver | n.d. | n.d. | 39.8 (9.6) | n.d. | n.d. | n.d. | n.d. |
| Left colon | n.d. | n.d. | 37.7 (8.6) | n.d. | n.d. | n.d. | n.d. |
| Abdominal adipose tissue | n.d. | n.d. | 54.6 (6.4) | n.d. | n.d. | n.d. | n.d. |
| Plasma (pre-mortem) | n.d. | n.d. | 9.0 (2.3) | n.d. | n.d. | n.d. | n.d. |

## 3. Discussion

The evaluation of opioid concentrations in tissues can give important insights about their distribution, tissue redistribution, possible tissue-specific effects and toxicity.

In this work, we described a multi-matrix, multiplexed method for the simultaneous quantification of the commonly used opioids in palliative care, which will be useful for studying drug distribution in autoptic specimens. The choice of the drugs to be included in this method was based on the most common opioid prescriptions [5,6] or pain management in terminal disease and on the drugs which were included in the treatment protocol from the "Last Gift" study. To assess the capability of this method to work in widely different matrices, we performed the validation considering several different tissues with widely different chemical compositions: heart, as a model for muscle tissue, subcutaneous fat, for fatty tissues, lung (alveolar tissue), liver (parenchymatous organ), intestine (one of main targets of opioid toxicity), kidney and plasma.

The results in terms of IS-nREC and IS-nME, both by post-extraction addition and standard curve slopes methods, confirmed the good robustness of this method. Nevertheless, it is important to note that the absolute REC and ME data, without the corrective effect of SIL-ISs, were widely variable: this highlights the importance of a rigorous internal standardization for a multi-matrix method as the one presented in this work. This aspect was particularly important considering the extremely variable chemical–physical composition of some of these tissues (e.g., fat, liver, and heart).

The developed method provided fast (7 min) and reliable results, confirming the eligibility for wider studies and, possibly, for medico-legal purposes. This multi-matrix validation, involving matrices with extremely different and variable composition, suggests the applicability of this method to other tissues with intermediate characteristics (e.g., brain tissue, smooth or striated muscles, etc.).

This method has been applied on seven sets of autoptic samples from four participants who lived with HIV. These samples confirmed the method's capability of quantifying opioid concentrations in human samples on a wide range of tissue amounts (from 10 to 100 mg).

Participant 1 presented a pancreatic tumor and had taken O-COD 10 mg/day with a history of FENT use.

Participant 2 took 15 g of MOR as a part of his legal right-to-die option, as extremely high concentrations were found in all the compartments, with the higher ones in the left colon, heart, kidney and lung.

Participant 3 had a squamous cell carcinoma of tongue and had taken MTD and FENT (previously took MOR): opioid concentrations in this participant were more homogeneous among tissues and plasma.

Participant 4 presented with rectal cancer and was treated with H-MOR, showing higher concentrations in heart and lung as compared with other tissues.

Concerning the analyzed samples, higher opioids concentrations were observed in tissues than in plasma. Particularly, O-MOR and O-COD showed higher concentrations in kidney and liver than other tissues (>15–20 times greater) and blood plasma (>100 times greater).

The possible limits of this method consist in the difficulty of estimating the real recovery in tissue samples, where the drugs could be less available for extraction. Nevertheless, the double replicate testing of samples from participants showed satisfactory reproducibility (CV < 15%), even with tissue sections of variable weight, suggesting that this simple extraction technique could be still considered reliable.

To our knowledge, this is the first method able to quantify a wide panel of opioids in different tissue matrices; in addition, this study grants deeper insights on opioid distribution in tissues, paving the way for a better understanding of concentration-related organ-specific toxicity.

## 4. Materials and Methods

### *4.1. Chemicals and Reagents*

The reference standard of FENT (purity 99.9%), MOR (purity 99.7%), MTD (purity 99.9%), O-MOR (purity 99.8%), O-COD (purity 99.8%), H-MOR (purity 99.6%), and H-COD (purity 99.8%) solutions in methanol (MeOH) were purchased from Sigma-Aldrich (Milan, Italy).

The compounds which were chosen as internal standards (IS) included (±)-FENT-D5 (purity 99.5%, isotopic purity 97.05%), MOR-D3 (purity 99.2%, isotopic purity 89.00%), (±)-MTD-D3 (purity 99.7%, isotopic purity 98.7%), O-MOR-D3 (purity 99.8%, isotopic purity 99.55%), O-COD-D3 (purity 99.9%, isotopic purity 93.30%), H-MTD-D3 (purity 99.7%, isotopic purity 89.70%), and H-COD-D3 (purity 99.8%, isotopic purity 86.49%) in MeOH were purchased from Sigma-Aldrich (Milan, Italy). HPLC-grade acetonitrile (ACN) and MeOH were purchased from VWR International (Radnor, PA, USA). HPLC grade water was produced with a Milli-DI system coupled with a Synergy 185 system by Millipore (Milan, Italy). Formic acid was purchased from Sigma-Aldrich (Milan, Italy). Blank plasma from healthy donors was kindly supplied by the Blood Bank of the "Città della Salute e della Scienza" of Turin, while blank tissue for method validation was obtained from meat for commercial use.

### *4.2. Preparation of Calibrators and Quality Control Samples*

Standard solutions at the concentration of 1 mg/mL in MeOH were used to independently spike MeOH:$H_2O$ (70:30 $v/v$) to obtain the highest calibrating solution, which were used, in turn, for the preparation of the highest standard point of the calibration curve (STD 7). The same procedure was performed for the preparation of the quality control (QCs) solutions at 3 different concentrations: high, medium and low (QC H, M and L, respectively).

At each analytical session, other calibration standards (STD 6 to STD 1) were obtained by 1:3 ($v/v$) serial dilution of the STD 7 with MeOH:$H_2O$ (70:30 $v/v$). Then, these calibration standards were used to independently spike blank tissue aliquots to obtain calibration curves in different matrices: 100 µL of calibrating solutions were added to each aliquot of tissue (weight range 10–100 mg). Exact concentrations for each standard (STD), calibration ranges and QC values are reported in Tables 2 and 3, respectively.

The type of matrices tested were as follows: heart, lung, kidney, liver, abdominal adipose tissue, intestine and plasma (post-mortem and pre-mortem, where available).

### *4.3. Sample Preparation*

Before each analytical session, an internal standard (IS) working solution was prepared in MeOH:$H_2O$ (70:30 $v/v$) at the concentration of 10 ng/mL for MTD, 20 ng/mL for MOR, 5 ng/mL for FENT, O-COD, O-MOR, H-COD and H-MOR.

After thawing at room temperature, each sample was treated as follows: 40 µL of IS working solution and 100 µL of calibration standard were added to an amount of pestled weighed tissue.

Then, samples were vortex-mixed for at least 10 s and sonicated for 10 min.

Subsequently, in the extraction process, 360 µL of a precipitant solution (ACN:MeOH 50:50 *v*/*v*) was added, and then, samples were vortex-mixed for at least 10 s.

All samples were subsequently centrifuged at 10,000× *g* for 5 min, without brake, at 10 °C, and the obtained supernatants (400 µL) were transferred in glass shots and dried in a vacuum centrifuge at 50 °C for about 1.5 h. Finally, the dry extracts were dissolved first with 50 µL of $H_2O$:MeOH 50:50 *v*/*v* and vortex-mixed; then, they were dissolved with 150 µL of $H_2O$ and vortex-mixed again; lastly, they were transferred in total recovery vials: 5 µL were injected in the chromatographic system.

### 4.4. LC-MS Analysis

The chromatographic separation was carried out with a LX50 UHPLC (Perkin Elmer, Milan, Italy) composed of an Integrity® autosampler, a SPH1299® Dual UHPLC Pump and a Mistral® column oven. The chromatographic column was a Kinetex® Biphenyl LC column, 2.1 × 100 mm, 2.6 µm (Phenomenex, Torrance, CA, USA) at 40 °C. The autosampler temperature was set at 15 °C.

The flow rate was settled at 0.4 mL/min with a gradient of two mobile phases: A (0.1% *v*/*v* formic acid in HPLC grade $H_2O$) and B (0.1% formic acid *v*/*v* in MeOH and ACN 60:40 *v*/*v*). Briefly, the chromatographic gradient started with 10% Mobile Phase B up to 0.5 min. Then, it was increased linearly to 95% at 4.7 min and held at the same percentage up to 5.60 min. After, a decrease to 10% Mobile Phase B was applied from 5.65 min and held to the end of analysis.

The total runtime was 7 min. $H_2O$:MeOH 95:5 vol:vol was used as weak washing solution, while $H_2O$:ACN 30:70 vol:vol was adopted as strong washing solution. Two strong washing and two weak washing cycles (250 µL each) were applied, sequentially, after each injection.

Tandem mass spectrometry detection was carried out with a QSight 220® (Perkin Elmer, Milan, Italy) tandem mass spectrometer with an electrospray ionization (ESI) interface.

The ESI source was set in positive ionization mode (ESI+) for all drugs.

"Zero-Air" (Dry air) was used as nebulizing and heating gas, while nitrogen was used as Drying and Collision gas: both these gasses were produced at high purity (>99.9%) with a Cinel Zefiro QS® (Cinel, Vigonza, Italy).

The general mass parameters for positive ionization were as follows: electrospray voltage 5.0 kV; source temperature 350 °C; nebulizing gas flow 350 L/h; drying gas flow 130 L/h; Heated Surface-Induced Desolvation (HSID) temperature, 300 °C.

Two mass transitions yielding the highest sensitivity were selected for all drugs: the first was quantification trace and the second was ion trace. All masses are reported in Table 2.

### 4.5. Method Validation

Once we obtained enough separation between the different analytes and enough selectivity/specificity, the method underwent a full validation in compliance with EMA and FDA guidelines for bioanalytical methods validation [33,34]. The validation covered the following: specificity and selectivity, accuracy and precision, linearity and sensitivity, carry-over, recovery and matrix effect. Since the stability data were already available for these drugs [38], analyte stability was tested only in particular conditions associated with the method.

#### 4.5.1. Results of Specificity and Selectivity Test

The method's specificity and selectivity were tested on 6–10 "blank" (without any trace of the analytes of interest) aliquots of each tissue after undergoing sample preparation. Good specificity and selectivity were interpreted as the absence of significant interfering peaks at the analytes retention times (with mean signal <20% of the LLOQ for the target analytes or 5% of the IS).

#### 4.5.2. Accuracy, Precision, Calibration, Limits of Quantification and Detection, Dilution Integrity

The accuracy and intra-day precision were evaluated in 5 intra-day replicates of each QC sample at 3 different concentrations. Inter-day precision was evaluated as the coefficient of variation (CV%) of the QC among 6 different validation sessions. The linearity of the calibration curve was evaluated in all the 6 sessions, following a linear fitting with 1/x weighing. The lower limit of quantification (LLOQ) and of detection (LOD) were considered as the lowest concentrations yielding signal-to-noise ratios of 5 and 3, respectively. Moreover, the bias% from the nominal value and CV% at the level of LLOQ had to be both <20%.

Finally, the dilution integrity was evaluated by quantifying in double replicate samples with a concentration twice higher than the highest point of the standard curve (STD 6) after a 3-fold dilution.

#### 4.5.3. Recovery

Recovery (REC) was estimated by comparing the signals of target analytes and their IS in 6 "blank" samples (without analytes) from different tissues spiked, after the extraction process, at the same concentration in vials of the QC samples (post-extraction addition) with the ones from the injection of QC samples (spiked before extraction). This was evaluated both as absolute REC and as IS-normalized-REC in order to assess the capability of the IS to mitigate the variability in REC data.

#### 4.5.4. Shelf-Life Results

The long- and short-term stability data at −20 and −80 °C were already extensively described in the literature [39–42]. However, a long-term stability study was performed up to 4 months at −80 °C on standards and QC samples (H and L) spiked in liver samples (from pig), since this tissue is expected to provide the highest potential for drug degradation, and in human plasma, in order to evaluate the feasibility of the samples' shipment and medium-term storage. Moreover, the short-term stability bench-top (at 2, 4 and 24 h) and in the autosampler (at 24 h) was investigated, again from pig liver samples and human plasma.

In addition, 2 freeze and thaw cycles were performed.

#### 4.5.5. Matrix Effect

The percentage of matrix effect (ME) was calculated by comparing the signals of the target analytes and QC of "post-extraction" spiked QCs (see Recovery) with the ones from the direct injection of pure solvents with the same analytes concentrations [37]. Moreover, ME and REC were also evaluated by comparing the standard curve slopes prepared in all the different tissues tested (heart, lung, kidney, liver, abdominal adipose tissue, intestine and plasma) with the ones obtained from the same curves prepared in pure solvent, as suggested by Matuszewsky et al. [35,36].

#### 4.5.6. Carry-Over results

Carry-over was investigated by injecting blank plasma extracts after the injection of a sample extract prepared at a twice higher concentration than the highest standard sample (STD 6).

The absence of significant carry-over was defined as a signal in these samples < 20% of the LLOQ and 5% of the IS signal.

#### 4.5.7. Application and Statistical Analysis

This method was applied on autoptic specimens obtained in the context of the observational clinical study "Last Gift" (this study was approved by the UCSD Office of Human Research Protections Program, protocol 160563), which aims to study drug concentrations in tissues and plasma in people who lived with HIV (PLWH) and used opioids for palliative care during terminal disease. Chromatographic and mass spectrometry data were processed through Simplicity® 3Q (Perkin Elmer, Milan, Italy) software. Drug concentrations were normalized by weight for each sample.

The Last Gift is a unique cohort that enrolls altruistic PLWH with a life-shortening illness who wish to participate in HIV cure research at the end-of-life (EOL), including tissue donation for a rapid research autopsy (completed within six hours of death), which greatly increases viable cell count, tissue and nucleic acid integrity, as well as maintaining the tissue distribution of drugs. The primary goal of the Last Gift study is to characterize the HIV reservoirs across various tissues and anatomic compartments. Opioid medications are used to relieve pain and suffering associated with terminal illness and might interact with HIV persistence. Thus, measuring levels of opioids in various tissues is a priority for the Last Gift study team.

#### 4.5.8. Samples Retesting

All the analyzed samples have been re-tested in the following analytical session. The CV% of these comparisons was calculated as a marker of "real-samples" precision.

## 5. Conclusions

The fast and simple method was fully validated in compliance with EMA and FDA guidelines for bioanalytical methods validation.

The obtained results gave evidence that the method could be useful for research purposes: it could be used for post-mortem pharmacological/toxicological studies, allowing to complement the lack of information in this field.

**Supplementary Materials:** The following supporting information can be downloaded at: https://www.mdpi.com/article/10.3390/ph16060903/s1, Supplementary Table S1: Short-term and long-term stability results.

**Author Contributions:** Conceptualization, A.M. and A.D.N.; methodology, A.M. and A.D.N.; software, J.M., A.P. and J.C.; validation, A.M. and A.D.N.; formal analysis, E.D.D.V.; investigation, A.M. and A.D.N.; resources, S.G.; data curation, M.F., S.O., S.K., N.H. and R.G.D.; writing—original draft preparation, A.M. and A.D.N.; writing—review and editing, A.D., S.B. and S.G.; visualization, J.M. and A.P.; supervision, A.D. and S.G.; project administration, A.M. and A.D.N.; funding acquisition, S.G. All authors have read and agreed to the published version of the manuscript.

**Funding:** This study was supported by NIH P01 fund (AI169609, Prof. David Smith) and the NIDA "Avenir" award (DA051915, Prof. Sara Gianella).

**Institutional Review Board Statement:** This study was approved by the UCSD Office of Human Research Protections Program (protocol 160563).

**Informed Consent Statement:** Informed consent was obtained from all subjects involved in the study.

**Data Availability Statement:** Data will be provided on request.

**Acknowledgments:** We thank David Smith for his support and assistance in the preparation and execution of this study.

**Conflicts of Interest:** A.D.A. and A.M. declare to be shareholders of CoQua Lab. The other authors declare no conflict of interest.

## References


1. Bruera, E.; Paice, J.A. Cancer pain management: Safe and effective use of opioids. *Am. Soc. Clin. Oncol. Educ. Book.* **2015**, *35*, e593–e599. [CrossRef] [PubMed]
2. McDonald, J.; Lambert, D.G. Opioid receptors. *Contin. Educ. Anaesth. Crit. Care Pain* **2005**, *5*, 22–25. [CrossRef]
3. Trescot, A.M.; Datta, S.; Lee, M.; Hansen, H. Opioid pharmacology. *Pain. Physician* **2008**, *11*, S133–S153. [CrossRef] [PubMed]
4. Waldhoer, M.; Bartlett, S.E.; Whistler, J.L. Opioid receptors. *Annu. Rev. Biochem.* **2004**, *73*, 953–990. [CrossRef] [PubMed]
5. CDC. Clinical Practice Guideline for Prescribing Opioids for Pain—United States. Available online: https://www.cdc.gov/mmwr/volumes/71/rr/rr7103a1.htm?s_cid=rr7103a1.htm_w (accessed on 18 April 2023).
6. Nora, D.; Volkow, M.D. Oida Opioid Facts for Teens. Available online: https://nida.nih.gov/sites/default/files/nida_opioidfactsforteens_brochure_final_web.pdf (accessed on 18 April 2023).
7. Wootton, M. Morphine is not the only analgesic in palliative care: Literature review. *J. Adv. Nurs.* **2004**, *45*, 527–532. [CrossRef]
8. Wasan, A.D.; Michna, E.; Janfaza, D.; Greenfield, S.; Teter, C.J.; Jamison, R.N. Interpreting urine drug tests: Prevalence of morphine metabolism to hydromorphone in chronic pain patients treated with morphine. *Pain. Med.* **2008**, *9*, 918–923. [CrossRef]
9. Inturrisi, C.E. Clinical pharmacology of opioids for pain. *Clin. J. Pain.* **2002**, *18*, S3–S13. [CrossRef]
10. Vallner, J.J.; Stewart, J.T.; Kotzan, J.A.; Kirsten, E.B.; Honigberg, I.L. Pharmacokinetics and bioavailability of hydromorphone following intravenous and oral administration to human subjects. *J. Clin. Pharm.* **1981**, *21*, 152–156. [CrossRef]
11. Vandenbosch, M.; Somers, T.; Cuypers, E. Distribution of Methadone and Metabolites in Skeletal Tissue. *J. Anal. Toxicol.* **2018**, *42*, 400–408. [CrossRef]
12. Mercadante, S. Opioid metabolism and clinical aspects. *Eur. J. Pharm.* **2015**, *769*, 71–78. [CrossRef]
13. Hutchinson, M.R.; Menelaou, A.; Foster, D.J.; Coller, J.K.; Somogyi, A.A. CYP2D6 and CYP3A4 involvement in the primary oxidative metabolism of hydrocodone by human liver microsomes. *Br. J. Clin. Pharm.* **2004**, *57*, 287–297. [CrossRef] [PubMed]
14. Zhou, S.F. Polymorphism of human cytochrome P450 2D6 and its clinical significance: Part II. *Clin. Pharm.* **2009**, *48*, 761–804. [CrossRef] [PubMed]
15. Caplehorn, J.R.; Drummer, O.H. Methadone dose and post-mortem blood concentration. *Drug. Alcohol. Rev.* **2002**, *21*, 329–333. [CrossRef]
16. Kinnunen, M.; Kokki, H.; Hautajarvi, H.; Lantto, J.; Rasanen, J.; Voipio, H.M.; Kokki, M. Oxycodone concentrations in the central nervous system and cerebrospinal fluid after epidural administration to the pregnant ewe. *Basic. Clin. Pharmacol. Toxicol.* **2019**, *125*, 430–438. [CrossRef] [PubMed]
17. Guillot, E.; de Mazancourt, P.; Durigon, M.; Alvarez, J.C. Morphine and 6-acetylmorphine concentrations in blood, brain, spinal cord, bone marrow and bone after lethal acute or chronic diacetylmorphine administration to mice. *Forensic Sci. Int.* **2007**, *166*, 139–144. [CrossRef] [PubMed]
18. Nakhla, D.S.; Hussein, L.A.; Magdy, N.; Abdallah, I.A.; Hassan, H.E. Precise simultaneous quantification of methadone and cocaine in rat serum and brain tissue samples following their successive i.p. administration. *J. Chromatogr. B Analyt Technol. Biomed. Life Sci.* **2017**, *1048*, 19–29. [CrossRef] [PubMed]
19. Colucci, A.P.; Aventaggiato, L.; Centrone, M.; Gagliano-Candela, R. Validation of an extraction and gas chromatography-mass spectrometry quantification method for cocaine, methadone, and morphine in postmortem adipose tissue. *J. Anal. Toxicol.* **2010**, *34*, 342–346. [CrossRef]
20. Frost, J.; Løkken, T.N.; Helland, A.; Nordrum, I.S.; Slørdal, L. Post-mortem levels and tissue distribution of codeine, codeine-6-glucuronide, norcodeine, morphine and morphine glucuronides in a series of codeine-related deaths. *Forensic Sci. Int.* **2016**, *262*, 128–137. [CrossRef]
21. Platosz, N.A.; Binz, T.M.; Baumgartner, M.R.; Lendoiro, E.; de Castro, A.; Concheiro, M. Quantification of Classic, Prescription and Synthetic Opioids in Hair by LC-MS-MS. *J. Anal. Toxicol.* **2021**, *45*, 943–949. [CrossRef]
22. Liu, P.; Liu, W.; Qiao, H.; Jiang, S.; Wang, Y.; Chen, J.; Su, M.; Di, B. Simultaneous quantification of 106 drugs or their metabolites in nail samples by UPLC-MS/MS with high-throughput sample preparation: Application to 294 real cases. *Anal. Chim. Acta* **2022**, *1226*, 340170. [CrossRef]
23. Shen, M.; Chen, H.; Xiang, P. Determination of opiates in human fingernail—comparison to hair. *J. Chromatogr. B Anal. Technol. Biomed. Life Sci.* **2014**, *967*, 84–89. [CrossRef] [PubMed]
24. Maurer, H.H. Advances in analytical toxicology: The current role of liquid chromatography-mass spectrometry in drug quantification in blood and oral fluid. *Anal. Bioanal. Chem.* **2005**, *381*, 110–118. [CrossRef]
25. Chatterton, C.N.; Scott-Ham, M. The distribution and redistribution of fentanyl & norfentanyl in post mortem samples. *Forensic Sci. Int.* **2018**, *284*, 146–152. [CrossRef] [PubMed]
26. Chesser, R.; Pardi, J.; Concheiro, M.; Cooper, G. Distribution of synthetic opioids in postmortem blood, vitreous humor and brain. *Forensic Sci. Int.* **2019**, *305*, 109999. [CrossRef] [PubMed]
27. Jakobsson, G.; Larsson, R.; Pelle, L.; Kronstrand, R.; Green, H. Oxycodone findings and CYP2D6 function in postmortem cases. *Forensic Sci. Int. Genet.* **2021**, *53*, 102510. [CrossRef]
28. Poklis, J.; Poklis, A.; Wolf, C.; Mainland, M.; Hair, L.; Devers, K.; Chrostowski, L.; Arbefeville, E.; Merves, M.; Pearson, J. Postmortem tissue distribution of acetyl fentanyl, fentanyl and their respective nor-metabolites analyzed by ultrahigh performance liquid chromatography with tandem mass spectrometry. *Forensic Sci. Int.* **2015**, *257*, 435–441. [CrossRef]

29. Gosetti, F.; Mazzucco, E.; Gennaro, M.C.; Marengo, E. Ultra high performance liquid chromatography tandem mass spectrometry determination and profiling of prohibited steroids in human biological matrices. A review. *J. Chromatogr. B Anal. Technol. Biomed. Life Sci.* **2013**, *927*, 22–36. [CrossRef]
30. Thomas, S.N.; French, D.; Jannetto, P.J.; Rappold, B.A.; Clarke, W.A. Liquid chromatography-tandem mass spectrometry for clinical diagnostics. *Nat. Rev. Methods Prim.* **2022**, *2*, 96. [CrossRef]
31. Perez de Souza, L.; Alseekh, S.; Scossa, F.; Fernie, A.R. Ultra-high-performance liquid chromatography high-resolution mass spectrometry variants for metabolomics research. *Nat. Methods* **2021**, *18*, 733–746. [CrossRef]
32. Thakur, D.; Lal, U.R.; Kapoor, D.N.; Kumar, D. Identification and quantification of polyphenolic secondary metabolites in Stem Bark of Ficus Religiosa (Moraceae) using UPLC-HRMS and RP-HPLC-PDA. *Separations* **2023**, *10*, 338. [CrossRef]
33. EMA. Guideline on Bioanalytical Method Validation. Available online: http://www.ema.europa.eu/docs/en_GB/document_library/Scientific_guideline/2011/08/WC500109686.pdf (accessed on 4 March 2021).
34. FDA. Guidance for Industry: Bioanalytical Method Validation. Available online: https://www.fda.gov/files/drugs/published/Bioanalytical-Method-Validation-Guidance-for-Industry.pdf (accessed on 4 March 2021).
35. De Nicolò, A.; Cantu, M.; D'Avolio, A. Matrix effect management in liquid chromatography mass spectrometry: The internal standard normalized matrix effect. *Bioanalysis* **2017**, *9*, 1093–1105. [CrossRef] [PubMed]
36. Karinen, R.; Øiestad, E.L.; Andresen, W.; Smith-Kielland, A.; Christophersen, A. Comparison of the Stability of Stock Solutions of Drugs of Abuse and Other Drugs Stored in a freezer, refrigerator, and at Ambient Temperature for Up to One Year. *J. Anal. Toxicol.* **2011**, *35*, 583–590. [CrossRef] [PubMed]
37. Wehrfritz, A.; Schmidt, S.; Ihmsen, H.; Schuttler, J.; Jeleazcov, C. Long-Term Stability of Hydromorphone in Human Plasma Frozen at −20 °C for Three Years Quantified by LC-MS/MS. *Int. J. Anal. Chem.* **2022**, *2022*, 3645048. [CrossRef] [PubMed]
38. Papoutsis, I.; Nikolaou, P.; Pistos, C.; Dona, A.; Stefanidou, M.; Spiliopoulou, C.; Athanaselis, S. Stability of morphine, codeine, and 6-acetylmorphine in blood at different sampling and storage conditions. *J. Forensic Sci.* **2014**, *59*, 550–554. [CrossRef] [PubMed]
39. Palmquist, K.B.; Swortwood, M.J. Long-Term Stability of 13 Fentanyl Analogs in Blood. *J. Anal. Toxicol.* **1999**, *45*, 870–877. [CrossRef]
40. Moody, D.E.; Monti, K.M.; Spanbauer, A.C. Long-term stability of abused drugs and antiabuse chemotherapeutical agents stored at −20 degrees C. *J. Anal. Toxicol.* **1999**, *23*, 535–540. [CrossRef]
41. Matuszewski, B.K. Standard line slopes as a measure of a relative matrix effect in quantitative HPLC-MS bioanalysis. *J. Chromatogr. B Anal. Technol. Biomed. Life Sci.* **2006**, *830*, 293–300. [CrossRef]
42. Matuszewski, B.K.; Constanzer, M.L.; Chavez-Eng, C.M. Strategies for the assessment of matrix effect in quantitative bioanalytical methods based on HPLC-MS/MS. *Anal. Chem.* **2003**, *75*, 3019–3030. [CrossRef]

pharmaceuticals

MDPI

Article

# Application of Direct Thermal Desorption–Gas Chromatography–Mass Spectrometry for Determination of Volatile and Semi-Volatile Organosulfur Compounds in Onions: A Novel Analytical Approach

Ana V. González-de-Peredo, Mercedes Vázquez-Espinosa, Estrella Espada-Bellido, Marta Ferreiro-González *, Ceferino Carrera, Miguel Palma and Gerardo F. Barbero

Department of Analytical Chemistry, Faculty of Sciences, Wine and Agrifood Research Institute (IVAGRO), University of Cadiz, Agrifood Campus of International Excellence (ceiA3), 11510 Puerto Real, Spain
* Correspondence: marta.ferreiro@uca.es; Tel.: +34-956-01-6359

**Abstract:** The population is now more aware of their diets due to the connection between food and general health. Onions (*Allium cepa* L.), common vegetables that are minimally processed and grown locally, are known for their health-promoting properties. The organosulfur compounds present in onions have powerful antioxidant properties and may decrease the likelihood of developing certain disorders. It is vital to employ an optimum approach with the best qualities for studying the target compounds to undertake a thorough analysis of these compounds. In this study, the use of a direct thermal desorption–gas chromatography–mass spectrometry method with a Box–Behnken design and multi-response optimization is proposed. Direct thermal desorption is an environmentally friendly technique that eliminates the use of solvents and requires no prior preparation of the sample. To the author's knowledge, this methodology has not been previously used to study the organosulfur compounds in onions. Likewise, the optimal conditions for pre-extraction and post-analysis of organosulfur compounds were as follows: 46 mg of onion in the tube, a desorption heat of 205 °C for 960 s, and a trap heat of 267 °C for 180 s. The repeatability and intermediate precision of the method were evaluated by conducting 27 tests over three consecutive days. The results obtained for all compounds studied revealed CV values ranging from 1.8% to 9.9%. The major compound reported in onions was 2,4-dimethyl-thiophene, representing 19.4% of the total area of sulfur compounds. The propanethial S-oxide, the principal compound responsible for the tear factor, accounted for 4.5% of the total area.

**Keywords:** *Allium cepa* L.; direct thermal desorption–gas chromatography–mass spectrometry; multi-response optimization; onion; organosulfur compounds

**Citation:** V. González-de-Peredo, A.; Vázquez-Espinosa, M.; Espada-Bellido, E.; Ferreiro-González, M.; Carrera, C.; Palma, M.; F. Barbero, G. Application of Direct Thermal Desorption–Gas Chromatography–Mass Spectrometry for Determination of Volatile and Semi-Volatile Organosulfur Compounds in Onions: A Novel Analytical Approach. *Pharmaceuticals* **2023**, *16*, 715. https://doi.org/10.3390/ph16050715

Academic Editors: Jan Oszmianski and Sabina Lachowicz-Wiśniewska

Received: 7 March 2023
Revised: 27 April 2023
Accepted: 2 May 2023
Published: 8 May 2023



## 1. Introduction

The *Allium* genus includes the onion, a vegetable that holds significant value both in terms of its economic and nutritional contributions [1]. The use of onions in food, including as seasoning in many dishes, has been prevalent in most countries for hundreds of years [2]. However, onions are not only used in culinary contexts but also for medical purposes [3]. Many studies have reported that onion consumption helps to prevent the occurrence of several illnesses, such as inflammatory diseases [4], cancer [5], diabetes [6], and neurological disorders [7]. These biological effects are largely associated with the chemical components of onions, especially organosulfur and phenolic compounds [8]. Regarding the latter, different studies carried out by our research group have demonstrated the high levels of flavonols and anthocyanins in onion bulbs [9–11]. In addition, these compounds are responsible for the color of the bulb and have strong antioxidant activity. On the other hand, organosulfur compounds can make up as much as 5% of an onion's dry weight.

They are responsible for the flavor and aroma of onions and have powerful antioxidant properties [12]. Nrf2, nuclear erythroid 2-related Factor 2, is a key factor that plays a vital role in protecting cells against oxidative stress and inflammation [13]. It has been shown that activation of this transcription factor can have a significant impact on various diseases, including renal, pulmonary, cardiovascular, neurodegenerative diseases, and cancer [14,15]. Likewise, antioxidant compounds, such as organosulfur compounds, can increase the activity of Nrf2, thus contributing to the activation of the cellular antioxidant defense system.

S-alk(en)yl cysteine sulfoxides (SCs), R-SO-$CH_2$-CH($NH_2$)COOH, are the primary odorless, non-volatile organosulfur compounds found in onions and are kept in the cytoplasm of whole bulbs [16]. When onions are damaged by cutting, cooking, or chewing, the cell tissue ruptures and SCs are released, thus coming into contact with the enzyme alliinase (S-alk(en)yl-L-cysteine sulfoxide lyase) [17]. This release triggers enzymatic hydrolysis of SCs, thus generating pyruvate, ammonia, and sulfenic acids (R-S-OH) [18]. These highly reactive sulfenic acids are rearranged causing thiosulfinates (R-S-SO-R′), which in turn react with each other or with other sulfenic acids, thus leading to more stable forms such as thiosulfonates (R-$SO_2$-S-R′), monosulfides (R-S-R′), disulfides (R-S-S-R′), and trisulfides (R-S-S-S-R′), as well as to other thiosulfinates [19]. This heterogeneous mixture of aromatic organosulfur compounds is a consequence of both the nature of the -R and -R′ groups of their precursors (i.e., SCs) and their proportion in onion bulbs [20]. For example, after all these fast and reactivating reactions, the S-*trans*-prop-1-enyl cysteine sulfoxide (isoaliin, $CH_3$-CH=CH-SO-$CH_2$-CH($NH_2$)COOH), which is the main SC of the common onion, generates propanethial S-oxide ($CH_3$-$CH_2$-CH=SO); this is very interesting because it is responsible for the irritation of the eyes (usually generating tears) when cutting an onion [21]. Therefore, in *Allium* vegetables, SCs serve as the precursor to a remarkable array of sensory and health-promoting active chemicals [22].

Given the importance of organosulfur compounds, their content in onions must be known to select and develop cultivars with high nutraceutical value for the food and pharmaceutical industry [18]. A range of techniques has been employed to examine these compounds, ranging from basic semi-quantitative screening to state-of-the-art chromatographic methods. These methods may be categorized as either direct or indirect [22]. Direct methods determine SCs before their enzymatic decomposition, and indirect methods determine the products obtained after enzymatic conversion [23]. The profile study of these compounds, which are known as secondary aroma compounds [8], is very important, as they are responsible for the flavor and biological qualities of onions. In addition, many studies have shown that the concentration of organosulfur compounds in the volatile and semi-volatile fractions of onions affect the antibacterial activity of sulfides [24].

Frequently, trace amounts of volatile and semi-volatile organic compounds (VOCs and SVOCs) can be found in fruits and vegetables, so effective extraction techniques and sensitive methodologies are required to properly characterize them [25]. In terms of analytical methodologies, gas chromatography combined with mass spectrometry (GC-MS) is a useful tool to characterize the volatile and semi-volatile organosulfur composition of *Allium* [8,26–28]. Regarding the extraction technique, conventional solvent extraction requires a high volume of solvent, as well as consuming a lot of time and money [29]. As a result, many solvent-free sample preparation methods have been developed. Solid-phase microextraction (SPME) is one of the most widely used due to its versatility, simplicity of operation, availability of many coating materials, and its robustness [30]. However, the competition for adsorption/absorption sites caused by the fiber's low surface capacity could hinder the extraction of some VOCs and SVOCs [31]. The direct thermal desorption (DTD) method stands out as an exceptional alternative to SPME due to its capacity for a large sorbent phase. Unlike normal thermal desorption (TD), which focuses on gaseous samples, DTD is specifically designed for solid samples. Instead of using sorbents to absorb the sample, the TD tube is directly filled with the solid material, which eliminates the need for sample pre-treatment since analytes are directly desorbed at the appropriate temperature [32]. Additionally, this technique is renowned for its higher final sensitivity and lower detection limits [33,34]. However, for onions, the use of DTD or TD coupled with GC-MS has not been previously explored to analyze the profile of their organosulfur compounds.

Although DTD is a versatile preconcentration technique for GC-MS, various parameters such as the sample amount, the desorption tube temperature, or the trap heat temperature could affect the extraction. Therefore, it is crucial that the DTD method has the most suitable characteristics for the compounds to be analyzed. The use of experimental designs (DOEs) combined with the response surface methodology (RSM) is a widely adopted strategy for optimizing methodologies for the analysis of bioactive compounds. This approach enables the determination of the optimal values of the factors that minimize or maximize the response or achieve a specific goal [35].

Therefore, this work aims to develop and validate a direct thermal desorption–gas chromatography–mass spectrometry (DTD-GC-MS) method for the simultaneous determination of organosulfur compounds in onions using both a BBD and an RSM. This combination could allow the relationship between factors and their response to be fully understood, as well as methods to be systematically and effectively optimized. Likewise, the DTD-GC-MS method proposed in this study could be used by laboratories, researchers, and companies to better understand the organosulfur onion profile and select onion cultivars with the best nutraceutical value or sensory characteristics.

## 2. Results and Discussion

### 2.1. Qualitative Analyses

To have an overview of the profile of compounds present in onions, qualitative analyses were performed under unoptimized conditions as follows: 40 mg of onion sample; tube heating at 150 °C for 600 s; a trap temperature heating at 265 °C for 180 s. Table 1 shows the compounds divided into families.

**Table 1.** First qualitative analyses of the lyophilized red onion by using DTD-GC-MS.

| Name | Formula | N° CAS | Similarity (%) |
|---|---|---|---|
| **Carboxylic acid** | | | |
| Acetic acid | $C_2H_4O_2$ | 64-19-7 | 97 |
| Formic acid | $CH_2O_2$ | 64-18-6 | 98 |
| Propanoic acid | $C_3H_6O_2$ | 79-9-4 | 83 |
| **Ester** | | | |
| Oxiranylmethyl ester 2-propenoic acid | $C_6H_8O_3$ | 106-90-1 | 90 |
| Methyl ester hexadecanoic acid | $C_{17}H_{34}O_2$ | 112-39-0 | 90 |
| Ethyl ester 2-methyl-3-oxo-butanoic acid | $C_7H_{12}O_3$ | 609-14-3 | 83 |
| 2-Hydroxy-*gamma*-butyrolactone | $C_4H_6O_3$ | 19444-84-9 | 87 |
| 3,5-Dihydroxy-2-methyl-4H-pyran-4-one | $C_6H_6O_4$ | 1073-96-7 | 92 |
| **Alkane** | | | |
| 1-(1H-pyrrol-2-yl)-ethanone | $C_6H_7NO$ | 1072-83-9 | 94 |
| **Alcohol** | | | |
| Ethanol | $C_2H_6O$ | 64-17-5 | 98 |
| 2,3-Butanediol | $C_4H_{10}O_2$ | 513-85-9 | 92 |
| 3-Furanmethanol | $C_5H_6O_2$ | 4412-91-3 | 94 |
| 2-Furanmethanol | $C_5H_6O$ | 98-0-0 | 90 |
| 5-Methyl-2-furan methanol | $C_6H_8O_2$ | 3857-25-8 | 95 |
| 3-Butene-1,2-diol | $C_4H_8O_2$ | 497-6-3 | 88 |
| **Aldehydes** | | | |
| Acetaldehyde | $C_2H_4O$ | 75-7-0 | 98 |
| Propanal | $C_3H_6O$ | 123-38-6 | 96 |
| 2-Methyl-propanal | $C_4H_8O$ | 78-84-2 | 98 |
| 2-Methyl-butanal | $C_5H_{10}O$ | 96-17-3 | 96 |
| 2-Methyl-2-butenal | $C_5H_{10}O$ | 590-86-3 | 93 |
| 2-Methyl-2-pentenal | $C_6H_{10}O$ | 14250-96-5 | 96 |

**Table 1.** *Cont.*

| Name | Formula | N° CAS | Similarity (%) |
|---|---|---|---|
| Furfural | $C_5H_4O_2$ | 98-1-1 | 98 |
| 2,2-Diethylbutyraldehyde | $C_8H_{16}O$ | 26254-89-7 | 85 |
| **Ketones** | | | |
| 2-Butanone | $C_4H_8O$ | 78-93-3 | 86 |
| 2,3-Butanedione | $C_4H_6O_2$ | 431-03-8 | 96 |
| Acetoin | $C_4H_8O_2$ | 513-86-0 | 90 |
| 1-Hydroxy-2-propanone | $C_3H_6O$ | 116-9-6 | 98 |
| 4,5-Dimethyl-1,3-dioxol-2-one | $C_5H_6O_3$ | 37830-90-3 | 86 |
| **Furans** | | | |
| 3-Methyl-furan | $C_5H_6O$ | 930-27-8 | 95 |
| 2,4-Dimethylfuran | $C_6H_8O$ | 3710-43-8 | 95 |
| **Other compounds without S** | | | |
| Carbon dioxide | $CO_2$ | 124-38-9 | 98 |
| **Organosulfur compounds** | | | |
| Thiols | | | |
| Methanethiol | $CH_4S$ | 74-93-1 | 98 |
| 1-Propanethiol | $C_3H_8S$ | 107-3-9 | 95 |
| Allyl mercaptan | $C_3H_6S$ | 870-23-5 | 94 |
| **Monosulfide** | | | |
| Dimethyl sulfide | $C_2H_6S$ | 75-18-3 | 98 |
| 1,1′-Thiobis-1-propene | $C_6H_{10}S$ | 33922-80-4 | 89 |
| (*Z*)-Allyl(prop-1-en-1-yl)sulfane * | $C_6H_{10}S$ | 104324-69-8 | 87 |
| (*E*)-Allyl(prop-1-en-1-yl)sulfane * | $C_6H_{10}S$ | 104324-36-9 | 85 |
| **Disulfides** | | | |
| Dimethyl disulfide | $C_2H_6S_2$ | 624-92-0 | 95 |
| Methyl propyl disulfide | $C_4H_{10}S_2$ | 2179-60-4 | 93 |
| (*Z*)-1-Methyl-2-(prop-1-en-1-yl)disulfane * | $C_4H_8S_2$ | 23838-19-9 | 96 |
| (*E*)-1-Methyl-2-(prop-1-en-1-yl)disulfane * | $C_4H_8S_2$ | 23838-18-8 | 97 |
| **Trisulfide** | | | |
| Dimethyl trisulfide | $C_2H_6S_3$ | 3658-80-8 | 97 |
| **Episulfide** | | | |
| Methyl-thiirane | $C_3H_6S$ | 1072-43-1 | 94 |
| **Thiophene** | | | |
| 2-Methyl-thiophene | $C_5H_6S$ | 554-14-3 | 93 |
| 3-Methyl-thiophene | $C_5H_6S$ | 616-44-4 | 96 |
| 2,5-Dimethyl-thiophene | $C_6H_8S$ | 638-02-8 | 91 |
| 3,4-Dimethyl-thiophene | $C_6H_8S$ | 638-0-6 | 94 |
| 2,4-Dimethyl-thiophene | $C_6H_8S$ | 638-00-6 | 95 |
| **Sulfine** | | | |
| Propanethial S-oxide | $C_3H_6OS$ | 32157-29-2 | 96 |
| **Other S-Compounds** | | | |
| Sulfur dioxide | $SO_2$ | 7446-09-5 | 96 |

* The identification of the isomer is tentative.

A total of fifty-one VOCs and SVOCs were tentatively identified, according to the Wiley Library, with a matching factor greater than 80%. Among these compounds, twenty were organosulfur compounds, and the remaining were eight aldehydes, three carboxylic acids, six alcohol, five ketones, five esters, two furans, one alkane, and one carbon dioxide. The distribution of the VOCs and SVOC in the red onion is graphically represented by a pie chart (Figure 1).

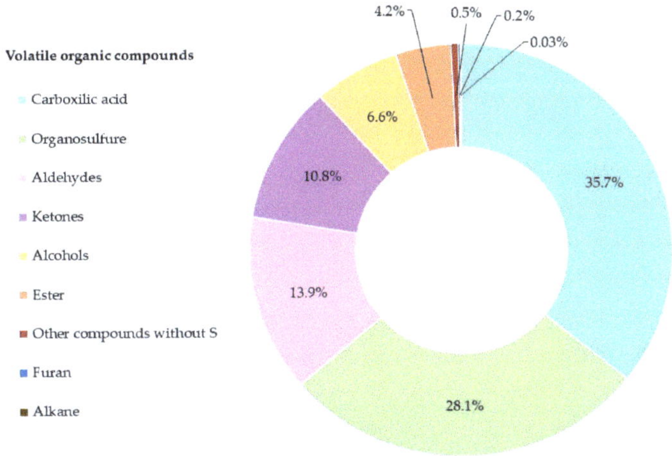


**Figure 1.** Distribution of the fifty-one compounds. Each color shows the percentage composition of the fifty-one VOCs and SVOCs following the normalization method of the chromatographic peak areas.

Carboxylic acids (35.7%), organosulfur (28.1%), and aldehydes (13.9%) accounted for more than 77.6% of the composition of red onion. This information agreed with the data reported by other authors who indicated that red onion is mainly characterized by these three families of compounds [27].

As for aldehydes, 2-methyl-2-pentenal was produced by the sequential transformation of 1-propenyl sulfenic acid into thiopropanal-S-oxide [36,37]. In addition, the condensation of propanal and acetaldehyde could produce (E)-2-methyl-2-butenal, which in turn could be reduced to 2-methyl-butanal [38]. Moreover, some aldehydes characteristics of heated onions were detected because of the temperatures applied during DTD. Likewise, acetaldehyde, propanal, 2-methyl-butanal, and 2-methyl-propanal have been recognized as by-products of the Maillard process, resulting from the Strecker degradation of the respective amino acids [39]. As for carboxylic acids, acetic, propanoic, and hexanoic acids have already been described in the volatile composition of roasted onions [36].

This study focused on the twenty organosulfur compounds identified in onions (Table 2) and aimed to analyze the impact of DTD conditions on their extraction and analysis. Figure 2 shows their chemical structures, and the characteristic mass spectral ions are shown in the Supplementary Materials (Table S1).

**Table 2.** ANOVA of the quadratic model for the total area.

| Source | Source Code | Coefficients | Sum of Squares | DF | Mean Square | *F*-Value | *p*-Value |
|---|---|---|---|---|---|---|---|
| A: Onion sample | $X_1$ | $-5.97 \times 10^8$ | $5.70 \times 10^{18}$ | 1 | $5.70 \times 10^{18}$ | 91.32 | <0.001 |
| B: Tube desorption temperature | $X_2$ | $9.83 \times 10^8$ | $1.55 \times 10^{19}$ | 1 | $1.55 \times 10^{19}$ | 247.71 | <0.001 |
| C: Tube desorption time | $X_3$ | $7.29 \times 10^7$ | $8.50 \times 10^{16}$ | 1 | $8.50 \times 10^{16}$ | 1.36 | 0.296 |
| D: Trap heat temperature | $X_4$ | $9.92 \times 10^7$ | $1.57 \times 10^{17}$ | 1 | $1.57 \times 10^{17}$ | 2.52 | 0.173 |
| E: Trap heat time | $X_5$ | $-4.91 \times 10^6$ | $3.86 \times 10^{14}$ | 1 | $3.86 \times 10^{14}$ | 0.01 | 0.940 |

**Table 2.** *Cont.*

| Source | Source Code | Coefficients | Sum of Squares | DF | Mean Square | *F*-Value | *p*-Value |
|---|---|---|---|---|---|---|---|
| AA | $X_1^2$ | $-6.81 \times 10^8$ | $4.05 \times 10^{18}$ | 1 | $4.05 \times 10^{18}$ | 64.89 | <0.001 |
| AB | $X_1X_2$ | $-1.30 \times 10^8$ | $6.77 \times 10^{16}$ | 1 | $6.77 \times 10^{16}$ | 1.08 | 0.345 |
| AC | $X_1X_3$ | $1.88 \times 10^8$ | $1.42 \times 10^{17}$ | 1 | $1.42 \times 10^{17}$ | 2.27 | 0.192 |
| AD | $X_1 X_4$ | $-1.37 \times 10^7$ | $7.52 \times 10^{14}$ | 1 | $7.52 \times 10^{14}$ | 0.01 | 0.917 |
| AE | $X_1X_5$ | $-3.59 \times 10^7$ | $5.15 \times 10^{15}$ | 1 | $5.15 \times 10^{15}$ | 0.08 | 0.786 |
| BB | $X_2^2$ | $-4.13 \times 10^8$ | $1.49 \times 10^{18}$ | 1 | $1.49 \times 10^{18}$ | 23.8 | 0.00460 |
| BC | $X_2X_3$ | $5.12 \times 10^8$ | $1.05 \times 10^{18}$ | 1 | $1.05 \times 10^{18}$ | 16.78 | 0.00940 |
| BD | $X_2X_4$ | $-7.35 \times 10^7$ | $2.16 \times 10^{16}$ | 1 | $2.16 \times 10^{16}$ | 0.35 | 0.582 |
| BE | $X_2X_5$ | $-1.65 \times 10^8$ | $1.08 \times 10^{17}$ | 1 | $1.08 \times 10^{17}$ | 1.74 | 0.245 |
| CC | $X_3^2$ | $-6.09 \times 10^8$ | $3.24 \times 10^{18}$ | 1 | $3.24 \times 10^{18}$ | 51.89 | <0.001 |
| CD | $X_3X_4$ | $-1.12 \times 10^8$ | $4.98 \times 10^{16}$ | 1 | $4.98 \times 10^{16}$ | 0.8 | 0.413 |
| CE | $X_3X_5$ | $3.62 \times 10^8$ | $5.25 \times 10^{17}$ | 1 | $5.25 \times 10^{17}$ | 8.41 | 0.0338 |
| DD | $X_4^2$ | $-5.44 \times 10^8$ | $2.58 \times 10^{18}$ | 1 | $2.58 \times 10^{18}$ | 41.4 | 0.00130 |
| DE | $X_4X_5$ | $2.90 \times 10^8$ | $3.37 \times 10^{17}$ | 1 | $3.37 \times 10^{17}$ | 5.4 | 0.0678 |
| EE | $X_5^2$ | $-6.77 \times 10^8$ | $4.00 \times 10^{18}$ | 1 | $4.00 \times 10^{18}$ | 64.12 | <0.001 |
| Lack-of-fit | | | $5.31 \times 10^{18}$ | 20 | $2.65 \times 10^{17}$ | 4.25 | 0.0577 |
| Residual | | | $5.62 \times 10^{18}$ | 25 | $2.25 \times 10^{17}$ | | |
| Pure Error | | | $3.12 \times 10^{17}$ | 5 | $6.24 \times 10^{16}$ | | |
| Cor Total | | | $3.71 \times 10^{19}$ | 45 | | | |
| Model | | $7.36 \times 1009$ | | | | | |

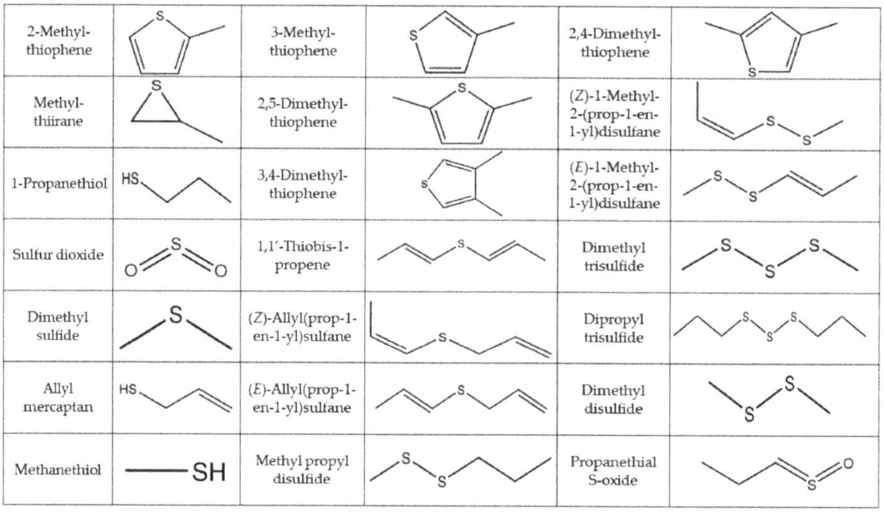


**Figure 2.** Identification of organosulfur compounds found in red onions: chemical structures.

### 2.2. Individual Box–Behnken Designs and Analysis of Variances

Two Box-Behnken designs were carried out for the development and optimisation of the DTD-GC-MS method for the pre-extraction and post-analysis of both the total sulfur compounds area and the area of the propanethial S-oxide. The ANOVA was applied to each experimental matrix, and the results are shown in Tables 2 and 3.

**Table 3.** ANOVA of the quadratic model for the area of the propanethial S-oxide.

| Source | Source Code | Coefficients | Sum of Squares | DF | Mean Square | *F*-Value | *p*-Value |
|---|---|---|---|---|---|---|---|
| A: Onion sample | $X_1$ | $-1.16 \times 10^8$ | $2.15 \times 10^{17}$ | 1 | $2.15 \times 10^{17}$ | 68.8 | 0.0004 |
| B: Tube desorption temperature | $X_2$ | $-4.60 \times 10^7$ | $3.39 \times 10^{16}$ | 1 | $3.39 \times 10^{16}$ | 10.83 | 0.0217 |
| C: Tube desorption time | $X_3$ | $4.71 \times 10^7$ | $3.55 \times 10^{16}$ | 1 | $3.55 \times 10^{16}$ | 11.33 | 0.0200 |
| D: Trap heat temperature | $X_4$ | $1.74 \times 10^7$ | $4.86 \times 10^{15}$ | 1 | $4.86 \times 10^{15}$ | 1.55 | 0.268 |
| E: Trap heat time | $X_5$ | $-3.29 \times 10^7$ | $1.73 \times 10^{16}$ | 1 | $1.73 \times 10^{16}$ | 5.53 | 0.0655 |
| AA | $X_1{}^2$ | $-1.27 \times 10^8$ | $1.40 \times 10^{17}$ | 1 | $1.40 \times 10^{17}$ | 44.6 | 0.00110 |
| AB | $X_1X_2$ | $1.22 \times 10^8$ | $5.92 \times 10^{16}$ | 1 | $5.92 \times 10^{16}$ | 18.89 | 0.00740 |
| AC | $X_1X_3$ | $3.05 \times 10^7$ | $3.72 \times 10^{15}$ | 1 | $3.72 \times 10^{15}$ | 1.19 | 0.325 |
| AD | $X_1X_4$ | $-2.01 \times 10^7$ | $1.61 \times 10^{15}$ | 1 | $1.61 \times 10^{15}$ | 0.52 | 0.505 |
| AE | $X_1X_5$ | $-4.36 \times 10^7$ | $7.62 \times 10^{15}$ | 1 | $7.62 \times 10^{15}$ | 2.43 | 0.180 |
| BB | $X_2{}^2$ | $-1.78 \times 10^8$ | $2.77 \times 10^{17}$ | 1 | $2.77 \times 10^{17}$ | 88.41 | <0.001 |
| BC | $X_2X_3$ | $4.26 \times 10^7$ | $7.27 \times 10^{15}$ | 1 | $7.27 \times 10^{15}$ | 2.32 | 0.188 |
| BD | $X_2X_4$ | $1.38 \times 10^8$ | $7.61 \times 10^{16}$ | 1 | $7.61 \times 10^{16}$ | 24.28 | 0.00440 |
| BE | $X_2X_5$ | $-5.43 \times 10^7$ | $1.18 \times 10^{16}$ | 1 | $1.18 \times 10^{16}$ | 3.76 | 0.110 |
| CC | $X_3{}^2$ | $-8.15 \times 10^7$ | $5.80 \times 10^{16}$ | 1 | $5.80 \times 10^{16}$ | 18.53 | 0.00770 |
| CD | $X_3X_4$ | $-7.83 \times 10^7$ | $2.45 \times 10^{16}$ | 1 | $2.45 \times 10^{16}$ | 7.83 | 0.0381 |
| CE | $X_3X_5$ | $9.73 \times 10^6$ | $3.79 \times 10^{14}$ | 1 | $3.79 \times 10^{14}$ | 0.12 | 0.742 |
| DD | $X_4{}^2$ | $-1.38 \times 10^8$ | $1.66 \times 10^{17}$ | 1 | $1.66 \times 10^{17}$ | 52.95 | <0.001 |
| DE | $X_4X_5$ | $1.06 \times 10^7$ | $4.47 \times 10^{14}$ | 1 | $4.47 \times 10^{14}$ | 0.14 | 0.721 |
| EE | $X_5{}^2$ | $-3.83 \times 10^7$ | $1.28 \times 10^{16}$ | 1 | $1.28 \times 10^{16}$ | 4.09 | 0.0992 |
| Lack-of-fit | | | $2.64 \times 10^{17}$ | 20 | $1.32 \times 10^{16}$ | 4.21 | 0.0586 |
| Residual | | | $2.80 \times 10^{17}$ | 25 | $1.12 \times 10^{16}$ | | |
| Pure Error | | | $1.57 \times 10^{16}$ | 5 | $3.13 \times 10^{15}$ | | |
| Cor Total | | | $1.18 \times 10^{18}$ | 45 | | | |
| Model | | $5.26 \times 10^8$ | | | | | |

The models effectively described the observed data for both responses, as well as the total area ($Y_{TA}$) and the area of the propanethial S-oxide ($Y_{C3H6OS}$), explaining 84.84% and 76.35% of their variability, respectively. Additionally, both models fit well as lack-of-fit tests with *p*-values greater than 0.05 (0.0577 and 0.0586, respectively). So, 2 s-order equations can be constructed to predict each response value as a function of the independent variables (Equations (1) and (2)).

$$\begin{aligned} Y_{TA}\ (g^{-1}) = {} & 7.36 \times 10^9 - 5.97 \times 10^8 \cdot X_1 + 9.83 \times 10^8 \cdot X_2 + 7.29 \times 10^7 \cdot X_3 + 9.92 \times 10^7 \cdot X_4 \\ & - 4.91\text{E+06} \cdot X_5 - 6.81 \times 10^8 \cdot X_1{}^2 - 1.30 \times 10^8 \cdot X_1X_2 + 1.88 \times 10^8 \cdot X_1X_3 - 1.37 \times 10^7 \cdot X_1X_4 \\ & - 3.59 \times 10^7 \cdot X_1X_5 - 4.13 \times 10^8 \cdot X_2{}^2 + 5.12 \times 10^8 \cdot X_2X_3 - 7.35 \times 10^7 \cdot X_2X_4 - 1.65 \times 10^8 \cdot X_2X_5 \\ & - 6.09 \times 10^8 \cdot X_3{}^2 - 1.12 \times 10^8 \cdot X_3X_4 + 3.62 \times 10^8 \cdot X_3X_5 - 5.44 \times 10^8 \cdot X_4{}^2 + 2.90 \times 10^8 \cdot X_4X_5 \\ & - 6.77 \times 10^8 \cdot X_5{}^2, \end{aligned} \quad (1)$$

$$\begin{aligned} Y_{C3H6OS}\ (g^{-1}) = {} & 5.26 \times 10^8 - 1.16 \times 10^8 \cdot X_1 - 4.60 \times 10^7 \cdot X_2 + 4.71 \times 10^7 \cdot X_3 + 1.74 \times 10^7 \cdot X_4 \\ & - 3.29 \times 10^7 \cdot X_5 - 1.27 \times 10^8 \cdot X_1{}^2 + 1.22 \times 10^8 \cdot X_1X_2 + 3.05 \times 10^7 \cdot X_1X_3 - 2.01 \times 10^7 \cdot X_1X_4 \\ & - 4.36 \times 10^7 \cdot X_1X_5 - 1.78 \times 10^8 \cdot X_2{}^2 + 4.26 \times 10^7 \cdot X_2X_3 + 1.38 \times 10^8 \cdot X_2X_4 - 5.43 \times 10^7 \cdot X_2X_5 \\ & - 8.15 \times 10^7 \cdot X_3{}^2 - 7.83 \times 10^7 \cdot X_3X_4 + 9.73 \times 10^6 \cdot X_3X_5 - 1.38 \times 10^8 \cdot X_4{}^2 + 1.06\text{E} \times 10^7 \cdot X_4X_5 \\ & - 3.83 \times 10^7 \cdot X_5{}^2, \end{aligned} \quad (2)$$

Likewise, ANOVA showed the significance of each factor and its interaction with the response variable. Only variables and interactions with a *p*-value less than 0.05 were significantly affected at a 95% level of significance. This statistical information was graphically represented using the Pareto chart (Figure 3).

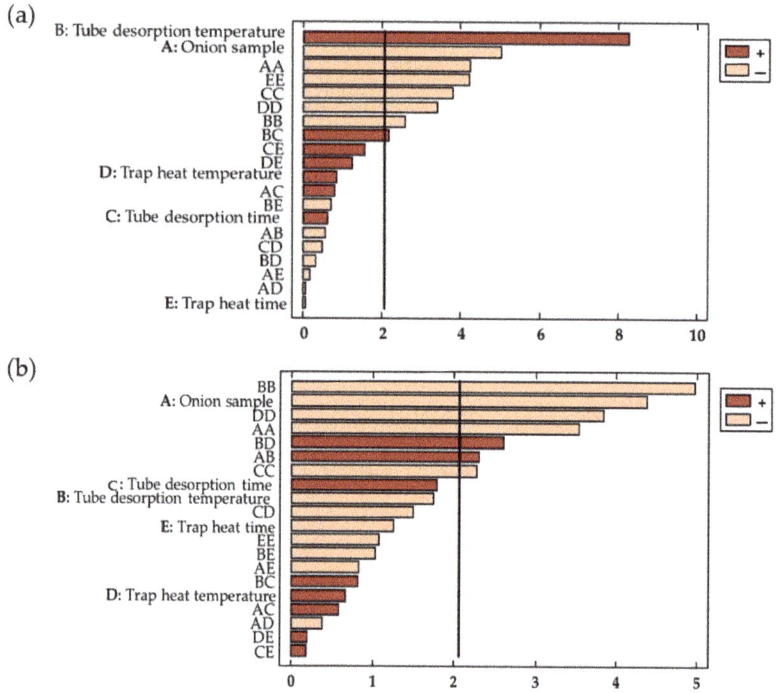


**Figure 3.** Pareto Chart of: (**a**) the total area; and (**b**) the area of the propanethial S-oxide.

Onion sample weight ($X_1$) significantly impacted the pre-extraction and post-analysis of the organosulfur compounds of the red onions ($p$-value < 0.05). The onion sample amount used negatively affected the total area of the organosulfur compounds ($b_1 = -5.97 \times 10^8$) and the area of the propanethial S-oxide area ($b_1 = - 1.16 \times 108$).

The tube desorption temperature positively affected the total area of the organosulfur compounds because greater temperatures increased the efficiency of the pre-extraction process ($b_2 = 9.83 \times 10^8$). However, it is worth noting that, if the desorption temperature is too high, the peak area could show a negative trend due to the degradation of the compounds [38]. This could be observed by the effect of the desorption temperature on the area of the propanethial S-oxide—the effect was not significant, but negative. The conditions in the tube oven during the desorption stage were crucial for the effective extraction of the compounds [37].

Finally, for a clear understanding of the interactive and main effects, 3D surface plots were represented using the designed model. Figure 4a–d show the combined impact of the onion sample–tube desorption temperature, the tube desorption temperature–tube desorption time, and the tube desorption temperature–trap heat temperature, on the response variables.

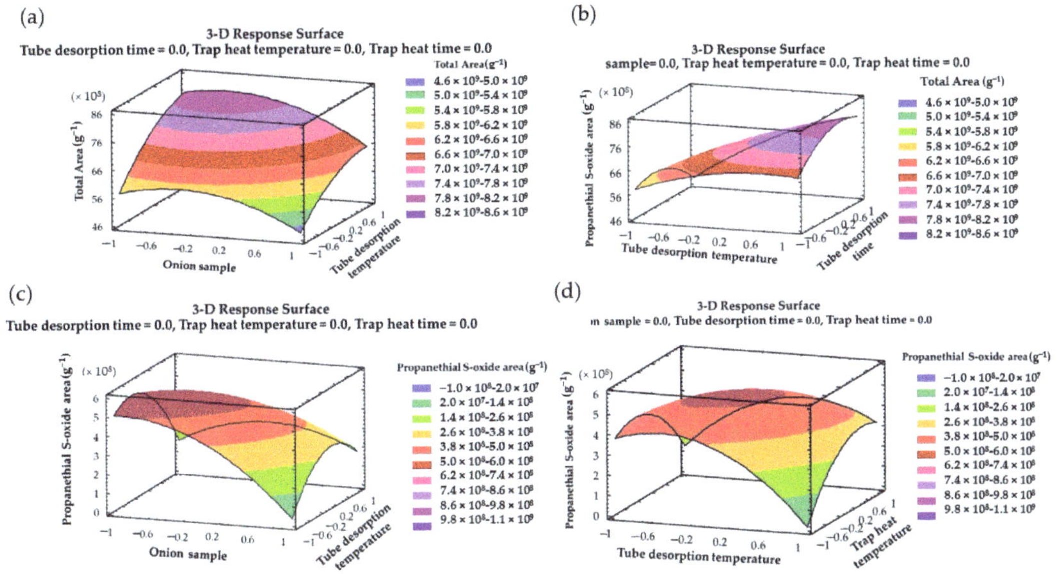


**Figure 4.** Three-dimensional (3D) surface plots of BBD using polynomial equations: (**a**) the effect of both the onion sample and the tube desorption temperature on the total organosulfur area; (**b**) the effect of both the tube desorption temperature and the tube desorption time on the total organosulfur area; (**c**) the effect of both the onion sample and the tube desorption temperature on the area of the propanethial S-oxide; and (**d**) the effect of both the tube desorption temperature and the trap heat temperature on the area of the propanethial S-oxide.

## 2.3. Multi-Response Optimization

Finally, RSM provided details on the optimal values that each factor should assume to achieve maximum response. Specifically, the values required to optimize the extraction of total organosulfur and the extraction of propanethial S-oxide are included in Table 4. On the other hand, to identify the best conditions not only for the total area but also for the area of the propanethial S-oxide, the MRO was applied. As Table 4 shows, the optimal conditions obtained through both individual experiments and MRO achieved a desirable value of 84.8%.

**Table 4.** Results of the individual optimization of the total area and the area of the propanethial S-oxide, as well as MRO of both response variables.

| Factors | Individual Optimization | | MRO |
|---|---|---|---|
| | Total Area | Propanethial S-Oxide Area | |
| $X_1$: Onion sample (mg) | 46 | 47 | 46 |
| $X_2$: Tube desorption temperature (°C) | 211 | 189 | 205 |
| $X_3$: Tube desorption time (s) | 16 | 9 | 16 |
| $X_4$: Trap heat temperature (°C) | 265 | 252 | 267 |
| $X_5$: Trap heat time (s) | 3 | 2 | 3 |

The precision of the MRO method was also validated through conducting 27 experiments carried out on 3 consecutive days. The approach exhibited desirable reproducibility and intermediate precision, as evidenced by a CV below 10% for all organosulfur compounds (Supplementary Material, Table S2). These results were considered acceptable, as the CVs were below the commonly accepted threshold of 10% [37].

The validated method was applied to the red onion sample, obtaining the results of the area shown in Table 5 and the total ion chromatogram (TIC) shown in Figure 5.

**Table 5.** Organosulfur compounds extracted and analyzed using the DTD-GC-MS MRO optimized method. Compositional percentages were computed by the normalization method from the GC peak areas.

| Code | Compound | Individual Relative Area ($g^{-1}$) | Percentage Composition (%) |
|---|---|---|---|
| 1 | Methanethiol | 79,946,258 ± 5,037,254 | 0.8 ± 0.1 |
| 2 | Dimethyl sulfide | 368,819,241 ± 26,379,719 | 3.9 ± 0.3 |
| 3 | 1-Propanethiol | 520,814,238 ± 25,833,947 | 5.9 ± 1.2 |
| 4 | Sulfur dioxide | 741,589,894 ± 68,150,056 | 8.1 ± 0.3 |
| 5 | Methyl-thiirane | 356,083,962 ± 6,570,787 | 3.8 ± 0.2 |
| 6 | Allyl mercaptan | 925,374,280 ± 53,675,232 | 10.7 ± 0.6 |
| 7 | Dimethyl disulfide | 734,818,197 ± 68,447,885 | 7.0 ± 0.4 |
| 8 | 2-Methyl-thiophene | 79,399,050 ± 7,255,625 | 0.7 ± 0.04 |
| 9 | 3-Methyl-thiophene | 161,784,317 ± 15,999,654 | 1.6 ± 0.1 |
| 10 | 2,5-Dimethyl-thiophene | 46,771,965 ± 4,109,263 | 0.4 ± 0.03 |
| 11 | 3,4-Dimethyl-thiophene | 266,951,634 ± 26,402,475 | 3.2 ± 0.3 |
| 12 | 1,1′-Thiobis-1-propene | 12,137,629 ± 982,646 | 0.1 ± 0.01 |
| 13 | (*Z*)-Allyl(prop-1-en-1-yl)sulfane | 47,080,601 ± 3,942,928 | 0.5 ± 0.03 |
| 14 | (*E*)-Allyl(prop-1-en-1-yl)sulfane | 34,336,629 ± 1,522,835 | 0.4 ± 0.03 |
| 15 | Methyl propyl disulfide | 127,532,708 ± 1,832,161 | 1.4 ± 0.1 |
| 16 | Propanethial S-oxide | 385,303,062 ± 14,922,610 | 4.5 ± 0.3 |
| 17 | 2,4-Dimethyl-thiophene | 1,728,974,335 ± 94,208,493 | 19.4 ± 0.1 |
| 18 | (*Z*)-1-Methyl-2-(prop-1-en-1-yl)disulfane | 457,771,521 ± 11,134,133 | 5.2 ± 0.5 |
| 19 | (*E*)-1-Methyl-2-(prop-1-en-1-yl)disulfane | 983,648,059 ± 82,014,566 | 10.3 ± 0.4 |
| 20 | Dimethyl trisulfide | 839,379,417 ± 73,906,935 | 9.6 ± 1.0 |
| | Total | 9,389,028,511 ± 908,282,182 | - |

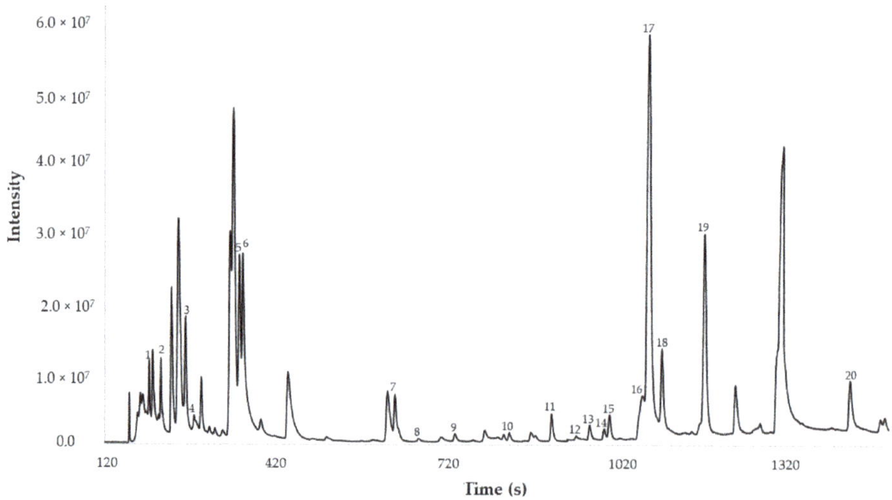


**Figure 5.** DTD-GC-MS chromatogram of the organosulfur compounds present in red onions. The code compounds are shown in Table 5.

### 2.4. Distribution of the Organosulfur Compound in Red Onion by MRO DTD-GC-MS Method

The optimized method for analyzing the distribution of 20 sulfur compounds in a red onion sample was then used, and the results are included in Figure 6.

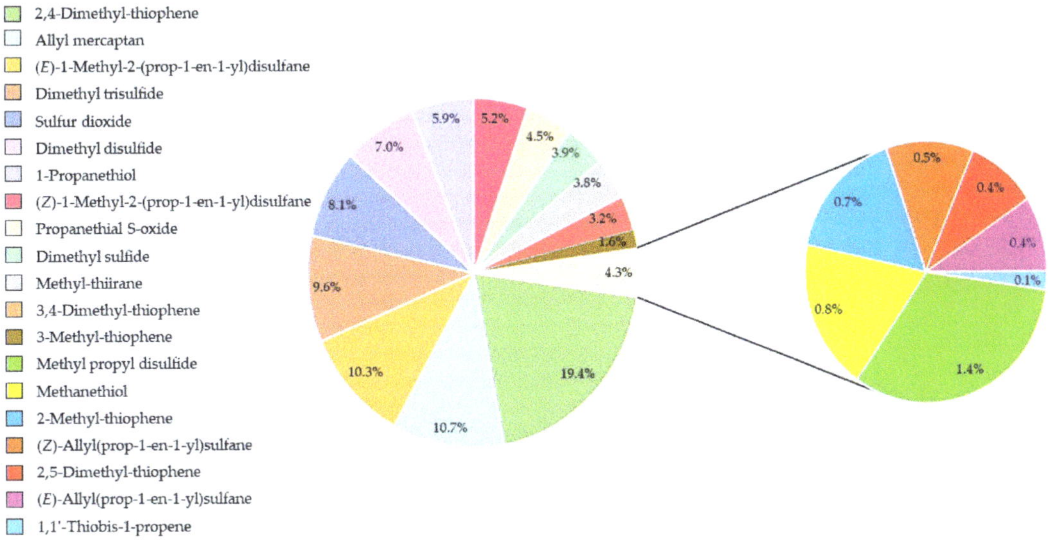


**Figure 6.** Distribution of the twenty organosulfur compounds. Each color shows its percentage composition calculated by using the normalization method from the GC peak areas.

When alliinase and cysteine sulfoxides came together, they generated a mixture of sulfenic acids, ammonia, and pyruvate. The major onion cysteine sulfoxide, i.e., S-1-propenyl-L-cysteine sulfoxide, was transformed into the 1-propenyl sulfenic acid, which was then turned into propanethial S-oxide. This organosulfur compound, which is known as the onion lachrymatory factor (LF), accounted for 4.5% of the composition of red onions and was stable during the gas chromatography analysis, making it easy to trap and measure.

The other products of the condensation reaction, i.e., thiosulfinates, were degraded during trapping and GC analysis, thus generating most of the organosulfur compounds identified, i.e., the sulfides [39], which were involved in further transformations and showed biological activity [34]. Disulfides were observed in frozen onions, while the drying of onions increased trisulfides [40]. The resulting mixture of monosulfides, disulfides, and trisulfides accounted for 4.9%, 23.9%, and 9.6% (i.e., 38.4% of carbon sulfide $CS_n$) of the red onion composition, respectively. Additionally, propanothiol was identified as a significant source of flavor in fresh onions [41], accounting for 5.9% of the red onion composition.

In addition, high temperatures during thermal desorption can trigger the thermolysis of alkyl-1-propenyl disulphides and di(1-alkenyl) disulphides to form thiophene [42]. This family of compounds accounted for 25.3% of the red onion composition, so it was one of the main components.

### *2.5. Comparison of DTD Methodology with Another Extraction Techniques*

Finally, the developed MRO DTD-GC-MS method was compared with other extraction techniques to highlight its advantages.

Concerning solvent extraction techniques for volatile compounds, DTD offers multiple advantages as reported by other authors [34,43]. Firstly, it is an automated extraction technique in which there is hardly any sample handling, and it does not use solvents, which means that it is a more environmentally friendly technique. In addition, the yields and the sensitivity offered by this technique are much higher than conventional solvent extractions.

However, it should be noted that it is not the only solvent-free technique currently available in the literature. SPME combined with HS has been used on several occasions for the study of sulfur compounds in onion samples [8,43]. Nonetheless, the choice of the adsorbent material is consistently a crucial aspect of the process because it must adsorb a diverse range of molecules with varying chemical properties, molecular weights, and polarities. With SPME, the fiber's surface capacity is relatively restricted, which can lead to analytes competing for adsorption sites, potentially causing a greater bias towards specific volatile compounds [18]. Overall, TD typically has a higher surface capacity than SPME due to the larger size of the trap used in TD compared to the SPME fiber. To facilitate an experimental comparison of both techniques, an analysis of the organosulfur compounds in the same onion matrix using SPME (AOC-6000 Plus Multifunctional Autosampler, Shimadzu, Kyoto, Japan) was carried out. The SPME analysis conditions used were as follows: 0.5 g of onion was pre-incubated at 150 °C with pulsed agitation for 10 min at a speed of 500 rpm. The headspace above the samples in a vial was exposed to a DVB/Carbon WR/PDMS SPME fiber (manufactured by Shimadzu, Kyoto, Japan) at a depth of 22 mm for 20 min. After extraction, the SPME fiber was withdrawn and introduced into the gas chromatograph inlet, where it was desorbed for 5 min using a split mod at a temperature of 220 °C. The results obtained showed that the number of sulfur compounds identified by SPME (14 sulfur compounds) was lower than that identified by the MRO DTD-GC-MS method (20 sulfur compounds). In addition, the extraction of these compounds also showed worse results, with a total area by SPME of 78,595,746 ± 612,8231 $g^{-1}$ —lower than the total area obtained by DTD (9,389,028,511 ± 908,282,182 $g^{-1}$). This shows that, in the case of sulfur compounds in onion, DTD correctly optimized by BBD and MRO and combined with GC-MS provides a suitable extraction and analysis method for the compounds of interest, yielding better results than other more common techniques, such as SPME.

## 3. Materials and Methods

### 3.1. Onion Samples

The onion samples that were the focus of this investigation were procured from a nearby marketplace located in the province of Cadiz, Spain. Specifically, this variety has been used by the research group in previous studies about the evaluation of the content of phenolic compounds [9–11,44]. For all the experiments carried out, the onions used were in freeze-dried form to avoid the presence of water and to have a more homogeneous onion matrix [45]. The type of onion used for our investigation is a variety of Spanish origin that is cultivated from June to December. The bulbs were characterized by their globose/conical shape with red outer skins in different shades (depending on the variety). The flesh has an intense purple color, a strong taste, and high pungency.

### 3.2. DTD-GC-MS Procedure and Conditions

The DTD equipment used was a TD-20 System (Shimadzu, Kyoto, Japan). Specifically, the onion sample was placed in a sample tube (with an outer diameter of 1/4″ (6.35 mm) and a length of 90 mm) and was secured at both ends with silica wool to prevent leakage. The flow rate of the carrier gas (helium) was adjusted to 1 mL $s^{-1}$. The cold trap, which concentrated the desorbed compounds to a bandwidth compatible with the capillary column, was set to −15 °C. The compounds were injected into the GC module in a split mode with a split ratio of 1:50. The sample amount collected in the tube, the temperature and time of the sample-tube heating block, and the temperature and time of the trap desorption were chosen according to the results of the response surface design of experiments. A detailed outline of the process is shown in the Supplementary Materials (Figure S1).

The GC-MS equipment used was GCMS-TQ8040 (Shimadzu, Kyoto, Japan). The chromatographic separation was performed on a Suprawax-280 capillary column (Teknokroma, Barcelona, Spain; 60 m length × 0.25 mm column I.D. × 0.25 μm film thickness). The injector was set at 25 °C. Moreover, the temperature program of the oven was as follows: 40 °C isothermal for 300 s; from 40 °C to 200 °C at a rate of 0.05 °C $s^{-1}$; 200 °C isothermal for 300 s; from 200 °C to 270 °C at a rate of 0.67 °C $s^{-1}$; and 270 °C isothermal for 120 s. Likewise, helium (99.999%) was used as the carrier gas at both a constant linear velocity of 35 cm $s^{-1}$ and a flow rate of 0.031 mL $s^{-1}$. Regarding the mass spectrometer, the ionization mode was electron impact with a voltage of 70 eV. The mass spectrometer worked in a full-scan mode in the range of 40–400 $m/z$. The ion source temperature was 200 °C.

Compounds were identified by comparing their mass spectra using the Wiley library (Wiley Registry of Mass Spectral Data, 7th Edition, 2000) and the criterion of at least 80% similarity [46]. The area of the chromatographic signal produced by the largest mass fragment (base peak) was measured to determine the area of each compound. Furthermore, a normalizing approach was used to obtain the percentage composition (%) from the peak area of each compound: the area of the base peak/total area.

### 3.3. Box–Behnken Design

The parameters that affected the analysis of the organosulfur compounds by using DTD-GC-MS were optimized with a BBD [47]. As aforementioned, several factors were evaluated to determine their optimum levels: the onion sample amount placed in the sample tube ($X_1$), the tube desorption temperature ($X_2$), the tube desorption time ($X_3$), the trap heat temperature ($X_4$), and the trap heat time ($X_5$). In a BBD, each factor was studied at 3 levels: at a lower level (−1), at an intermediate level (0), and at a higher level (1). Due to the lack of prior studies on the application of DTD to analyze the organosulfur compounds in onions, the operating range of each factor for the BBD was selected according to the outcome of the OFAT experiments. The OFAT experiments are summarized in the Supplementary Materials (Table S3 and Figure S2). According to these results, the range for each factor was chosen to consider not only the greatest total areas but also the area of the individual propanethial S-oxide: 40-50-60 mg for $X_1$; 180-200-220 °C for $X_2$; 600-900-1200 s for $X_3$; 250-265-280 °C for $X_4$; 120-180-240 s for $X_5$. Two response variables were defined: the total area ($Y_{TA}$) of the base peak of each compound, and the area of the propanethial S-oxide ($Y_{C3H6OS}$). The latter is of interest as it is obtained from the main SCs of onions and plays the role of lacrimator [48]. Both response variables were expressed as relative areas, i.e., as a function of the accurate mass of onion weighed for each experiment: the total area per gram of onion ($g^{-1}$) and the area of the propanethial S-oxide per gram of onion ($g^{-1}$), respectively. The study involved two separate BBD experiments, each design consisted of 46 treatments, including 6 repetitions at the center point to calculate the error. The complete matrix with the experimental and predicted values for each response variable is included in Supplementary Materials (Table S4).

### 3.4. Response Surface Methodology

In combination with BBD, RSM comprises a series of statistical and mathematical methods utilized to develop and optimize processes. RSM allows for the modelling of the curvature relationship between factors and their response by employing a second-order polynomial equation (Equation (3)) [49].

$$Y = \beta_0 + \sum_{i=1}^{k} \beta_i X_i + \sum_{i=1}^{k} \beta_{ii} X_i^2 + \sum\sum_{i<j}^{k} \beta_{ij} X_i X_j + r, \quad (3)$$

where $Y$ represents the predicted responses (i.e., $Y_{TA}$ and $Y_{C3H6OS}$); $X_i$ and $X_j$ the factors involved; $X_iX_j$ the interactions between factors; $X_i^2$ the quadratic interaction between factors; $\beta_0$ the intercept; $\beta_i$ the linear coefficient; $\beta_{ij}$, the coefficient of interaction between factors; $\beta_{ii}$ the quadratic coefficient; and $r$ the random error.

Utilizing the Statgraphics Centurion version XVI software (Warrenton, VA, USA) and Design Expert software (Version 13, Stat-Ease Inc., Minneapolis, MN, USA), an analysis of variance (ANOVA) was executed.

After optimizing each of the two response variables separately, a multi-response optimization (MRO) was performed to determine their optimal conditions simultaneously. The desirability optimization methodology was employed, which combines the desirability function analysis with the design of experiments [50]. Each response was assigned a desirability score ($d_i$) ranging from 0 to 1, where 0 indicated an unacceptable response, and 1 indicated an ideal response. The individual desirability scores were then geometrically averaged to obtain an overall desirability score (D). Ultimately, the multi-optimization process aimed to maximize the value of D.

## 4. Conclusions

Among all vegetables, onions have perhaps the biggest market niche not only because of their great use in cooking but also because of their sulfur compounds, which can have enormously beneficial properties for one's health. Particularly for the latter reason, it is necessary to develop methods of analysis and extraction that allow for the study of these sulfur compounds in an efficient way. In this work, a DTD-GC-MS methodology has been developed for the pre-extraction and subsequent determination of the organosulfur compounds present in onions. A BBD, together with MRO, was used for optimization, considering the total area (sum of the individual area of each of the 20 identified sulfur compounds) and the individual area of the propanethial S-oxide as response variables. The MRO conditions were as follows: 46 mg of onion in the tube, a desorption heat of 205 °C for 960 s, and a trap heat of 267 °C for 180 s. In addition, the efficacy of the technique has been confirmed by demonstrating that all organosulfur compounds exhibit high levels of repeatability and intermediate precision, with coefficients of variation (CV) lower than 10%. The distribution of the twenty organosulfur compounds showed 25.3% of thiophenes, 38.4% of carbon sulfides (mixture of monosulfides, disulfides, and trisulfides), and 4.5% of propanethial S-oxide. Overall, the optimized DTD-GC-MS method has significant practical implications for laboratories, researchers, and companies hoping to determine the organosulfur content accurately and reliably in onions. To evaluate the limitations of this study, the developed method was compared with SPME, the most widely used pre-extraction technique, to study these compounds in onions. The developed method showed better extractions, with a higher amount of extracted sulfur compounds. In the future, high-resolution analytical techniques will allow us to know more about the content of these compounds in different types of onions, which could be useful for assessing how factors such as variety, origin, and cultivation method affect the composition and activity of these compounds.

**Supplementary Materials:** The following supporting information can be downloaded at: https://www.mdpi.com/article/10.3390/ph16050715/s1, Figure S1: Schematic diagram of the DTD system; Figure S2: OFAT experiments concerning the effect of the factor: (a) onion sample amount placed in the sample tube ($X_1$), (b) tube desorption temperature ($X_2$), (c) tube desorption time ($X_3$), (d) trap heat temperature ($X_4$), and (e) trap heat time ($X_5$), on both the relative total area ($g^{-1}$) and the relative area of the propanethial S-oxide ($g^{-1}$); Table S1: Characteristic mass spectral ions of the organosulfur compounds identified in red onion; Table S2: Validation of the developed DTD-GC-MS MRO method; Table S3: OFAT experiments to choose the range of BBD ($n$ = 3); Table S4: BBD for the total area and the area of the propanethial S-oxide. The results corresponded to experimental and predicted values.

**Author Contributions:** Conceptualization, M.F.-G. and G.F.B.; methodology, A.V.G.-d.-P., M.V.-E. and C.C.; software, M.F.-G.; validation, A.V.G.-d.-P. and M.V.-E.; formal analysis, A.V.G.-d.-P., M.V.-E. and C.C.; investigation, A.V.G.-d.-P. and M.V.-E.; resources, M.P and G.F.B.; data curation, E.E.-B., M.F.-G. and G.F.B.; writing—original draft preparation, A.V.G.-d.-P. and M.V.-E.; writing—review and editing, G.F.B. and E.E.-B.; visualization, G.F.B.; supervision, M.F.-G. and G.F.B.; project administration, G.F.B.; funding acquisition, G.F.B. and M.P. All authors have read and agreed to the published version of the manuscript.

**Funding:** This work has been supported by the project "EQC2018-005135-P" (Equipment for liquid chromatography using mass spectrometry and ion chromatography), of the State Subprogram of Research Infrastructures and Technical Scientific Equipment. This research was funded by the University of Cadiz, the INIA (National Institute for Agronomic Research), and FEDER (European Regional Development Fund, within the framework of the Operational Program under the Investment for Growth 2014–2020) who provided financial support (RTA2015-00042-C02-01), and by the Ministry of Science and Innovation of Spain (FPU grant AP-2018-03811 to Ana Velasco González de Peredo).

**Institutional Review Board Statement:** Not applicable.

**Informed Consent Statement:** Not applicable.

**Data Availability Statement:** The data presented in this study are contained within the article.

**Acknowledgments:** The authors are grateful to the "Instituto de Investigación Vitivinícola y Agroalimentaria" (IVA-GRO) for providing the necessary facilities to carry out the research. A special acknowledgement is extended to the Mass Spectrometry Division of the Central Research Services for Science and Technology (SC-ICYT) from the University of Cadiz for their collaboration throughout the analysis of the samples.

**Conflicts of Interest:** The authors declare no conflict of interest.

## References

1. Böttcher, C.; Krähmer, A.; Stürtz, M.; Widder, S.; Schulz, H. Comprehensive Metabolite Profiling of Onion Bulbs (*Allium cepa*) Using Liquid Chromatography Coupled with Electrospray Ionization Quadrupole Time-of-Flight Mass Spectrometry. *Metabolomics* **2017**, *13*, 35. [CrossRef]
2. Crowther, T.; Collin, H.A.; Smith, B.; Tomsett, A.B.; O'Connor, D.; Jones, M.G. Assessment of the flavour of fresh uncooked onions by taste-panels and analysis of flavour precursors, pyruvate and sugars. *J. Sci. Food Agric.* **2005**, *85*, 112–120. [CrossRef]
3. Goncharov, N.V.; Belinskaia, D.A.; Ukolov, A.I.; Jenkins, R.O.; Avdonin, P.V. Chapter 41—Organosulfur Compounds as Nutraceuticals. In *Nutraceuticals: Efficacy, Safety and Toxicity*; Ramesh, C.G., Ed.; Elsevier Inc.: Amsterdam, The Netherlands, 2016; pp. 555–568. [CrossRef]
4. Arshad, M.S.; Sohaib, M.; Nadeem, M.; Saeed, F.; Imran, A.; Javed, A.; Amjad, A.; Batool, S.M.; Yildiz, F. Status and trends of nutraceuticals from onion and onion by-products: A critical review. *Cogent Food Agric.* **2017**, *3*, 1280254. [CrossRef]
5. Nicastro, H.L.; Ross, S.A.; Milner, J.A. Garlic and Onions: Their Cancer Prevention Properties. *Cancer Prev. Res.* **2015**, *8*, 181–189. [CrossRef] [PubMed]
6. Abouzed, T.K.; del Mar Contreras, M.; Sadek, K.M.; Shukry, M.; Abdelhady, H.D.; Gouda, W.M.; Abdo, W.; Nasr, N.E.; Mekky, R.H.; Segura-Carretero, A.; et al. Red Onion Scales Ameliorated Streptozotocin-Induced Diabetes and Diabetic Nephropathy in Wistar Rats in Relation to Their Metabolite Fingerprint. *Diabetes Res. Clin. Pract.* **2018**, *140*, 253–264. [CrossRef] [PubMed]
7. Ferioli, F.; D'Antuono, L.F. Evaluation of Phenolics and Cysteine Sulfoxides in Local Onion and Shallot Germplasm from Italy and Ukraine. *Genet. Resour. Crop Evol.* **2016**, *63*, 601–614. [CrossRef]
8. Fredotović, Ž.; Soldo, B.; Šprung, M.; Marijanović, Z.; Jerković, I.; Puizina, J. Comparison of Organosulfur and Amino Acid Composition between Triploid Onion *Allium cornutum clementi* Ex Visiani, 1842, and Common Onion *Allium cepa* L., and Evidences for Antiproliferative Activity of Their Extracts. *Plants* **2020**, *9*, 98. [CrossRef]
9. González-de-peredo, A.V.; Vázquez-espinosa, M.; Espada-bellido, E.; Carrera, C.; Ferreiro-gonzález, M.; Barbero, G.F.; Palma, M. Flavonol Composition and Antioxidant Activity of Onions (*Allium cepa* L.) Based on the Development of New Analytical Ultrasound-assisted Extraction Methods. *Antioxidants* **2021**, *10*, 273. [CrossRef]
10. González-de-Peredo, A.V.; Vázquez-Espinosa, M.; Espada-Bellido, E.; Ferreiro-González, M.; Carrera, C.; Barbero, G.F.; Palma, M. Extraction of Antioxidant Compounds from Onion Bulb (*Allium cepa* L.) Using Individual and Simultaneous Microwave-Assisted Extraction Methods. *Antioxidants* **2022**, *11*, 846. [CrossRef]
11. González-de-Peredo, A.V.; Vázquez-Espinosa, M.; Espada-Bellido, E.; Ferreiro-González, M.; Carrera, C.; Barbero, G.F.; Palma, M. Development of Optimized Ultrasound-Assisted Extraction Methods for the Recovery of Total Phenolic Compounds and Anthocyanins from Onion Bulbs. *Antioxidants* **2021**, *10*, 1755. [CrossRef]
12. Kim, S.; Kim, D.B.; Jin, W.; Park, J.; Yoon, W.; Lee, Y.; Kim, S.; Lee, S.; Kim, S.; Lee, O.H.; et al. Comparative Studies of Bioactive Organosulphur Compounds and Antioxidant Activities in Garlic (*Allium sativum* L.), Elephant Garlic (*Allium ampeloprasum* L.) and Onion (*Allium cepa* L.). *Nat. Prod. Res.* **2017**, *32*, 1193–1197. [CrossRef]
13. Cardozo, L.F.M.F.; Pedruzzi, L.M.; Stevinkel, P.; Stockler-Pinto, M.B.; Daleprane, J.B.; Leire, M., Jr.; Mafra, D. Nutritional strategies to modulate inflammation and oxidative stress pathways via activation of the master antioxidant switch Nrf2. *Biochimie* **2013**, *95*, 1525–1533. [CrossRef]
14. Wu, S.; Liao, X.; Zhu, Z.; Huang, R.; Chen, M.; Huang, A.; Zhang, J.; Wu, Q.; Wang, J.; Ding, Y. Antioxidant and anti-inflammation effects of dietary phytochemicals: The Nrf2/NF-κB signalling pathway and upstream factors of Nrf2. *Phytochemistry* **2022**, *204*, 113429. [CrossRef]
15. Esmaeili, Y.; Yarjanli, Z.; Pakniya, F.; Bridam, E.; Los, M.J.; Eshraghi, M.; Klionsky, D.J.; Ghavami, S.; Zarrabi, A. Targeting autophagy, oxidative stress, and ER stress for neurodegenerative disease treatment. *J. Control. Release* **2022**, *345*, 147–175. [CrossRef]
16. Kato, M.; Masamura, N.; Shono, J.; Okamoto, D.; Abe, T.; Imai, S. Production and Characterization of Tearless and Non-Pungent Onion. *Sci. Rep.* **2016**, *6*, 23779. [CrossRef]
17. Yoshimoto, N.; Saito, K. S-Alk(En)Ylcysteine Sulfoxides in the Genus *Allium*: Proposed Biosynthesis, Chemical Conversion, and Bioactivities. *J. Exp. Bot.* **2019**, *70*, 4123–4137. [CrossRef]
18. Ramirez, D.A.; Locatelli, D.A.; González, R.E.; Cavagnaro, P.F.; Camargo, A.B. Analytical Methods for Bioactive Sulfur Compounds in *Allium*: An Integrated Review and Future Directions. *J. Food Compos. Anal.* **2017**, *61*, 4–19. [CrossRef]

19. Tocmo, R.; Liang, D.; Lin, Y.; Huang, D. Chemical and Biochemical Mechanisms Underlying the Cardioprotective Roles of Dietary Organopolysulfides. *Front. Nutr.* **2015**, *2*, 1. [CrossRef]
20. Bahram-Parvar, M.; Lim, L.T. Fresh-Cut Onion: A Review on Processing, Health Benefits, and Shelf-Life. *Compr. Rev. Food Sci. Food Saf.* **2018**, *17*, 290–308. [CrossRef]
21. Nandakumar, R.; Eyres, G.T.; Burritt, D.J.; Kebede, B.; Leus, M.; Oey, I. Impact of Pulsed Electric Fields on the Volatile Compounds Produced in Whole Onions (*Allium cepa* and *Allium fistulosum*). *Foods* **2018**, *7*, 183. [CrossRef]
22. Cazzola, R.; Cestaro, B. Chapter 9 - Antioxidant Spices and Herbs Used in Diabetes. In *Diabetes: Oxidative Stress Diet Antioxidants*; Preedy, V.R., Ed.; Elsevier Inc.: Amsterdam, The Netherlands, 2014; pp. 89–97. [CrossRef]
23. Rektorisova, M.; Hrbek, V.; Jiru, M.; Ovesna, J.; Hajslova, J. Variability in S-Alk(En)Yl-l-Cysteine Sulfoxides in Garlic within a Seven-Month Period Determined by a Liquid Chromatography—Tandem Mass Spectrometry Method. *Plant Foods Hum. Nutr.* **2020**, *75*, 376–382. [CrossRef] [PubMed]
24. Cecchi, L.; Ieri, F.; Vignolini, P.; Mulinacci, N.; Romani, A. Characterization of Volatile and Flavonoid Composition of Different Cuts of Dried Onion (Allium cepa L.) by HS-SPME-GC-MS, HS-SPME-GC×GC-TOF and HPLC-DAD. *Molecules* **2020**, *25*, 408. [CrossRef] [PubMed]
25. Ma, J.-h.; Zhong, Y.; Zhou, Y.; Zhang, Y.; Feng, X.-s. Organosulfur in Food Samples: Recent Updates on Sampling, Pretreatment and Determination Technologies. *J. Chromatogr. A* **2023**, *1689*, 463769. [CrossRef]
26. Gîtin, L.; Dinică, R.; Neagu, C.; Dumitrascu, L. Sulfur Compounds Identification and Quantification from *Allium* Spp. Fresh Leaves. *J. Food Drug Anal.* **2014**, *22*, 425. [CrossRef] [PubMed]
27. Aguiar, J.; Gonçalves, J.L.; Alves, V.L.; Câmara, J.S. Relationship between Volatile Composition and Bioactive Potential of Vegetables and Fruits of Regular Consumption—An Integrative Approach. *Molecules* **2021**, *26*, 3653. [CrossRef]
28. Colina-Coca, C.; González-Peña, D.; Vega, E.; De Ancos, B.; Sánchez-Moreno, C. Novel Approach for the Determination of Volatile Compounds in Processed Onion by Headspace Gas Chromatography-Mass Spectrometry (HS GC-MS). *Talanta* **2013**, *103*, 137–144. [CrossRef]
29. Flores, R.M.; Mertoglu, E.; Durenne, B.; Blondel, A.; Druart, P.; Fauconnier, M.L.; Duan, Z.; Kjeldsen, P.; Scheutz, C. Optimization of a Thermal Desorption-Gas Chromatography/Mass Spectrometry Method for Characterization of Semi-Volatile Organic Compounds in High Time Resolved PM2.5. *Atmos. Pollut. Res.* **2020**, *11*, 619–629. [CrossRef]
30. Gouda, M.; Ma, M.; Sheng, L.; Xiang, X. SPME-GC-MS & Metal Oxide E-Nose 18 Sensors to Validate the Possible Interactions between Bio-Active Terpenes and Egg Yolk Volatiles. *Food Res. Int.* **2019**, *125*, 108611. [CrossRef]
31. Cheng, Z.; Mannion, D.T.; O'Sullivan, M.G.; Miao, S.; Kerry, J.P.; Kilcawley, K.N. Comparison of Automated Extraction Techniques for Volatile Analysis of Whole Milk Powder. *Foods* **2021**, *10*, 2061. [CrossRef]
32. Shiea, C.; Huang, Y.L.; Liu, D.L.; Chou, C.C.; Chou, J.H.; Chen, P.Y.; Shiea, J.; Huang, M.Z. Rapid Screening of Residual Pesticides on Fruits and Vegetables Using Thermal Desorption Electrospray Ionization Mass Spectrometry. *Rapid Commun. Mass Spectrom.* **2015**, *29*, 163–170. [CrossRef]
33. Humbert, K.; Debret, M.; Morin, C.; Cosme, J.; Portet-Koltalo, F. Direct Thermal Desorption-Gas Chromatography-Tandem Mass Spectrometry versus Microwave Assisted Extraction and GC-MS for the Simultaneous Analysis of Polyaromatic Hydrocarbons (PAHs, PCBs) from Sediments. *Talanta* **2022**, *250*, 123735. [CrossRef]
34. Elorduy, I.; Elcoroaristizabal, S.; Durana, N.; García, J.A.; Alonso, L. Diurnal Variation of Particle-Bound PAHs in an Urban Area of Spain Using TD-GC/MS: Influence of Meteorological Parameters and Emission Sources. *Atmos. Environ.* **2021**, *138*, 87–98. [CrossRef]
35. Vera Candioti, L.; De Zan, M.M.; Cámara, M.S.; Goicoechea, H.C. Experimental Design and Multiple Response Optimization. Using the Desirability Function in Analytical Methods Development. *Talanta* **2014**, *124*, 123–138. [CrossRef]
36. Villière, A.; Le Roy, S.; Fillonneau, C.; Guillet, F.; Falquerho, H.; Boussely, S.; Prost, C. Evaluation of Aroma Profile Differences between Sué, Sautéed, and Pan-Fried Onions Using an Innovative Olfactometric Approach. *Flavour* **2015**, *4*, 24. [CrossRef]
37. Elorduy, I.; Durana, N.; García, J.A.; Gómez, M.C.; Alonso, L. Optimization and Validation of Thermal Desorption Gas Chromatography-Mass Spectrometry for the Determination of Polycyclic Aromatic Hydrocarbons in Ambient Air. *J. Anal. Methods Chem.* **2018**, *2018*, 8734013. [CrossRef]
38. Gao, S.; Zuom, G.; Chen, J.; Zhou, J.; Liu, J. Study on the TD-GC-MS to Detect HD in Exhaust Gas. In *Proceedings from the ICERP 2016*; De Gruyter: Warsaw, Poland, 2017; pp. 13–22. [CrossRef]
39. Poojary, M.M.; Putnik, P.; Bursać Kovačević, D.; Barba, F.J.; Lorenzo, J.M.; Dias, D.A.; Shpigelman, A. Stability and Extraction of Bioactive Sulfur Compounds from *Allium* Genus Processed by Traditional and Innovative Technologies. *J. Food Compos. Anal.* **2017**, *61*, 28–39. [CrossRef]
40. Choi, S.M.; Lee, D.J.; Kim, J.Y.; Lim, S.T. Volatile Composition and Sensory Characteristics of Onion Powders Prepared by Convective Drying. *Food Chem.* **2017**, *231*, 386–392. [CrossRef]
41. Taglienti, A.; Araniti, F.; Piscopo, A.; Tiberini, A. Characterization of Volatile Organic Compounds in 'Rossa Di Tropea' Onion by Means of Headspace Solid-Phase Microextraction Gas Chromatography–Mass Spectrometry (Hs/Spme Gc–Ms) and Sensory Analysis. *Agronomy* **2021**, *11*, 874. [CrossRef]
42. Boelends, H.; Brandsma, L. Formation of dialkylthiophenes by thermolysis of di(1-alkenyl) disulfides and alkyl 1-propenyl disulfides. *Recueil des Travaux Chimiques des Pays-Bas* **1972**, *91*, 141–145. [CrossRef]

43. Woolfenden, E. Chapter 10—Thermal Desorption Gas Chromatography. In *Handbooks in Separation Science*; Poole, C.F., Ed.; Elsevier Inc.: Amsterdam, The Netherlands, 2021; pp. 267–323. [CrossRef]
44. González-de-Peredo, A.V.; Vázquez-Espinosa, M.; Espada-Bellido, E.; Ferreiro-González, E.; Barbero, G.F.; Palma, M.; Carrera, C. Optimization of a Microwave Assisted Extraction Method for Maximum Flavonols and Antioxidant Activity of Onion Extracts. *Antioxidants* **2022**, *11*, 2393. [CrossRef]
45. González-de-Peredo, A.V.; Vázquez-Espinosa, M.; Carrera, C.; Espada-Bellido, E.; Ferreiro-González, E.; Barbero, G.F.; Palma, M. Development of a Rapid UHPLC-PDA Method for the Simultaneous Quantification of Flavonol Contents in Onions (*Allium cepa* L.). *Pharmaceuticals* **2021**, *14*, 310. [CrossRef] [PubMed]
46. Machová, M.; Bajer, T.; Šilha, D.; Ventura, K.; Bajerová, P. Release of Volatile Compounds from Sliced Onion Analysed by Gas Chromatography Coupled to Mass Spectrometry and Its Antimicrobial Activity. *J. Food Nutr. Res.* **2019**, *58*, 393–400.
47. Maran, J.; Manikandan, S.; Thirugnanasambandham, K.; Vigna Nivetha, C.; Dinesh, R. Box-Behnken Design Based Statistical Modeling for Ultrasound-Assisted Extraction of Corn Silk Polysaccharide. *Carbohydr. Polym.* **2013**, *92*, 604–611. [CrossRef]
48. Eshel, D.; Teper-Bamnolker, P.; Vinokur, Y.; Saad, I.; Zutahy, Y.; Rodov, V. Fast Curing: A Method to Improve Postharvest Quality of Onions in Hot Climate Harvest. *Postharvest Biol. Technol.* **2014**, *88*, 34–39. [CrossRef]
49. Marinković, V. A Novel Desirability Function for Multi-Response Optimization and Its Application in Chemical Engineering. *Chem. Ind. Chem. Eng. Q.* **2020**, *26*, 309–319. [CrossRef]
50. Myers, R.H.; Montgomery, D.C.; Anderson-Cooj, C.M. *Response Surface Methodology: Process and Products Optimization Using Designed Experiments*; Wiley: Hoboken, NJ, USA, 2015; pp. 1–821.

*pharmaceuticals*

*Article*

# Identification of Compounds of *Crocus sativus* by GC-MS and HPLC/UV-ESI-MS and Evaluation of Their Antioxidant, Antimicrobial, Anticoagulant, and Antidiabetic Properties

**Aziz Drioiche [1,2,*], Atika Ailli [1], Nadia Handaq [1,3], Firdaous Remok [1], Mohamed Elouardi [1], Hajar Elouadni [1], Omkulthom Al Kamaly [4], Asmaa Saleh [4], Mohamed Bouhrim [5], Hanane Elazzouzi [1], Fadoua El Makhoukhi [1] and Touriya Zair [1,*]**

1 Research Team of Chemistry of Bioactive Molecules and the Environment, Laboratory of Innovative Materials and Biotechnology of Natural Resources, Faculty of Sciences, Moulay Ismaïl University, B.P. 11201 Zitoune, Meknes 50070, Morocco
2 Medical Microbiology Laboratory, Mohamed V. Hospital, Meknes 50000, Morocco
3 Research Team of Enhancement and Protection of Plants, Laboratory of Environmental Biology and Sustainable Development, Higher Normal School, Abdelmalek Essaadi University, Tetouan 93000, Morocco
4 Department of Pharmaceutical Sciences, College of Pharmacy, Princess Nourah Bint Abdulrahman University, P.O. Box 84428, Riyadh 11671, Saudi Arabia
5 Laboratory of Biological Engineering, Team of Functional and Pathological Biology, Faculty of Sciences and Technology Beni Mellal, University Sultan Moulay Slimane, Beni-Mellal 23000, Morocco
* Correspondence: a.drioiche@edu.umi.ac.ma (A.D.); t.zair@umi.ac.ma (T.Z.); Tel.: +212-600394841 (A.D.)

**Abstract:** In order to valorize the species *Crocus sativus* from Morocco and to prepare new products with high added value that can be used in the food and pharmaceutical industry, our interest was focused on the phytochemical characterization and the biological and pharmacological properties of the stigmas of this plant. For this purpose, the essential oil of this species, extracted by hydrodistillation and then analyzed by GC-MS, revealed a predominance of phorone (12.90%); (R)-(-)-2,2-dimethyl-1,3-dioxolane-4-methanol (11.65%); isopropyl palmitate (9.68%); dihydro-$\beta$-ionone (8.62%); safranal (6.39%); *trans*-$\beta$-ionone (4.81%); 4-keto-isophorone (4.72%); and 1-eicosanol (4.55%) as the major compounds. The extraction of phenolic compounds was performed by decoction and Soxhlet extraction. The results of the determination of flavonoids, total polyphenols, condensed tannins, and hydrolyzable tannins determined by spectrophotometric methods on aqueous and organic extracts have proved the richness of *Crocus sativus* in phenolic compounds. Chromatographic analysis by HPLC/UV-ESI-MS of *Crocus sativus* extracts revealed the presence of crocin, picrocrocin, crocetin, and safranal molecules specific to this species. The study of antioxidant activity by three methods (DPPH, FRAP, and total antioxidant capacity) has proved that *C. sativus* is a potential source of natural antioxidants. Antimicrobial activity of the aqueous extract ($E_0$) was investigated by microdilution on a microplate. The results have revealed the efficacy of the aqueous extract against *Acinetobacter baumannii* and *Shigella* sp. with MIC $\leq$ 600 µg/mL and against *Aspergillus niger*, *Candida kyfer*, and *Candida parapsilosis* with MIC = 2500 µg/mL. Measurements of pro-thrombin time (PT) and activated partial thromboplastin time (aPTT) in citrated plasma obtained from routine healthy blood donors were used to determine the anticoagulant activity of aqueous extract ($E_0$). The anticoagulant activity of the extract ($E_0$) studied showed that this extract can significantly prolong the partial thromboplastin time ($p < 0.001$) with a 359 µg/mL concentration. The antihyperglycemic effect of aqueous extract was studied in albino Wistar rats. The aqueous extract ($E_0$) showed strong in vitro inhibitory activity of $\alpha$-amylase and $\alpha$-glucosidase compared with acarbose. Thus, it very significantly inhibited postprandial hyperglycemia in albino Wistar rats. According to the demonstrated results, we can affirm the richness of *Crocus sativus* stigmas in bioactive molecules and its use in traditional medicine.

**Keywords:** *Crocus sativus* L.; phorone; crocins; picrocrocin; safranal; DPPH; FRAP; total antioxidant capacity; antimicrobial activity; anticoagulant activity; antidiabetic activity

**Citation:** Drioiche, A.; Ailli, A.; Handaq, N.; Remok, F.; Elouardi, M.; Elouadni, H.; Al Kamaly, O.; Saleh, A.; Bouhrim, M.; Elazzouzi, H.; et al. Identification of Compounds of *Crocus sativus* by GC-MS and HPLC/UV-ESI-MS and Evaluation of Their Antioxidant, Antimicrobial, Anticoagulant, and Antidiabetic Properties. *Pharmaceuticals* **2023**, *16*, 545. https://doi.org/10.3390/ph16040545

Academic Editor: Jan Oszmianski

Received: 21 January 2023
Revised: 17 March 2023
Accepted: 27 March 2023
Published: 5 April 2023

## 1. Introduction

The overproduction of reactive oxygen species generates damage in biomolecules (DNA, proteins, amino acids, and lipids) and can produce oxidative stress, which subsequently can induce various pathologies, such as neurodegenerative diseases, cardiovascular diseases, diabetes, and cancer [1,2]. Currently, there are therapies against these various pathologies, but they present serious side effects, most often undesirable. To reduce these side effects, several works have focused on the research of effective bioactive molecules from natural plant resources, in particular aromatic and medicinal plants.

As humankind has been progressing in its development, we have developed interactions with plants to take better advantage of them for our well-being. We have empirically learned some of the virtues of plants and have improved them with the development of science. However, plants in their natural state continue to be used by most of the world's population. In particular, the Moroccan population benefits from a rich and diversified flora characterized by high endemism of 7000 existing species, subspecies, and varieties with 940 genera and 135 families of which 537 are endemic to the country and 1625 are rare or threatened [3]. Its diversity is the backbone of Morocco's industry for aromatic and medicinal plants, which is regarded as the world's 12th largest exporter of these plants.

The Fez-Meknes area, with its rich and varied ecosystems, offers an ideal setting for the research of new bioactive molecules of agro-food and pharmaceutical interest. Saffron is one of the plants of this region which plays a very important role at the economic and social level, and its conservation constitutes a major and preponderant priority.

The species *Crocus sativus* has been widely used as a dyeing agent, as well as for medication in the traditional medicinal systems of India, Thailand, the Philippines, Malaysia, China, Australia, Africa, etc. [4,5]. The Moroccan flora is also abundant in this species, which is used as a condiment plant and in traditional medicine to treat a variety of illnesses [6].

Saffron (*C. sativus* L.) is an autumnal herbal flowering plant of the family Iridaceae. It is currently ranked as the most expensive spice in the world. It is the most important plant of the genus *Crocus*. With an estimated 418 tons produced per year, this plant is commonly cultivated in Iran, Morocco, India, Afghanistan, Italy, Greece, and Spain [7]. The most commonly used plant parts of *C. sativus* are the stigmas, flowers, and petals. Saffron has many traditional applications, and it is used to treat dysmenorrhea, gastric ulcers [8], nervous system disorders, asthma, whooping cough, and inflammations [9]. It can be used as an aperitif, stimulant, and tonic [8]. In addition, phytochemical studies of saffron have identified multiple active components such as *cis*- and *trans*-crocin (dye), safranal (aroma), and glycoside carotenoids [10]. In addition, the petals and stigmas of this plant contain various flavonoids, phenylpropanoids, phenolic glycosides, alkaloids, coumarins, terpenoids, fatty alcohols, and fatty acids of which glycosylated flavonoids and anthocyanins are considered important antioxidant and anti-inflammatory compounds [11–13].

Currently, the literature on saffron provides a deep scope on plant varieties, major constituents, and culinary uses, mainly based on traditional Chinese, Indian, and Iranian medicine. To enlarge the scope of the future implications of this Moroccan plant species for medicinal and culinary purposes, the present study aims to valorize *C. sativus* cultivated in the Boulemane area, through phytochemical investigations and evaluation of biological and pharmacological properties of this plant species. Certainly, the phytochemical profile, antioxidant, antimicrobial, and anticoagulant properties, and antidiabetic activities of the stigmas of *C. sativus* cultivated in the Boulemane area are highlighted by taking into account in vitro and in vivo investigations.

## 2. Results

### 2.1. The Yield of *C. sativus* EO

The average yield of EO was calculated based on the dry plant material obtained from the stigmas of *C. sativus*.

The yield of the obtained essential oil is very low, about 0.25 ± 0.01% (Table 1). This oil is characterized by a yellowish color with an aromatic spicy and woody smell.

**Table 1.** The yield of the essential oil of *C. sativus*.

| Species | Properties | | |
|---|---|---|---|
| | Yield (%) | Color | Smell |
| *C. sativus* | 0.25 ± 0.01 | Deep yellow | Spicy and woody aromatics |

## 2.2. GC-MS Analysis of *C. sativus* Essential Oil

The GC/MS analyses of EO from *C. sativus* stigmas were used to determine the chromatographic profile (Figure 1), identify the different constituents, and measure their relative abundances in the EO analyzed (Table 2).

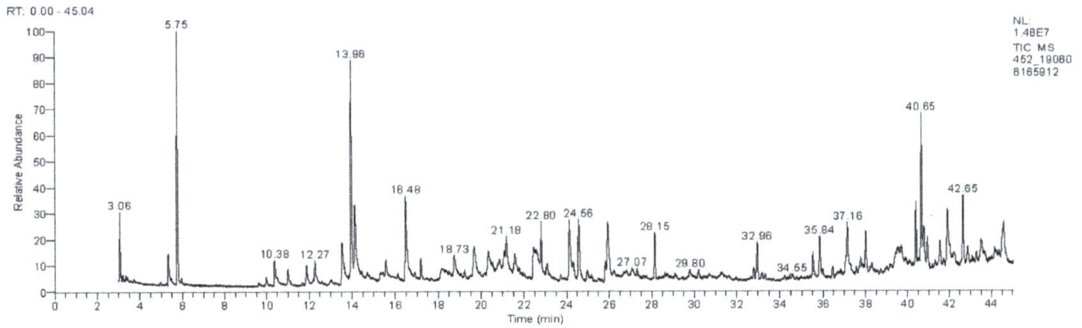


**Figure 1.** Chromatographic profile of the GC/MS analysis of *C. sativus* EO studied.

**Table 2.** Chemical profile (GC/MS) of the essential oil of *C. sativus*.

| KI | Compounds | Area (%) |
|---|---|---|
| 856 | (R)-(-)-2,2-dimethyl-1,3-dioxolane-4-methanol | 11.65 |
| **912** | 2-acetylfuran | 1.42 |
| **1017** | γ-terpinene | 1.32 |
| **1050** | 2-acetylcyclopentanone | 0.86 |
| **1070** | *Trans*-arbusculone | 0.91 |
| **1121** | Isophorone | 3.56 |
| 1095 | Phorone | 12.90 |
| **1145** | **4-keto-isophorone** | **4.72** |
| 1196 | Safranal | 6.39 |
| **1230** | Duroquinone | 1.10 |
| **1252** | Thymoquinone | 2.42 |
| **1282** | 2-ethyl menthone | 0.93 |
| **1289** | Cyclopent-2-en-1-one, 2-pentyl- | 2.03 |
| **1314** | 2,3,4-trimethyl benzaldehyde | 1.25 |

**Table 2.** *Cont.*

| KI | Compounds | Area (%) |
|---|---|---|
| 1436 | Dihydro-β-ionone | 8.62 |
| **1455** | β-ionone epoxide | 0.78 |
| **1488** | ***Trans*-β-ionone** | **4.81** |
| **1668** | (Z)-Coniferyl alcohol | 3.62 |
| **1800** | Octadecane | 1.20 |
| **1812** | Isopropyl myristate | 2.38 |
| **1900** | Nonadecane | 1.91 |
| **2000** | Eicosane | 3.27 |
| 2010 | Isopropyl palmitate | 9.68 |
| **2049** | 4-tert-Octyl-*o*-cresol | 1.29 |
| **2060** | 13-epi-manool | 2.46 |
| **2065** | **1-eicosanol** | **4.55** |
| **2100** | Heneicosane <n-> | 3.96 |
| **Monoterpene hydrocarbons (%)** | | **1.32** |
| **Oxygenated monoterpenes (%)** | | **67.97** |
| **Hydrocarbon sesquiterpenes (%)** | | **3.11** |
| **Oxygenated sesquiterpenes (%)** | | **13.35** |
| **Hydrocarbon diterpenes (%)** | | **7.23** |
| **Oxygenated diterpenes (%)** | | **7.01** |
| **Total (%)** | | **99.99** |

The analysis of the chemical composition of the EO identified 27 chemical compounds for *C. sativus* representing 99.99% of the total composition. By examining the EO's major chemical classes (Figure 1), we found that the saffron sample presents an EO rich in oxygenated monoterpenes (67.97%), oxygenated sesquiterpenes (13.35%), hydrocarbon diterpenes (7.23%), and oxygenated diterpenes (7.01%). The hydrocarbon sesquiterpenes and hydrocarbon monoterpenes come last, with levels of 3.11 and 1.32%, respectively. In addition, the majority of compounds of this EO are phorone (12.90%); (R)-(-)-2,2-dimethyl-1,3-dioxolane-4-methanol (11.65%); isopropyl palmitate (9.68%); dihydro-β-ionone (8.62%); safranal (6.39%); *trans*-β-ionone (4.81%); 4-keto-isophorone (4.72%); and 1-eicosanol (4.55%).

The EO of *C. sativus* is composed essentially of ketones (44.28%) and non-aromatic alcohols (18.66%), followed by esters (12.06%), hydrocarbons (11.66%), and aldehydes (7.64%). In addition, we discovered additional chemical families, such as weakly proportioned phenols and epoxides, that describe the species of saffron (Figure 2).

*2.3. Phytochemical Screening*

Phytochemical tests showed that *C. sativus* stigmas contain important secondary metabolites (Table 3). A strongly positive reaction was observed showing the presence of sterols and triterpenes, flavonoids, gallic tannins, alkaloids, and saponosides. In addition, a negative reaction of anthocyanins, catechic tannins, and anthracene derivatives was observed.

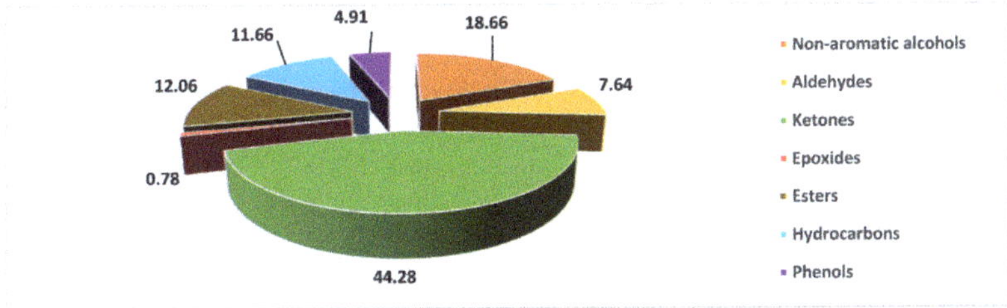


**Figure 2.** Distribution of chemical classes identified in *C. sativus* EO (%).

**Table 3.** Results of phytochemical tests.

| Compounds/Species | | *C. sativus* |
|---|---|---|
| Part used | | *Stigma* |
| **Sterols and triterpenes** | | +++ |
| **Flavonoids** | | ++ |
| **Anthocyanins** | | - |
| **Tannins** | Catechic tannins | - |
| | Gallic tannins | + |
| **Anthracene derivatives** | Quinones | - |
| | O-Heterosides | - |
| | C-Heterosides | - |
| **Saponosides** | | + |
| **Alkaloids** | Dragendorff | ++ |
| | Mayer | ++ |

Category: Strong presence: +++; average presence: ++; low presence: + and absent: -.

### 2.4. Extraction and Quantitative Analysis of Phenolic Compounds

#### 2.4.1. Extraction Yields

We performed the solid–liquid extraction by decoction ($E_0$), by Soxhlet water ($E_1$), and by Soxhlet ethanol–water ($E_2$) extraction, to compare the yields and the contents of phenolic compounds obtained by these two methods. From Figure 3, it appears that the yield of extracts obtained by decoction for *C. sativus* stigmas is higher than those obtained by Soxhlet extraction.

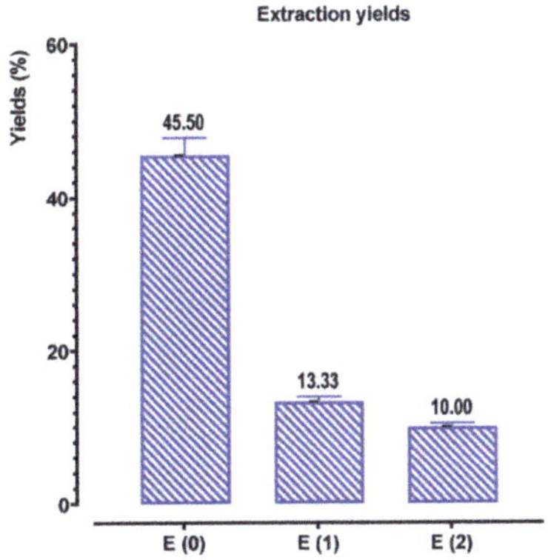


**Figure 3.** Extraction yields of phenolic compounds from *C. sativus*.

### 2.4.2. Determination of Phenolic Compounds

In order to evaluate the contents of polyphenols, flavonoids, condensed tannins, and hydrolyzable tannins in the aqueous extracts and the hydroethanolic extract of *C. sativus*, we established calibration curves for gallic acid (Y = 0.095X + 0.003; $R^2$ = 0.998), quercetin (Y = 0.073X − 0.081; $R^2$ = 0.995), catechin (Y = 0.7421X + 0.0318; $R^2$ = 0.998), and tannic acid (Y = 0.1700X − 0.0006718; $R^2$ = 0.996). The quantities of total polyphenols, flavonoids, condensed tannins, and hydrolyzable tannins in the extracts were estimated by milligram equivalents of gallic acid, quercetin, vanillin, and tannic acid per gram of extract, respectively.

The results of the total polyphenol assay show that the contents of these molecules vary considerably from one extract to another (Figure 4). The best contents recorded for *C. sativus* stigmas were noted for the hydroethanolic extract (45.474 mg EAG/g), followed by decocted and then the aqueous extract obtained by Soxhlet extraction. Flavonoid content values (Figure 4) show that *C. sativus* extracts are visibly rich in flavonoids with the best contents recorded for the hydroethanolic extract (33.497 mgEQ/g), followed by decocted (113.877 mgEQ/g) then the aqueous extract obtained by Soxhlet extraction (5.639 mgEQ/g). An analysis of the tannin results shows that *C. sativus* stigmas are rich in tannins (condensed and hydrolyzable tannins). The extracts obtained by Soxhlet extraction of *C. sativus* represent the highest tannin contents, while the decocted one has a lower content of condensed tannins and hydrolyzable tannins (0.558 mg EC/g and 0.364 mg EAT/g extracts).

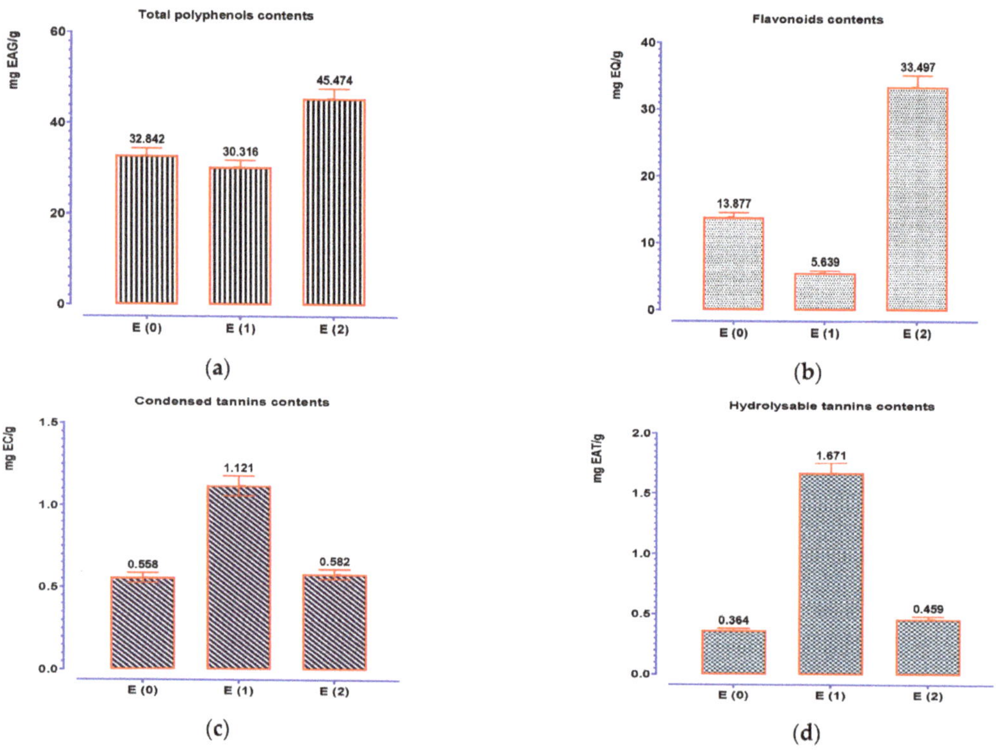


**Figure 4.** Contents of phenolic compounds: (**a**) total polyphenols; (**b**) flavonoids; (**c**) condensed tannins; (**d**) hydrolyzable tannins. Mean values ± standard deviations of determinations performed in triplicate are reported. Means are significantly different ($p < 0.001$).

2.4.3. Analysis and Identification of Polyphenols in *C. sativus* Extract by High-Pressure Liquid Chromatography–Mass Spectrometry (HPLC/UV-ESI-MS)

The extracts of *C. sativus* were analyzed by HPLC/UV-ESI-MS, and the chromatogram in Figure 5 presents the different compounds detected in the stigmas of *C. sativus*. The analysis of the mass spectra in addition to the chromatogram allowed us to propose 27 molecules which are recorded in Table 4.

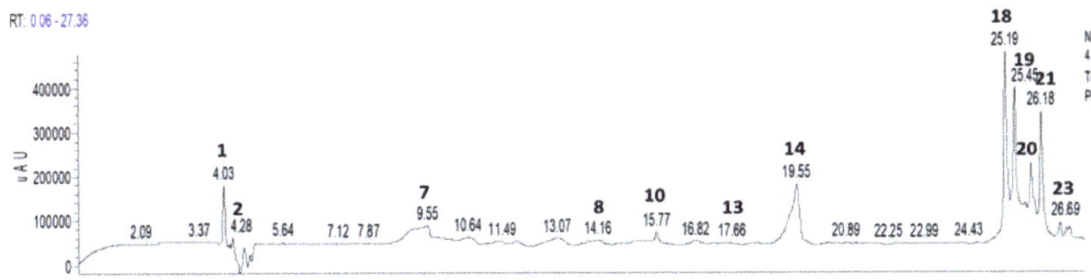


(a)

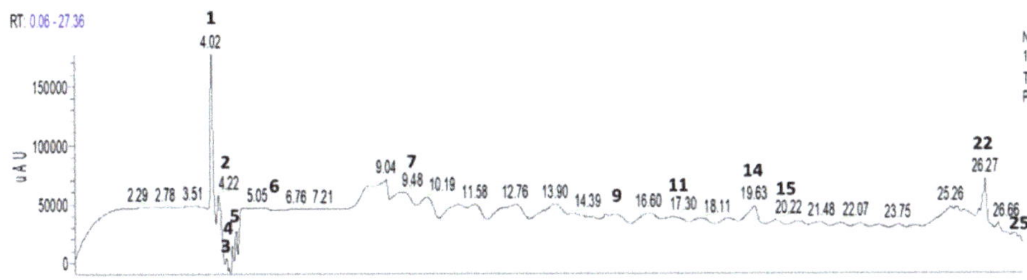


(b)

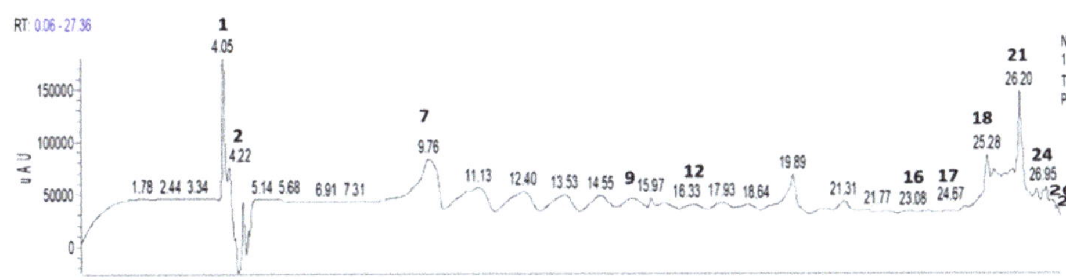


(c)

**Figure 5.** HPLC chromatograms of *C. sativus* compounds: (**a**) decocted; (**b**) aqueous extract obtained by Soxhlet extraction; (**c**) hydroethanolic extract obtained by Soxhlet extraction.

The analytical study of the mass spectra of the aqueous and hydroethanolic extracts of *C. sativus* in negative mode shows the presence of several carotenoids, flavonoids, and phenolic acids. The principal carotenoids identified in the decocted extract ($E_0$), aqueous extract obtained by Soxhlet extraction ($E_1$), and hydroethanolic extract obtained by Soxhlet extraction ($E_2$) are *cis*-crocin-3 (0.32%, 1.64%, and 1.48%), *cis*-crocin-4 (1.06%, 2.58%, and 4.12%), crocetin (0.57%, 0.16%, and 2.47%), crocin-2 (7.8%, 11.25%, and 24.58%), crocin-3 (25.65%, 10.1%, and 3.58%), crocin-4 (10.52%, 1.32%, and 17.99%), picrocrocin (18.78%, 30.23%, and 0.5%), *trans*-crocin-1 (0.62%, 2.34%, and 4.3%), and $\beta$-carotene (1.32%, 4.44%, and 7%), respectively. We also recorded the presence of safranal in the three extracts ($E_0$, $E_1$, and $E_2$) with respective percentages in the range of 0.78%; 3.83%; and 0.69%.

**Table 4.** List of compounds identified by mass spectrometry in the stigmas of *C. sativus*.

| N° Pics | TR (min) | Molecules | Classes | MS $[M\text{-}H]^-$ *m/z* | Exact Weights | Area (%) | | |
|---|---|---|---|---|---|---|---|---|
| | | | | | | E (0) | E (1) | E (2) |
| 1 | **4.04** | Safranal | Other | 149 | 150 | 0.78 | 3.83 | 0.69 |
| 2 | **4.28** | Caffeoyl coumaroyl methyl citric acid | Phenolic acid | 551-387 | 552 | 0.76 | 2.08 | 0.59 |
| 3 | **4.63** | Quinic malonyl glucoside acid | Other | 439 | 440 | 0 | 0.17 | 0.07 |
| 4 | **4.75** | Hydroxybenzoic acid hexoside | Polyphenol | 299-239-179 | 300 | 0.01 | 0.29 | 0.02 |
| 5 | **4.90** | Caffeic acid glycoside dimmer | Phenolic acid | 683 | 684 | 0.02 | 0.34 | 0.03 |
| 6 | **5.68** | Isorhamnetin 3-*O*-neohesperidoside | Flavonoide | 623 | 624 | 0.01 | 0.15 | 0.02 |
| 7 | **9.59** | Kaempferol 3-(6″-acetylglucoside) | Flavonoid | 489 | 490 | 0.63 | 1.77 | 1.02 |
| 8 | **14.16** | Syringetin hexoside | Flavonoid | 345-507 | 508 | 5.86 | 3.09 | 2.8 |
| 9 | **15.14** | Xanthoangelol | Flavonoid | 391 | 392 | 1.57 | 2.77 | 1.3 |
| 10 | **15.77** | Secoisolariciresinol | Polyphenol | 361 | 362 | 2.52 | 0 | 1.53 |
| 11 | **17.30** | Caffeic acid | Phenolic acid | 179 | 180 | 0.01 | 0.13 | 0.01 |
| 12 | **17.25** | Vanillic acid-dihexoside | Phenolic acid | 537 $[M + HCOOH\text{-}H]^-$ | 492 | 1.65 | 0 | 1.05 |
| 13 | **17.66** | α-tocopherol | Other | 429 | 430 | 0.66 | 0 | 0 |
| 14 | **19.55** | Picrocrocin | Carotenoid | 375 $[M\text{-}H + HFA]^-$ | 330 | 18.78 | 30.23 | 0.5 |
| 15 | **20.22** | Oleanolic acid | Fatty acid | 409 | 456 | 4.26 | 16 | 6.85 |
| 16 | **23.08** | Apigenin-*O*-rhamnoside | Flavonoid | 415 | 416 | 1.06 | 0.52 | 0.64 |
| 17 | **24.67** | Gallocatechin-pgd-3-*O*-glucoside | Flavonoid | 737 | 738 | 1.63 | 1.1 | 3.68 |
| 18 | **25.19** | *Trans*-crocin-4 (*trans*-crocetin di(b-*D*-gentiobiosyl) ester) | Carotenoid | 1021 $[M\text{-}H + HFA]^-$ | 976 | 10.52 | 1.32 | 17.99 |
| 19 | **25.45** | *Trans*-crocin-3 (*trans*-crocetin (b-*D*-glucosyl)-(b-*D*-gentiobiosyl) ester) | Carotenoid | 859 $[M\text{-}H + HFA]^-$ | 814 | 25.65 | 10.1 | 3.58 |
| 20 | **25.90** | Quercetin | Flavonoid | 347 $[M\text{-}H + HFA]^-$ | 301 | 11.1 | 2.18 | 4.12 |
| 21 | **26.18** | *Trans*-crocin-2 (*trans*-crocetin (b-*D*-gentibiosyl) ester) | Carotenoid | 651 | 652 | 7.8 | 11.25 | 24.58 |
| 22 | **26.27** | β-Carotene | Carotenoid | 581 $[M\text{-}H + HFA]^-$ | 536 | 1.32 | 4.44 | 7 |
| 23 | **26.69** | Coumaric acid-hexoside | Phenolic acid | 325 | 326 | 0.8 | 0 | 0 |
| 24 | **26.95** | *Cis*-crocin-4 | Carotenoid | 975 | 976 | 1.06 | 2.58 | 4.12 |
| 25 | **27.15** | *Cis*-crocin-3 | Carotenoid | 813 | 814 | 0.32 | 1.64 | 1.48 |
| 26 | **27.26** | *Trans*-crocin-1 | Carotenoid | 489 | 490 | 0.62 | 2.34 | 4.3 |
| 27 | **27.57** | Crocetin | Carotenoid | 327 | 328 | 0.57 | 0.16 | 2.47 |

Each compound can be mass fragmented using electrospray mass spectrometry (ESI-MS) detection in the negative ion mode, which is an additional technique for determining their structure.

The ESI-MS spectrum of picrocrocine ($C_{16}$ $H_{26}$ $O_7$, *m/z* = 330) showed an addition ion (with formic acid (HFA)) at *m/z* = 375 $[M\text{-}H + HFA]^-$.

The crocin-4, crocetin esterified with a gentiobiose at each end, *trans* and *cis* isomers (calculated as $C_{44}$ $H_{64}$ $O_{24}$, M = 976) produced the pseudo-molecular ion $[M\text{-}H]^-$ at *m/z* = 975 at 25.19 and 26.96 min, respectively. Upon adduct formation with formic acid in the mass spectrum of crocin-4, the peak formed with *m/z* = 1021 was identified as $[M\text{-}H + HFA]^-$.

The crocin-3 ($C_{38}$ $H_{54}$ $O_{19}$, M = 814), crocetin esterified with a gentiobiose at one end and glucose at the other end, *trans* and *cis* isomers showed the $[M\text{-}H]^-$ ion at *m/z* = 813. The use of formic acid as a mobile phase modifier gave an intense signal at *m/z* = 859 corresponding to the presence of adduct ion with crocin-3 $[M\text{-}H + HFA]^-$.

The crocin-2 ($C_{33}$ $H_{44}$ $O_{14}$, M = 652), a crocetin esterified with a gentiobiose at one extreme and a free carboxyl group at the other, showed the $[M\text{-}H]^-$ ion at *m/z* = 651 at 26.18 min. The compound has ion fragmentation similar to that of crocin-4 and crocin-3.

The $\beta$-carotene showed the $[M\text{-}H]^-$ ion at *m/z* = 535 at 26.27 min. On adduct formation with formic acid in the mass spectrum of $\beta$-carotene, the peak formed with *m/z* = 581 was identified as $[M\text{-}H + HFA]^-$.

The other ions at *m/z* = 489 $[M\text{-}H]^-$ correspond to the pseudomolecular ion of *trans*-crocin-1 and at *m/z* = 327 $[M\text{-}H]^-$ to the quasi-molecular ion of crocetin. Thus, the pseudo-molecular ions at *m/z* = 149 $[M\text{-}H]^-$ and *m/z* = 409 $[M\text{-}H]^-$ correspond to the molecules of safranal and oleanolic acid, respectively.

### 2.5. Heavy Metal Contents

The determination of heavy metal contents in the *crocus* genus has been the subject of very little work. In our study, a total of seven elements (As, Fe, Cd, Sb, Cr, Pb, and Ti) were determined. The *Crocus sativus* sample showed values below the FAO/WHO regulated limit values (Table 5).

**Table 5.** Heavy metal concentration (mg/L) and maximum limits FAO/WHO (2009).

| Species | Arsenic (As) | Cadmium (Cd) | Chromium (Cr) | Iron (Fe) | Lead (Pb) | Antimony (Sb) | Titanium (Ti) |
|---|---|---|---|---|---|---|---|
| *C. sativus* | 0.013 | ≤0.001 | 0.005 | 0.513 | 0.305 | 0.001 | 0.017 |
| *Maximum limits (FAO/WHO)* | 1 | 0.3 | 2 | 20 | 3 | 1 | - |

### 2.6. Antioxidant Activity

Three techniques—DPPH, FRAP, and TAC—were used to assess the antioxidant properties of the aqueous and hydroethanolic extracts of C. sativus and the reference (ascorbic acid). The calibration curves of ascorbic acid by DPPH (Y = 1.013X − 8.032; $R^2$ = 0.9893), FRAP (Y = 0.004760X + 0.09740; $R^2$ = 0.8963), and TAC (Y = 0.04066X + 0.02110; $R^2$ = 0.9949) methods were determined.

The extracts are referred to as natural antioxidants, due to their ability to reduce and/or prevent free radical formation. The results in Figure 6A show that the aqueous extracts and the hydroethanolic extract of *C. sativus* are endowed with antiradical power. The hydroethanolic extract ($E_2$) and the aqueous extract obtained by Soxhlet extraction ($E_1$) exerted remarkable antioxidant activity with $EC_{50}$ values equal to 565.590 and 627.003 μg/mL, respectively. The $EC_{50}$ of ascorbic acid was 19.378 μg/mL. The $EC_{50}$ of iron reduction in the *C. sativus* extracts is quite evident, according to the results of the FRAP method. Figure 6B shows significant differences between the reducing power of the extracts from *C. sativus* stigmas and the positive control (ascorbic acid). The aqueous extract obtained by Soxhlet extraction of *C. sativus* showed a better reducing capacity than the decocted extract (112.530 µg/mL) or the aqueous extract obtained by Soxhlet extraction (113.914 µg/mL). However, these results fall short of the ascorbic acid standard, which had a concentration of 0.470 g/mL. As the total antioxidant activity (TAA) is measured in terms of ascorbic acid equivalents, the phosphomolybdenum technique is quantitative. According to Figure 6C, the tested extracts showed good total antioxidant capacity. The results showed that the organic extracts presented the highest TAC compared to the decocted extract, with a high activity recorded for the aqueous extract (193.356 mg EAA/g) followed by the hydroethanolic extract (175.666 mg EAA/g). Moreover, the results grouped in Figure 7 present a perfect correlation between the phenolic compound contents of the studied extracts and the different antioxidant activities carried out, in particular between the contents of flavonoids, condensed tannins, and hydrolyzable tannins concerning DPPH, FRAP, and TAC assays.

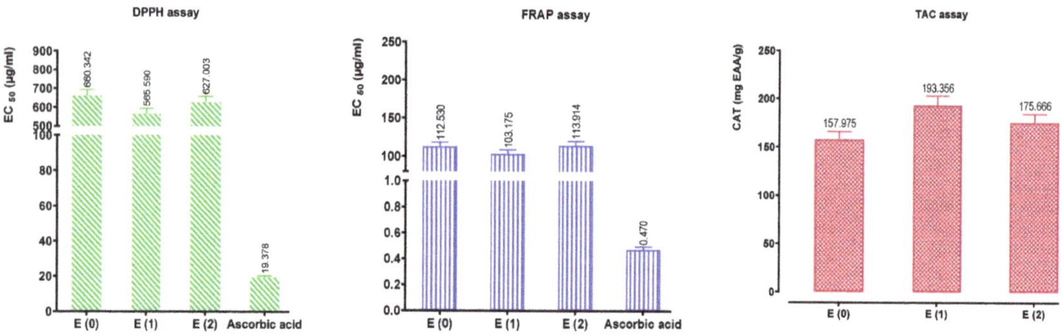


(**a**) (**b**) (**c**)

**Figure 6.** Antioxidant activity of ascorbic acid and extracts by: (**a**) DPPH; (**b**) FRAP; (**c**) TAC assays. Mean values ± standard deviations of determinations performed in triplicate are reported. Means are significantly different ($p < 0.001$).

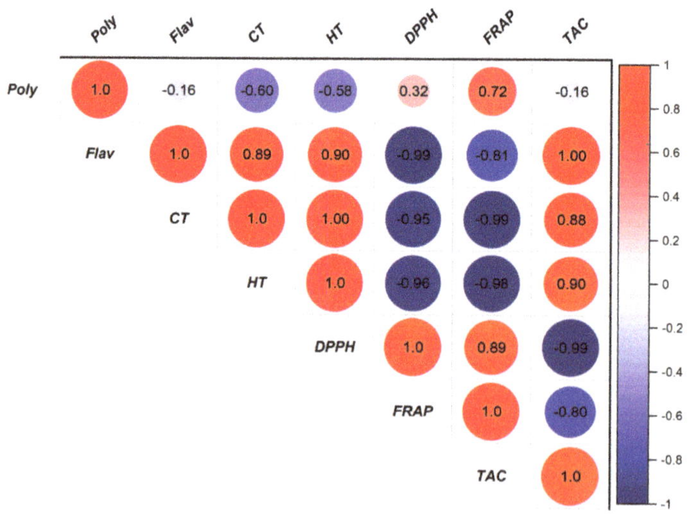


**Figure 7.** Correlation between antioxidant activities and phenolic compound contents of *C. sativus* stigma extracts. (Poly: polyphenols; Flav: flavonoids; CT: condensed tannins; HT: hydrolyzable tannins).

### 2.7. Antimicrobial Activity

The results of the antimicrobial activity of decocted *C. sativus* extract are represented in Table 6. The MIC of the extract was categorized using the standards put forth by Sartoratto, Duarte, Wang, Oliveira, and their associates [14–17]. Antimicrobial activity was classified as high (MIC < 600 µg/mL), moderate (MIC between 600 and 2500 µg/mL), and low (MIC > 2500 µg/mL). The MIC, BMC, and FMC analyses of *C. sativus* show a high bactericidal quality of the decocted extract against enterobacterial species. The other strains tested show some resistance. Indeed, the evaluated decocted extract is powerful against Gram-negative bacilli and, in particular, against *A. baumannii* and *Shigella* sp. with MIC lower than or equal to 600 µg/mL. Moreover, the decocted extract of *C. sativus* is more active against strains causing candidiasis, in particular against *C. kefyr* and *C. parapsilosis* with a MIC of about 2500 µg/mL. The mold (*A. niger*) was also sensitive to the decocted extract studied (MIC = 2500 µg/mL).

### 2.8. Anticoagulant Activity

According to the data in Figure 8, it is apparent that the studied extract ($E_0$) does not affect prothrombin time (PT). However, the results of the coagulation test (aPPT) showed a remarkable anticoagulant activity by inhibiting the intrinsic pathway in a dose-dependent manner. Indeed, *C. sativus* extract ($E_0$) can significantly prolong the partial thromboplastin time ($p < 0.001$) with a 359 µg/mL concentration.

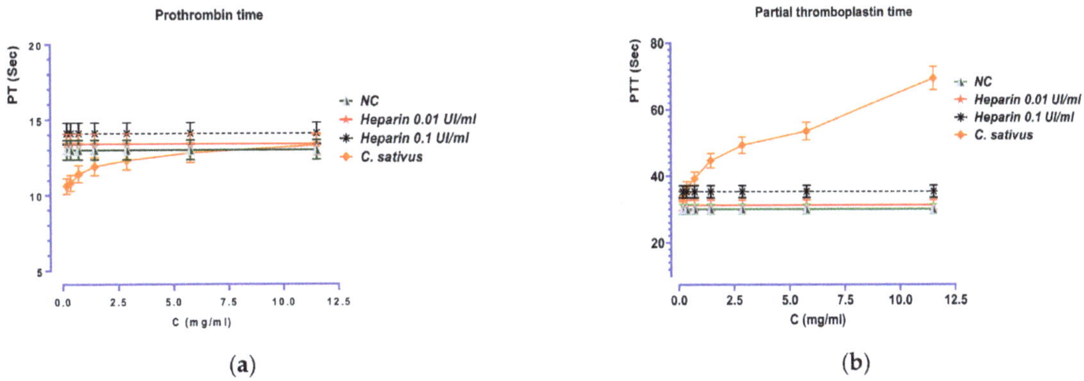


**Figure 8.** Effect of *C. sativus* decocted extract ($E_0$), normal control (NC), and heparin on prothrombin time (**a**) and partial thromboplastin time (**b**).

### 2.9. Antidiabetic Activity

#### 2.9.1. Evaluation of the Inhibitory Effect of Decocted Extract on the Activity of α-Amylase and α-Glucosidase, In Vitro

Regarding the in vitro inhibition tests of α-amylase and α-glucosidase by aqueous extract of *C. sativus*, the results presented in Figure 9 show that the extract exerts a variable and dose-dependent inhibitory effect. *C. sativus* decocted extract as well as acarbose showed strong inhibitory activity on α-amylase and α-glucosidase. The effect of acarbose against α-amylase and α-glucosidase was increased in a concentration-dependent manner, with $EC_{50}$ of 364.446 and 17.269 µg/mL, respectively. The decocted extract's effect was significantly higher than that of acarbose on α-amylase and α-glucosidase (86.326 mg/mL; 10.957 µg/mL, respectively).

**Table 6.** The MIC, MBC, and MFC (µg/mL) of *C. sativus* decocted extract, and the MIC of antibiotics and antifungal agent.

| | Microorganism | *C. sativus* | | *Antibiotics* * | | | | | *Antifungals* # |
|---|---|---|---|---|---|---|---|---|---|
| | | *MIC* | *MBC or MFC* | *Gentamycin* | *Amoxicillin–Clavulanate* | *Vancomycin* | *Trimethoprim–Sulfamethoxazole* | *Penicillin G* | *Terbinafine* |
| GPC | *S. epidermidis* | >5000 | >5000 | 2 | | >8 | >4/76 | | |
| | *S. aureus BLACT* | 5000 | 5000 | <0.5 | | 2 | <10 | | |
| | *S. aureus STAIML/MRS/mecA/HLMUP/BLACT* | >5000 | >5000 | 2 | | >8 | >4/76 | | |
| | *S. acidominimus* | 5000 | 5000 | ≤250 | | <0.5 | | 0.03 | |
| | *S. group D* | >5000 | >5000 | >1000 | | <0.5 | | 0.13 | |
| | *S. agalactiae (B)* | 5000 | 5000 | ≤250 | | >4 | | 0.06 | |
| | *S. porcinus* | >5000 | >5000 | ≤250 | | <0.5 | | 0.06 | |
| | *E. faecalis* | 2500 | 2500 | ≤500 | | 1 | ≤0.5/9.5 | | |
| | *E. faecium* | >5000 | >5000 | ≤500 | | >4 | >4/76 | | |
| GNB | *A. baumannii* | 600 | 1200 | ≤1 | ≤2/2 | | ≤1/19 | | |
| | *E. coli* | >5000 | >5000 | 2 | 8/2 | | ≤1/19 | | |
| | *E. coliESBL* | >5000 | >5000 | 2 | >8/2 | | >4/76 | | |
| | *E. aerogenes* | >5000 | >5000 | ≤1 | 8/2 | | ≤1/19 | | |
| | *E. cloacae* | >5000 | >5000 | >4 | >8/2 | | >4/76 | | |
| | *C. koseri* | 2500 | 2500 | <1 | >8/2 | | <20 | | |
| | *K. pneumoniae* | 5000 | 5000 | ≤1 | ≤2/2 | | ≤1/19 | | |
| | *P. mirabilis* | 2500 | 2500 | 2 | ≤2/2 | | >1/19 | | |
| | *P. aeruginosa* | >5000 | >5000 | 2 | >8/2 | | 4/76 | | |
| | *P. fluorescence* | 2500 | 2500 | 4 | >8/2 | | 4/76 | | |
| | *P. putida* | >5000 | >5000 | >4 | >8/2 | | >4/76 | | |
| | *S. marcescences* | 5000 | 5000 | 4 | >8/2 | | >4/76 | | |
| | *Sallemonella* sp. | 2500 | 2500 | >4 | 8/2 | | >4/76 | | |
| | *Shigella* sp. | 600 | 1200 | >4 | 8/2 | | >4/76 | | |
| | *Y. enterolitica* | 2500 | 2500 | ≤1 | 8/2 | | 2/38 | | |

**Table 6.** *Cont.*

| | Microorganism | C. sativus | | Antibiotics * | | | | | Antifungals # |
|---|---|---|---|---|---|---|---|---|---|
| | | MIC | MBC or MFC | Gentamycin | Amoxicillin–Clavulanate | Vancomycin | Trimethoprim–Sulfamethoxazole | Penicillin G | Terbinafine |
| Yeasts | *C. albicans* | >5000 | >5000 | | | | | | 12.500 |
| | *C. kefyr* | 2500 | 2500 | | | | | | 25.000 |
| | *C. krusei* | >5000 | >5000 | | | | | | 50.000 |
| | *C. parapsilosis* | 2500 | 2500 | | | | | | 6.250 |
| | *C. tropicalis* | >5000 | >5000 | | | | | | 12.500 |
| | *C. dubliniensis* | >5000 | >5000 | | | | | | 3.125 |
| | *S. cerevisiae* | >5000 | >5000 | | | | | | 3.125 |
| Molds | *A. niger* | 2500 | 2500 | | | | | | 3.125 |

*: the MIC (μg/mL) of the antibiotics was determined by the BD Phoenix™ identification and antibiogram instrument; #: the MIC (μg/mL) of terbinafine was determined on a microplate.

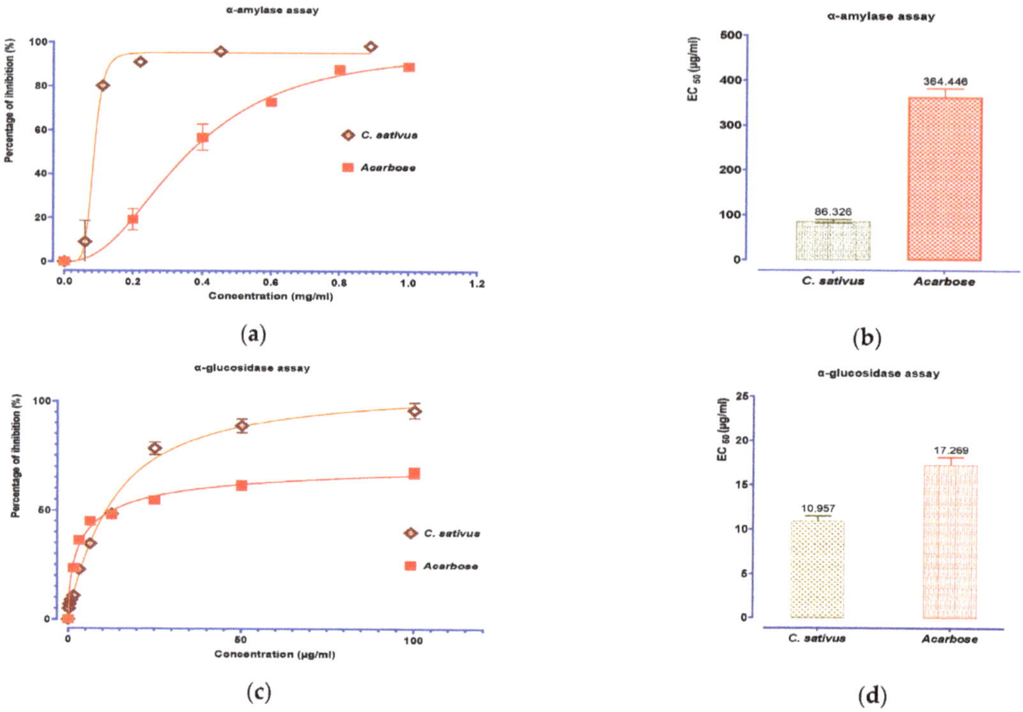


**Figure 9.** Percent inhibition and $EC_{50}$ of inhibitory effects on α-amylase (**a**,**b**) and α-glucosidase (**c**,**d**) activities by *C. sativus* decocted extract and acarbose, in vitro. Values are means ± SEM (n = 3).

### 2.9.2. Acute Toxicity Study of *C. sativus* Decocted Extract

The result of this acute toxicity assay shows that the decocted material is not toxic even at 2 g/kg. Neither toxicity (diarrhea, vomiting, abnormal mobility, etc.) nor mortality were caused by the decocted substance during the observation period.

### 2.9.3. Study of the Antihyperglycemic Activity of *C. sativus* Decocted Extract in Normal Rats In Vivo

Analysis of the glucose tolerance test and comparison of the total areas under the blood glucose curve during the 150 min period are shown in Figures 10 and 11.

- Oral glucose tolerance test
- Blood glucose levels in normal rats showed a high peak 30 min after glucose loading. A positive effect on the response of rats to glucose loading was noted in rats treated with decocted extract and glibenclamide. Oral administration of *C. sativus* extract at a dose of 400 mg/kg, 30 min before glucose overload, to normal rats significantly attenuated postprandial hyperglycemia for this decocted extract study (Figure 10), compared with the group of control rats pretreated with distilled water. However, glibenclamide very significantly inhibited postprandial hyperglycemia during the first hour (60 min) after glucose overload, ($p < 0.001$; 1.08 g/L) compared with the distilled water-pretreated rat group.
- Areas under the curve (AUCs) of postprandial glucose levels.
- The area under the curve was significantly smaller in the decocted extract-treated rats ($p \geq 0.001$; 56.11 g/L/h) than in the distilled water-treated rats (62.91 g/L/h). In addition, the area under the curve of glibenclamide was significantly ($p \geq 0.01$) smaller

(55.95 g/L/h) compared with the area under the curve of distilled water-treated rats (62.91 g/L/h) (Figure 11).

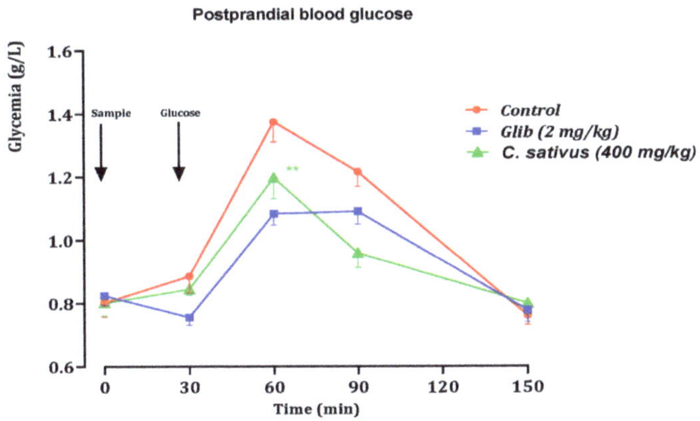


**Figure 10.** Variation in postprandial blood glucose in normal rats after administration of test products (decocted extract and glibenclamide). Values are means ± SEM. (n = 6). ** $p < 0.01$ in comparison with the control.

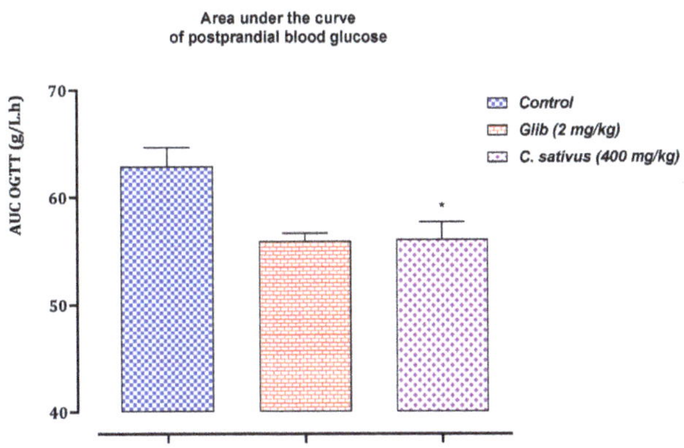


**Figure 11.** Variation in the area under the curve of postprandial blood glucose in normal rats after administration of tested products (decocted extract and glibenclamide). Values are means ± SEM. (n = 6). * $p < 0.05$ when compared with control.

## 3. Discussion

More than 100 volatile chemical compounds are found in saffron after processing the scent precursors and dehydrating the stigmas [18]. Monoterpenes and sesquiterpenes obtained from the isoprenoid synthesis pathway, phenylpropanoids and benzenoids derived from the shikimic acid pathway, and chemicals derived from the enzymatic conversion of lipids by $\beta$-oxidation are the several types of aromatic compounds found in saffron. A few volatile chemicals are also created by shortening or altering the skeletons of other molecules [19]. Safranal, which makes up about 0.001–0.006% of the dry matter and is responsible for 30–70% of the saffron's scent, is the main constituent [20]. Picrocrocine is converted into safranal through an enzyme reaction and/or dehydration. However, there is proof that this conversion can take place at low pH levels.

The yield obtained in EO from *C. sativus* stigmas in our work remains low and comparable to the yields of saffron harvested in Mexico by Cid-Pérez [21] and in Kashmir, Iran, and Turkey by Kafi [22]. Previous studies on the chemical composition of *C. sativus*

EO show similarities and variations in major constituents. Using GC-MS/FID, Anastasaki et al. [23] identified the following significant chemicals in samples of *C. sativus* from Italy and Spain: safranal, 4-hydroxy-2,6,6-trimethyl-3-oxocyclohexa-1,4-diene-1-carboxaldehyde, isophorone, and dihydrooxophorone. The same substances, albeit in a different sequence of concentration, were found in samples from Iran and Greece. Aliakbarzadeh et al. [24] used GC-MS to identify 77 volatile compounds of Iranian saffron, of which 10 are biomarkers, incuding 8 secondary metabolites (isophorone; phenylethyl alcohol; phorone; keto-isophorone; dihydrooxophorone; safranal; 2,6,6-trimethyl-4-oxo-2-cyclohexene-1-carbaldehyde; 2,4,4-trimethyl-3-carboxaldehyde-5-hydroxy-2,5-cyclohexadiene-1-one), 1 is a primary metabolite (linoleic acid), and 1 is a long chain fatty alcohol (nonacosanol). Azarabadi and Özdemir [25] used solid phase microextraction (SPME) to extract and determine the significant compounds from Iran, including acetic acid; 2-(5H)-furanone; isophorone; 4-keto-isophorone; 2,6,6-trimethyl-1,4-cyclohexanedione; eucarvone; and safranal. Similarly, Carmona et al. [26] determined high concentrations of acetic acid, dihydrooxophorone, and isophorone in samples from Morocco and Iran as markers to distinguish their origin. This polymorphism of the identified constituents varies according to the time of harvesting, seasonal factor, place of origin, and the condition of the plant, fresh or dry. Thus, the concentration of compounds of interest may vary according on the preparation, extraction, and characterization processes [24].

The results of phenolic compound assays indicated that the content of phenolic compounds in the studied extracts showed a significantly higher variation in the aqueous extract and hydroethanolic extract obtained by Soxhlet extraction compared to the decocted extract. The total contents of phenolic compounds, flavonoids, and tannins in *C. sativus* stigmas were higher compared to those reported by the work of Karimi et al. in Iran [27]. Additionally, this work offers the first account of a comparison of three distinct extracts made via decoction, Soxhlet extraction, and ethanol/water mixture. These data confirm that *C. sativus* stigmas harvested from the Boulemane region are rich in total polyphenols, flavonoids and tannins, and could therefore show important biological and pharmacological activities. In addition, the results of the chromatogram and mass spectral analysis (HPLC/UV-ESI-MS) obtained were in agreement with earlier studies in a significant way [28]. The provided information offers trustworthy confirmation of the mass spectral analysis-based structures of crocins, picrocrocin, $\beta$-carotene, crocetin, and safranal. Mineral elements are involved in important biological functions of cells. The contents of heavy metals (As, Cd, Cr, Fe, Pb, Sb, and Ti) determined in *C. sativus* powder were found to be below the standards described by the FAO/WHO. This will allow safe and healthy uses of the studied plant in the fields of pharmacy, food, and cosmetics.

Concerning the evaluation of the antioxidant potential of *C. sativus* extracts, we measured the antioxidant activities using three methods (DPPH, FRAP, and TAC). The results of these activities showed the richness of the studied extracts of *C. sativus* in antioxidant molecules. Several previous works have focused on the antioxidant effects of *C. sativus* flower, petal, and leaf extracts, however, we found that few studies have been reported on the antioxidant activity of stigma extracts [29–31]. The outcomes of our research demonstrate that the antioxidant potential of the *C. sativus* stigma extracts from the Boulemane region is comparable to the information gathered by Assimopoulou et al. [32], who presented studies on the antioxidant activity of saffron. They estimated that the antioxidant activity of saffron could be attributed to two bioactive compounds, crocin and safranal. According to Chen et al. [33], who compared the antioxidant activity of Chinese saffron (*C. sativus*) with that of Gardenia jasminoides, Chinese saffron had greater free radical scavenging abilities than Gardenia, with corresponding values between 107 and 421 mg tocopherol/g. In the present work, the antioxidant activity of saffron stigma extracts could be attributed to the presence and synergic effects of phenolic compounds (total polyphenols, flavonoids, and tannins) or other bioactive molecules, as affirmed by the correlation test between antioxidant activities and phenolic compound contents described in Figure 7. Moreover, the difference between the antioxidant activity of the extracts studied in this work and those from China could be due to the difference in their polyphenol, flavonoid, and tannin contents and experimental conditions (plant, standard, etc.). Indeed, many studies have previously mentioned that phenolic compounds have remarkable antioxidant properties in both in vitro and in vivo systems [34–36].

Up to now, some global studies have reported work on the antimicrobial activity of *C. sativus* extracts. Jadouali and co-workers demonstrated that *C. sativus* leaf extracts had strong antibacterial activity against *Listeria* sp. in a prior investigation carried out in Morocco [29]. Vahidi and his team investigated the antimicrobial efficacy of *C. sativus* extracts prepared with various solvents and harvested in Khorasan, Iran, against various bacterial and fungal strains [37]. Asgarpanah and co-workers studied the antibacterial properties of methanolic extract made from the various portions of *C. sativus* that were harvested from the southern Iranian province of Khorasan. They focused on the extract's ability to inhibit different types of bacteria and fungi [38]. Aqueous and ethanolic extracts of *C. sativus* against mammitis pathogens in Turkey showed an intriguing antibacterial action [39]. Additionally, Muzaffar demonstrated the potent antibacterial properties of *C. sativus* extracts made from India-sourced methanol and petroleum ether [40]. Furthermore, our findings demonstrated that *C. sativus* decocted extract is superior to Gram-positive bacteria in its ability to combat *Aspergillus niger*, candidiasis, and Gram-negative pathogens. This difference could be due to the difference in the cell wall structure of the tested microorganisms. Moreover, our results showed that the antibacterial activity of *C. sativus* extracts from Morocco is similar to those from Iran and Turkey [37,38].

By forming a platelet thrombus, the normal hemostatic process halts a cut or wound on the blood vessels; once healing is complete, the thrombus is eventually removed. This complex multiphasic mechanism depends on platelets and coagulation factors interacting with blood arteries. Thrombosis or hemorrhage can result from a flaw in any of these stages [41]. In this study, the decocted extract showed a clear difference between coagulation and anticoagulation by prolonging the partial thromboplastin time (aPPT) in a more significant manner. Earlier research with saffron demonstrated that crocin might inhibit thrombosis in rats, delay blood clotting time and relieve respiratory discomfort during pulmonary thrombosis in mice, and decrease platelet aggregation in rabbits [42]. Hence, crocetin effectively inhibited platelet aggregation caused by collagen and ADP, but not by arachidonic acid [43]. Furthermore, while neither platelet adherence to collagen nor cyclic AMP levels were impacted by crocetin, it greatly reduced the release of dense granules [44]. After this, administration of saffron tablets (200 and 400 mg/day) failed to significantly affect the coagulation and anticoagulation system after one month in a double-blind, placebo-controlled clinical investigation with a large sample size. The authors suggested that the case reports of bleeding complications could be due to the high dose of saffron, the long period of consumption, or idiosyncratic activities [45].

The results of the antidiabetic activity of the decocted extract studied against digestive enzymes showed higher inhibitory activities against α-glucosidase and α-amylase compared to acarbose and compared to other published studies [46]. The traditional usage of *C. sativus* aqueous extract in the treatment of diabetes is supported by its capacity to inhibit the enzymes α-glucosidase and α-amylase, which suggests that substances with antidiabetic activity are extracted into the water. Since limiting the rate of glucose absorption from the intestines into the bloodstream was thought to be the only method that could help prevent diabetes, numerous prior studies have shown the hypoglycemic effect of phenolic compounds, improving postprandial blood glucose, acute insulin secretion, and insulin sensitivity [47].

## 4. Materials and Methods

### *4.1. Vegetal Material*

The studied sample of *Crocus sativus* was collected from the cultivated stands in the Boulemane region. The data on the origin, the part collected, and the site of harvesting are shown in Table 7. The Scientific Institute of Rabat's Laboratory of Botany and Plant Ecology is where identification of this species was completed.

**Table 7.** Origin, part used, location, and harvesting period of *Crocus sativus*.

| Scientific Name | Part Collected | Type of Extract Used | Harvesting Area | | | | | | |
|---|---|---|---|---|---|---|---|---|---|
| | | | Region | Province | Municipality | Latitude (x) | Longitude (y) | Altitude (m) | Collection Period |
| *Crocus sativus* L. | Stigma | EO and extract | Fez-Meknes | Boulemane | Serghina | 33°20′44″ N | 4°24′11″ W | 1496 m | September 2019 |

### 4.2. Microbial Materials

The determination of the antimicrobial activity of the aqueous extract of *C. sativus* was performed on twenty-four bacterial and eight fungal strains (Table 8). These particular microorganisms are harmful and well-known for their strong resistance, ability to invade, and toxicity toward humans. They are frequently encountered in many infections in Morocco that represent a clinical and therapeutic problem. These strains were isolated from the hospital environment: Mohamed V-Meknes Provincial Hospital. All strains were kept in a 20% glycerol stock at −80 °C, rejuvenated in Mueller–Hinton and Sabouraud broths, and subcultured before use.

**Table 8.** List of bacterial and fungal strains tested with their references.

| | Strains | Abbreviations | References |
|---|---|---|---|
| Gram-positive cocci | *Staphyloccocus epidermidis* | *S. epidermidis* | 5994 |
| | *Staphyloccocus aureus BLACT* | *S. aureus BLACT* | 4IH2510 |
| | *Staphyloccocus aureus STAIML/MRS/mecA/HLMUP/BLACT* | *S. aureus STAIML/MRS/mecA/HLMUP/BLACT* | 2DT2220 |
| | *Streptococcus acidominimus* | *S. acidominimus* | 7DT2108 |
| | *Streptococcus group D* | *S. group D* | 3EU9286 |
| | *Streptococcus agalactiae* | *S. agalactiae* | 7DT1887 |
| | *Streptococcus porcinus* | *S. porcinus* | 2EU9285 |
| | *Enterococcus faecalis* | *E. faecalis* | 2CQ9355 |
| | *Enterococcuss faecium* | *E. faecium* | 13EU7181 |
| Gram-negative bacilli | *Acinetobacter baumannii* | *A. baumannii* | 7DT2404 |
| | *Escherichia coli* | *E. coli* | 3DT1938 |
| | *Escherichia coliESBL* | *E. coli ESBL* | 2DT2057 |
| | *Enterobacter aerogenes* | *E. aerogenes* | 07CQ164 |
| | *Enterobacter cloacae* | *E. cloacae* | 02EV317 |
| | *Citrobacter koseri* | *C. koseri* | 3DT2151 |
| | *Klebsiella pneumonie ssp. pneumonie* | *K. pneumonie* | 3DT1823 |
| | *Proteus mirabilis* | *P. mirabilis* | 2DS5461 |
| | *Pseudomonas aerogenosa* | *P. aerogenosa* | 2DT2138 |
| | *Pseudomonas fluorescence* | *P. fluorescence* | 5442 |
| | *Pseudomonas putida* | *P. putida* | 2DT2140 |
| | *Serratia marcescens* | *S. marcescens* | 375BR6 |
| | *Salmonella* sp. | *Salmonella* sp. | 2CG5132 |
| | *Shigella* sp. | *Shigella* sp. | 7DS1513 |
| | *Yersinia enterocolitica* | *Y. enterocolitica* | ATCC27729 |

**Table 8.** *Cont.*

| | Strains | Abbreviations | References |
|---|---|---|---|
| Yeasts | *Candida albicans* | *C. albicans* | Ca |
| | *Candida kefyr* | *C. kefyr* | Cky |
| | *Candida krusei* | *C. krusei* | Ckr |
| | *Candida parapsilosis* | *C. parapsilosis* | Cpa |
| | *Candida tropicalis* | *C. tropicalis* | Ct |
| | *Candida dubliniensis* | *C. dubliniensis* | Cd |
| | *Saccharomyces cerevisiae* | *S. cerevisiae* | Sacc |
| **Fungi** | *Aspergillus niger* | *A. niger* | AspN |

### 4.3. Animal Selection for Research

The acute toxicity study was performed on albino mice (male and female). The study of the antidiabetic activity in vivo was carried out on Wistar rats (males and females). The animals were kept in an environment with a photoperiod of 12 h of light and 12 h of darkness and a temperature of 22 ± 2 °C in the Biology Department's animal home at the Faculty of Sciences in Oujda. The animals were maintained under favorable rearing conditions with free access to water and food. The animals were cared for, used, and handled in full compliance with internationally accepted standard guidelines and the institutional animal ethics committee (02/2019/LBEAS) [48].

### 4.4. The Qualitative and Quantitative Study of Essential Oils

#### 4.4.1. Extraction of Essential Oils from *C. sativus* and Determination of Yield

Essential oils (EOs) were extracted from *C. sativus* stigmas by hydrodistillation using a Clevenger-type apparatus. Briefly, 20 g of plant material was boiled for three hours with 200 mL of distilled water to produce three distillations. Then, the obtained oil was dried by adding anhydrous sodium sulfate ($Na_2SO_4$) and stored at a temperature of −4 °C in a dark bottle until its use. The yield of EO was calculated from 20 g of the plant material by Formula (1) [49]:

$$\text{Yield (\%)} = \frac{\text{W (EO)}}{\text{W0}} \times 100 \quad (1)$$

where: W (EO): weight of HE recovered (g); W0: weight of plant material (20 g).

#### 4.4.2. Analysis and Identification of the Chemical Composition of *C. sativus* EO

The sample of the studied essential oil was subjected to chromatographic analysis using a gas chromatograph of the Thermo Electron (Trace GC Ultra) type in conjunction with a mass spectrometer of the Thermo Electron Trace MS system (Thermo Electron: Trace GC Ultra; Polaris Q MS). The fragmentation was accomplished by an electron impact of intensity of 70 eV. The chromatograph was outfitted with a flame ionization detector (FID) fed by an H2/air–gas mixture and a DB-5 type column (5% phenyl-methyl-siloxane) (30 m × 0.25 mm × 0.25 m film thickness). The column temperature was set to increase by 4 °C/min for 5 min, from 50 to 200 °C. The carrier gas utilized was nitrogen at a flow rate of 1 mL/min in the split injection mode (leak ratio: 1/70).

The identification of the chemical composition of the EO was performed by determining and comparing the Kovats indices (KIs) of the compounds with those of the known standard products described in the databases of Kovats [50], Adams [51], and Hübschmann [52]. The retention times of the peaks were compared to those of known authentic standards kept in the authors' lab, and their reported KI and MS data were compared to those kept in the WILEY and NIST 14 standard mass spectral database and published literature to identify every compound. This was carried out using the Kovats index. The retention times of any products were compared using Kovats indices to the retention times of linear alkanes with the same number of carbons. They were determined by co-injecting a mixture of alkanes (standard $C_7$-$C_{40}$) under the same operating conditions.

### *4.5. Phytochemical Screening*

In this qualitative investigation, chemical families were discovered using tests for the solubility of compounds, precipitation, and turbidity responses. It can also be carried out by looking for a specific color change or by carrying out an examination under UV light. The stigmas of *C. sativus* were used for this phytochemical investigation. Dry samples of the plant were ground into a fine powder, followed by characterization tests of different chemical groups, carried out according to the protocols of Dohou et al., Judith, Mezzoug et al., Bekro et al., Bruneton, and N'Guessan et al. [53–58].

### *4.6. Study of Phenolic Compounds*

#### 4.6.1. Extraction of Phenolic Compounds

Extraction of phenolic compounds was performed using two methods: decoction and solid–liquid extraction by Soxhlet apparatus. The first sample of 30 g was added to 600 mL of distilled water, heated, and boiled at 80 °C for 1 h. The mixture was allowed to sit for five minutes before being filtered at a lower pressure. The decocted extract was dried in an oven at 70 °C and then recovered as a powder in a glass vial and stored until use. The Soxhlet apparatus was used to extract the second and third samples, each weighing 30 g, using 300 mL of water or an ethanol/water solution (70/30, *v*/*v*) as the extraction solvent. After different extraction cycles, the extracts were concentrated using a rotary evaporator. Table 9 below describes the coding adopted for the extracts prepared in this work.

**Table 9.** Extraction coding.

| Extraction Methods | Solvents | Codification |
|---|---|---|
| **Soxhlet** | Ethanol/Water (70/30; *v*/*v*) | E (2) |
| | Water | E (1) |
| **Decoction** | Water | E (0) |

#### 4.6.2. Determination of Total Polyphenols

The Folin–Ciocalteu method, as described by Singleton and Rossi [59], was used to determine the total polyphenol content of the plant extracts under study. The methods of their determination are generally based on the oxidation of these compounds and the development of color. The Folin–Ciocalteu technique is the most popular. It consists of the reduction of a mixture of phosphotungstic ($H_3PW_{12}O_{40}$) and phosphomolybdic ($H_3PMo_{12}O_{40}$) acids (Folin–Ciocalteu reagent) into a mixture of blue oxides of tungsten ($W_8O_{23}$) and molybdenum ($Mo_8O_3$). The assay of the compounds is carried out by colorimetry with an optical density reading. Using a spectrophotometer (UV mini-1240) set to 760 nm, the absorbance reading was taken in comparison to a blank (a reaction mixture without extract). Gallic acid was used as a positive control from a concentration range of 50 µg/mL in a parallel calibration curve that was made under the same operating circumstances. The results are expressed as the milligram equivalent of gallic acid per gram of extract (mg GAE/g) and were determined by the equation of type Y = a x + b obtained with the calibration curve. Each test was repeated three times.

#### 4.6.3. Determination of Flavonoids

The flavonoid content was determined by the colorimetric method with aluminum trichloride by adaptation of the methods of Djeridane [60] and Hung [61] and their coworkers. Aluminum chloride ($AlCl_3$) forms a complex with the hydroxyl groups (OH) of flavonoids. The flavonoids were estimated by UV spectroscopy at a wavelength of 433 nm. Quercetin, a standard flavonoid that experienced the same analytical conditions as the samples and had concentrations between 5 and 30 µg/mL, was used to generate a calibration curve (type Y = a x + b) for the measurement of flavonoids. The flavonoid content is expressed as milligram equivalent of quercetin per gram of extract (mg EQ/g). Each test was repeated three times.

#### 4.6.4. Determination of Condensed Tannins

Condensed tannin contents were estimated by the vanillin method [62]. Briefly, different prepared concentrations of (+)-catechin solution (2 mg/mL) were added to a volume of 3 mL of vanillin/methanol solution (4%; *m*/*v*). The mixture was manually stirred. Then, 1.5 mL of strong hydrochloric acid was added to each concentration. The resulting mixes were allowed to react for 20 min at room temperature. With the aid of a UV–visible spectrophotometer, the absorbance at 499 nm was determined in comparison to a blank. By substituting our samples for the catechin in the calibration curve plotting process, the quantity of condensed tannins in our samples was determined. The calibration curve was used to quantify the tannin concentration in milligram equivalent of catechin per gram of dry matter weight.

#### 4.6.5. Determination of Hydrolyzable Tannins

Hydrolyzable tannins were determined by the method of Willis and Allen [63], with minimal modifications. The extract was vortexed for ten seconds with 5 mL of 2.5% $KIO_3$ in a mixture of 10 µL of the extract. After 2 min for the extract and the reaction's optimum, and 4 min for the standard tannic acid solution, the maximum absorbance was reached. A spectrophotometer was used to measure the absorbance at 550 nm (UV–visible). The results are expressed as mg of tannic acid per gram of dry plant, and 11 different tannic acid concentrations (ranging from 100 to 2000 g/mL) were used to elaborate the calibration curve.

#### 4.6.6. HPLC/UV ESI-MS Analysis of *C. sativus* Stigma Extracts

Analysis of phenolic compounds of *C. sativus* decocted by high-performance liquid chromatography coupled to Q Exactive Plus mass spectrometry with electrospray as a molecular ionization method (HPLC/UV-ESI-MS) was performed with an UltiMate 3000 HPLC (Thermo Fisher Scientific, Sunnyvale, CA, USA) equipped with a sample changer, in which the samples were stored at 5 °C. This HPLC system was equipped with a reverse phase $C_{18}$ column (250 × 4 mm, id 5 µm, Lichro CART, Lichrospher, Merck, Darmstadt, Germany). During the analysis, the column temperature was set at 40 °C. The mobile phase degassed by ultrasonic treatment was: solvent A: 0.1% formic acid in water (*v*/*v*) and solvent B: 0.1% formic acid in acetonitrile (*v*/*v*). The gradient composition was: 2% B at the beginning (0 min) and then changed to 30%, 95%, 2%, and 2% B at 20, 25, 26, and 30 min, respectively. The flow rate was 1 mL/min and the injection volume was 20 µL.

Detection was performed on a Maxis Impact HD (Bruker Daltonik, Bremen, Germany) in MS/MS mode (broadband collision-induced dissociation (bbCID)) after negative electrospray ionization. In addition, UV detection by an L-2455 diode array detector (Merck-Hitachi, Darmstadt, Germany) was performed by scanning in the wavelength range of 190–600 nm and then at three acquisition wavelengths, 280 nm, 320 nm, and 360 nm. The values of the other parameters were set as follows: capillary voltage of 3000 V; drying gas temperature 200 °C; dry gas flow rate 8 L/min; a nebulizing gas pressure of 2 bar; and an offset plate of −500 V. Nitrogen was used as the desolvation gas and nebulizer gas. MS data were acquired over an *m/z* range of 100 to 1500. A Thermo Scientific™ Chromeleon™ 7.2 Chromatography Data System (CDS) was used as software for data acquisition and evaluation. The analysis of the eluted compounds was carried out by analyzing the mass spectra of the eluted molecules.

### 4.7. *Heavy Metal Assays: Inductively Coupled Plasma Atomic Emission Spectrometry (ICP-AES)*

A number of heavy metals, including arsenic (As), cadmium (Cd), chromium (Cr), iron (Fe), lead (Pb), antimony (Sb), and titanium (Ti), were studied. Each metal has a set of somewhat lax contamination requirements. Drugs whose primary ingredients are known to accumulate significant amounts of cadmium are exempt from this rule. The standard mineralization protocol (AFNOR, 1999) using aqua regia ($HNO_3$ + 3 HCl) was chosen for the analysis of the concentrations of the key elements (As, Cd, Cr, Fe, Pb, Sb, and Ti). The latter permits large sample sizes, which reduce issues with sample representativeness. The method involves mixing the pulverized plant material (0.1 g) with 3 mL of aqua regia, which is made from 1 mL of nitric acid ($HNO_3$; 99%) and 2 mL of hydrochloric acid (HCl; 37%). The mixture was then placed in a reflux setup at 200 °C for two hours, following which it was cooled and allowed to settle. The supernatant was then taken, filtered through a 0.45 m membrane, and diluted to a final volume of 15 mL using distilled water. The ICP-AES (Ultima 2 Jobin Yvon) at the UATRS laboratory (Technical Support Unit to Scientific Research) at CNRST in Rabat measured the concentrations of heavy metals [64].

### 4.8. Antioxidant Activities

#### 4.8.1. Antiradical Activity by the DPPH• Test

The evaluation of the antiradical activity was based on the ability of an antioxidant (phenolic compound) to donate a single electron to the synthetic radical DPPH• (chemical compound 2,2-diphenyl-1-picrylhydrazyl), with purple coloring, to stabilize it into DPPH with yellow coloring [65]. At a wavelength of 515 nm, the experiment was carried out in a UV–visible spectrophotometer. The $6 \times 10^{-5}$ M DPPH• solution was obtained by dissolving 2.4 mg of DPPH• in 100 mL of ethanol. In order to prepare the extract samples, pure ethanol was used to dissolve them. The test was carried out by combining 200 µL of extract (sample) or standard antioxidant (ascorbic acid) at various concentrations with 2.8 mL of the preceding DPPH• solution.

Following 30 min of incubation at room temperature in the dark, the absorbance was measured at 515 nm in comparison to a blank that solely contained ethanol. The DPPH• solution without extract served as the negative control, and the values were subsequently converted into percent inhibition using the formula below (2) [66]:

$$\%\ AA = \frac{\text{Abs control} - \text{Abs sample}}{\text{Abs control}} \times 100 \tag{2}$$

where: % AA: percentage of antiradical activity; Abs control: absorbance of the blank (optical density of the solution consisting of DPPH• and ethanol); Abs sample: absorbance of the test compound (extracts).

#### 4.8.2. Ferric Reducing Antioxidant Power (FRAP) Method

According to Oyaizu's approach from 1986, the reduction of $Fe^{3+}$ in the $K_3Fe(CN)_6$ complex to $Fe^{2+}$ was used to assess the iron reducing activity of our extracts. To carry out the measurements of antioxidant activity by the FRAP method, samples from the different extracts were studied, proceeding as described below. A quantity of extract underwent the same treatment as the samples described above. Using distilled water to calibrate the device, the absorbance value was read at 700 nm against a prepared blank (UV–Vis spectrophotometer). A typical antioxidant solution was used as the positive control (ascorbic acid whose absorbance was measured under the same conditions as the samples). Increases in absorbance were accompanied by increases in the extracts' tested reducing power [67,68].

#### 4.8.3. Total Antioxidant Capacity (TAC)

The TAC of the extracts was evaluated by the phosphomolybdenum method described by Khiya [69]. According to this method, molybdenum Mo (VI), which is present as molybdate ions ${MoO_4}^{2-}$, is reduced to molybdenum Mo (V) $MoO^{2+}$ in the presence of the extract to form a green phosphate/Mo(V) complex at an acidic pH. Each extract was combined with 3 mL of the reagent solution and a 0.3 mL aliquot (0.6 M sulfuric acid, 28 mM sodium phosphate, and 4 mM ammonium molybdate). The tubes were secured with screws and left to sit at 95 °C for 90 min. Following cooling, the solutions' absorbance at 695 nm was measured in comparison to a produced blank that was incubated under identical circumstances as the sample. Several ascorbic acid concentrations were prepared as a standard range. Ascorbic acid milligram equivalents per gram of crude extract (mg EAA/g EB) are used to express the TAC.

### *4.9. Determination of the Minimum Inhibitory Concentration, Minimum Bactericidal Concentration, and Minimum Fungicidal Concentration*

Using 96-well microplates and the reference microdilution method, the minimum inhibitory concentration (MIC) was determined [70]. The MIC is the lowest amount of EO necessary to completely stop the tested microorganism's development during incubation, as measured by how much growth is visible to the naked eye. Thus, a series of dilutions were carried out from a stock solution of the essential oil produced in 10% DMSO to obtain concentrations of 5 to $0.93 \times 10^{-2}$ mg/mL of each EO. These dilutions were made with a final volume of 100 µL for each concentration in Sabouraud broth for fungi and Mueller–Hinton medium for bacteria. Thereafter, 100 µL of microbial inoculum with a final concentration of $10^6$ or $10^4$ CFU/mL for bacteria or fungi, respectively, was added to the various stages of the dilution series. Ten microliters of resazurin was added to each well as a measure of bacterial growth after a 24 h incubation period at 37 °C. The coloring changed from purple to pink after a second incubation at 37 °C for two hours, indicating microbial development. The MIC value was determined as the lowest concentration that prevents resazurin from changing color. The growth and sterility controls, respectively, were the eleventh and twelfth wells. For this oil, the test was conducted twice. Terbinafine 250 mg, the typical antifungal used in the study, was mixed with 2 mL of 10% DMSO after being ground. The minimum bactericidal concentration (MBC) and minimum fungicidal concentration (MFC) were determined by taking 10 µL from each well that had no apparent growth and plating it for 24 h at 37 °C on Mueller–Hinton (MH) agar for bacteria or in Sabouraud broth for fungi. The lowest sample concentration that resulted in a 99.99% reduction in CFU/mL relative to the control was designated as the MBC and MFC. Additionally, it was possible to calculate the MBC/MIC or MFC/MIC ratio of each extract to evaluate its antimicrobial potency. Accordingly, if the ratio is less than 4, the essential oil has a bactericidal/fungicidal effect, and if it is greater than 4, the sample has a bacteriostatic/fungistatic effect [71].

### *4.10. Anticoagulant Activity*

The anticoagulant effect was evaluated by chronometric coagulation tests involving the prothrombin time and partial thromboplastin time by adopting the method described by Hmidani et al. [72]. The prepared decocts were studied for possible investigation of anticoagulant agents. The extract concentrations used in the coagulation mixtures were 11.500, 5.750, 2.875, 1.438, 0.719, 0.359, and 0.179 mg/mL. In a polypropylene container, 3.8% trisodium citrate tubes were used to collect blood samples. They were then separated and pooled to create a plasma pool after being immediately centrifuged at 25,000 rpm for 10 min. Before use, the freshly made plasma pool was kept at −10 °C.

By combining the citrated normal plasma pool (50 µL) with a plant extract solution (50 µL) and incubating for 10 min at 37 °C, the partial thromboplastin time (aPTT) of the various samples examined was determined. After that, the mixture was mixed with 100 µL of the PTT reagent (CKPREST®) provided by Diagnostica Stago and incubated for 5 min at 37 °C. Coagulation was induced by adding 25 mmol/L $CaCl_2$ (100 µL) and the coagulation time was recorded. In contrast, the prothrombin time (PT) experiment was performed by combining 50 µL of the citrated normal plasma pool with 50 µL of a plant extract solution, and incubating the mixture for 10 min. After that, 200 µL of Neoplastin ® Cl reagent was added and preincubated for 10 min at 37 °C. The clotting time was then noted. The anticoagulant activity of the various aqueous plant extracts was measured in seconds at various doses. A coagulometer (MC4Plus MERLIN Medical®) was used to automatically carry out each measurement six times.

### *4.11. Antidiabetic Activity*

#### 4.11.1. Study of the Inhibitory Effect of Aqueous Extracts on the Activity of Pancreatic α-Amylase, In Vitro

The following concentrations of acarbose were used for testing: 1, 0.8, 0.6, 0.4, and 0.2 mg/mL. The following concentrations were used to test the aqueous extract: 0.89, 0.45, 0.22, 0.11, and 0.06 mg/mL. The inhibitory effect of the aqueous extract on α-amylase enzymatic activity was measured according to the method described by Daoudi et al. [73]. Phosphate buffer solution was mixed with 200 µL of the aqueous extract solution or 200 µL of the acarbose solution (positive control). All tubes received an addition of 200 µL of the enzyme solution, with the exception of the blank tube, which received 200 µL of phosphate buffer instead. The tubes underwent a 10 min preincubation at 37 °C. Starch solution was then added to each tube in a volume of 200 µL. The entire set was incubated for 15 min at 37 °C. The tubes were filled with 600 µL of DNSA to halt the enzymatic process.

The tubes were then submerged for 8 min in a pot of boiling water. Heat shock was then used to inhibit this process. Before adding 1 mL of diluted water to each tube, the tubes were submerged in an ice water bath. Using a spectrophotometer and a blank made up of the buffer solution rather than the enzyme solution, the absorbance was measured at 540 nm. The following Equation (3) was used to compute the % inhibition of each extract or acarbose:

$$\%\ \text{Inhibition} = \frac{\text{A control} - \text{A sample}}{\text{A control}} \times 100 \tag{3}$$

where: A control: absorbance of enzyme activity without inhibitor; A sample: absorbance of enzymatic activity in the presence of extract or acarbose.

#### 4.11.2. Study of the Inhibitory Effect of Aqueous Extracts on the Activity of α-Glucosidase, In Vitro

The α-glucosidase inhibitory activity of the extract was determined using pNPG substrate according to the method described by Chatsumpun et al. [74], with some modifications. The extracts were tested using the concentration range from 0.488 to 100 µmg/mL. DMSO (5%) was used to prepare each sample, and phosphate buffer was used to prepare the α-glucosidase enzyme (pH 6.8). As a solvent control, DMSO (5%) was employed, while acarbose served as the positive control. Step by step, 40 µL of the α-glucosidase enzyme (0.1 U/mL) and 10 µL of each sample were added to a 96-well plate. This mixture was then preincubated for 10 min at 37 °C. Thereafter, 50 µL of pNPG (1 mM) was added, and the mixture was incubated at 37 °C for 20 min. To end this reaction, 100 µL of $Na_2CO_3$ (0.1 M) solution was added. In a microplate reader, the mixture's absorbance was determined at a wavelength of 405 nm. Equation (3) was used to compute the percentage of α-glucosidase inhibition.

#### 4.11.3. Acute Toxicity Study

This experiment aimed to demonstrate that normal mice do not experience any short-term toxicity from the therapeutic dose. We investigated acute toxicity by the oral pathway, as this is the usual pathway implicated under normal conditions for humans. This study was conducted according to the guidelines of the Organization for Economic Cooperation and Development (OECD) [75]. The products tested were *C. sativus* extract ($E_0$) selected for the pharmacological study, at doses of 0.5, 1, and 2 g/kg.

Two lots of albino mice (20–35 g) in a fasted state (14 h) were randomly divided into 4 groups (n = 6; ♂/♀= 1):

- Control: distilled water (10 mL/kg).
- Group 1: aqueous extract $E_0$ (0.5 g/kg).
- Group 2: aqueous extract $E_0$ (1 g/kg).
- Group 3: aqueous extract $E_0$ (2 g/kg).

The mice were weighed before the test began. They were then given a single dose of the aqueous extract right away. After that, we continuously observed them for 10 h to look for indications of apparent toxicity. The mice were observed every day for the following 14 days for any new clinical or behavioral indications of harm.

4.11.4. Study of the Antihyperglycemic Activity of the Aqueous Extract of *C. sativus* in Normal Rats In Vivo

The ex vivo oral glucose tolerance test (OGTT) or oral sucrose tolerance test (OSTT) was performed by administering the different test products to normoglycemic mice. This study aimed to investigate whether the extract ($E_0$) had a postprandial antihyperglycemic effect in normal rats overloaded with D-glucose.

The normal rats (200–250 g) in a fasted state (14 h) were grouped into 3 groups (n = 6; ♂/♀= 1):

- Control: administration of distilled water (10 mL/kg).
- Extract: administration of aqueous extract $E_0$ (2 mL/kg).
- Glib: administration of glibenclamide (2 mg/kg).

First, normal rats were anesthetized with ether (inhalation), then a blood sample was taken from the queue to measure blood glucose at $t_0$ and, immediately after that, the test material (distilled water, aqueous extract, or glibenclamide) was administered orally. Thirty minutes afterward, another blood glucose measurement was performed, and immediately afterward the rats were overloaded with D-glucose (2 mg/kg). Thereafter, the change in blood glucose was monitored for 3 h, at 30, 60, 90, and 150 min.

*4.12. Statistical Analysis*

The results were expressed as mean $\pm$ standard error of mean. The results were analyzed using the one-factor ANOVA test of variance and followed by the Tukey posttest using GraphPad Prism 9 software (version 9.5.1) (San Diego, CA, USA). Values with $p < 0.05$ are considered significant. Correlations between phenolic compound contents and antioxidant activities were investigated using R software (version 4.1.3).

## 5. Conclusions

Morocco is a country with abundant plant resources, and there is a wide variety of medicinal plants there that are used to treat and prevent a number of diseases. From this study, we conclude that essential oils and extracts of *Crocus sativus* stigmas from the Boulemane region contain the same bioactive molecules required by European and American norms for the export of this spice, knowing that Morocco became the 3rd world producer of saffron in 2021, in front of Greece and India. The studied extracts are rich in phenolic compounds and carotenoids. Moreover, the extracts of *Crocus sativus* stigmas studied have great antioxidant, antimicrobial, anticoagulant, and antidiabetic properties. As a result, in addition to its numerous uses in the food and medical industries, it can be employed as a source of bioactive molecules in the treatment of infectious disorders brought on by bacteria that are multidrug resistant.

**Author Contributions:** T.Z. and A.D.; methodology: A.A., N.H. and F.R. Software and analysis: A.D. and M.E. Validation: F.E.M. and O.A.K. Investigation: H.E. (Hajar Elouadniand) and A.S. Writing—original draft preparation: A.D. Writing—revision and editing: A.D., M.B., and H.E. (Hanane Elazzouzi). Supervision: T.Z. All authors have read and agreed to the published version of the manuscript.

**Funding:** This study was funded by Princess Nourah bint Abdulrahman University Researchers Supporting Project number (PNURSP2023R141), Princess Nourah bint Abdulrahman University, Riyadh, Saudi Arabia.

**Institutional Review Board Statement:** The Institutional Ethics Committee for Care and Use of Laboratory Animals of the Faculty of Sciences of Oujda, University Sidi Mohamed Ben Abdallah Oujda, Morocco, has reviewed and approved this study #04/2019/LBEAS.

**Informed Consent Statement:** Not applicable.

**Data Availability Statement:** Not applicable.

**Acknowledgments:** The authors extend their appreciation for Princess Nourah bint Abdulrahman University Researchers Supporting Project number (PNURSP2023R141), Princess Nourah bint Abdulrahman University, Riyadh, Saudi Arabia.

**Conflicts of Interest:** The authors declare that they have no known competing financial interest or personal relationship that could have appeared to influence the work reported in this paper.

## References

1. Ren, X.; Santhosh, S.M.; Coppo, L.; Ogata, F.T.; Lu, J.; Holmgren, A. The combination of ascorbate and menadione causes cancer cell death by oxidative stress and replicative stress. *Free Radic. Biol. Med.* **2019**, *134*, 350–358. [CrossRef] [PubMed]
2. Wu, Q.; Wang, X.; Nepovimova, E.; Wang, Y.; Yang, H.; Kuca, K. Mechanism of cyclosporine A nephrotoxicity: Oxidative stress, autophagy, and signalings. *Food Chem. Toxicol.* **2018**, *118*, 889–907. [CrossRef] [PubMed]
3. Benabid, A. *Flore et Ecosystèmes du Maroc. Évaluation et Préservation de la Biodiversité*; Ibis Press: Paris, France; Librairies et Éditions Kalila Wa Dimna: Rabat, Morocco, 2000; p. 359.
4. Van Wyk, B.-E.; Wink, M. *Medicinal Plants of the World: An Illustrated Scientific Guide to Important Medicinal Plants and Their Uses*; Timber Press: Portland, OR, USA, 2004.
5. Pickersgill, B. *The Cultural History of Plants*; Prance, S.G., Nesbit, M., Eds.; Routledge: London, UK, 2005; ISBN 0415927463.
6. Drioiche, A.; Benhlima, N.; Kchibale, A.; Boutahiri, S.; Ailli, A.; El Hilali, F.; Moukaid, B.; Zair, T. Ethnobotanical Investigation of Herbal Food Additives of Morocco Used as Natural Dyes. *Ethnobot. Res. Appl.* **2021**, *21*, 1–43. [CrossRef]
7. Cardone, L.; Castronuovo, D.; Perniola, M.; Cicco, N.; Candido, V. Saffron (*Crocus sativus* L.), the King of Spices: An Overview. *Sci. Hortic.* **2020**, *272*, 109560. [CrossRef]
8. González, J.A.; García-Barriuso, M.; Amich, F. Ethnobotanical Study of Medicinal Plants Traditionally Used in the Arribes Del Duero, Western Spain. *J. Ethnopharmacol.* **2010**, *131*, 343–355. [CrossRef] [PubMed]
9. Idolo, M.; Motti, R.; Mazzoleni, S. Ethnobotanical and phytomedicinal knowledge in a long-history protected area, the Abruzzo, Lazio and Molise National Park (Italian Apennines). *J. Ethnopharmacol.* **2010**, *127*, 379–395. [CrossRef]
10. Maghsoodi, V.; Kazemi, A.; Akhondi, E. Effect of Different Drying Methods on Saffron (*Crocus sativus* L.) Quality. *Iran. J. Chem. Chem. Eng.* **2012**, *31*, 85–89.
11. Goupy, P.; Vian, M.A.; Chemat, F.; Caris-Veyrat, C. Identification and quantification of flavonols, anthocyanins and lutein diesters in tepals of *Crocus sativus* by ultra performance liquid chromatography coupled to diode array and ion trap mass spectrometry detections. *Ind. Crops Prod.* **2013**, *44*, 496–510. [CrossRef]
12. Alam, M.W.; Najeeb, J.; Naeem, S.; Usman, S.M.; Nahvi, I.; Alismail, F.; Abuzir, A.; Farhan, M.; Nawaz, A. Electrochemical Methodologies for Investigating the Antioxidant Potential of Plant and Fruit Extracts: A Review. *Antioxidants* **2022**, *11*, 1205. [CrossRef]
13. Aldughaylibi, F.S.; Raza, M.A.; Naeem, S.; Rafi, H.; Alam, M.W.; Souayeh, B.; Farhan, M.; Aamir, M.; Zaidi, N.; Mir, T.A. Extraction of Bioactive Compounds for Antioxidant, Antimicrobial, and Antidiabetic Applications. *Molecules* **2022**, *27*, 5935. [CrossRef]
14. Duarte, M.C.T.; Leme, E.E.; Delarmelina, C.; Soares, A.A.; Figueira, G.M.; Sartoratto, A. Activity of essential oils from Brazilian medicinal plants on *Escherichia coli*. *J. Ethnopharmacol.* **2007**, *111*, 197–201. [CrossRef] [PubMed]
15. Wang, H.; Wang, J.; Qiu, C.; Ye, Y.; Guo, X.; Chen, G.; Li, T.; Wang, Y.; Fu, X.; Liu, R.H. Comparison of phytochemical profiles and health benefits in fiber and oil flaxseeds (*Linum usitatissimum* L.). *Food Chem.* **2017**, *214*, 227–233. [CrossRef] [PubMed]
16. Pedro, R.O.; Takaki, M.; Gorayeb, T.C.C.; Del Bianchi, V.L.; Thomeo, J.C.; Tiera, M.J.; Tiera, V.A.O. Synthesis, Characterization and antifungal activity of quaternary derivatives of chitosan on *Aspergillus flavus*. *Microbiol. Res.* **2013**, *168*, 50–55. [CrossRef] [PubMed]
17. Sartoratto, A.; Machado, A.L.M.; Delarmelina, C.; Figueira, G.M.; Duarte, M.C.T.; Rehder, V.L.G. Composition and antimicrobial activity of essential oils from aromatic plants used in Brazil. *Braz. J. Microbiol.* **2004**, *35*, 275–280. [CrossRef]
18. Carmona, M.; Zalacain, A.; Salinas, M.; Alonso, G.L. Une nouvelle approche de l'arôme du safran. *Crit. Rev Food Sci. Nutr.* **2007**, *47*, 145–159. [CrossRef] [PubMed]
19. Moraga, Á.R.; Rambla, J.L.; Ahrazem, O.; Granell, A.; Gómez-Gómez, L. Metabolite and target transcript analyses during *Crocus sativus* stigma development. *Phytochemistry* **2009**, *70*, 1009–1016. [CrossRef]
20. Rahaiee, S.; Moini, S.; Hashemi, M.; Shojaosadati, S.A. Evaluation of antioxidant activities of bioactive compounds and various extracts obtained from saffron (*Crocus sativus* L.): A review. *J. Food Sci. Technol.* **2015**, *52*, 1881–1888. [CrossRef]
21. Cid-Pérez, T.S.; Nevárez-Moorillón, G.V.; Ochoa-Velasco, C.E.; Navarro-Cruz, A.R.; Hernández-Carranza, P.; Avila-Sosa, R. The relation between drying conditions and the development of volatile compounds in saffron (*Crocus sativus*). *Molecules* **2021**, *26*, 6954. [CrossRef]
22. Kafi, M.; Koocheki, A.; Rashed, M.H.; Nassiri, M. (Eds.) *Saffron (Crocus sativus) Production and Processing*, 1st ed.; Science Publishers: Washington, DC, USA, 2016; ISBN 978-1-57808-427-2.
23. Anastasakis, E.; Kanakis, C.; Pappas, C.; Maggi, L.; Del Campo, C.P.; Carmona, M.; Alonso, G.L.; Polissiou, M.G. Geographicaldifferentiation of saffron by GC–MS/FID and chemometrics. *Eur. Food Res. Technol.* **2009**, *229*, 899–905. [CrossRef]

24. Aliakbarzadeh, G.; Sereshti, H.; Parastar, H. Pattern recognition analysis of chromatographic fingerprints of *Crocus sativus* L. secondary metabolites towards source identification and quality control. *Anal. Bioanal. Chem.* **2016**, *408*, 3295–3307. [CrossRef]
25. Azarabadi, N.; Özdemir, F. Determination of Crocin Content and Volatile Components in Different Qualities of Iranian Saffron. *Gida J. Food* **2018**, *43*, 476–489. [CrossRef]
26. Carmona, M.; Martínez, J.; Zalacain, A.; Rodríguez-Méndez, M.L.; De Saja, J.A.; Alonso, G.L. Analysis of saffron volatile fraction by TD–GC–MS and e-nose. *Eur. Food Res. Technol.* **2006**, *223*, 96–101. [CrossRef]
27. Karimi, E.; Oskoueian, E.; Hendra, R.; Jaafar, H.Z. Evaluation of *Crocus sativus* L. stigma phenolic and flavonoid compounds and its antioxidant activity. *Molecules* **2010**, *15*, 6244–6256. [CrossRef]
28. Lech, K.; Witowska-Jarosz, J.; Jarosz, M. Saffron yellow: Characterization of carotenoids by high performance liquid chromatography with electrospray mass spectrometric detection. *J. Mass Spectrom.* **2009**, *44*, 1661–1667. [CrossRef] [PubMed]
29. Jadouali, S.M.; Atifi, H.; Bouzoubaa, Z.; Majourhat, K.; Gharby, S.; Achemchem, F.; Elmoslih, A.; Laknifli, A.; Mamouni, R. Chemical characterization, antioxidant and antibacterial activity of Moroccan *Crocus sativus* L. petals and leaves. *J. Mater. Environ. Sci.* **2018**, *9*, 113–118. [CrossRef]
30. Jadouali, S.M.; Atifi, H.; Mamouni, R.; Majourhat, K.; Bouzoubaâ, Z.; Laknifli, A.; Faouzi, A. Chemical Characterization and Antioxidant Compounds of Flower Parts of Moroccan *Crocus sativus* L. *J. Saudi Soc. Agric. Sci.* **2019**, *18*, 476–480. [CrossRef]
31. Huang, W.-Y.; Cai, Y.-Z.; Zhang, Y. Natural phenolic compounds from medicinal herbs and dietary plants: Potential use for cancer prevention. *Nutr. Cancer.* **2009**, *62*, 1–20. [CrossRef]
32. Assimopoulou, A.N.; Sinakos, Z.; Papageorgiou, V.P. Radical scavenging activity of *Crocus sativus* L. extract and its bioactive constituents. *Phytother. Res.* **2005**, *19*, 997–1000. [CrossRef]
33. Chen, Y.; Zhang, H.; Tian, X.; Zhao, C.; Cai, L.; Liu, Y.; Jia, L.; Yin, H.-X.; Chen, C. Antioxidant potential of crocins and ethanol extracts of Gardenia jasminoides ELLIS and *Crocus sativus* L.: A relationship investigation between antioxidant activity and crocin contents. *Food Chem.* **2008**, *109*, 484–492. [CrossRef]
34. Naceiri Mrabti, H.; Marmouzi, I.; Sayah, K.; Chemlal, L.; El Ouadi, Y.; Elmsellem, H.; Cherrah, Y.; Faouzi, M.A. *Arbutus unedo* L. aqueous extract is associated with *in vitro* and *in vivo* antioxidant activity. *J. Mater. Environ. Sci.* **2017**, *8*, 217–224.
35. Eman, R.E.; Abdelaziz, E.; Emad, M.A.; Ahmed, M.H.A. Antioxidant, antimicrobial and antifeedant activity of phenolic compounds accumulated in *Hyoscyamus muticus* L. *Afr. J. Biotechnol.* **2018**, *17*, 311–321. [CrossRef]
36. Belkamel, A.; Bammi, J.; Belkamel, A. *Origanum compactum* (Benth.). *J. Anim. Plant Sci.* **2013**, *19*, 2880–2887.
37. Vahidi, H.; Kamalinejad, M.; Sedaghati, N. Antimicrobial properties of *Croccus sativus* L. *Iran. J. Pharm. Res.* **2010**, *1*, 33–35.
38. Asgarpanah, J.; Darabi-Mahboub, E.; Mahboubi, A.; Mehrab, R.; Hakemivala, M. In-vitro evaluation of *Crocus sativus* L. petals and stamens as natural antibacterial agents against food-borne bacterial strains. *Iran. J. Pharm. Sci.* **2013**, *9*, 69–82.
39. Okmen, G.; Vurkun, M.; Arslan, A.; Ceylan, O. The antibacterial activities of *Piper nigrum* L. against mastitis pathogens and its antioxidant activities. *Indian J. Pharm. Educ. Res.* **2017**, *51*, S170–S175. [CrossRef]
40. Muzaffar, S.; Rather, S.A.; Khan, K.Z. *In vitro* bactericidal and fungicidal activities of various extracts of saffron (*Crocus sativus* L.) stigmas from Jammu & Kashmir, India. *Cogent Food Agric.* **2016**, *2*, 1158999.
41. Ogedegbe, H.O. An Overview of Hemostasis. *Lab. Med.* **2002**, *33*, 948–953. [CrossRef]
42. Shiping, M.; Baolin, L.; Sudi, Z.; Xiangwei, X.; Qiaoqiao, Y.; Jinxiang, Z. Pharmacological studies of glycosides of saffron Crocus (*Crocus sativus*). II. Effects on blood coagulation, platelet aggregation and thromobosis. *Zhong Cao Yao Chin. Tradit. Herb. Drugs.* **1999**, *30*, 196–198.
43. Khan, A.; Muhamad, N.A.; Ismail, H.; Nasir, A.; Khalil, A.A.K.; Anwar, Y.; Khan, Z.; Ali, A.; Taha, R.M.; Al-Shara, B.; et al. Potential Nutraceutical Benefits of In Vivo Grown Saffron (*Crocus sativus* L.) As Analgesic, Anti-inflammatory, Anticoagulant, and Antidepressant in Mice. *Plants* **2020**, *9*, 1414. [CrossRef]
44. Yang, L.; Qian, Z.; Yang, Y.; Sheng, L.; Ji, H.; Zhou, C.; Kazi, H.A. Involvement of Ca2+ in the inhibition by crocetin of platelet activity and thrombosis formation. *J. Agric. Food Chem.* **2008**, *56*, 9429–9433. [CrossRef]
45. Ayatollahi, H.; Javan, A.O.; Khajedaluee, M.; Shahroodian, M.; Hosseinzadeh, H. Effect of *Crocus sativus* L. (Saffron) on Coagulation and Anticoagulation Systems in Healthy Volunteers. *Phytother. Res.* **2013**, *28*, 539–543. [CrossRef] [PubMed]
46. Anis, S.; Ali, T.; Seiied, S.S.; Razieh, K.; Haniye, T.; Mohammadhossein, T.; Nasim, S.E.; Seyyed, M.M.A.D.; Niloofar, D. A review of the anti-diabetic potential of saffron. *Nutr. Metab. Insights.* **2022**, *15*, 11786388221095224. [CrossRef]
47. Dada, F.A.; Oyeleye, S.I.; Ogunsuyi, O.B.; Olasehinde, T.A.; Adefegha, S.A.; Oboh, G.; Boligon, A.A. Phenolic constituents and modulatory effects of Raffia palm leaf (*Raphia hookeri*) extract on carbohydrate hydrolyzing enzymes linked to type-2 diabetes. *J. Tradit. Complement. Med.* **2017**, *7*, 494–500. [CrossRef] [PubMed]
48. Institute for Laboratory Animal Research (ILAR). *Guide for the Care and Use of Laboratory Animals*, 8th ed.; National Academies Press (US): Washington, DC, USA, 2011.
49. Akrout, A.; Chemli, R.; Chreïf, I.; Hammami, M. Analysis of the essential oil of *Artemisia campestris* L. *Flavour Fragr. J.* **2001**, *16*, 337–339. [CrossRef]
50. Kovats, E.S. Gas chromatographic characterization of organic substances in the retention index system. *Adv Chromatogr.* **1965**, *1*, 229–247.
51. Adams, R.P. *Identification of Essential Oil Components by Gas. Chromatography/Mass Spectrometry*, 4th ed.; Allured Publishing Corporation: Carol Stream, IL, USA, 2007.
52. Hübschmann, H.J. *Handbook of GC-MS: Fundamentals and Applications*; John Wiley & Sons: Hoboken, NJ, USA, 2015.

53. Dohou, R.; Yamni, K.; Tahrouch, S.; Hassani, L.I.; Badoc, A.; Gmira, N. Screening Phytochimique d'une Endémique Iberomarocaine, *Thymelaea lythroides*. *Bull.-Société Pharm. Bordx.* **2003**, *142*, 61–78.
54. Judith, M.D. Etude Phytochimique et Pharmacologique de *Cassia nigricans* Vhal (Caeslpiniaceae) Utilisée Dans le Traitement des Dermatoses au Tchad. Ph.D. Thesis, Thèse de Pharmacie, Université de Bamako, Bamako, Mali, 2005; pp. 136–140.
55. Mezzoug, N.; Elhadri, A.; Dallouh, A.; Amkiss, S.; Skali, N.S.; Abrini, J.; Zhiri, A.; Baudoux, D.; Diallo, B.; El Jaziri, M.; et al. Investigation of the Mutagenic and Antimutagenic Effects of *Origanum Compactum* Essential Oil and Some of Its Constituents. *Mutat. Res. Toxicol. Environ. Mutagen.* **2007**, *629*, 100–110. [CrossRef]
56. Bekro, Y.A.; Mamyrbekova, J.A.; Boua, B.B.; Bi, F.H.T.; Ehile, E.E. Etude ethnobotanique et screening phytochimque de *Caesalpina benthamiana* (Ball). Herend et Zarucchi (Caesalpiniaceae). *Sci. Nat.* **2007**, *4*, 217–225.
57. Bruneton, J. *Pharmacognosie, Phytochimie et Plantes Médicinales*, 4th ed.; Technique et Documentation; Lavoisier: Paris, France, 2009; p. 1268.
58. N'Guessan, K.; Kadja, B.; Zirihi, G.; Traoré, D.; Ake-Assi, L. Screening phytochimique de quelques plantes médicinales ivoiriennes utilisées en pays Krobou (Agboville, Côte-d'Ivoire). *Sci. Nat.* **2009**, *6*. [CrossRef]
59. Singleton, V.L.; Rossi, J.A. Colorimetry of total phenolics with phosphomolybdic-phosphotungstic acid reagents. *Am. J. Enol. Vitic.* **1965**, *16*, 144–158.
60. Djeridane, A.; Yousfi, M.; Nadjemi, B.; Boutassouna, D.; Stocker, P.; Vidal, N. Antioxidant activity of some Algerian medicinal plants extracts containing phenolic compounds. *Food. Chem.* **2006**, *97*, 654–660. [CrossRef]
61. Hung, P.V.; Maeda, T.; Miyatake, K.; Morita, N. Total phenolic compounds and antioxidant capacity of wheat graded flours by polishing method. *Food Res. Int.* **2009**, *42*, 185–190. [CrossRef]
62. Price, M.L.; Van Scoyoc, S.; Butler, L.G. A critical evaluation of the vanillin reaction as an assay for tannin in sorghum grain. *J. Agric. Food Chem.* **1978**, *26*, 1214–1218. [CrossRef]
63. Willis, R.B.; Allen, P.R. Improved method for measuring hydrolysable tannins using potassium iodate. *Analyst* **1998**, *123*, 435–439. [CrossRef]
64. Skujins, S. *Handbook for ICP-AES (Varıan-Vista). A Short Guide to Vista Series ICP-AES Operation*; Varian International AG: Steinhausen, Switzerland, 1998.
65. Brand-Williams, W.; Cuvelier, M.E.; Berset, C. Use of a free radical method to evaluate antioxidant activity. *LWT—Food Sci. Technol.* **1995**, *28*, 25–30. [CrossRef]
66. Tagnaout, I.; Zerkani, H.; Mahjoubi, M.; Bourakhouadar, M.; Alistiqsa, F.; Bouzoubaa, A.; Zair, T. Phytochemical study, antibacterial and antioxidant activities of extracts of *Capparis spinosa* L. *Int. J. Pharmacol. Phytochem. Res.* **2016**, *8*, 1993–2006.
67. Končić, M.Z.; Kremer, D.; Karlović, K.; Kosalec, I. Evaluation of antioxidant activities and phenolic content of *Berberis vulgaris* L. and *Berberis croatica* Horvat. *Food Chem. Toxicol.* **2009**, *48*, 2176–2180. [CrossRef]
68. Oyaizu, M. Studies on Products of Browning Reaction—Antioxidative Activities of Products of Browning Reaction Prepared from Glucosamine. *Jpn. J. Nutr.* **1986**, *44*, 307–315. [CrossRef]
69. Khiya, Z.; Oualcadi, Y.; Gamar, A.; Berrekhis, F.; Zair, T.; Hilali, F.E. Correlation of Total Polyphenolic Content with Antioxidant Activity of Hydromethanolic Extract and Their Fractions of the *Salvia officinalis* Leaves from Different Regions of Morocco. *J. Chem.* **2021**, *2021*, e8585313. [CrossRef]
70. Balouiri, M.; Sadiki, M.; Ibnsouda, S.K. Methods for in vitro evaluating antimicrobial activity: A review. *J. Pharm. Anal.* **2016**, *6*, 71–79. [CrossRef] [PubMed]
71. Bissel, S.J.; Winkler, C.C.; DelTondo, J.; Wang, G.; Williams, K.; Wiley, C.A. Coxsackievirus B4 myocarditis and meningoencephalitis in newborn twins. *Neuropathology* **2014**, *34*, 429–437. [CrossRef] [PubMed]
72. Hmidani, A.; Bouhlali, E.d.T.; Khouya, T.; Ramchoun, M.; Filali-zegzouti, Y.; Benlyas, M.; Alem, C. Effect of extraction methods on antioxidant and anticoagulant activities of *Thymus atlanticus* aerial part. *Sci. Afr.* **2019**, *5*, e00143. [CrossRef]
73. Daoudi, N.E.; Bouhrim, M.; Ouassou, H.; Legssyer, A.; Mekhfi, H.; Ziyyat, A.; Aziz, M.; Bnouham, M. Inhibitory effect of roasted/unroasted *Argania spinosa* seeds oil on α-glucosidase, α-amylase and intestinal glucose absorption activities. *South Afr. J. Bot.* **2020**, *135*, 413–420. [CrossRef]
74. Chatsumpun, N.; Sritularak, B.; Likhitwitayawuid, K. New biflavonoids with α-glucosidase and pancreatic lipase inhibitoryactivities from *Boesenbergia rotunda*. *Molecules* **2017**, *22*, 1862. [CrossRef] [PubMed]
75. Weng, L.; Chen, T.H.; Zheng, Q.; Weng, W.H.; Huang, L.; Lai, D.; Fu, Y.S.; Weng, C.F. Syringaldehyde promoting intestinal motility with suppressing alpha-amylase hinders starch digestion in diabetic mice. *Biomed. Pharm.* **2021**, *141*, 111865. [CrossRef]

*pharmaceuticals*

*Article*

# Analysis of Transfer of Tiamulin to Animal Tissue after Oral Administration: An Important Factor for Ensuring Food Safety and Environmental Protection

**Viviana Carmen Ciucă, Carmen Otilia Rusănescu * and Victor Viorel Safta ***

Faculty of Biotechnical Systems Engineering, Polytechnic University of Bucharest, 313 Spl. Independentei, 060042 Bucharest, Romania

* Correspondence: rusanescuotilia@gmail.com (C.O.R.); saftavictorviorel@yahoo.com (V.V.S.)

**Abstract:** The administration of veterinary medicinal products containing tiamulin hydrogen fumarate (THF) leads to the appearance of the following residues in animal tissues: THF and metabolites that can be hydrolyzed to 8-α-hydroxymutilin. The marker residue for tiamulin, according to Regulation EEC 2377/90, is the sum of the metabolites that can be hydrolyzed to 8-α-hydroxymutilin. The main aim of this study was to analyze the depletion of tiamulin residues and metabolites that can be hydrolyzed to 8-α-hydroxymulinin by liquid chromatography with tandem mass spectrometry (LC-MS/MS) in pig, rabbit and bird tissues after tiamulin administration and to determine minimum withdrawal times for products of animal origin intended for human consumption. Tiamulin was administered orally as follows: 12,000 μg/kg body weight/day for 7 days to pigs and rabbits and 20,000 μg tiamulin/kg body weight/day for 7 days to broiler chickens and turkeys. The values found for tiamulin marker residues were 3 times higher in liver than in muscle in pigs, 6 times in rabbits and 8–10 times in birds. The content of tiamulin residues in eggs from laying hens was below 1000 μg/kg at all times of analysis. The minimum withdrawal times for animal products intended for human consumption, resulting from this study, are 5 days for pigs, rabbits and turkeys, 3 days for broiler chickens and 0 days for eggs.

**Keywords:** tiamulin; withdrawal time; LC-MS/MS

**Citation:** Ciucă, V.C.; Rusănescu, C.O.; Safta, V.V. Analysis of Transfer of Tiamulin to Animal Tissue after Oral Administration: An Important Factor for Ensuring Food Safety and Environmental Protection. *Pharmaceuticals* **2023**, *16*, 387. https://doi.org/10.3390/ph16030387

Academic Editors: Jan Oszmianski and Sabina Lachowicz-Wiśniewska

Received: 30 January 2023
Revised: 25 February 2023
Accepted: 27 February 2023
Published: 2 March 2023



## 1. Introduction

Monitoring the use of antibiotics in animals is essential to ensure food safety and environmental protection. The administration of veterinary medicinal products containing tiamulin hydrogen fumarate (THF) leads to the appearance of the following residues in animal tissues: THF and metabolites that can be hydrolyzed to 8-α-hydroxymutilin. The marker residue for tiamulin, according to Regulation EEC 2377/90, is the sum of the metabolites that can be hydrolyzed to 8-α-hydroxymutilin. Tiamulin residues in animal tissue can lead to many adverse biological effects and allergic reactions in consumers.

Tiamulin hydrogen fumarate, discovered in 1950, is a semi-synthetic derivative of the antibiotic pleuromutilin, produced by the basidiomycete *Clitopilus scyphoides* (formerly *Pleurotus mutilis*) and is chemically similar to valnemulin. The mechanism of action of tiamulin consists in inhibiting the synthesis of microbial proteins; this is achieved by binding with rRNA in the peptidyl-transferase slot on the ribosome, in which it prevents the correct positioning of CCA on the ends of tRNA for the transfer of peptides and the subsequent production of specific proteins [1]. Tiamulin hydrogen fumarate has a pleuromutilin chemical structure similar to that of valnemulin HCl. The hydroxyacetyl side chain is replaced by a larger diethylaminoethylthioacetyl moiety. This gives tiamulin hydrogen fumarate better hydrophobicity. The hemi-fumarate moiety provides a stable salt with improved water solubility (Figure 1) [2].

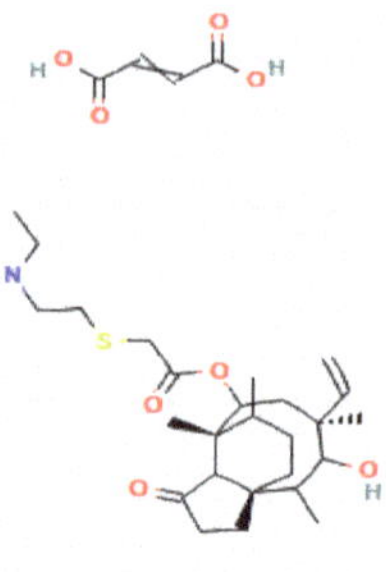

**Figure 1.** Structure of tiamulin hydrogen fumarate (adapted from [2]).

Tiamulin is not used in human medicine. It is used exclusively for animals. Tiamulin has demonstrated in vitro activity against porcine and avian Mycoplasma species, as well as against gram-positive aerobic species (streptococci and staphylococci), gram-positive anaerobes (clostridia), gram-negative anaerobes (*Brachyspira hyodysenteriae, Brachyspira pilosicoli*) and gram-negative aerobes (*Actinobacillus pleuropneumoniae* and *Pasteurella multocida*). Tiamulin is effective against mycoplasmas resistant to tylosin and erythromycin.

For veterinary medicinal products, the recommended dose for prevention or metaphylaxis when not associated with treatment is between 2000 and 5000 µg tiamulin as hydrogen fumarate/kg b.w. for 5 days to 6 weeks and for treatment-related metaphylaxis it is between 4000 and 10,000 µg for up to 10 days.

Analysis of the distribution of tiamulin in animal tissue is essential for ensuring food safety as well as preventing the spread of resistant bacteria and the acquisition of bacterial resistance.

The withdrawal time required for safe consumption of meat products from birds treated with tiamulin at therapeutic levels is 72 h in several states of the European Union. No waiting time is required for the removal of tiamulin residues from the egg in several states of the European Union [1,3].

In one study, tiamulin residues in meat were determined by liquid chromatography/mass spectrometry. Extraction of tiamulin was performed with acetone-tetrahydrofuran. The organic layer, separated from water with dichloromethane, was evaporated to dryness and the dry residue was diluted in methanol-1-heptane sulfonic acid. The method is advantageous due to the low quantification limit of 0.002 µg/g [4]. Tiamulin content was also successfully determined in medicated feed using a different gradient of 0.1% formic acid in acetonitrile and 0.1% formic acid in ultrapure water and a biphenyl column [5].

Studies on the metabolism of tiamulin have resulted in the following findings. Approximately 85–90% of the drug is absorbed. Maximum blood levels were reached 2–4 h after its oral administration. The target tissue for residues is the liver. The antibiotic is rapidly absorbed and the metabolites are eliminated through bile in feces (2/3) and urine (1/3). Only 0.3 to 0.5% of the parent compound is excreted unchanged in the urine. At least 25 metabolites have been found in urine and bile, of which 16 have been structurally identified. In animals, tiamulin is transformed by N-dealkylation, hydroxylation, oxidation and sulfoxidation. Most metabolites have a substantially lower antibacterial activity and the others have no antibacterial activity. In pigs, tiamulin is readily absorbed and extensively metabolized. None of the 14 metabolites found in pig urine exceeded 6% of the dose and none of the 16 metabolites in pig bile exceeded 7% of the administered drug dose. All individual metabolites found in pig urine, bile or liver (target tissue for residues) represent up to 10% of the administered dose [6].

The distribution of tiamulin in animal tissues was studied by gas chromatography with electrochemical detection in pigs and turkeys, by liquid scintillation counting (LSC) in broilers and by gas chromatography (GLC) with an electron capture detector in rabbits.

Under the same treatment conditions, similar results were obtained by the LC-MS/MS method described in this study, an equally efficient and accurate method [7–19].

The main aim of this study was to analyze the depletion of tiamulin residues and metabolites that can be hydrolyzed to 8-α-hydroxymulinin by an effective method, LC-MS/MS, in pig, rabbit and bird tissues after tiamulin administration and to determine minimum withdrawal times for products of animal origin intended for human consumption, according to European Medicines Agency (EMEA) requirements and EEC Regulation 2377/90 [20,21]. Compliance with the withdrawal period for residues to reach concentrations below the limits of safety protects humans from exposure to medicinal substances added to food. It is the responsibility of veterinarians and animal breeders to follow and comply with the withdrawal period in treated animals.

## 2. Results

### 2.1. Calculation of the Recovery Percentage from Samples

The calibration curve was drawn with the standard calibration solutions (0.05; 0.1; 0.25; 0.5; 1.0 μg/mL) in the control sample extract (matrix), which under working conditions (2 g sample/1 mL final purified extract) corresponds to 25–500 μg/kg; the linearity of the response was checked, then the percent recovery of tiamulin was calculated.

### 2.2. Calculation of Residue Content

The residue concentration was calculated using the calibration curve, taking into account the percentage recovery, or using the following formula:

$$Tiamulin(8\text{-}\alpha\text{-}hydroxymutilin) = c_{pr} \cdot \frac{V_{extract\ final}}{m} \cdot \frac{100}{R} \cdot 1000\ \mu g/kg \quad (1)$$

where $c_{pr}$ is the concentration of tiamulin (8-α-hydroxymutilin) in the final sample extract, read from the calibration curve, in μg/mL; $V_{extract\ final}$ is the final sample extract volume, in mL; $m$ is the mass of sample taken in work, in g; and $R$ is the recovery percentage.

### 2.3. Quality Control

A negative control without traces of tiamulin and a positive control obtained by fortifying the control sample to maximum residue limits (MRLs) were added to each series of samples analyzed.

### 2.4. Identity Confirmation

The ESI-MS/MS technique performed with the acquisition of the spectrum is good evidence of the identity and quantification. The spectrum obtained from the extract must match the spectrum of the standard solution analyzed in the same series of samples. The base ion must be the same. Qualifier ions must be present and have a relative abundance comparable to that of the standard.

The precursor ion and two MRM transitions corresponding to two fragmentary ions were selected, which means 4 identification points (Table 1 of Directive 2002/657/EC) [20,22]. The identity confirmation is presented in Table 1.

**Table 1.** Confirmation of identity.

| Analyte | Precursor Ion (*m/z*) | Fragment Ions, (*m/z*) | Ion Ratio, (%) | Max. Tolerance Allowed (%) |
|---|---|---|---|---|
| Tiamulin | 494.5 | 191.9; 118.6 | 17.6 | 25 |
| 8-alphahydroxymutilin | 337.4 | 282.8; 300.8 | 3.7 | 25 |

The results obtained for tiamulin hydrogen fumarate (THF) marker residues are shown in Tables 2–6. The values shown are corrected with recovery coefficients. The content of tiamulin and 8-α-hydroxymutilin for the samples taken from untreated control animals was below the detection limit of the analysis method.

**Table 2.** Values found in pigs for tiamulin marker residues (dose 12,000 µg tiamulin/kg body weight/day for 7 days).

| Animal Number | Days after Treatment | Muscle (µg/kg) | Liver (µg/kg) |
|---|---|---|---|
| 1 | 1 | 114 | 322 |
| 2 | 1 | 125 | 488 |
| 3 | 1 | 95 | 330 |
| 4 | 1 | 104 | 608 |
| 5 | 3 | 67 | 112 |
| 6 | 3 | 41 | 224 |
| 7 | 3 | 53 | 194 |
| 8 | 3 | 77 | 334 |
| 9 | 7 | 38 | 93 |
| 10 | 7 | 23 | 73 |
| 11 | 7 | 35 | 51.6 |
| 12 | 7 | 29 | 119 |

**Table 3.** Values found in rabbits for tiamulin marker residues (dose 12,000 µg tiamulin/kg body weight/day for 7 days).

| Animal Number | Days after Treatment | Muscle (µg/kg) | Liver (µg/kg) |
|---|---|---|---|
| 1 | 1 | 54 | 419 |
| 2 | 1 | 46.7 | 425 |
| 3 | 1 | 72 | 372 |
| 4 | 1 | 16.9 | 120 |
| 5 | 3 | 12.4 | 85 |
| 6 | 3 | 15.5 | 116 |
| 7 | 3 | 31.3 | 141 |
| 8 | 3 | 10 | 46.6 |
| 9 | 7 | 15 | 92.5 |
| 10 | 7 | 13.8 | 65 |
| 11 | 7 | 11.9 | 111 |

**Table 4.** Values found in broiler chickens for tiamulin marker residues (dose 20,000 µg tiamulin/kg body weight/day 7 days).

| Animal Number | Days after Treatment | Muscle (µg/kg) | Liver (µg/kg) | Skin/Fat (µg/kg) |
|---|---|---|---|---|
| 1 | 1 | 107 | 622 | 122 |
| 2 | 1 | 78 | 544 | 118 |
| 3 | 1 | 88.8 | 255 | 127 |
| 4 | 1 | 91 | 430 | 117 |
| 5 | 3 | 37.5 | 239 | 52.8 |
| 6 | 3 | 44.6 | 123 | 41.9 |
| 7 | 3 | 34.1 | 119 | 47 |
| 8 | 3 | 35 | 178 | 55 |
| 9 | 7 | 25 | 150 | 28.8 |
| 10 | 7 | 18.7 | 86 | 21.4 |
| 11 | 7 | 20 | 110 | 18 |
| 12 | 7 | 16.8 | 57 | 16 |

**Table 5.** Values found in turkeys for tiamulin marker residues (dose 20,000 µg THF/kg body weight/day 7 days).

| Animal Number | Days after Treatment | Muscle (µg/kg) | Liver (µg/kg) | Skin/Fat (µg/kg) |
|---|---|---|---|---|
| 1 | 1 | 62 | 400 | 77.6 |
| 2 | 1 | 36 | 356 | 37.5 |

**Table 5.** *Cont.*

| Animal Number | Days after Treatment | Muscle (µg/kg) | Liver (µg/kg) | Skin/Fat (µg/kg) |
|---|---|---|---|---|
| 3 | 1 | 53.5 | 460 | 64 |
| 4 | 3 | 29.5 | 183 | 34.2 |
| 5 | 3 | 17.2 | 192 | 31.8 |
| 6 | 3 | 25.7 | 226 | 27.8 |
| 7 | 7 | 12 | 88 | 25.5 |
| 8 | 7 | 15 | 80 | 20.3 |
| 9 | 7 | 20 | 110 | 22 |

**Table 6.** Content of tiamulin residues in eggs from laying hens (dose 20,000 µg THF/kg body weight /day for 7 days).

| Days after Treatment | Residues of Tiamulin in Egg (µg/kg) | | | | |
|---|---|---|---|---|---|
| 1 | 412 | 391 | 279 | 366 | 408 |
| 2 | 552 | 445 | 332 | 519 | 450 |
| 3 | 469 | 426 | 416 | 387 | 321 |
| 5 | 269 | 257 | 273 | 166 | 261 |
| 8 | 129 | 136 | 133 | 185 | 96 |
| 10 | 100 | 118 | 90 | 72 | 70 |
| 12 | 90 | 83 | 70 | 50 | 55 |

The highest concentrations of tiamulin (and 8-alphahydroxymutilin) residues were found in the liver in all species tested. In turkeys, at 7 days after treatment (20,000 µg tiamulin/kg body weight/day), the residues in the liver samples were below 100 µg/kg.

The residues from pig liver (608 µg/kg) were about three times higher than those from muscle. In rabbits, the residues in the liver (425 µg/kg) were about six times higher than those in the muscles and in birds (chicken and turkey broilers) about four times larger in liver than in muscle and skin/fat. Tiamulin residues in the analyzed eggs were below the MRL (1000 µg/kg) at all times of analysis.

The concentration of tiamulin (and 8-alphahydroxymutilin) in samples from untreated control animals was below the detection limit of the analytical method (Appendix A, Figures A1–A4) [23–25].

## 3. Discussion

Tiamulin use in intensively raised animals can lead to the appearance of residues in tissues and other products of animal origin intended for human consumption, in higher concentrations than those allowed. The presence of tiamulin in animal products intended for human consumption can often lead to the removal and destruction of a significant amount of meat, causing a serious economic consequences. For this reason, monitoring tiamulin residues is necessary.

For the analysis of tiamulin residues in animal tissues, an analytical HPLC-MS/MS method was developed in this paper. The presented method ensures specificity and selectivity for the determined compounds and efficient recovery and involves simple and fast sample preparation. The precision and accuracy of the described method are validation parameters that meet all the requirements for residue analysis. The results obtained for recoveries on samples fortified at the 0.5, 1 and 1.5× MRL levels exceeded the value of 75.0%, and coefficients of variability were lower than 15%. So, for tiamulin, the mean recovery for muscle was 84.2 ± 5.6% with a coefficient of average CV variability of 6.65%; for the liver, the average recovery was 80.5 ± 7.34% with an average CV coefficient of variability of 9.12%; and for the egg, the average recovery was 79.6 ± 8.80% with an average CV coefficient of variation of 11.06%.

In addition to the fragmentations (m/z) specific for tiamulin, the specific fragmentations were also monitored for 8-α-hydroxymutilin. The limit of quantification based on acceptable precision and accuracy (25 μg/kg) is below the MRL for different tissues, which is a necessary condition for the residue analysis methods. Values for MRL according to the EMA documents for the amount of metabolites that can be hydrolyzed to 8-α-hydroxymutilin are the following: in pigs and rabbits, 100 μg/kg in muscle and 500 μg/kg in liver, in broilers, 100 μg/kg in muscle and skin and 1000 μg/kg in liver, in turkeys, 100 μg/kg in muscle and skin and 300 μg /kg in liver and in eggs from laying hens, 1000 μg/kg.

According to a study in pigs, concentrations of liver metabolites that could be hydrolyzed to form 8-α-hydroxymutilin were determined by gas chromatography with electrochemical detection. In this paper, the LC-MS/MS method was used to determine the concentrations of liver metabolites that could be hydrolyzed to form 8-α-hydroxymutilin under the same conditions as in that study. The concentrations detected by gas chromatography with electrochemical detection in pigs with access to feed containing tiamulin at a concentration of 39,000 μg/kg for 10 days were 247 μg/kg 12 h after administration, whereas, in animals fed likewise for 18 consecutive days, mean liver concentrations of 8-α-hydroxymutilin were 184 μg/kg 12 h after administration. Using LC-MS, under the same conditions, mean concentrations of 272 μg/kg and 202 μg/kg, respectively, 12 h after administration were detected.

According to another study, in turkeys that had access to drinking water containing 0.025% *w*/*v* for 5 consecutive days, the concentrations of metabolites that could be hydrolyzed to form 8-α-hydroxymutilin, detected by gas chromatography with electrochemical detection, were less than 50 μg/kg in muscle. The mean liver concentrations of 8-α-hydroxymutilin were 905, 518, 527, 253 and 228 μg/kg at 0 h, 8 h, 1 day, 2 days and 3 days after treatment, respectively. Using the LC-MS/MS method described in this work, under the same experimental conditions, the average concentrations in the liver were 883, 489, 398, 203 and 188 μg/kg at 0 h, 8 h, 1 day, 2 days and 3 days, respectively, after treatment.

In broilers given 50,000 μg 3H-tiamulin/kg body weight/day for 5 consecutive days, average total residue concentrations in liver, muscle, skin and fat, determined by LSC (liquid scintillation counting), were 108,000, 550 and 6500 μg equivalents/kg, respectively, 2 h after dosing and the average tiamulin residues in liver, fat and muscle were 15,500, 1400 and 2200 μg/kg, respectively. 8-α-hydroxymutilin metabolite residues in broiler tissues represented approximately 7%, 3% and 2%, respectively, of the total residue from the liver, muscle and skin and fat [8–11].

## 4. Materials and Methods

The method went through the following steps: sample extraction (muscle, liver, kidney) with McIlvaine–EDTA buffer, precipitation of proteins with trichloroacetic acid, purification on an Oasis HLB cartridge and elution with methanol, evaporation of the eluate and reconstitution in 1 mL mobile phase (water–acetonitrile 90:10 *v*/*v* with 0.2% formic acid), LC analysis and MS/MS detection by ESI + technique with MRM monitoring of two transitions characteristic for tiamulin (494.5 > 191.9/494.5 > 118.6) and for 8-α-hydroxymutilin (337.4 > 282.8/337.4 > 300.8) [7].

### 4.1. Reagents Used

The following reagents were used: tiamulin, standard substance (Dr. Ehrenstorfer GmbH—Augsburg, Germany), disodium salt of ethylenediaminetetraacetic acid ($Na_2$EDTA dihydrate; Sigma—Livonia, MI, USA), anhydrous citric acid p.a. (Panreac—Barcelona, Spain), pure trichloroacetic acid, 99%, (Fluka -France), disodium phosphate anhydrous ($Na_2HPO_4$) p.a. (Panreac), acetonitrile, methanol, formic acid, 98–100% (Riedel-de Haen—Nottingham, UK), trichloroacetic acid, 20% (*m*/*v*).

The extraction solution is McIlvaine–EDTA buffer 0.1 M at pH 4, stable for a week. Standard stock solutions (1000 μg/mL) in methanol were prepared, stable for 6 months at

−20 °C. On the day of use, 5 standard calibration solutions (0.05; 0.1; 0.25; 0.5; 1.0 μg/mL) were prepared in control sample extract (matrix) corresponding to 25–500 μg/kg.

### 4.2. Animal Experiments and Sample Collection

Studies have been carried out on the depletion of tiamulin residues in pigs, rabbits and birds (broiler chickens and turkeys).

Depletion studies on tiamulin marker residues were performed on: 12 young pigs, weight 25–35 kg, 11 rabbits, weight 2.5–3 kg, 12 broiler chickens, Cob hybrid, average weight 750 g, and 9 turkeys, average weight 2.3 kg.

To eliminate interference, the animals and birds (broiler chickens and turkeys) used in the study were not treated before the experiment with another antibiotic. Birds and rabbits had not previously been treated with an ionophore coccidiostat (monensin, amproli, narasin or salinomycin).

The animals (pigs, rabbits, broiler chickens and turkeys) in the study were divided into two groups: an untreated control group and an experimental group in which the animals were treated according to the scheme of the experimental protocol with an oral solution containing 100,000 μg tiamulin hydrogen fumarate/mL. The administered doses were as follows: in pigs, 12,000 μg tiamulin/kg body/day, for 7 days, in rabbits, 12,000 μg tiamulin/kg body/day, for 7 days, and in birds (broiler chickens and turkeys), 20,000 μg tiamulin/kg body weight/day for 7 days. At specified time points after the end of treatment, animals were sacrificed for tissue sampling (muscle, liver and kidney).

### 4.3. Equipment and Materials

The chromatographic system used in the study was a Waters 2695 high-performance liquid chromatography (HPLC) system (Waters Corporation, Milford, MA, USA), consisting of a vacuum degasser, a binary pump, and an automatic sampler. An analytical Xbridge RP 18 column (2.1 × 150 mm, 3.5 μm) was used for the elution of the analyte. Using a triple-quadrupole mass spectrometer (Quatro micro MS-MS detector Micromass Waters Corporation, Milford, MA, USA) connected to an electrospray ion source; positive MRM was carried out with this mass spectrometer equipped with an ESI interface, MassLynx software.

The following equipment were also used in this study: KERN Abj analytical balance, REAX control vortex stirrer (Heidolph, Germany), Ultrathurax IKA T25, refrigerated centrifuge Centra MP 4R, ultrapure water SG GmbH production system, SPE purification station with pump vacuum (Supelco, Germany), Oasis HLB cartridge (60,000 μg/3 cc), Turbo-Vap evaporator (Zymark, Germany), Moulinex "Moulinette" type mixer, DD55210, 600 W, and Sonorex RK 100 H ultrasonic bath.

### 4.4. Sample Preparation

The collected samples were placed in a knife mixer for fine grinding and stored in a freezer (−20 °C) before analysis. Fresh and frozen samples were analyzed. If not analyzed immediately, the samples can be kept for up to 1 month at −20 °C.

#### 4.4.1. Sample Extraction

For extraction, approximately 2 g of minced and homogenized tissue was transferred to a 50 mL centrifuge tube with a cap. Then, 10 mL of McIlvaine–EDTA buffer was added and vortexed for 1 min at high speed followed by 10 min on a platform shaker at medium speed, then sonicated for 5 min. After centrifugation at 4000 rpm for 10 min, the supernatant was transferred to a test tube. The extraction was repeated with 5 mL buffer solution. For protein precipitation, 1 mL of 20% trichloroacetic acid solution was added to the combined supernatants, then homogenized for 1 min with a vortex mixer and centrifuged for 10 min at 4000 rpm (1685× *g*) at room temperature.

#### 4.4.2. Purification

After preconditioning the Oasis HLB cartridge with 6 mL methanol and 6 mL water, the sample supernatant was transferred to the cartridge. The cartridge was washed with 6 mL water methanol (5 + 95, *v*/*v*) and dried under air current for 10 min. Elution was done with 6 mL of methanol. After evaporating the eluate to dryness under a stream of nitrogen, at 40 °C, the residue, taken up in 1 mL mobile phase water–acetonitrile (90 + 10, *v*/*v*) with 0.2% formic acid, was analyzed by LC-MS/MS.

### *4.5. LC-MS—MS Analysis*

#### 4.5.1. LC Working Parameters

The mobile phase consisted of formic acid 0.2% (A) and acetonitrile with 0.2% formic acid (B). Gradient elution was conducted as indicated in Table 7.

**Table 7.** Mobile phase concentration gradient.

| Time (min) | % A | % B |
|---|---|---|
| 0 | 90 | 10 |
| 1 | 90 | 10 |
| 12 | 66 | 34 |
| 12.5 | 20 | 80 |
| 15 | 20 | 80 |
| 15.5 | 90 | 10 |
| 25 | 90 | 10 |

The flow rate was 0.2 mL/min and the injection volume was 20 μL. The temperature of the column thermostat was 30 °C.

Mass spectrometry detection was operated in positive ESI mode with multiple reaction monitoring (MRM) at two specific transitions: for tiamulin, 494.5 > 191.9/494.5 > 118.6, and for 8-α-hydroxymutilin, 337.4 > 282.8/337.4 > 300.8.

#### 4.5.2. MS Analysis Parameters

The mass spectrometry detection was operated in the positive ESI mode with multiple reaction monitoring (MRM). The operating parameters were set as follows: desolvation gas flow (nitrogen)—400 L/h, dissolving temperature—400 °C and ionization source temperature—120 °C. Fragmentation of molecular ions was done in the collision cell with argon ($3.0 \times 10^{-3}$ mbar). Tuning of the device was performed with standard solutions (concentration 10 μg/mL), by direct infusion (flow rate 10 μL/min). The retention time and optimal parameters for tiamulin hydrogen fumarate and 8-α-hydroxymutilin are given below (Table 8).

**Table 8.** Retention time and MS/MS parameters for tiamulin hydrogen fumarate and 8-α-hydroxymutilin.

| Component | Retention Time (min) | Precursor Ion (m/z) | Fragmentary Ions (m/z) | Cone Voltage (V) | Collision Energy (eV) |
|---|---|---|---|---|---|
| Tiamulin | 9.85 | 494.5 | 191.9 *; 118.6 | 26 | 29 |
| 8-α-hydroxymutilin | 9.10 | 337.4 | 282.8 *; 300.8 | 24 | 13 |

* Quantification ion (most abundant) (dwell time 20 ms).

Voltage at the extractor was 1 V and capillary voltage was 3.5 kV. The evaluations were done with the QuanLynx V4.1 program.

### *4.6. Method Validation*

Method validation and measurement uncertainty calculation for the determination of tiamulin residues in tissues by LC-MS/MS were performed in accordance with Food and Drug Administration (FDA) and European Medicines Agency (EMA) guidance requirements, with Directive 2002/657/EC and according to specific internal procedure PS-IP-CD-16.

All admissibility conditions for all validation parameters (identity, repeatability, reproducibility, specificity, interferences, limit of detection and quantification, linearity, precision and accuracy, decision limit (CCα) and detection capability (CCβ), stability and measurement uncertainty) were met. According to Directives 2002/657/EC [20] and 96/23/EC [22], tiamulin belongs to group B in Annex I, a minimum of 3 identification points being required for the LC-MS/MS technique to ensure the specificity of the method. In our case, the precursor ion and two transitions were selected, corresponding to two fragmented ions, which means 4 identification points [13]. The specific ions chosen for tiamulin were the precursor ion 494.5 and two fragment ions 191.9 and 118.6 and, for 8-α-hydroxymutilin, the precursor ion 337.4 and fragment ions 282.8 and 300.8. The results obtained by comparing the relative intensities of the fragment ions for the samples with those obtained for the standard solution at comparable concentrations, measured under the same conditions, fell within the allowed tolerance limits. In order to determine the linearity of the detector, a series of standard solutions with concentrations between 0.05 and 1.0 μg/mL were analyzed, which corresponds to the conditions described elsewhere in the work (2 g sample/1 mL final purified extract) 25–500 μg/kg (Appendix B, Figure A5). The correlation coefficient ($r^2$) was greater than 0.9950. The limits of detection and quantification were calculated by the method based on the calibration curve. The obtained values were LOD = 8.0 μg/kg and LOQ = 25 μg/kg. The limit of quantification was confirmed by fortifying the arrays at the level of quantification obtained and comparing the relative standard deviations with a limit of 20%. No interferences were observed with retention times of tiamulin and 8-α-hydroxymutilin greater than 20% of the response obtained at the limit of quantification of the method. The values found for the relative standard deviations of the repeatability were below the maximum value of 15% established by the legislation. The recovery efficiency and precision of the method were estimated. Control samples treated at the 3 levels, 0.5× maximum recovery limit (MRL), MRL and 1.5× MRL were analyzed in six replicates. The mean recoveries obtained for muscle, liver and egg were 84.2 ± 5.60%, 80.5 ± 7.34% and 79.6 ± 8.8%, respectively, and were within the limits, from 70% to 110%, allowed by legislation [12]. Decision limit ($CC_\alpha$) and detection capability ($CC_\beta$) are shown in Table 9.

**Table 9.** Decision limit ($CC_\alpha$) and detection capability ($CC_\beta$).

| Tissue | Quantity Added (x) (μg/kg) | Quantity Found (y) (μg/kg) | y = ax + b | $CC_\alpha$ (μg/kg) | $CC_\beta$ (μg/kg) |
|---|---|---|---|---|---|
| Muscle | 100 | 85.2 | y = 0.935x − 8.27 | 93.6 | 102 |
| Liver | 500 | 388.4 | y = 0.668x + 54.4 | 440.0 | 491.5 |
| Egg | 1000 | 776 | y = 0.696x + 80 | 929.0 | 1082 |

Quantification of measurement uncertainty was based on precision studies, accuracy studies and the identification and assessment of other contributions to uncertainty.

The calculated values for the expanded uncertainty were between 26 and 36%. The expanded uncertainty of the method was muscle—2 × 0.105 = 0.21 or 21%, liver—2 × 0.132 = 0.264 or 26.4% and egg—2 × 0.145 = 0.29 or 29% [21–26].

## 5. Conclusions

The control of antibiotics in the tissues of treated animals is an important factor that ensures the high quality of animal husbandry, consumer protection, as well as environmental protection. A simple and precise method was developed and validated in this study for the determination of tiamulin residues in animal tissue, which is useful for calculating the minimum withdrawal time for animal products intended for human consumption.

Protecting the population from exposure to tiamulin residues in the edible tissues of treated animals involves the determination and observance of withdrawal times. The method for analyzing tiamulin residues by the technique described in this work ensures specificity and selectivity for the determined compounds, fulfills all validation conditions for all analyzed performance parameters and is fast and accurate. By applying the statistical calculation of the European Medicines Agency (EMA) it was found that the minimum withdrawal times for meat and offal from animals treated with an approved oral solution containing 100,000 μg tiamulin hydrogen fumarate/mL are 5 days for pigs, rabbits and turkeys, 3 days for broilers and 0 days for eggs. The withdrawal times obtained as a result of this study fall within the limits stipulated and reconfirmed by Commission Regulation (EU) no. 37/2010. Veterinarians monitor withdrawal times for medicinal products to ensure that illegal levels of residues do not occur in animal products intended for human consumption from treated animals. This study is important because it establishes the withdrawal time and informs the consumer about the risk of consuming meat, organs and eggs that can harm human health.

The withdrawal time calculated and recommended in this study are in accordance with the data published in the specialized literature for products with similar indications of use. The withdrawal times determined in the paper have been confirmed in several European Union states (72 h for birds and zero for eggs) [27–29].

In the monitoring of tiamulin residues, although the LC-MS/MS method is well designed, demonstrated by its accuracy, sensitivity and precision, it is not without analytical problems. In addition to limitations in selectivity due to the occurrence of "isobaric" interferences, unpredictable attenuations of ion yield, known as the "ion suppression effect", must be considered. Lack of traceability to reference materials can lead to inaccuracy in testing as well as inaccurate reporting of results if the basic rules of method validation are ignored.

**Author Contributions:** Conceptualization: V.C.C.; methodology: V.C.C.; investigation: V.C.C., V.V.S.; validation: V.C.C., V.V.S.; review: V.C.C., C.O.R. All authors have read and agreed to the published version of the manuscript.

**Funding:** This work has been funded by the European Social Fund from the Sectorial Operational Programme Human Capital 2014–2020, through the Financial Agreement with the title "Training of PhD students and postdoctoral researchers in order to acquire applied research skills-SMART". project number: 13530/16.062022-SMIS code: 153734. The APC was funded by Polytechnic University of Bucharest, Romania, within the PubArt Program.

**Institutional Review Board Statement:** Not applicable.

**Informed Consent Statement:** Not applicable.

**Data Availability Statement:** Not applicable.

**Conflicts of Interest:** The authors declare no conflict of interest.

## Appendix A. Chromatograms

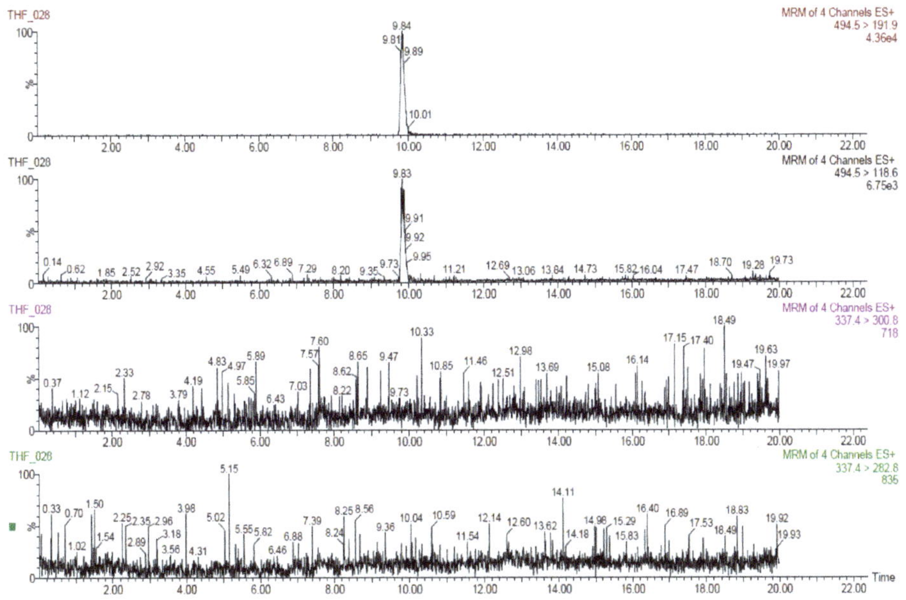

**Figure A1.** Chromatogram of a muscle sample fortified with tiamulin at 25 μg/kg.

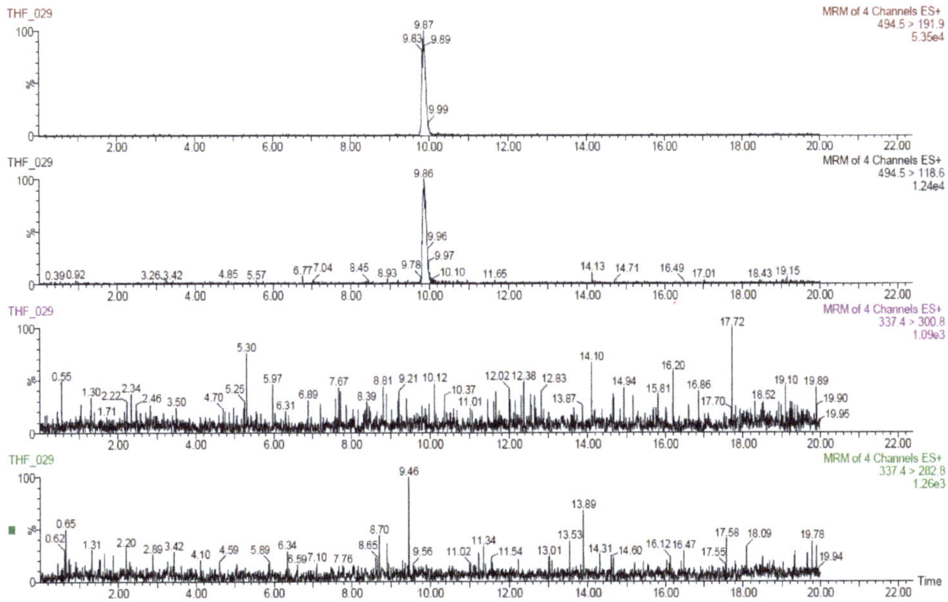

**Figure A2.** Chromatogram of a liver sample fortified with tiamulin at a level of 25 μg/kg.

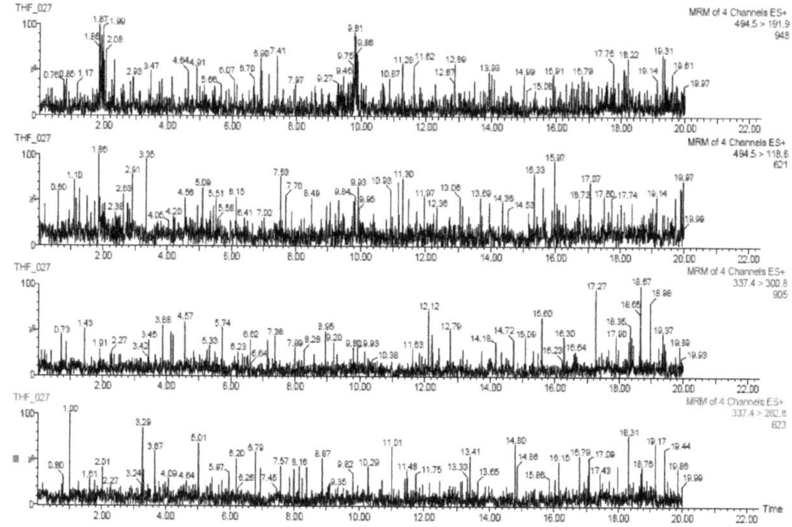

**Figure A3.** Chromatogram of a blank liver sample.

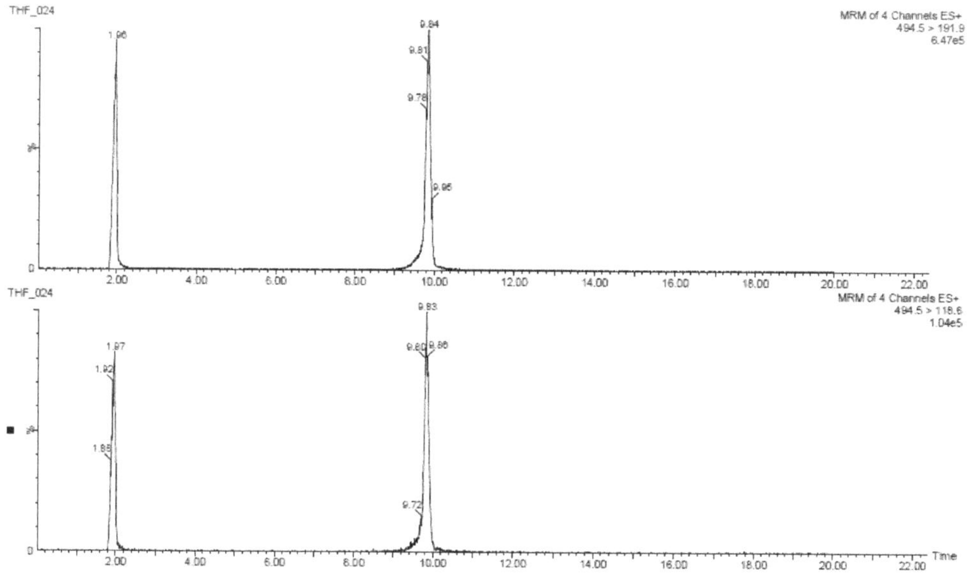

**Figure A4.** Chromatogram of the reference solution with a concentration of 0.5 µg tiamulin/mL in the matrix.

## Appendix B

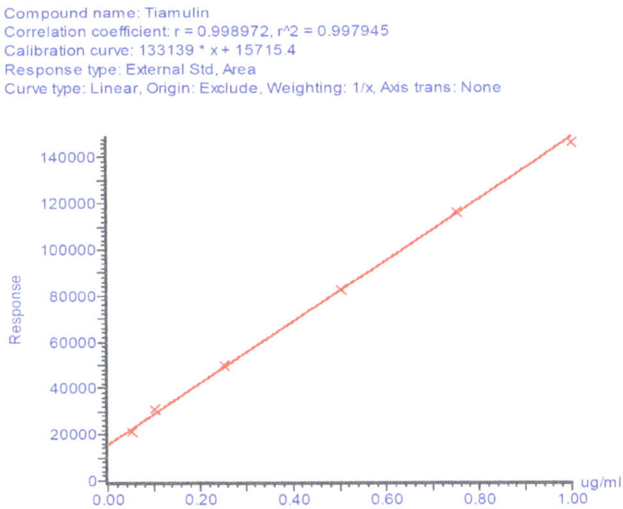


**Figure A5.** Calibration curve for tiamulin (0.05–1 µg/kg).

## References


1. Islam, K.M.S.; Klein, U.; Burch, D.G.S. The activity and compatibility of the antibiotic tiamulin with other drugs in poultry medicine-A review. *Poult. Sci.* **2009**, *88*, 2353–2359. [PubMed]
2. Available online: https://pubchem.ncbi.nlm.nih.gov/compound/Tiamulin (accessed on 1 November 2022).
3. Available online: https://www.ema.europa.eu/en/documents/referral/tiamulin-referral-article-35-annex-i-iii_ro.pdf (accessed on 10 November 2022).
4. Hormazábal, V.; Yndestad, M.; Ostensvik, O. Determination of Flunixin and Tiamulin Hydrogen Fumarate in Meat and Toltrazuril and the Metabolite Toltrazurilsulfon in Meat and Eggs Using LC/MS. *J. Liq. Chromatogr. Relat. Technol.* **2003**, *26*, 791–801. [CrossRef]
5. Patyra, E.; Nebot, C.; Gavilán, R.E.; Cepeda, A. Development and validation of an LC-MS/MS method for the quantification of tiamulin, trimethoprim, tylosin, sulfadiazine and sulfamethazine in medicated feed. *Food Addit. Contam. Part A* **2018**, *35*, 882–891. [CrossRef] [PubMed]
6. Date of Approval: November 26, Freedom of Information Summary Supplemental New Animal Drug Application NADA 139–472 for the Use of DENAGARD®(Tiamulin) Type A Medicated Article 2001. Available online: https://animaldrugsatfda.fda.gov/adafda/app/search/public/document/downloadFoi/459 (accessed on 13 November 2022).
7. Bohm, D.A.; Stachel, C.S.; Gowick, P. Validated determination of eight antibiotic substance groups in cattle and pig muscle by HPLC/MS/MS. *J. AOAC Int.* **2011**, *94*, 407–419. [CrossRef] [PubMed]
8. Chen, H.C.; Cheng, S.H.; Tsai, Y.H.; Hwang, D.F. Determination of tiamulin residue in pork and chicken by solid phase extraction and HPLC. *J. Food Drug Anal.* **2006**, *14*, 80–83. [CrossRef]
9. Nozal, M.A.J.; Bernal, J.L.; Martín, M.A.T.; Bernal, J.; Alvaro, A.; Martín, R.; Higes, M. Trace analysis of fumagillin in honey by liquid chromatography-diode array-electrospray ionization mass spectrometry. *J. Chromatogr. A* **2008**, *1190*, 224–231. [CrossRef] [PubMed]
10. *Methods of Analysis of the AOAC*; AOAC International: Washington, DC, USA, 1980.
11. Pickles, R.W. Tiamulin water medication in the treatment of swine dysentery under farm conditions. *Vet. Rec.* **1982**, *110*, 403–405. [CrossRef] [PubMed]
12. Available online: http://www.aca.org.ro/content/media/pagini/REGULAMENTUL%20(CEE)%20NR.%20237790%20rezziduuri.pdf (accessed on 13 November 2022).
13. Barros, A.; Novo, C.; Feddern, V.; Coldebella, A.; Scheuermann, G. Determination of Eleven Veterinary Drugs in Chicken Meat and Liver. *Appl. Sci.* **2021**, *11*, 8731. [CrossRef]
14. Daeseleire, E.; Van Pamel, E.; Van Poucke, C.; Croubels, S. Chapter 6—Veterinary Drug Residues in Foods. In *Chemical Contaminants and Residues in Food*, 2nd ed.; Woodhead Publishing Series in Food Science, Technology and Nutrition; Woodhead Publishing: Sawston, UK, 2017; pp. 117–153. Available online: https://biblio.ugent.be/publication/8575746 (accessed on 17 November 2022).

15. Dutta, T.K.; Yadav, S.K.; Chatterjee, A. Antibiotics as feed additives for livestock: Human health concerns. *Indian J. Anim. Health* **2019**, *58*, 121–136. [CrossRef]
16. Saettone, V.; Biasato, I.; Radice, E.; Schiavone, A.; Bergero, D.; Meineri, G. State-of-the-Art of the Nutritional Alternatives to the Use of Antibiotics in Humans and Monogastric Animals. *Animals* **2020**, *10*, 2199. [CrossRef] [PubMed]
17. Ciucă, V.C.C.; Rusănescu, C.O.; Safta, V.V. Research on the Removal of Levamisole Residues in Bovine, Ovine, Caprine, Porcine and Poultry Tissues. *Separations* **2022**, *9*, 261. [CrossRef]
18. Luo, R.; Wang, Y.; Guo, Q.; Fan, C.; Jiang, G.; Wang, L.; Zou, M.; Wang, T.; Sun, Y.; Peng, X. Andrographolide attenuates *Mycoplasma gallisepticum*-induced inflammation and apoptosis by the JAK/PI3K/AKT signal pathway in the chicken lungs and primary alveolar type II epithelial cells. *Int. Immunopharmacol.* **2022**, *109*, 108819. [CrossRef] [PubMed]
19. Chang, E.; Yamashita, K.; Arora, R. *Multi-Residue Screening of Veterinary Drugs (I) and (II) in Meat According to the Japan Positive List Using Cartridge-Based SPE and LC/MS/MS*; Application Note SI-02112; Varian, Inc.: Palo Alto, CA, USA, 2010.
20. Directive 2002/657/EC: Commission Decision of 12 August 2002 Implementing Directive 96/23/EC on the Performance of Analytical Methods and the Interpretation of Results. Available online: https://eur-lex.europa.eu/legal-content/EN/ALL/?uri=CELEX%3A32002D0657 (accessed on 13 November 2022).
21. Available online: https://www.ema.europa.eu/en/documents/scientific-guideline/note-guidance-approach-towards-harmonisation-withdrawal-periods_en.pdf (accessed on 13 November 2022).
22. Council Directive 96/23/EC of 29 April 1996 on Measures to Monitor Certain Substances and Residues Thereof in Live Animals and Animal Products and Repealing Directives 85/358/EEC and 86/469/EEC and Decisions 89/187/EEC and 91/664/EEC. Available online: https://eur-lex.europa.eu/legal-content/en/ALL/?uri=CELEX%3A31996L0023 (accessed on 19 November 2022).
23. Heitzman, R.J. *Veterinary Drug Residues-Residues in Food Producing Animals and Their Products: Reference Materials and Methods*, 2nd ed.; EUR 15127 EN; EU Commission: Brussels, Belgium, 1994; Available online: http://aei.pitt.edu/44343/ (accessed on 23 November 2022).
24. Ellison, S.L.R.; Roesslein, M.; Williams, A. Quantifying uncertainty in analytical measurement. In *Eurachem/Citac Guide*; Eurachem: Gembloux, Belgium, 2000.
25. *ISO 5725-2:1994*; Accuracy (Trueness and Precision) of Measurement Methods and Results, Part 1: General Principles and Definitions. American National Standards Institute: Washington, DC, USA, 1994.
26. *ISO 5725-2:1994*; Accuracy (Trueness and Precision) of Measurement Methods and Results—Part 2: Basic Method for the Determination of Repeatability and Reproducibility of a Standard Measurement Method. American National Standards Institute: Washington, DC, USA, 1994.
27. Krasucka, D.; Mitura, A.; Cybulski, W.; Kos, K.; Pietro, W. Tiamulin hydrogen fumarate ñ veterinary uses and hplc method of determination in premixes and medicated feeding stuffs. *Acta Pol. Pharm. Ñ Drug Res.* **2010**, *67*, 682–685.
28. HACCP Manual, Tiamulin in Veterinary Medicine ñ Overview. Available online: http://www.bioagrimix.com/haccp/html/tiamulin.htm (accessed on 29 November 2022).
29. Shimaa, E.; Aziz, A.; Nermin, S.; Ahmed Sawsan, A.; Razeq, A. Stability-indicating UHPLC and TLC-densitometric, Methods for the Determination of Tiamulin Fumarate. *Int. Res. J. Pure Appl. Chem.* **2021**, *22*, 8–21.

*pharmaceuticals*

*Article*

# Determination of Tyrosine Kinase Inhibitors via Capillary Electrophoresis with Tandem Mass Spectrometry and Online Stacking Preconcentration

**Jan Petr**

Department of Analytical Chemistry, Faculty of Science, Palacký University Olomouc, 17. listopadu 12, 77146 Olomouc, Czech Republic; jan.petr@upol.cz

**Abstract:** Capillary electrophoresis connected with tandem mass spectrometry was employed for the development of a method for determination of various tyrosine kinase inhibitors in plasma samples. A stacking online preconcentration with a 120 cm-long capillary was used for the determination of bosutinib, dasatinib, canertinib, and erlotinib at physiologically relevant concentrations. The optimization included both capillary electrophoresis and mass spectrometry steps. Under optimal conditions, 50 mM formic acid pH 2.5, an injection time of 120 s, and an optimized mass spectrometry set-up (as sheath liquid composition 75:24.9:0.1 ($v/v$) methanol, water, formic acid, and appropriate conditions for ion transitions), LODs in a range of 3.9–23.0 nmol·L$^{-1}$ were observed. The method was validated in terms of linearity, limit of detection, limit of quantification, repeatability of migration times and peak area, and recovery using plasma as a matrix for analytes. The results showed that this method has great promise for use in many analytical tasks, e.g., therapeutic drug monitoring.

**Keywords:** capillary electrophoresis; mass spectrometry; online preconcentration; stacking; tyrosine kinase inhibitors

**Citation:** Petr, J. Determination of Tyrosine Kinase Inhibitors via Capillary Electrophoresis with Tandem Mass Spectrometry and Online Stacking Preconcentration. *Pharmaceuticals* **2023**, *16*, 186. https://doi.org/10.3390/ph16020186

Academic Editors: Jan Oszmianski and Sabina Lachowicz-Wiśniewska

Received: 31 December 2022
Revised: 16 January 2023
Accepted: 20 January 2023
Published: 25 January 2023



## 1. Introduction

Tyrosine kinase inhibitors (TKIs) are a family of small molecules or peptides with the ability to inhibit either cytosolic or receptor tyrosine kinases. Inhibition by this class of agents occurs through different types of actions, where the direct competition for ATP binding to tyrosine kinase is often described [1,2]. Generally, tyrosine kinases have the function of catalyzing the transfer of a phosphoryl group from a nucleoside triphosphate donor to the hydroxyl group of tyrosine residues on protein substrates and then triggering the activation of downstream signaling cascades [3]. Abnormal activation of tyrosine kinases due to mutations, translocations, or amplifications is implicated in tumorigenesis, progression, invasion, and metastasis of malignancies [4]. As a result, tyrosine kinases have emerged as major targets for drug discovery. In 2001, the FDA approved imatinib for the treatment of chronic myeloid leukemia. However, the majority of TKIs are currently not in clinical use. The exceptions are the first-used imatinib, erlotinib, and dasatinib, which show promise as "targeted" therapeutics in the treatment of various cancers as well as leukemia [1,2].

Therapeutic drug monitoring (TDM) is the clinical practice of measuring a drug's concentration in blood or plasma, or in other biological fluids that can be linked to blood drug concentrations. The measured drug concentration is then used to adjust a drug dosing regimen by targeting a predefined concentration or exposure interval, called a therapeutic range [5]. The WHO has issued specific guidelines on how a drug should be monitored due to TDM's clinical importance, which is defined as individualizing a drug's dose by keeping a drug's concentrations in the plasma or blood within a target range to act as a guide for healthcare staff [6]. Such guidelines deal with large interpatient pharmacokinetic variability, adverse effects, therapeutic concentration-related effects, undefined therapeutic concentration ranges, and difficult-to-manage desired therapeutic effects. The WHO report also stressed certain criteria, such as an increased drug concentration in the blood, being related to increased efficacy and/or toxicity in the organism, the difficulty in monitoring a target drug's pharmacological effects, and drug concentration-related adverse effects [7]. Likewise, the report suggested a list of a pharmacological groups requiring monitoring, i.e., antibiotics (aminoglycosides and glycopeptides), anticonvulsants (e.g., valproic acid, phenobarbital, carbamazepine), cytotoxic drugs (e.g., metotrexate), antiarrhythmics (e.g., digoxin), and immunosuppressants (e.g., cyclosporine), which are indispensable drugs for the treatment of a myriad of diseases in current clinical practice [8].

Different techniques have been necessarily employed in TDM due to the nature of the investigated drugs to be quantified. Some of the most common techniques are high-performance liquid chromatography (HPLC) and its connection with mass spectrometry (HPLC-MS), gas chromatography–mass spectrometry (GC–MS), and immunoassays. The chromatographic techniques are the most robust and specific reference techniques; however, these methods need trained personnel, involve long sample processing times, and require costly reagents. Moreover, a sample is processed in the chromatograph each time (unlike other techniques that usually do not need expensive reagents after the validation of the technique) and requires a specialized laboratory for processing [6,9]. In contrast, immunoassays have been of great use as these techniques operate in less time than HPLC and/or GC-MS (chromatographic techniques require samples and mobile-phase preparations, extractions and/or filtrations, derivatizations, and continuous control of the equipment for correct operation). However, these techniques require trained personnel and a clinical laboratory with the necessary equipment and reagents. Such techniques include radioimmunoassays, enzyme-linked immunosorbent assays (known as ELISA), and fluorescence polarization immunoassays, which have been used for quantifying antibiotics as well as anticancer/antineoplastic, antiarrhythmic, and biological drugs [6,10].

Last but not least, the capillary electrophoresis (CE) technique represents an alternative to HPLC. CE has a unique separation mechanism, speed, efficiency, and versatility. CE separation depends on the different migration of solutes in an electric field. CE is performed in narrow-bore capillaries filled with a background electrolyte (BGE). The driving forces in CE are electrophoretic migration and electro-osmotic flow (EOF). Compared with other techniques, the instrumentation of CE is simple and consists of electrodes, sample-introduction systems, a capillary, a power supply, a detector, and a liquid-handling system. Detection can be achieved with online (diode-array spectrophotometric, spectrofluorimetric, and electrochemical) or external detectors (mass spectrometer, MS) [11–13]. CE has been noted as a "green" technique as it "consumes" an ultralow amount of chemicals and samples. Usually, nanoliter volumes of samples are analyzed (microliters are necessary for injection from commercial instruments). In the case of chemicals, typically only a few milliliters of BGE is needed. Hence, the use of substances potentially harmful to the environment, e.g., organic solvents, is limited, especially in comparison with HPLC [14]. Moreover, since CE suffers from less-sensitive UV detection, so-called online sample-concentration techniques have been developed in the past. These include four basic approaches: stacking, transient isotachophoresis, dynamic pH junction, and sweeping. The use of such techniques can improve LOD values from tens to millions, especially when the electrokinetic injection of samples is employed. Compared to "traditional" offline sample pretreatment steps, such as liquid–liquid extraction or solid-phase extraction, online sample preconcentration techniques are user-friendly, as most enrichment processes are performed in the separation capillary. In general, the composition of zones in the capillary is "programmed" to focus analytes of interest. The simplest mechanism (stacking) is based on the analyte velocity change (e.g., slowing down) in a two-discontinuous-solutions system (e.g., at the boundary formed by diluted BGE and non-diluted BGE) [15–18].

As noted, TDM is considered a very useful tool in helping clinicians with individual dose adjustment. Many studies have highlighted the clinical benefit of TDM for the first tyrosine kinase inhibitor—imatinib [19–21]. Like many other TKIs, imatinib exhibits large interpatient pharmacokinetic variability with more than ten-fold differences in drug plasma concentration, which sometimes results in therapeutic failures [22,23]. This variation has been linked to multiple genetic factors but may also be influenced by other physiological and environmental factors, such as drug–drug/food interactions and patient adherence [24]. Previous studies have shown that about 95% of imatinib is bound to albumins and $\alpha$-1-acid glycoproteins in plasma. The large variability in the concentration of the latter results in inconsistent concentrations of unbound and active imatinib [25]. Clinical data have also shown that the plasma concentration of imatinib is directly linked to therapeutic success; for example, in the case of chronic myeloid leukemia, the concentration needs to be higher than 1000 ng·mL$^{-1}$, while for GISTs, the minimum concentration is 1100 ng·mL$^{-1}$. Current literature data recommend that the therapeutic dose should not exceed 3000 ng·mL$^{-1}$; however, a systematic analysis of available patient data, where side effects were reported, appears to suggest that a much lower value (closer to 1500 ng·mL$^{-1}$) should be used [22].

Given the research discussed above, it is of great interest to develop new tools to measure plasma concentrations of different TKIs. Several methods have already been developed and published, mainly those connecting the use of liquid chromatography with mass spectrometry, as described previously [26–32]. For example, Merienne et al. presented a UPLC-MS/MS method for determination of 17 TKIs in one run using C18 column and LOD in 0.1 ng·mL$^{-1}$ levels [30]. Koller et al. developed a new clean-up procedure for analysis of 11 TKIs with Poroshell C18 column and similar LOD levels [31]. Further, supercritical fluid chromatography with MS was used for quantitation of 11 TKIs on a DIOL column with a gradient program [32].

The use of capillary electrophoresis, a technique for analysis of TKIs with a different separation principle based on the migration of ions in an electric field, has also been described in the literature. For example, Horská et al. [33] separated seven TKIs in 7 min using phosphate buffer pH 2.75. Rodriguez et al. [34] developed a non-aqueous CE method for analysis of imatinib with its metabolite and two analgesics using a background electrolyte containing ammonium acetate and acetic acid in methanol. Gonzales et al. [35] compared analysis of dasatinib via UHPLC and CE, both with UV detection. An LOD more than one-hundred-times better was obtained by UHPLC. These papers only described the use of low-sensitivity UV detection, as the optical path for UV detection is extremely low in CE. This problem can be overcome by using different type of detection, such as mass spectrometry (which also offers identification of the compounds of interest), or by using offline or online preconcentration techniques. Here, Rodriguez et al. [36] achieved a 70 ng·mL$^{-1}$ LOD for sunitinib using CE connected to TOF-MS. Forough et al. [37] used nanocomposite-based electromembrane extraction followed by field-amplified sample injection online preconcentration in CE with UV detection, resulting in a determination of imatinib with an LOD of 6.24 ng·mL$^{-1}$. Nanomaterials, especially multi-walled carbon nanotubes, were also used for dispersive solid-phase extraction prior to CE-UV of dabrafenib and trametinib in serum samples (with LODs from 8 ng·mL$^{-1}$) [38]. The Perrin group published a set of papers [39–41] dealing with salting-out procedures for determination of TKIs in plasma samples with LODs higher than 16 ng·mL$^{-1}$. Acetonitrile (ACN) was used to precipitate proteins from plasma samples followed by the addition of NaCl to the ACN–plasma mixture to introduce two-phase separation. Next, the high ACN content allowed for stacking of the analytes in CE. Finally, in 2020, Zhao et al. [42] used CE-MS with a field-amplified sample stacking method for the determination of imatinib and its metabolites with an LOD of 0.2 ng·mL$^{-1}$, which is fully comparable with those obtained by UPLC-MS/MS.

Given the above, CE-MS seems to be a good alternative to HPLC-MS or UHPLC-MS for TKIs. Moreover, in contrast to HPLC (and UHPLC), CE is considered a green (environment-friendly) technique, as it uses smaller amounts of solvents and samples. Considering this, the aim of the current work was to develop a CE-MS/MS method for the determination of multiple TKIs, dasatinib, erlotinib, canertinib, and bosutinib (Figure 1), using an online preconcentration strategy. These four TKIs were chosen because of their use in clinical praxis and their different structural features. In addition, they had not been previously analyzed by CE-MS.

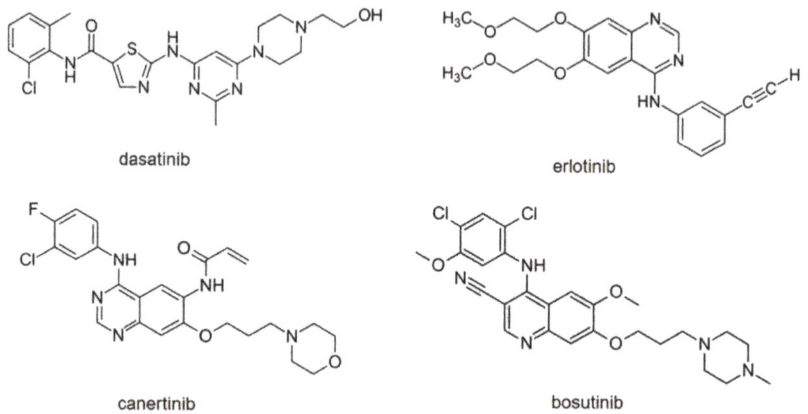


**Figure 1.** Structures of model TKIs.

## 2. Results

### 2.1. Method Development

In our previous work [33], the model TKIs were separated by CZE with UV detection under acidic conditions with non-volatile phosphate buffer pH 2.75. The LODs were incomparable with the levels of these TKIs in plasma samples. However, the right pH adjustment was necessary to obtain separation. Hence, first, the effect of pH was studied using 200 mM, 100 mM, 50 mM, and 10 mM formic acid (pH 2.2, 2.4, 2.5, and 2.9, respectively) and 100 mM, 50 mM, and 10 mM acetic acid (pH 2.9, 3.0, and 3.4, respectively) as background electrolytes (BGEs). The decrease in pH led to an increase in the duration of analysis from 3 min (pH 3.4) to more than 60 min (pH 2.2). This was likely due to differences in the electroosmotic flow and sucking of MS. The best separation (the highest resolution between all the peaks) was achieved at pH 2.5. Hence, 50 mM formic acid was used in further experiments.

Next, MS conditions were optimized. First, the effect of sheath liquid composition on TKI signals as well as background noise was studied. Initially, water–methanol mixtures at ratios of 25:75, 50:50, and 75:25 ($v/v$) were evaluated (without the addition of formic acid). The TKI signals were the highest for the 25:75 (water–methanol, $v/v$) ratio, Figure 2a; the same results were observed for the signal/noise ratio. Then, the effect of formic acid presence (0%, 0.1%, 0.5%; $v/v$) was analyzed (75% methanol, 25% or 24.9% or 24.5% water; $v/v$). The addition of 0.1% ($v/v$) had a dramatic effect on the erlotinib signal while the other TKIs had similar signals to those without any formic acid (Figure 2b). The higher concentration (0.5%; $v/v$) did not have any positive impact on the signal intensity. Similar effects were observed for the signal/noise ratio. Hence, the addition of 0.1% ($v/v$) formic acid was used for further optimization. Additional parameters included MS operational parameters, such as electrospray voltage (3.0–4.5 kV), drying gas temperature (200–300 °C), nebulizing gas flow rate (5–13 L min$^{-1}$), nebulizing gas pressure (10–20 psi), and sheath liquid flow rate (0.4–1.0 μL·min$^{-1}$). The highest TKI signals were observed at an electrospray voltage of 4.0 kV, drying gas temperature 250 °C, nebulizing gas flow rate 5 L·min$^{-1}$, nebulizing gas pressure 15 psi, and sheath liquid flow rate 0.6 μL·min$^{-1}$.

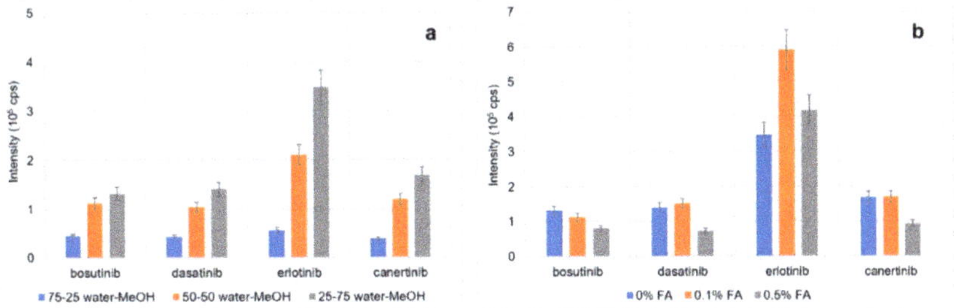


**Figure 2.** Optimization of the sheath liquid composition. (**a**) Optimization of water–methanol ratio, (**b**) effect of formic acid content; error bars represent SD values (n = 3).

To obtain the correct single reaction monitoring (SRM) transitions for both determination and identification, all the TKIs were fragmented by increasing the collision energy from 0 eV to 50 eV. The SRM transitions used for the quantitation and identification are listed in Table 1. They also correspond with previously published literature [30,43].

**Table 1.** SRM transitions of model TKIs.

| Compound | Parent Ion (*m*/*z*) | Quantification Transition Ion (*m*/*z*) | Collision Energy for Quantification Transition (eV) | Confirmation Transition Ion (*m*/*z*) | Collision Energy for Confirmation Transition (eV) |
|---|---|---|---|---|---|
| Dasatinib | 488 | 401 | 20 | 232 | 40 |
| Erlotinib | 394 | 336 | 20 | 278 | 30 |
| Canertinib | 486 | 128 | 15 | 100 | 40 |
| Bosutinib | 530 | 141 | 10 | 113 | 50 |

### 2.2. Stacking Preconcentration

To decrease the LOD values, sample stacking preconcentration was applied. This technique is based on an extended injection of analytes in a diluted (low-conductivity) electrolyte to the background electrolyte (higher conductivity). The differences in conductivities, reflected in differences in the electric field strengths, lead to the slowdown of analytes at the boundary between the sample and electrolyte plugs and their preconcentration [44,45]. In this work, a methanol–BGE mixture (90:10, $v/v$) was used as the TKI background to ensure some minimal conductivity of the sample zone when using an extended injection. Next, the effect of injection time of TKIs was studied in a range of 5 to 120 s. A linear increase in the TKI peak area was observed up to 60 s. LOD values were roughly estimated to be about $1 \times 10^{-7}$ mol·L$^{-1}$. This is quite high given practical (therapeutic drug monitoring) applications. In addition, the resolution between the TKIs rapidly decreased (with an 85 cm-long capillary) and there was not enough time for stacking. Roughly, a length of about 6 cm, representing about 7% of the capillary length, is injected within 60 s. Theoretically, about 40 cm of the capillary length is needed for the separation of the TKIs (the mobilities of TKIs are: bosutinib 29.7, canertinib 27.5, dasatinib 22.7, and erlotinib 19.6, all $10^{-9}$ m$^2$V$^{-1}$s$^{-1}$). The motion of TKIs in the rest of the capillary is likely caused by the suction of MS.

Therefore, using a longer capillary may be the solution to improving the LOD values. If a longer capillary is used, the preconcentration ratio should be the same or better, and the TKIs should separate. Hence, a capillary of 120 cm was applied. Here, an injection time of 120 s, two-times longer than that used with an 85 cm-long capillary, represents the same portion of the capillary length (7%; 8.5 cm). With the new capillary (120 cm), the effect of the injection time of TKIs was studied in a range of 60 to 150 s (150 s represents an injection of 8.9% of the capillary length). A linear increase in the TKI peak area was observed up to 120 s for bosutinib, canertinib, and erlotinib, and up to 150 s for dasatinib. Moreover, the effect of stacking on the peak height was studied (Figure 3). The peak height linearly increased with the injection time. Hence, an injection time of 120 s was chosen for further experiments.

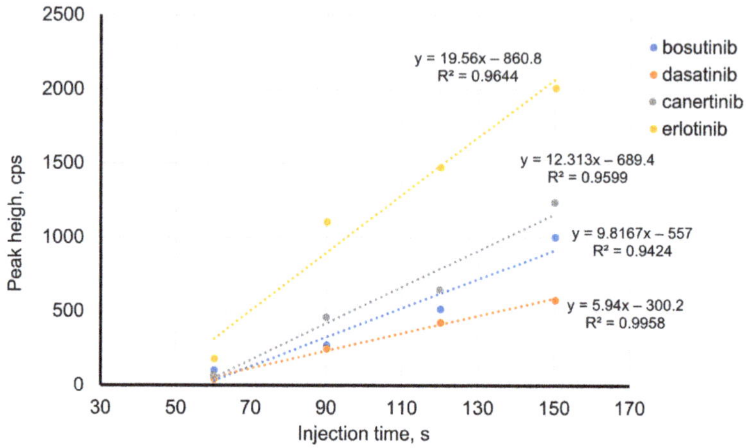


**Figure 3.** Effect of injection time on TKIs' peak height BGE: 50 mM formic acid, pH 2.5; 120 cm-long capillary; injection pressure 50 mbar; voltage of 20 kV; concentration of TKIs: $10^{-7}$ mol·L$^{-1}$.

### 2.3. *Analysis of Spiked Plasma Samples*

Finally, the CE-MS/MS method was applied for the analysis of spiked plasma samples. The spiked plasma was deproteinated using trichloroacetic acid. The final sample was diluted with methanol to decrease the conductivity of the sample, allowing for the stacking preconcentration. An example of the TKI analysis at a concentration level of $10^{-7}$ mol·L$^{-1}$ is given in Figure 4. It can be seen that the method allows for determination of TKIs at low concentration levels with high resolution.

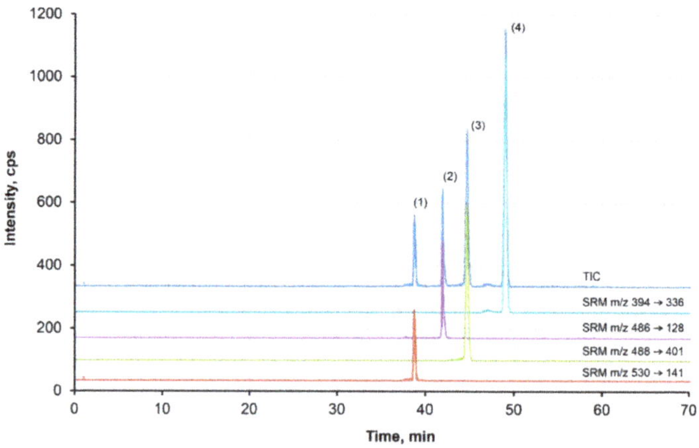


**Figure 4.** Example of analysis of the model TKIs by CE-MS/MS Peaks: (1) bosutinib, (2) canertinib, (3) dasatinib, (4) erlotinib; BGE: 50 mM formic acid, pH 2.5; 120 cm-long capillary; injection 120 s by 50 mbar; voltage of 20 kV; MS conditions: $U_{ESI}$ = 4.0 kV, T = 250°C, 5 μL/min, 15 psi, sheath liquid: 75:24.9:0.1 MeOH, H2O, formic acid (*v*/*v*), flow rate 0.6 μL·min$^{-1}$; SRM for quantification: bosutinib *m*/*z* 530 → *m*/*z* 141, 10 eV; dasatinib *m*/*z* 488 → *m*/*z* 401, 20 eV; erlotinib *m*/*z* 394 → *m*/*z* 336, 20 eV; canertinib *m*/*z* 486 → *m*/*z* 128, 15 eV; sample: deproteinated plasma with addition of 1·$10^{-7}$ mol·L$^{-1}$ TKIs.

Finally, the method for the determination of TKIs in plasma samples was validated in terms of linearity, LOD, LOQ, repeatability, selectivity, and recovery (Table 2). All calibrations were linear with correlation coefficients higher than 0.985. The LOD and LOQ values were in a range of 3.9–23.0 nmol·L$^{-1}$ (1.5–11.5 ng·mL$^{-1}$) and 11.9–69.7 nmol·L$^{-1}$ (4.7–34.7 ng·mL$^{-1}$), respectively. The selectivity was evaluated by comparing analyses of extracted blank plasma (without the addition of TKIs) with the analyses of TKIs at LOQ levels. There was no signal overlap. The intraday and interday repeatability (precision) of migration times varied between 0.5 and 3.2%, and 1.5 and 3.9%, respectively. The repeatability of peak heights was less than 3.2% (intraday) and 8.1% (interday). The trueness, expressed as the recovery, ranged from 96% to 103%. As can be seen in Table 2, all the data are fully acceptable for trace analysis; therefore, the method could be employed in routine therapeutic drug monitoring.

**Table 2.** Summary of the method validation.

| Parameter | Analyte | | | |
|---|---|---|---|---|
| | **Bosutinib** | **Dasatinib** | **Canertinib** | **Erlotinib** |
| Calibration range (mol × L$^{-1}$) | $1 \times 10^{-8}$–$1 \times 10^{-5}$ | $3\times10^{-9}$–$1 \times 10^{-5}$ | $3\times10^{-9}$–$1 \times 10^{-5}$ | $3\times10^{-9}$–$1 \times 10^{-5}$ |
| Calibration equation | $y = 1.528 \times 10^8x + 145$ | $y = 1.148 \times 10^8x + 94$ | $y = 4.977 \times 10^8x + 167$ | $y = 8.417 \times 10^8x + 385$ |
| Correlation coefficient | 0.9921 | 0.9849 | 0.9976 | 0.9943 |
| LOD (nmol·L$^{-1}$) | 21.6 | 23.0 | 8.0 | 3.9 |
| LOQ (nmol·L$^{-1}$) | 65.4 | 69.7 | 24.1 | 11.9 |
| Intraday repeatability of migration time (%) | 0.55 | 2.11 | 1.01 | 3.18 |
| Interday repeatability of migration time (%) | 1.76 | 3.76 | 1.57 | 3.85 |
| Intraday repeatability of peak heights (%) | 3.17 | 1.94 | 2.62 | 2.20 |
| Interday repeatability of peak heights (%) | 5.89 | 5.21 | 8.11 | 3.60 |
| Recovery (%) | 103.2 | 96.4 | 96.1 | 101.5 |

## 3. Discussion

In this work, a novel CE-MS/MS method for the determination of four model TKIs, namely bosutinib, dasatinib, canertinib, and erlotinib, was developed. The stacking online preconcentration technique was used to achieve LOD and LOQ values comparable with HPLC-MS methods, allowing for determination of all drugs at physiological levels in plasma samples. A simple extension of the capillary length (from 85 cm to 120 cm, representing a 41% increase) led to a significant decrease in LOD and LOQ values to ng·mL$^{-1}$ levels (LODs of 1.5–11.5 ng·mL$^{-1}$; LOQs of 4.7–34.7 ng·mL$^{-1}$). Theoretically, the LOD and LOQ values can be improved by an extension of the capillary length. However, the increase in separation distance also led to an increase in analysis time, in some cases. In this work, the analysis lasted for 60 min; in contrast, LC-MS could finish in approx. 10 min (without column equilibration) [30,46]. This is, of course, a drawback of the CE-MS/MS method. However, this weakness is balanced by the reduced consumption of chemicals and samples by the "environmental friendliness" of CE-MS. Subsequently, LOD values were compared with a CE-MS/MS method without any preconcentration (5 s injection by 50 mbar); the preconcentration factors ranged between 12.6 (dasatinib) and 14.4 (bosutinib). Since the levels of these drugs are in ng·mL$^{-1}$ concentrations in plasma [47,48], the method could be employed in routine therapeutic drug monitoring. In conclusion, a new CE-MS/MS method for determination of certain TKIs in plasma samples was developed and validated.

## 4. Materials and Methods

### *4.1. Chemicals and Materials*

Chemicals (mainly acetic acid, formic acid, trichloroacetic acid, sodium hydroxide solution (0.1 mol·L$^{-1}$), methanol, isopropanol, water) and standards (dasatinib, erlotinib, canertinib, and bosutinib), all of analytical grade or higher (solvents of HPLC-grade) purity, were bought from Sigma-Aldrich (St. Louis, MO, USA). Deionized water with resistivity of 18.2 MΩ.cm was prepared by the MilliQ system from Millipore (Molsheim, France).

Background electrolytes (BGEs were prepared by dissolving corresponding volumes of acids in HPLC-grade water. The pH was measured using an inoLab (WTW, Weilheim, Germany) pH meter. The ionic strength was calculated using Peakmaster software [49]. Finally, all the BGEs were filtered using 0.45 µm PTFE syringe filters (Labicom, Czech Republic).

The blood plasma sample was obtained from Sigma-Aldrich (St. Louis, MO, USA). The sample was spiked with the model TKIs and deproteinated as follows: 100 µL of the sample was mixed with 15 µL of cold trichloroacetic acid and shaken for 15 min. Then, it was centrifuged at 12,000× *g* for 5 min. Finally, the supernatant (50 µL) was carefully transferred to the sample vial for CE analysis and diluted with BGE–methanol mixture (80:20, *v*/*v*, 50 µL).

### *4.2. CE-MS*

All the experiments were performed using an Agilent 7100 capillary electrophoresis instrument, which was connected to an Agilent 6460 triple quadrupole mass spectrometer (Waldbronn, Germany). The sheath liquid was delivered into the electrospray interface via an isocratic LC pump Agilent 1260 with a 1:100 flow splitter.

Separations were performed in fused silica capillaries of 85 cm or 120 cm in length (the effective length was the same) 50 µm ID, from Molex (Lisle, IL, USA). Prior to first use, the capillaries were initially conditioned by rinsing them with 0.1 mol·L$^{-1}$ NaOH for 20 min and then deionized water for 30 min, out of the MS. Between each sample run, the capillary was flushed with 0.1 mol·L$^{-1}$ NaOH for 10 s, HPLC-grade water for 3 min, and BGE for 3 min. All the rinsing was carried out at a pressure drop of 935 mbar. The capillary cassette was thermostated at 25 °C, except for the part of the cassette leading to the MS interface. Each experiment was conducted in triplicate, unless stated otherwise.

### *4.3. Validation*

The method was validated using the following performance characteristics: linearity, limit of detection (LOD), limit of quantification (LOQ), repeatability of migration time and peak area, and recovery. Linearity was tested using calibration within a concentration range of $1 \times 10^{-9}$–$1 \times 10^{-5}$ mol·L$^{-1}$ for all the TKIs. LODs and LOQs were calculated according to equations: LOD = 3.3 SD/s and LOQ = 10 SD/s, where SD is the standard deviation of the signal intensity and s is the slope of the calibration curve. The selectivity was investigated by comparing analyses of blank plasma without the addition of TKIs with the analyses of TKIs at LOQ levels. Acceptance criteria for any interference included a signal response lower than 5% of the LOQ. The reproducibility of migration times and peak areas was calculated from repeated analyses at the $1 \times 10^{-7}$ mol·L$^{-1}$ level; the intraday repeatability was calculated from three repetitions within one day; and the interday repeatability was calculated from repetitive analyses on three consecutive days (with three repetitions each day). The recovery was evaluated using analyses of TKIs in blood samples spiked at the $1 \times 10^{-7}$ mol·L$^{-1}$ level; the recovery was calculated as the ratio of the determined TKI concentration and the true (added) concentration.

## 5. Conclusions

In this work, a novel CE-MS/MS method was employed for the determination of four model TKIs, namely bosutinib, dasatinib, canertinib, and erlotinib. The stacking online preconcentration technique was used with a 120 cm-long capillary to achieve LOD and LOQ values comparable with HPLC-MS methods, allowing for determination of all drugs at physiological levels in plasma samples. Under optimal conditions, 50 mM formic acid pH 2.5, an injection time of 120 s, and an optimized mass spectrometry set-up (sheath liquid composition 75:24.9:0.1 ($v/v$) methanol, water, formic acid, and appropriate conditions for ion transitions), LODs in a range of 3.9–23.0 nmol·$L^{-1}$ were observed.

**Funding:** This research was funded by the Czech Science Foundation (project 19-23033S).

**Institutional Review Board Statement:** Not applicable.

**Informed Consent Statement:** Not applicable.

**Data Availability Statement:** The data are available from the author upon reasonable request.

**Acknowledgments:** The author thanks Michaela Kotková for her pilot measurements on CE-MS.

**Conflicts of Interest:** The author declares no conflict of interest.

## References

1. Cismowski, M.J. *Tyrosine Kinase Inhibitors, in xPharm: The Comprehensive Pharmacology Reference*; Enna, S.J., Bylund, D.B., Eds.; Elsevier: Amsterdam, The Netherlands, 2007; pp. 1–4.
2. Shawver, L.K.; Slamon, D.; Ullrich, A. Smart drugs: Tyrosine kinase inhibitors in cancer therapy. *Cancer Cell* **2002**, *1*, 117–123. [CrossRef] [PubMed]
3. Zámečníková, A. Novel approaches to the development of tyrosine kinase inhibitors and their role in the fight against cancer. *Expert. Opin. Drug. Discov.* **2014**, *9*, 77–92. [CrossRef] [PubMed]
4. Huang, L.; Jiang, S.; Shi, Y. Tyrosine kinase inhibitors for solid tumors in the past 20 years (2001–2020). *J. Hematol. Oncol.* **2020**, *13*, 143. [CrossRef]
5. Ates, H.C.; Roberts, J.A.; Lipman, J.; Cass, A.E.G.; Urban, G.A.; Dincer, C. On-site therapeutic drug monitoring. *Trends Biotechnol.* **2020**, *38*, 1262–1277. [CrossRef] [PubMed]
6. Garzón, V.; Pinacho, D.G.; Bustos, R.-H.; Garzón, G.; Bustamante, S. Optical biosensors for therapeutic drug monitoring. *Biosensors* **2019**, *9*, 132. [CrossRef] [PubMed]
7. Figueras, A. *Review of the Evidence to Include TDM in the Essential In Vitro Diagnostics List and Prioritization of Medicines to be Monitored*; Fundació Institut Català de Farmacologia: Barcelona, Spain, 2019.
8. Kang, J.-S.; Lee, M.-H. Overview of therapeutic drug monitoring. *Korean J. Intern. Med.* **2009**, *24*, 2687654. [CrossRef]
9. Paal, M.; Zoller, M.; Schuster, C.; Vogeser, M.; Schütze, G. Simultaneous quantification of cefepime, meropenem, ciprofloxacin, moxifloxacin, linezolid and piperacillin in human serum using an isotope-dilution HPLC–MS/MS method. *J. Pharm. Biomed. Anal.* **2018**, *152*, 102–110. [CrossRef]
10. Bustos, R.; Zapata, C.; Esteban, E.; García, J.-C.; Jáuregui, E.; Jaimes, D. Label-free quantification of anti-TNF-α in patients treated with adalimumab using an optical biosensor. *Sensors* **2018**, *18*, 691. [CrossRef]
11. Suntornsuk, L. Recent advances of capillary electrophoresis in pharmaceutical analysis. *Anal. Bioanal. Chem.* **2010**, *398*, 29–52. [CrossRef]
12. Torano, J.S.; Ramautar, R.; de Jong, G. Advances in capillary electrophoresis for the life sciences. *J. Chromatogr. B* **2019**, *1118–1119*, 116–136. [CrossRef]
13. Voeten, R.L.C.; Ventouri, I.K.; Haselberg, R.; Somsen, G.W. Capillary electrophoresis: Trends and recent advances. *Anal. Chem.* **2018**, *90*, 1464–1481. [CrossRef] [PubMed]
14. Ranasinghe, M.; Quirino, J.P. Can we replace liquid chromatography with the greener capillary electrophoresis? *Curr. Opin. Green Sustain. Chem.* **2021**, *31*, 100515. [CrossRef]
15. Kitagawa, F.; Otsuka, K. Recent applications of on-line sample preconcentration techniques in capillary electrophoresis. *J. Chromatogr. A* **2014**, *1335*, 43–60. [CrossRef] [PubMed]
16. Grochocki, W.; Markuszewski, M.J.; Quirino, J.P. Different detection and stacking techniques in capillary electrophoresis for metabolomics. *Anal. Methods* **2016**, *8*, 1216–1221. [CrossRef]
17. Gackowski, M.; Przybylska, A.; Kruszewski, S.; Koba, M.; Madra-Gackowska, K.; Bogacz, A. Recent applications of capillary electrophoresis in the determination of active compounds in medicinal plants and pharmaceutical formulations. *Molecules* **2021**, *26*, 4141. [CrossRef] [PubMed]
18. Petr, J.; Jiang, C.X.; Ševčík, J.; Tesařová, E.; Armstrong, D.W. Sterility testing by CE: A comparison of online preconcentration approaches in capillaries with greater internal diameters. *Electrophoresis* **2009**, *30*, 3870–3876. [CrossRef]
19. Mičová, K.; Friedecký, D.; Faber, E.; Polýnková, A.; Adam, T. Flow injection analysis vs. ultra high performance liquid chromatography coupled with tandem mass spectrometry for determination of imatinib in human plasma. *Clin. Chim. Acta* **2010**, *411*, 1957–1962. [CrossRef] [PubMed]
20. Friedecký, D.; Mičová, K.; Faber, E.; Hrdá, M.; Široká, J.; Adam, T. Detailed study of imatinib metabolization using high-resolution mass spectrometry. *J. Chromatogr. A* **2015**, *1409*, 173–181. [CrossRef]

21. Miura, M.; Takahashi, N. Routine therapeutic drug monitoring of tyrosine kinase inhibitors by HPLC-UV or LC-MS/MS methods. *Drug Metabol. Pharmacokin.* **2016**, *31*, 12–20. [CrossRef]
22. Pearce, C.M.; Resmini, M. Towards point of care systems for the therapeutic drug monitoring of imatinib. *Anal. Bioanal. Chem.* **2020**, *412*, 5925–5933. [CrossRef]
23. Picard, S.; Titier, K.; Etienne, G.; Teilhet, E.; Ducint, D.; Bernard, M.-A.; Lassalle, R.; Marit, G.; Reiffers, J.; Begaud, B.; et al. Trough imatinib plasma levels are associated with both cytogenetic and molecular responses to standard-dose imatinib in chronic myeloid leukemia. *Blood* **2007**, *109*, 3496–3499. [CrossRef] [PubMed]
24. Peng, B.; Lloyd, P.; Schran, H. Clinical pharmacokinetics of imatinib. *Clin. Pharmacokinet.* **2005**, *44*, 879–894. [CrossRef] [PubMed]
25. Haouala, A.; Widmer, N.; Guidi, M.; Montemurro, M.; Leyvraz, S.; Buclin, T.; Eap, C.B.; Decosterd, L.A.; Csajka, C. Prediction of free imatinib concentrations based on total plasma concentrations in patients with gastrointestinal stromal tumours. *Br. J. Clin. Pharmacol.* **2013**, *75*, 1007–1018. [CrossRef]
26. Pasquini, B.; Orlandini, S.; Furlanetto, S.; Gotti, R.; Del Bubba, M.; Boscaro, F.; Bertaccini, B.; Douša, M.; Pieraccini, G. Quality by design as a risk-based strategy in pharmaceutical analysis: Development of a liquid chromatography—Tandem mass spectrometry method for the determination of nintedanib and its impurities. *J. Chromatogr. A* **2020**, *1611*, 460615. [CrossRef]
27. Matsumoto, A.; Shiraiwa, K.; Suzuki, Y.; Tanaka, K.; Kawano, M.; Iwasaki, T.; Tanaka, R.; Tatsuta, R.; Tsumura, H.; Itoh, H. Sensitive quantification of free pazopanib using ultra-high performance liquid chromatography coupled to tandem mass spectrometry and assessment of clinical application. *J. Pharm. Biomed. Anal.* **2021**, *206*, 114348. [CrossRef] [PubMed]
28. Ferrer, F.; Solas, C.; Giocanti, M.; Lacarelle, B.; Deville, J.-L.; Gravis, G.; Ciccolini, J. A simple and rapid liquid chromatography—Mass spectrometry method to assay cabozantinib in plasma: Application to therapeutic drug monitoring in patients with renal cell carcinoma. *J. Chromatogr. B* **2020**, *1138*, 121968. [CrossRef]
29. Ezzeldin, E.; Iqbal, M.; Herqash, R.N.; ElNahhas, T. Simultaneous quantitative determination of seven novel tyrosine kinase inhibitors in plasma by a validated UPLC-MS/MS method and its application to human microsomal metabolic stability study. *J. Chromatogr. B* **2020**, *1136*, 121851. [CrossRef] [PubMed]
30. Merienne, C.; Rousset, M.; Ducint, D.; Castaing, N.; Titier, K.; Molimard, M.; Bouchet, S. High throughput routine determination of 17 tyrosine kinase inhibitors by LC–MS/MS. *J. Pharm. Biomed. Anal.* **2018**, *150*, 112–120. [CrossRef]
31. Koller, D.; Vaitsekhovich, V.; Mba, C.; Steegmann, J.L.; Zubiaur, P.; Abad-Santos, F.; Wojnicz, A. Effective quantification of 11 tyrosine kinase inhibitors and caffeine in human plasma by validated LC-MS/MS method with potent phospholipids clean-up procedure. Application to therapeutic drug monitoring. *Talanta* **2020**, *208*, 120450. [CrossRef]
32. Zhang, S.; Jin, W.; Yang, Y. Simultaneous identification and determination of eleven tyrosine kinase inhibitors by supercritical fluid chromatography—Mass spectrometry. *Anal. Methods* **2019**, *11*, 2211–2222. [CrossRef]
33. Horská, J.; Ginterová, P.; Ševčík, J.; Petr, J. CZE separation of new drugs for treatment of leukemia. *Chromatographia* **2014**, *77*, 1477–1482. [CrossRef]
34. Rodriguez, J.; Castaneda, G.; Munoz, L.; Lopez, S. Development and validation of a non-aqueous capillary electrophoresis method for the determination of imatinib, codeine and morphine in human urine. *Anal. Methods* **2014**, *6*, 3842–3848. [CrossRef]
35. Gonzales, A.G.; Taraba, T.; Hraníček, J.; Kozlík, P.; Coufal, P. Determination of dasatinib in the tablet dosage form by ultra high performance liquid chromatography, capillary zone electrophoresis, and sequential injection analysis. *J. Sep. Sci.* **2017**, *40*, 400–406. [CrossRef]
36. Rodriguez, J.; Castaneda, G.; Munoz, L.; Villa, J.C. Quantitation of sunitinib, an oral multitarget tyrosine kinase inhibitor, and its metabolite in urine samples by nonaqueous capillary electrophoresis time of flight mass spectrometry. *Electrophoresis* **2015**, *36*, 1580–1587. [CrossRef]
37. Forough, M.; Farhadi, K.; Eyshi, A.; Molaei, R.; Khalili, H.; Kouzegaran, V.J.; Matin, A.A. Rapid ionic liquid-supported nano-hybrid composite reinforced hollow-fibed electromembrane extraction followed by field-amplified sample injection-capillary electrophoresis: An effective approach for extraction and quantification of imatinib mesylate in human plasma. *J. Chromatogr. A* **2017**, *1516*, 21–34. [PubMed]
38. Sanz, I.L.; Bernardo, F.J.G.; Penalvo, G.C.; Flores, J.R. Determination of dabrafenib and trametinib in serum by dispersive solid phase extraction with multi-walled carbon nanotubes and capillary electrophoresis coupled to ultraviolet/visible detection. *Microchem. J.* **2021**, *165*, 160180.
39. Ahmed, O.S.; Ladner, Y.; Montels, J.; Philibert, L.; Perrin, C. Coupling of salting-our assisted liquid-liquid extraction with on-line stacking for the analysis of tyrosine kinase inhibitors in human plasma by capillary zone electrophoresis. *J. Chromatogr. A* **2018**, *1579*, 121–128. [CrossRef] [PubMed]
40. Ahmed, O.S.; Ladner, Y.; Xia, J.; Montels, J.; Philibert, L.; Perrin, C. A fully automated on-line salting-out assisted liquid-liquid extraction capillary electrophoresis methodology: Application to tyrosine kinase inhibitors in human plasma. *Talanta* **2020**, *208*, 120391. [CrossRef] [PubMed]
41. Ahmed, O.S.; Ladner, Y.; Bousquet, C.; Montels, J.; Dubský, P.; Philibert, L.; Perrin, C. Direct salting-out assisted liquid-liquid extraction (SALLE) from human blood: Application for the analysis of tyrosine kinase inhibitors. *Microchem. J.* **2020**, *155*, 104791. [CrossRef]
42. Zhao, T.; Wang, L.; Chen, D.D.Y. Quantification of imatinib and related compounds using capillary electrophoresis—Tandem mass spectrometry with field-amplified sample stacking. *Electrophoresis* **2020**, *41*, 1843–1850. [CrossRef]

43. Niessen, W.M.A.; Rosing, H.; Beijnen, J.H. Interpretation of MS-MS spectra of small-molecule signal transduction inhibitors using accurate-m/z data and m/z-shifts with stable-isotope-labeled analogues and metabolites. *Int. J. Mass. Spectrom.* **2021**, *464*, 116559. [CrossRef]
44. Beckers, J.L.; Boček, P. Sample stacking in capillary zone electrophoresis: Principles, advantages and limitations. *Electrophoresis* **2000**, *21*, 2747–2767. [CrossRef] [PubMed]
45. Simpson, S.L.; Quirino, J.P.; Terabe, S. On-line sample preconcentration in capillary electrophoresis: Fundamentals and applications. *J. Chromatogr. A* **2008**, *1184*, 504–541. [CrossRef] [PubMed]
46. Kehl, N.; Schlichtig, K.; Dürr, P.; Bellut, L.; Dörje, F.; Fietkau, R.; Pavel, M.; Mackensen, A.; Wullich, B.; Maas, R.; et al. An easily expandable multi-drug LC-MS assay for the simultaneous quantification of 57 oral antitumor drugs in human plasma. *Cancers* **2021**, *13*, 6329. [CrossRef]
47. Miura, M. Therapeutic drug monitoring of imatinib, nilotinib, and patients with chronic myeloid leukemia. *Biol. Pharm. Bull.* **2015**, *38*, 645–654. [CrossRef]
48. García-Ferrer, M.; Wojnicz, A.; Mejía, G.; Koller, D.; Zubiaur, P.; Abad-Santos, F. Utility of therapeutic drug monitoring of imatinib, nilotinib, and dasatinib in chronic myeloid leukemia: A systematic review and meta-analysis. *Clin. Ther.* **2019**, *41*, 2258–2570.e7. [CrossRef]
49. Available online: https://web.natur.cuni.cz/gas/peakmaster.html (accessed on 15 March 2021).

pharmaceuticals

Article

# Development and Validation of an LC-MS/MS Method for the Quantitative Determination of Contezolid in Human Plasma and Cerebrospinal Fluid

Guanxuanzi Zhang [1,2,†], Na Zhang [2,†], Liuhan Dong [1,2], Nan Bai [2] and Yun Cai [2,*]

1 Medical School of Chinese PLA, Graduate School of Chinese PLA General Hospital, Beijing 100853, China
2 Center of Medicine Clinical Research, Department of Pharmacy, Medical Supplies Center, PLA General Hospital, Beijing 100853, China
* Correspondence: caicai_hh@126.com; Tel.: +86-10-6693-7166
† These authors contributed equally to this work.

**Abstract:** To develop and verify a liquid chromatography-tandem mass spectrometry (LC-MS/MS) method for determining contezolid in plasma and cerebrospinal fluid (CSF). Protein precipitation was performed on samples using linezolid as the internal standard. We used an Agilent EclipsePlus C18 column operating at 0.4 mL/min in conjunction with acetonitrile and water mobile phases for the LC-MS/MS analysis. Using the precursor-product ion pairs 409.15→269.14 (contezolid) and 338.14→195.1 (linezolid), multiple reaction monitoring was used to quantify the compounds. Plasma linearity range was 50.0 to 5000 ng/mL, and CSF was 20.0 to 1000 ng/mL ($r^2$ = 0.999). The inter-batch and intra-batch precisions were ≤2.57% and ≤5.79%, respectively. Plasma recovered 92.94%, and CSF recovered 97.83%. Plasma, CSF, hemolytic plasma, and hyperlipidemic plasma all showed a coefficient of variation ≤ 7.44%. The stability and dilution integrity of this method were also acceptable. The study also demonstrated that artificial CSF can be used as a matrix for the preparation of standard curve samples. A simple and accurate method was developed and validated for the determination of contezolid concentrations in human plasma and CSF, which may be useful for monitoring the therapeutic effect of central nervous system medications.

**Keywords:** contezolid; CSF; LC-MS/MS; plasma

**Citation:** Zhang, G.; Zhang, N.; Dong, L.; Bai, N.; Cai, Y. Development and Validation of an LC-MS/MS Method for the Quantitative Determination of Contezolid in Human Plasma and Cerebrospinal Fluid. *Pharmaceuticals* **2023**, *16*, 32. https://doi.org/10.3390/ph16010032

Academic Editors: Jan Oszmianski and Sabina Lachowicz-Wiśniewska

Received: 1 December 2022
Revised: 21 December 2022
Accepted: 23 December 2022
Published: 26 December 2022



## 1. Introduction

Antimicrobial resistance in bacteria is becoming increasingly serious, especially in methicillin-resistant *Staphylococcus aureus* (MRSA), methicillin-resistant coagulase-negative *Staphylococci*, penicillin-resistant *Streptococcus pneumoniae* (PRSP), vancomycin-resistant *enterococci* (VRE), vancomycin-intermediate *Staphylococcus aureus*, and vancomycin-resistant *Staphylococcus aureus* [1,2]. The treatment and control of infectious diseases are faced with severe challenges [3].

Oxazolidinone is a kind of synthetic antimicrobial agent [4]. These drugs have a unique mode of action, targeting a unique region of 23S rRNA adjacent to the peptidyl transferase center of the 50s ribosomal subunit, inhibiting protein biosynthesis in bacteria and used to treat severe infections caused by Gram-positive pathogens [5–8]. Linezolid is the first member of the oxazolidinone antibiotics and has good antibacterial activity against many important Gram-positive bacteria [9,10]. However, potential neurotoxicity and hematotoxicity, as well as significant inhibition of monoamine oxidase, limited the wide use of linezolid [11]. The homology between the human cell mitochondrial protein synthesis pathway and the bacterial ribosomal target also inhibits human cell mitochondrial protein synthesis, which could lead to lactic acidosis, myelosuppression, anemia, thrombocytopenia, pancytopenia, and other serious adverse drug reactions [12]. Contezolid (S)-5-([isoxazol-3-ylamino]methyl)-3-(2,3,5-trifluoro-4-[4-oxo-3,4-dihydropyridin-1(2H)-yl]phenyl), a new oxazolidinone antimicrobial agent, is an innovative antibiotic developed in China and has independent intellectual property rights. Its main active structure is the same as that of linezolid. Studies in vitro showed that contezolid had good antibacterial activity against MRSA, PRSP, and VRE and had no cross-resistance with existing antimicrobial agents [13,14]. Compared with linezolid, it shows better safety as well as minimal myelosuppression and inhibition of monoamine oxidase, which are independent adverse reactions associated with linezolid therapy [13]. Therefore, contezolid may provide a promising alternative therapy for infections with multidrug-resistant Gram-positive bacteria [15].

The methodology for analyzing contezolid in human body fluids has not been extensively studied, and only in two articles is it described how ultra-performance liquid chromatography-mass spectrometry can be used to analyze contezolid concentrations in human plasma and urine [16,17]. The method for the analysis of contezolid in human cerebrospinal fluid (CSF) has not been studied. Infections of the central nervous system (CNS) can be extremely harmful. These lead to fatal outcomes and long-term neurological problems in survivors, including cognitive deficits and motor impairments [18]. Pathogens with Gram-positive characteristics are common etiologic agents of infections of the CNS, including brain abscesses [19]. Antibiotics should be present at therapeutic concentrations at the site of infection if they are to be effective against CNS infections. But the existence of the blood-brain barrier made it difficult for antibiotics to enter the CNS, and it was difficult to obtain effective antibacterial concentration in the CSF and brain tissue, which seriously affected the antibacterial effect and resulted in poor treatment results, even led to ineffective treatment results [20]. Therefore, for new antimicrobials, it is necessary to develop a method for quantitative determination of CSF concentration, which is helpful to explore their efficacy in the CNS.

Our previous study has successfully verified the feasibility of liquid chromatography-tandem mass spectrometry (LC-MS/MS) for the analysis of drug concentrations in human plasma and CSF [21]. The objective of this study is to develop a method for the analysis of contezolid in human plasma and CSF by LC-MS/MS and to verify the methodology to provide an analysis method for the study and application of the drug in treating CNS infections.

## 2. Results and Discussion

This section may be divided into subheadings. It should provide a concise and precise description of the experimental results, their interpretation, and the experimental conclusions that can be drawn.

### 2.1. Method Development

At present, there are no known LC-MS/MS methods for the determination of contezolid in CSF. In this study, we propose to develop a method for the quantitative determination of contezolid in CSF by LC-MS/MS, except for plasma.

#### 2.1.1. Chromatography and MS Conditions

Several C18 columns and several gradient programs were tested for the chromatographic separation of target drugs. The Agilent Eclipse Plus C18 column (100 × 2.1 mm, 3.5 μm) provided the best chromatographic performance for all compounds with appropriate peak shape and sharpness. Moreover, the addition of 0.1% formic acid (FA) to the mobile phase acetonitrile: water (90:10, $v/v$) and the compatibility of the chromatographic column enabled the target compounds to be effectively retained and separated by chromatography. The Agilent Eclipse Plus C18 column showed durability and robustness under these conditions.

For MS conditions, usually when the polarity of the analyte is high, electrospray ionization (ESI) is selected. Generally, acidic compounds, which are compounds containing -OH, -COOH, and phenolic compounds, are determined by the negative ESI mode, and basic compounds, which are compounds containing more heteroatoms and nucleosides, are determined by the positive ESI mode. Contezolid has high polarity and is a basic compound containing N, so the ESI and positive ion modes were selected.

In order to optimize chromatographic separation, a series of preliminary experiments were carried out to test different mobile phases, including ammonium acetate, acetonitrile, the mixture of acetonitrile and ammonium, or water, as well as different concentrations of mobile phase additives, such as FA and acetic acid. In the positive ESI mode, the addition of FA reduced the tailing of the peak and improved the response of the target compounds. Finally, 0.1% FA in 10 mM aqueous ammonium acetate (A) and 0.1% FA in 5 mM ammonium acetate acetonitrile: water (90:10, $v/v$) solution (B) were selected as the best mobile phase.

For the purpose of establishing the multiple reaction monitoring (MRM) scanning mode, the product ions of each analyte were characterized by product ion scanning under 30 eV collision energy. The product ion spectra of each analyte were obtained. For all analytes, the two most abundant product ions were used as quantifier ions and qualifier ions, respectively. Quantitative determination was carried out by the $m/z$ transitions 409.15→269.14 for contezolid and 338.14→195.1 for linezolid.

#### 2.1.2. Sample Preparation

The preparation of samples through effective extraction steps is a common step for the determination of compounds in biological matrices. Yet the cost of the nitrogen used is high, and liquid-liquid extraction is relatively time consuming and difficult to operate. In this work, we considered the method of protein precipitation for sample preparation. Using methanol and acetonitrile as protein precipitates, the matrix effect and extraction recovery were investigated. The results showed that the method had satisfactory recovery and almost no matrix effect and could be used for the determination of clinical biological samples. Of course, this method also had some shortcomings. For example, it was difficult to remove salts and lipids from the matrix, which was easy to interfere with and affect the reproducibility and accuracy of the results. Moreover, the non-specific precipitation reaction might cause the loss of trace analytes along with the co-precipitation of matrix proteins. But the method had a simple operation procedure, shortened the extraction time, and saved the cost. It was suitable for hospitals to analyze clinical biological samples with large size.

### 2.2. Method Validation

#### 2.2.1. Calibration Curve

The calibration standard curves of three different batches were prepared for linear evaluation. Two set of calibration curve samples were prepared freshly on the day of analysis. Each calibration curve consisted of a double blank sample (without analyte or internal standard, IS), and a single blank sample (only IS). Eight concentration levels (50.0, 100, 250, 500, 1000, 2500, 4000 and 5000 ng/mL) were used; the linearity range of plasma was between 50.0 and 5000 ng/mL. Eight concentration levels (20.0, 40.0, 100, 200, 400, 600, 800 and 1000 ng/mL) were used; the linearity range of CSF was between 20.0 and 1000 ng/mL. The calibration curve was generated by the response of peak area ratio (y) and analyte concentration (x), and the linear $1/X^2$ weighted relation was used for regression. We also evaluated the batch size, and the results showed that each batch of plasma could be injected 136 times and CSF could be injected 101 times.

#### 2.2.2. Specificity and Selectivity

The specificity and selectivity results showed that the analyte or the IS in the relevant mass channels contributed less than 20% of the mean of the lower limit of quantification (LLOQ) and less than 5% of the IS response. The method produced a specific signal for the analyte, which allowed for the separation of contezolid from the IS and other components in samples. In this study, endogenous substances and IS did not affect the determination of contezolid (Figures 1–3). The matrix effect (ME) of contezolid at low quality control (LQC) and high quality control (HQC) was 101.68% and 100.37% in hemolysis, as well as 104.20% and 101.00% in hyperlipidemia (Table 1).

I

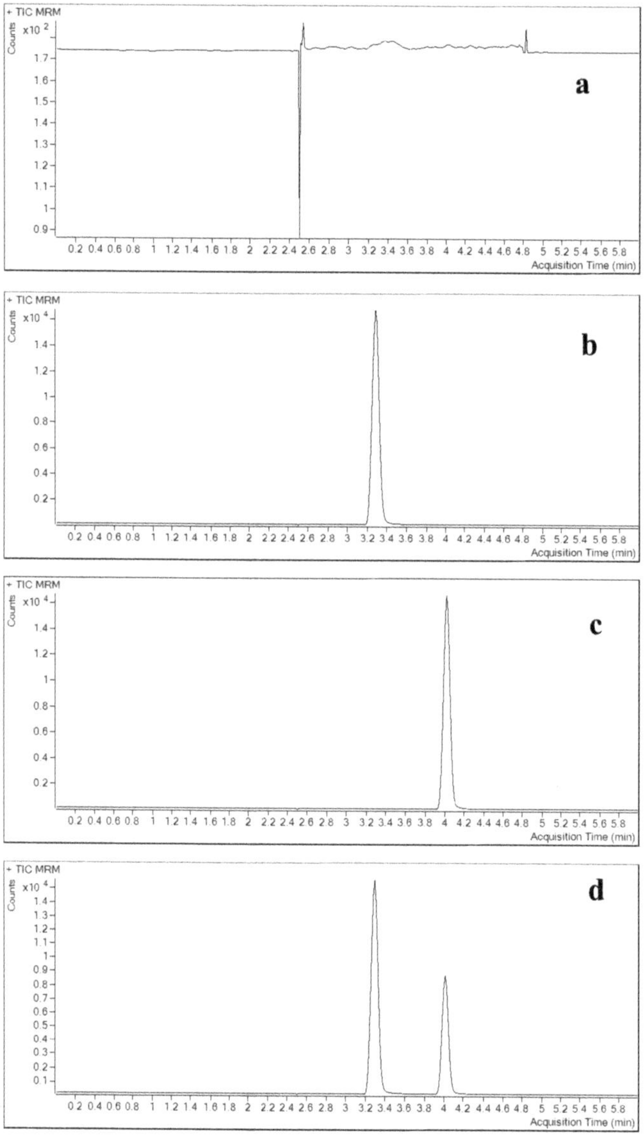


**Figure 1.** *Cont.*

II

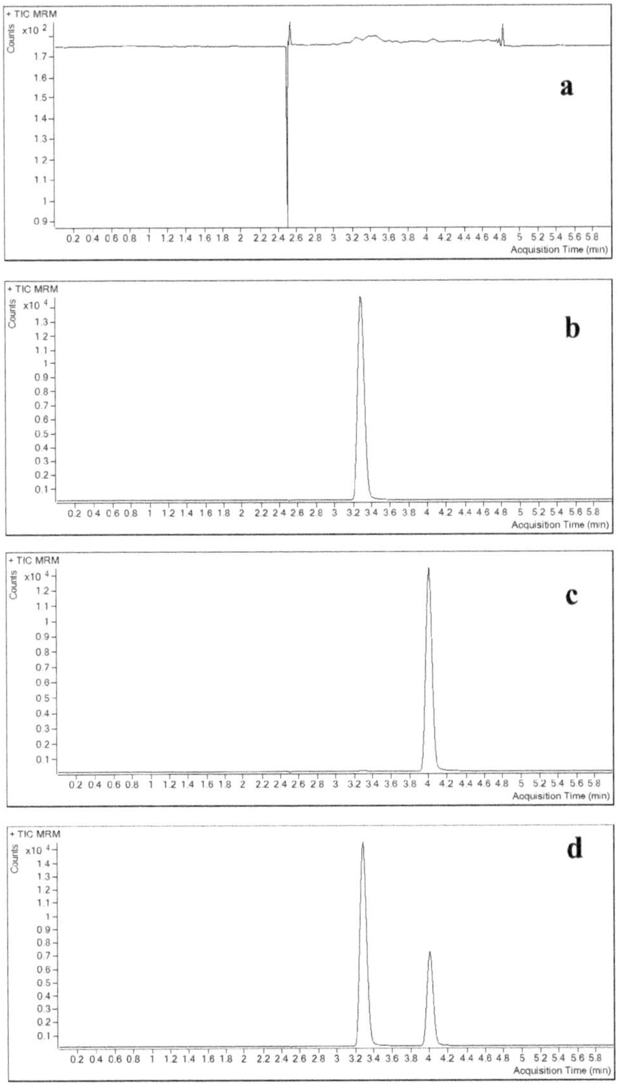

**Figure 1.** *Cont.*

**III**

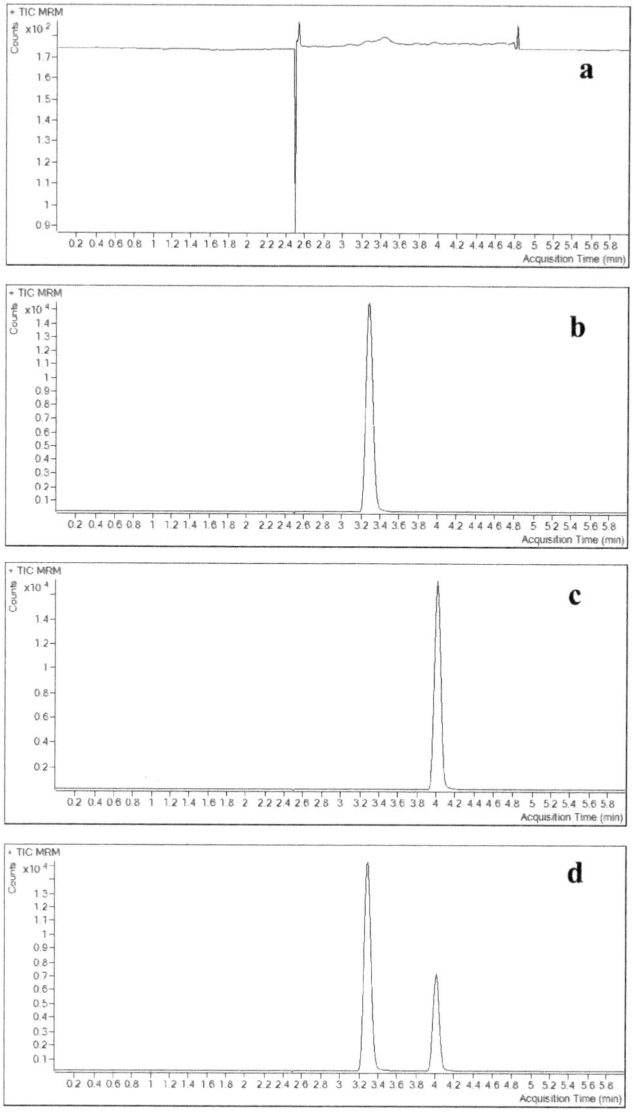


**Figure 1.** *Cont.*

**IV**

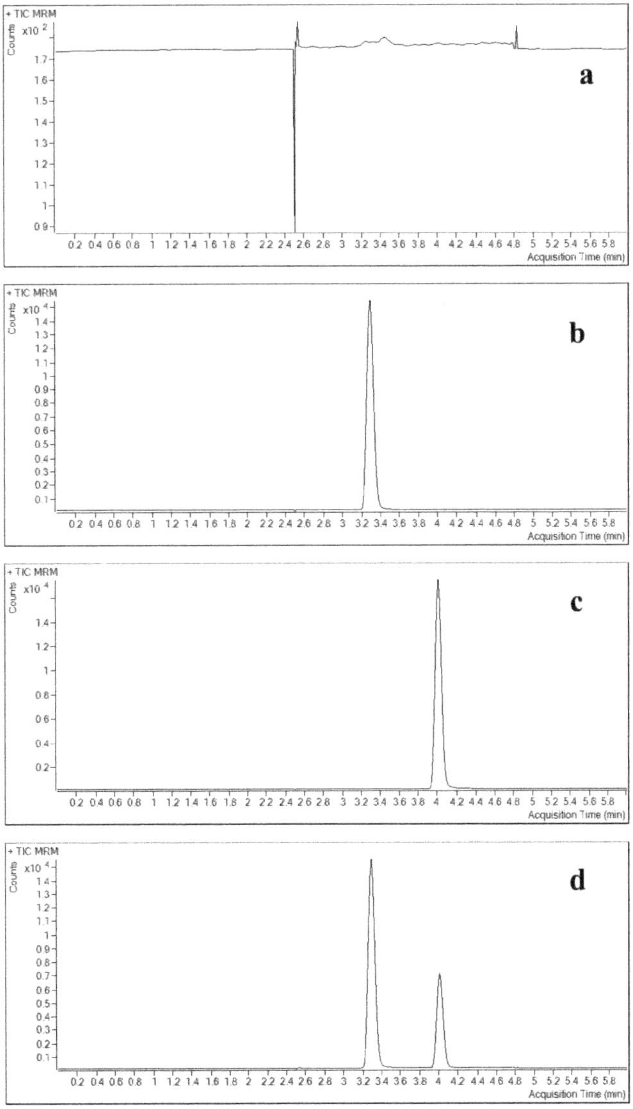


**Figure 1.** *Cont.*

**V**

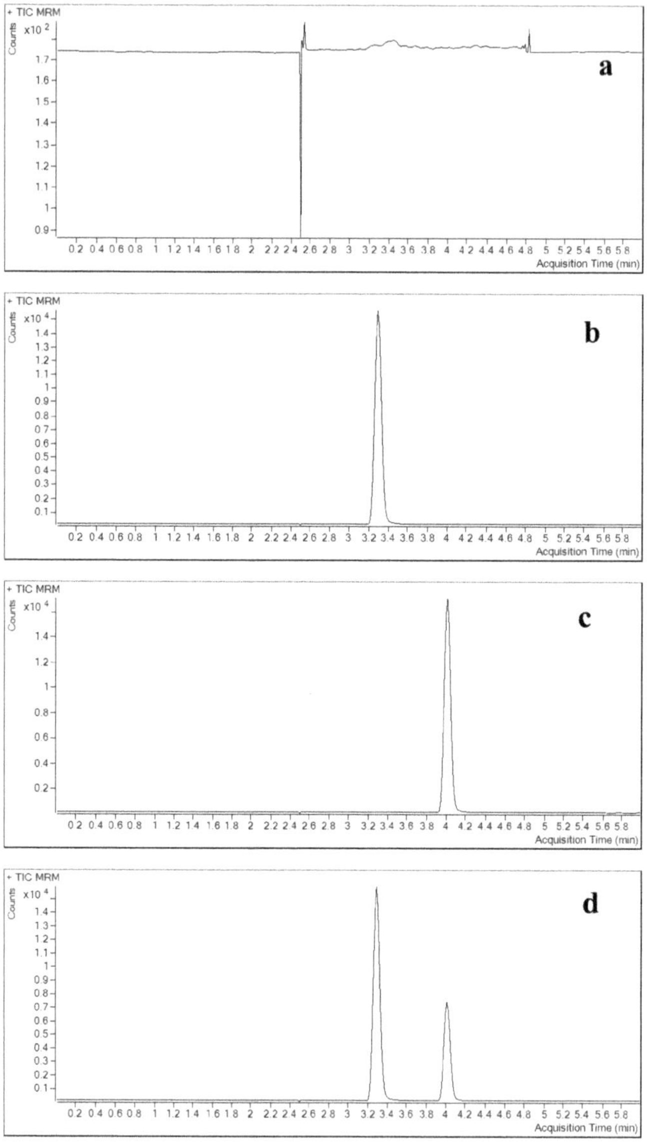


**Figure 1.** *Cont.*

VI

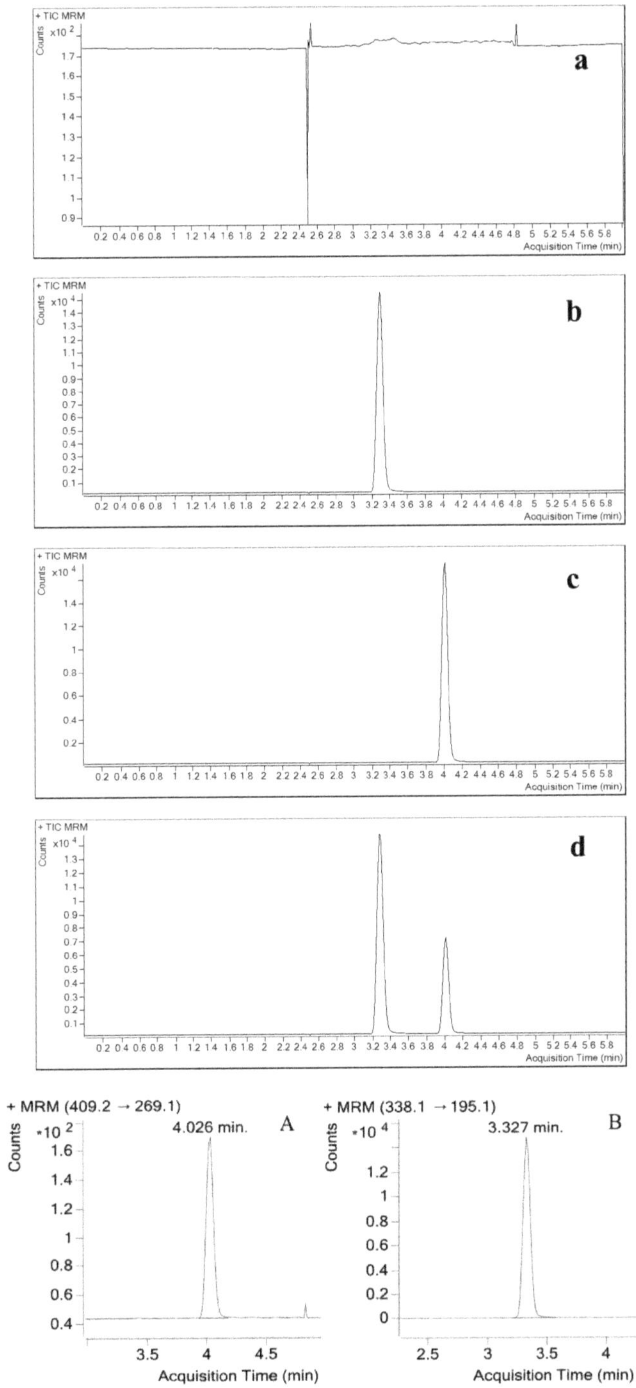


**Figure 1.** Specificity and selectivity of the method in plasma. (**a**) double blank, (**b**) single blank, (**c**) analyte only and (**d**) analyte and IS (in six different lots). Chromatograms of contezolid (**A**) and linezolid (**B**).

**I**

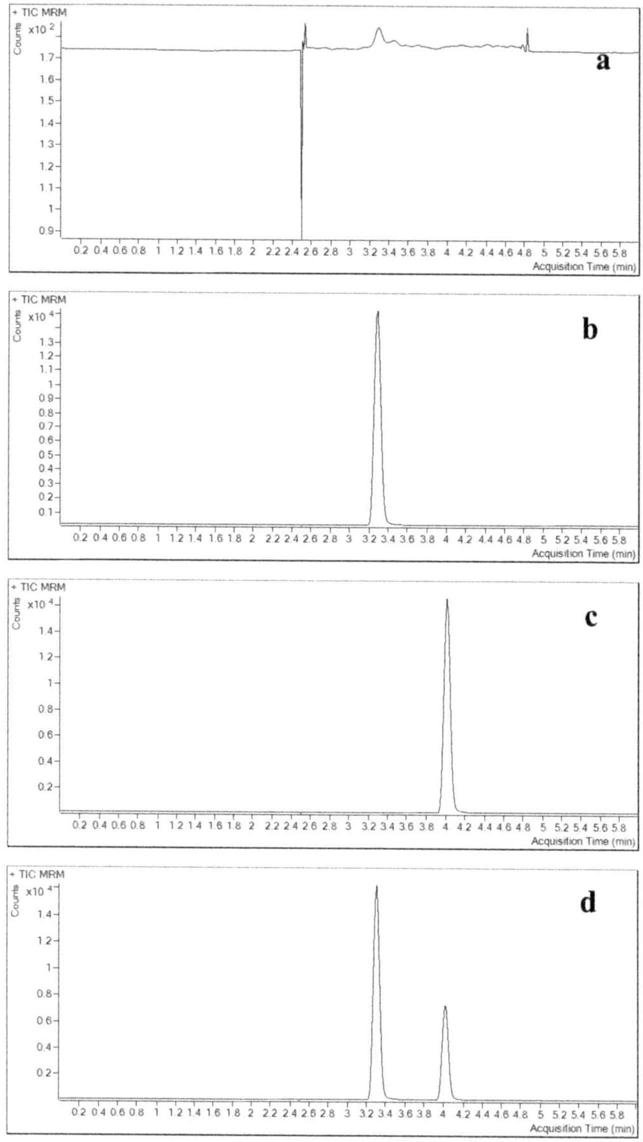


**Figure 2.** *Cont.*

**II**

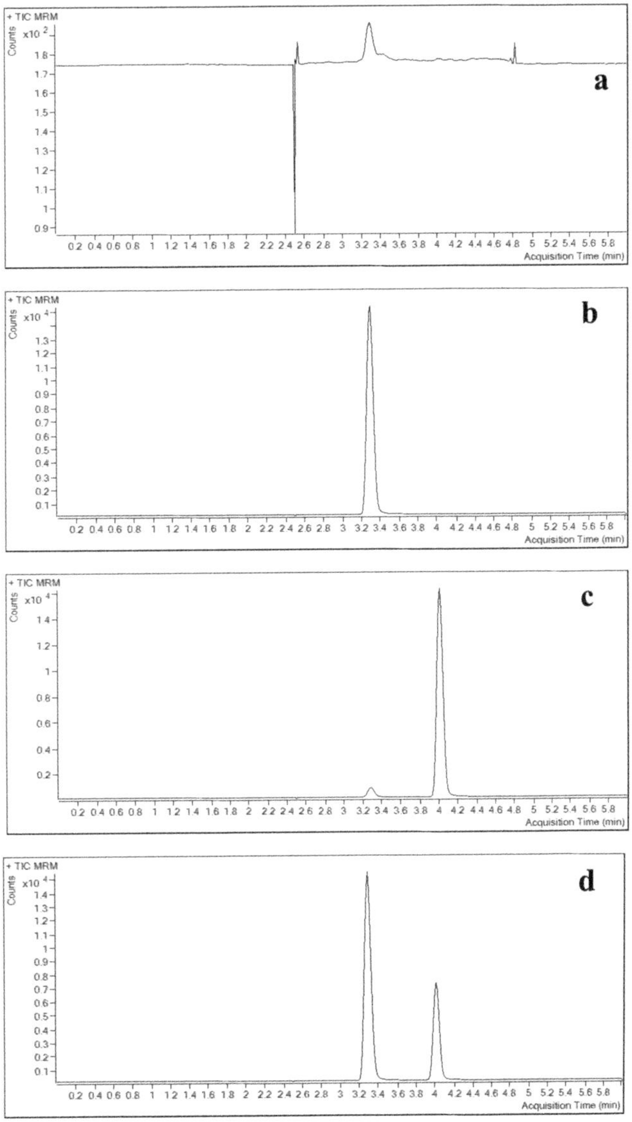


**Figure 2.** *Cont.*

III

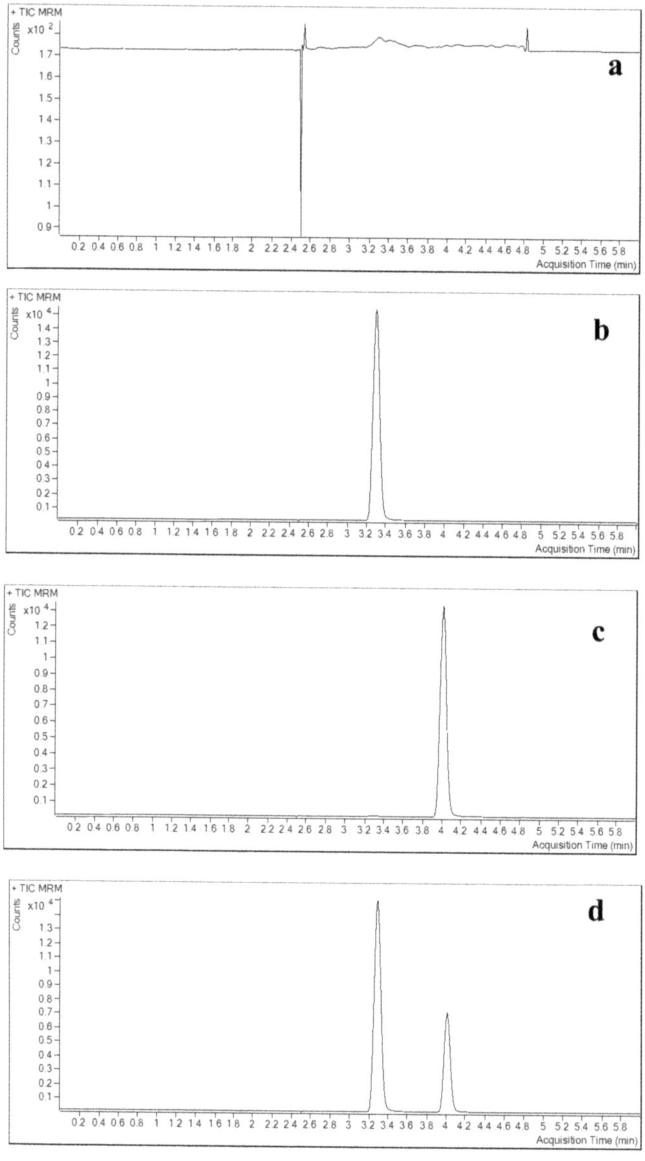


**Figure 2.** *Cont.*

**IV**

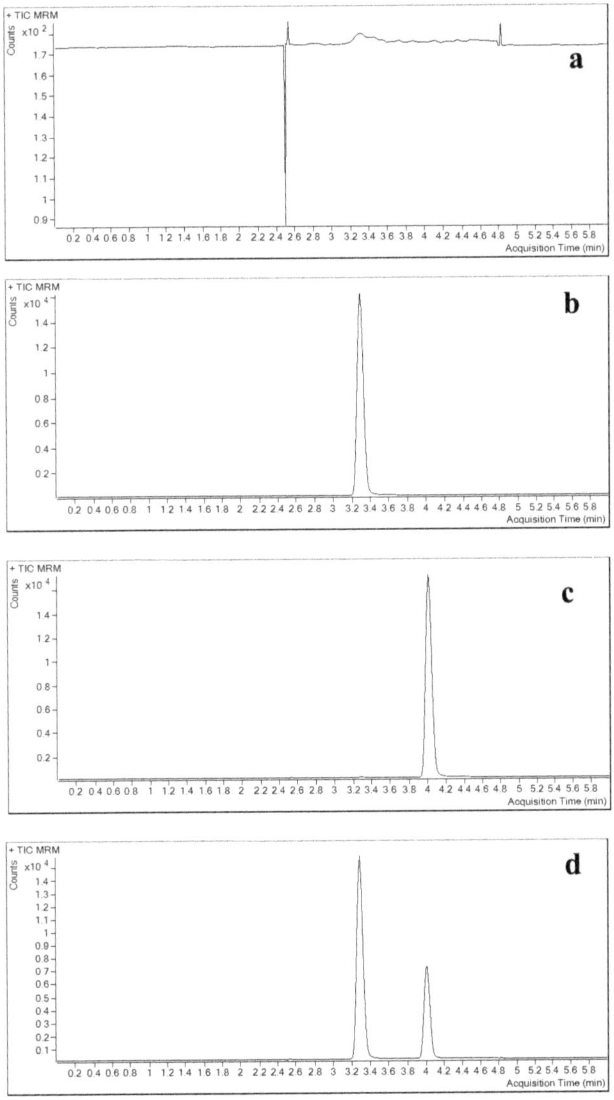


**Figure 2.** *Cont.*

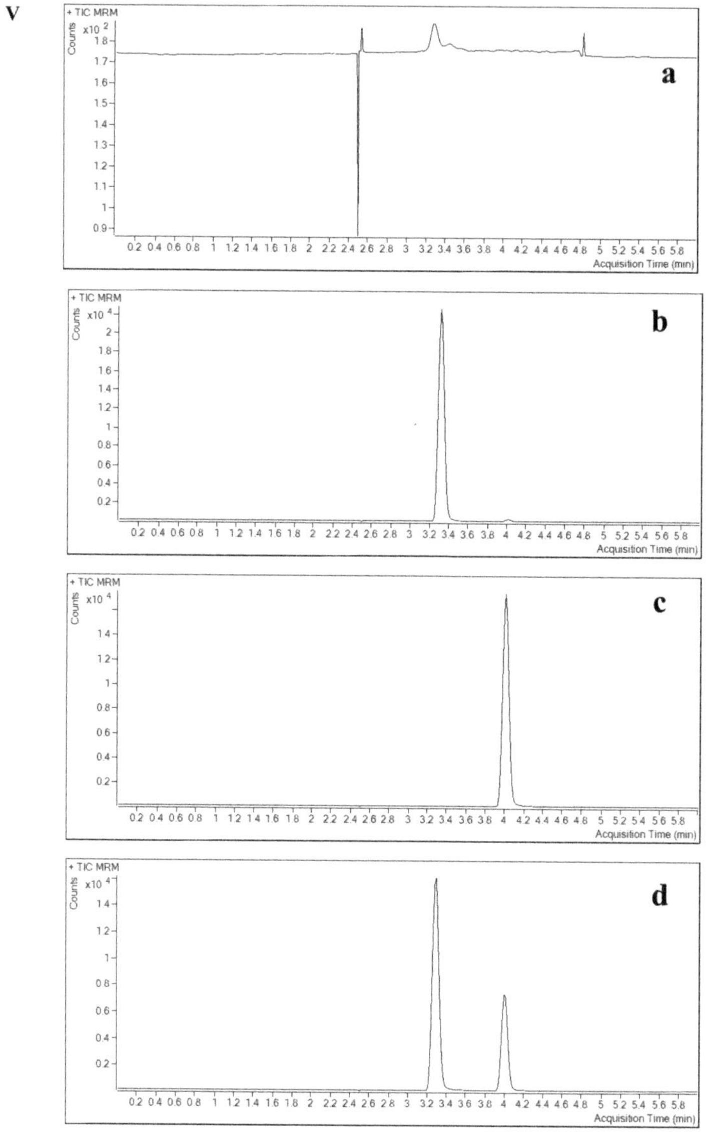


**Figure 2.** *Cont.*

**VI**

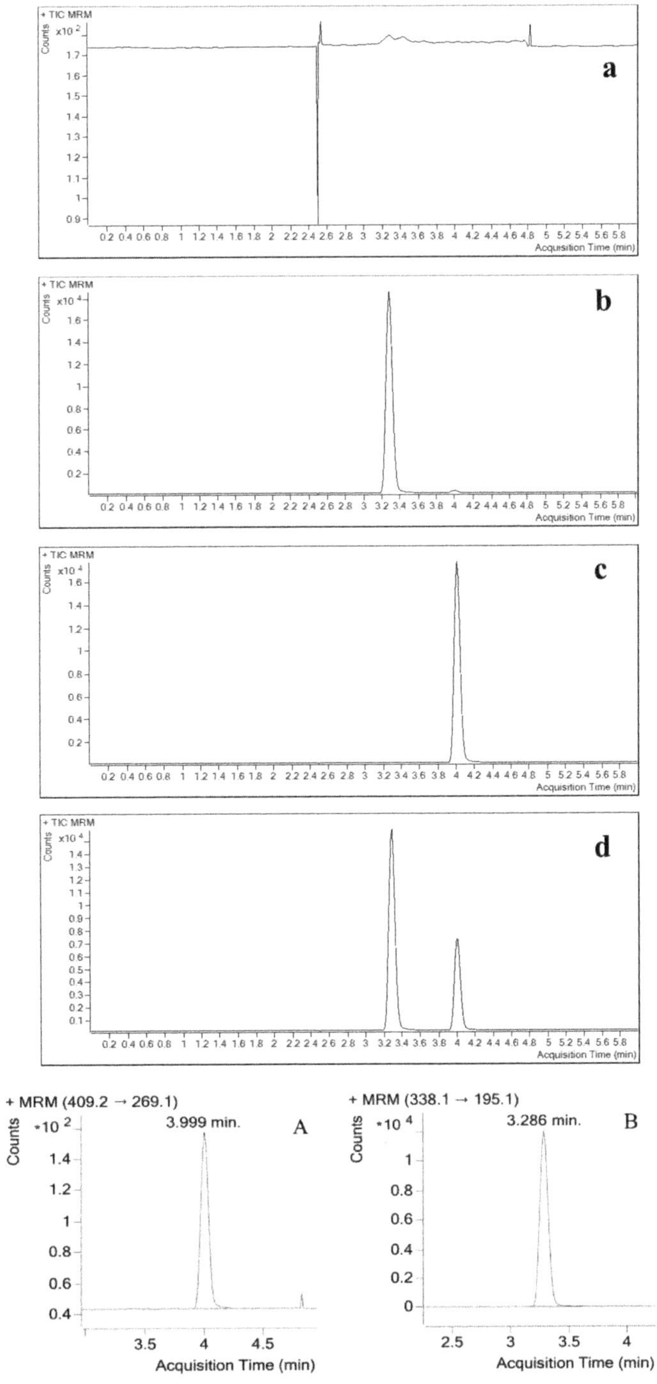


**Figure 2.** Specificity and selectivity of the method in plasma. (**a**) double blank, (**b**) single blank, (**c**) analyte only and (**d**) analyte and IS (in six different lots). Chromatograms of contezolid (**A**) and linezolid (**B**).

**Hemolytic plasma**

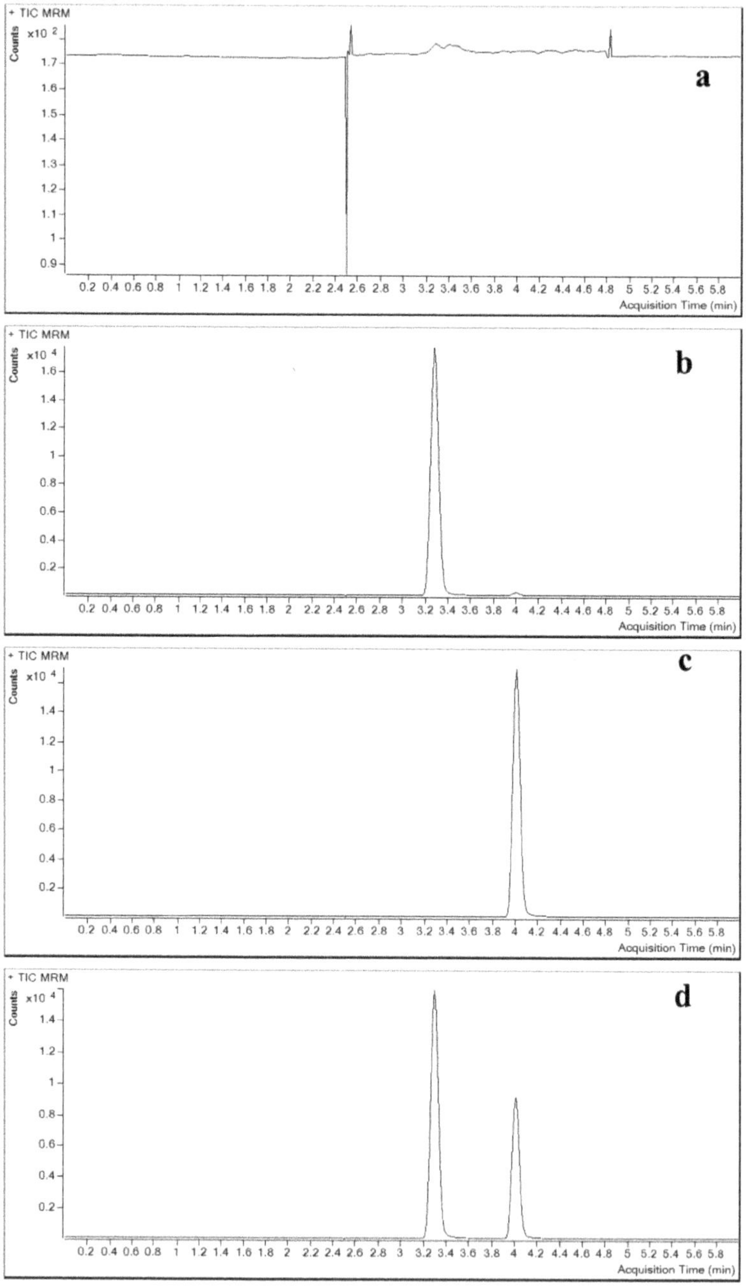


**Figure 3.** *Cont.*

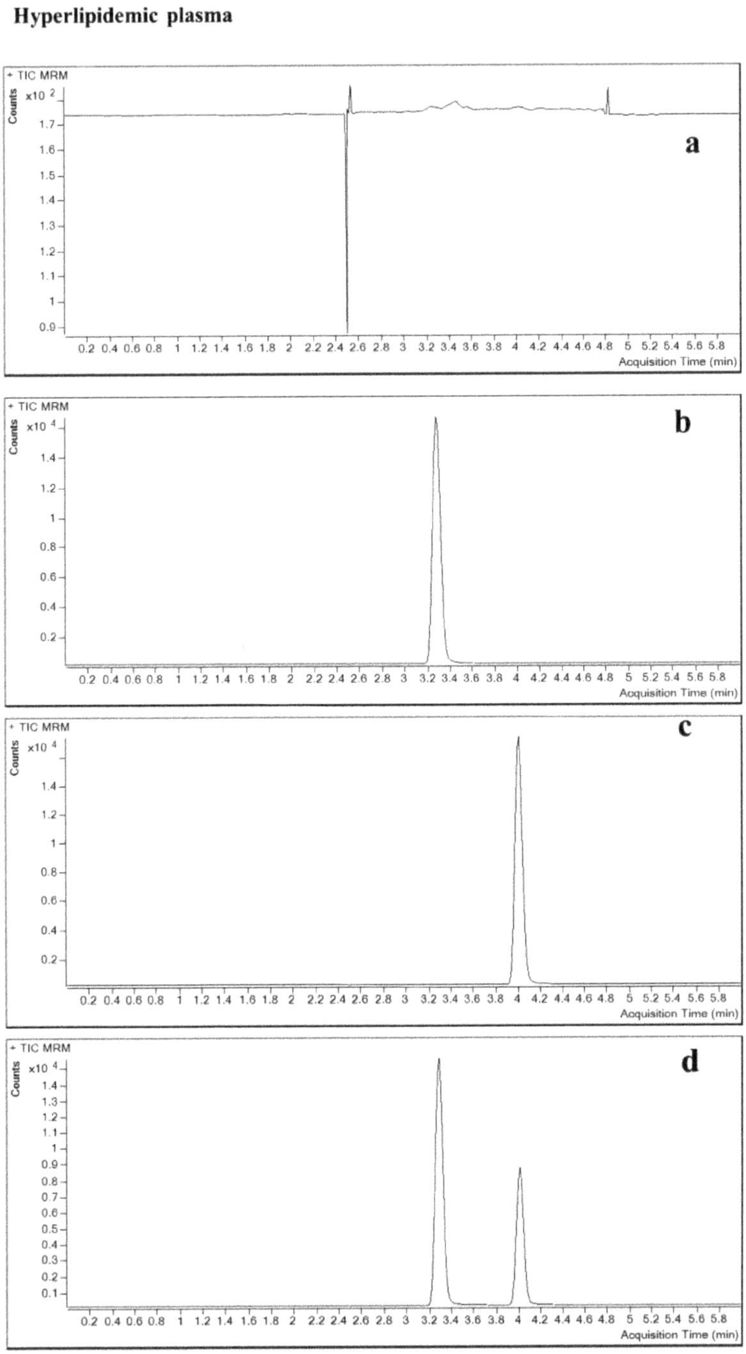


**Figure 3.** Specificity and selectivity of the method in plasma. (**a**) double blank, (**b**) single blank, (**c**) analyte only and (**d**) analyte and IS (in six different lots). Chromatograms of contezolid (**A**) and linezolid (**B**).

**Table 1.** ME of the method for determining contezolid in hemolytic and hyperlipidemic plasma.

| Concentration (ng/mL) | Number | Hemolytic Plasma | Hyperlipidemic Plasma |
|---|---|---|---|
| 150 | 1 | 153.3127 | 155.2132 |
| | 2 | 150.9690 | 156.3032 |
| | 3 | 153.2864 | 157.3617 |
| | Mean | 152.5227 | 156.2927 |
| | SD | ±1.3456 | ±1.0743 |
| | %CV | 0.8800 | 0.6900 |
| | Accuracy | 101.6800 | 104.2000 |
| | RE (%) | 1.6800 | 4.2000 |
| 3750 | 1 | 3767.9127 | 3792.1579 |
| | 2 | 3752.9170 | 3775.6060 |
| | 3 | 3770.4290 | 3794.5686 |
| | Mean | 3763.7529 | 3787.4442 |
| | SD | ±9.4681 | ±10.3228 |
| | %CV | 0.2500 | 0.2700 |
| | Accuracy | 100.3700 | 101.0000 |
| | RE (%) | 0.3700 | 1.0000 |

### 2.2.3. Sensitivity

The signal-to-noise ratio (SNR) of the LLOQs and the zero calibrators was greater than 15 in plasma and 35 in CSF. Among the LLOQs evaluated for this method, SNR reached 18.36 in plasma and 41.33 in CSF, which were considered sensitive. The mean concentration was 49.3196 (±1.3508) ng/mL in plasma and 19.9802 (±0.4460) ng/mL in CSF. The accuracy was 98.64% in plasma and 99.9% in CSF. The coefficients of variation (%CV) were 2.74% in plasma and 2.23% in CSF. The results showed that the lowest detection limits of this method were 50 ng/mL in plasma and 20 ng/mL in CSF. The results all met the criteria for acceptance.

### 2.2.4. Precision and Accuracy

The results of precision and accuracy validation are shown in Table 2. Inter-batch precision and accuracy of the plasma samples were 0.29% to 3.37% and 93.30% to 105.59%, respectively, and intra-batch precision and accuracy were 0.10% to 1.02% and 97.34% to 103.33%, respectively. The intra-batch accuracy and precision for CSF samples were 97.57% to 107.86% and 0.63% to 5.79%, respectively, and the inter-batch accuracy and precision were 99.41% to 105.62% and 1.35% to 2.57%, respectively.

### 2.2.5. Recovery

A simple protein precipitation method had been proved to be reliable and provided the cleanest samples. The comparison results of neat standards vs. plasma-extracted standards of contezolid and neat standards vs. CSF-extracted standards were evaluated.

The plasma recoveries of contezolid in LQC, medium quality control (MQC) and HQC were 90.14%, 95.04% and 93.63%, respectively. Their mean value was 92.94% and %CV was 2.52%. The CSF recoveries of contezolid in LQC, MQC and HQC were 102.21%, 98.05% and 93.50%, respectively. Their mean value was 97.68% and %CV was 1.47%.

### 2.2.6. Dilution Integrity

When the concentration of the unknown sample was higher than the standard curve, a dilution study was carried out to report the accuracy and precision of the diluted sample. The samples were diluted five-fold with plasma and ten-fold with human CSF. The samples showed that the accuracy relevant error (RE) in plasma and CSF were 4.08 and −2.82.

The %CV values were 0.41 and 3.07 (Table 3). If the verified dilution multiple could not meet the requirements of testing the clinical samples, the dilution multiple would be supplemented according to the actual need.

**Table 2.** Precision and accuracy of the method for determining contezolid.

| Concentration (ng/mL) | Batch | Intra-Batch (Mean ± SD) | Intra-Batch Accuracy (%) | Intra-Batch (%CV) | Inter-Batch (Mean ± SD) | Inter-Batch Accuracy (%) | Inter-Batch (%CV) |
|---|---|---|---|---|---|---|---|
| In plasma ($n$ = 6) | | | | | | | |
| 50.0 | 1 | 51.2670 ± 1.5182 | 102.53 | 2.96 | 51.6672 ± 0.5278 | 103.33 | 1.02 |
| | 2 | 50.9382 ± 1.7169 | 101.88 | 3.37 | | | |
| | 3 | 52.7963 ± 0.7198 | 105.59 | 1.36 | | | |
| 150 | 1 | 142.0762 ± 1.2834 | 94.72 | 0.90 | 146.0054 ± 0.8599 | 97.34 | 0.59 |
| | 2 | 139.9441 ± 2.4430 | 93.30 | 1.75 | | | |
| | 3 | 155.9958 ± 0.7632 | 104.00 | 0.49 | | | |
| 2000 | 1 | 2008.7019 ± 10.5603 | 100.44 | 0.53 | 2004.4452 ± 2.1689 | 100.22 | 0.11 |
| | 2 | 1973.4304 ± 7.4792 | 98.67 | 0.38 | | | |
| | 3 | 2031.2032 ± 11.6640 | 101.56 | 0.57 | | | |
| 3750 | 1 | 3712.0060 ± 14.6794 | 98.99 | 0.40 | 3721.1817 ± 3.8675 | 99.23 | 0.10 |
| | 2 | 3651.3176 ± 10.4087 | 97.37 | 0.29 | | | |
| | 3 | 3800.2216 ± 18.1291 | 101.34 | 0.48 | | | |
| In CSF ($n$ = 6) | | | | | | | |
| 20.0 | 1 | 20.7859 ± 0.6668 | 103.93 | 3.21 | 21.1237 ± 0.4009 | 105.62 | 1.90 |
| | 2 | 21.0129 ± 0.4486 | 105.06 | 2.13 | | | |
| | 3 | 21.5723 ± 1.2258 | 107.86 | 5.68 | | | |
| 60.0 | 1 | 62.7186 ± 1.6451 | 104.53 | 2.62 | 61.3609 ± 1.4473 | 102.27 | 2.36 |
| | 2 | 58.7197 ± 0.8075 | 97.87 | 1.38 | | | |
| | 3 | 62.6443 ± 3.6259 | 104.41 | 5.79 | | | |
| 350 | 1 | 359.8353 ± 3.2945 | 102.81 | 0.92 | 347.9318 ± 4.6878 | 99.41 | 1.35 |
| | 2 | 342.4519 ± 4.3888 | 97.84 | 1.28 | | | |
| | 3 | 341.5083 ± 11.9056 | 97.57 | 3.49 | | | |
| 750 | 1 | 738.0302 ± 17.8846 | 98.40 | 2.42 | | | |
| | 2 | 751.4551 ± 4.7026 | 100.19 | 0.63 | 747.1321 ± 19.1938 | 99.62 | 2.57 |
| | 3 | 751.9110 ± 42.5167 | 100.25 | 5.65 | | | |

**Table 3.** Dilution integrity of the method for determining contezolid.

| Number | Nominal Concentration (ng/mL) | Actual Concentration (ng/mL) | Accuracy (%) | RE (%) |
|---|---|---|---|---|
| In plasma ($n$ = 6, diluted five-fold) | | | | |
| 1 | 20,000 | 20,858.37 | 104.29 | 4.29 |
| 2 | | 20,820.85 | 104.10 | 4.10 |
| 3 | | 20,873.87 | 104.37 | 4.37 |
| 4 | | 20,757.35 | 103.79 | 3.79 |
| 5 | | 20,675.17 | 103.38 | 3.38 |
| 6 | | 20,909.58 | 104.55 | 4.55 |
| Mean | - | 20,815.87 | 104.08 | - |
| SD | - | ±86.25 | - | - |
| %CV | - | 0.41 | - | - |
| RE (%) | - | 4.08 | - | - |
| In CSF ($n$ = 6, diluted five-fold) | | | | |
| 1 | 1750 | 1739.23 | 99.38 | −0.62 |
| 2 | | 1747.11 | 99.84 | −0.16 |
| 3 | | 1757.69 | 100.44 | 0.44 |
| 4 | | 1654.12 | 94.52 | −5.48 |
| 5 | | 1657.21 | 94.70 | −5.30 |
| 6 | | 1648.88 | 94.22 | −5.78 |
| Mean | - | 1700.71 | 97.18 | - |
| SD | - | ±52.22 | - | - |
| %CV | - | 3.07 | - | - |
| RE (%) | - | −2.82 | - | - |

2.2.7. Stability

The stability of two concentrations of contezolid in human plasma (3.5 h at room temperature, 116 h in an automatic sampler, after three cycles from −20 °C and −80 °C to room temperature freeze-thawed, 28 days at −20 °C, and 28 and 118 days at −80 °C) and CSF (6 h at room temperature, 23.5 h in an automatic sampler, after three cycles from

−20 °C and −80 °C to room temperature freeze-thawed, 90 days at −20 °C and −80 °C), had been established. The %CV values ≤ 14.22% (Table 4). We also investigated the stability of contezolid stock solution for 24 h at room temperature and 90 days at −20 °C, and the stability of IS stock solution for 24 h at room temperature and 62 days at −20 °C. The %CV values ≤ 1.96%. The results showed that the samples had good stability under different storage conditions.

**Table 4.** Sample stability investigation.

| Nominal Concentration (ng/mL) | | 150 | | | | 3750 | | | |
|---|---|---|---|---|---|---|---|---|---|
| | | Mean | SD | %CV | RE (%) | Mean | SD | %CV | RE (%) |
| | | In plasma ($n$ = 3) | | | | | | | |
| 0 h | Determined concentration (ng/mL) | 154.68 | ±1.51 | 0.98 | | 3827.04 | ±13.58 | 0.35 | |
| | Accuracy (%) | 103.12 | | | 3.12 | 102.05 | | | 2.05 |
| 3.5 h—at room temperature | Determined concentration (ng/mL) | 154.03 | ±2.92 | 1.90 | | 3812.67 | ±32.46 | 0.85 | |
| | Accuracy (%) | 102.68 | | | 2.68 | 101.67 | | | 1.67 |
| 116 h—in the automatic sampler * | Determined concentration (ng/mL) | 161.21 | ±7.12 | 4.42 | | 3819.02 | ±12.55 | 0.33 | |
| | Accuracy (%) | 107.47 | | | 7.47 | 101.84 | | | 1.84 |
| Repeated freeze-thaw three times from −20 °C to room temperature | Determined concentration (ng/mL) | 153.76 | ±1.53 | 1.00 | | 3791.53 | ±21.06 | 0.56 | |
| | Accuracy (%) | 102.50 | | | 2.50 | 101.11 | | | 1.11 |
| Repeated freeze-thaw three times from −80 °C to room temperature | Determined concentration (ng/mL) | 152.80 | ±1.61 | 1.05 | | 3793.19 | ±9.28 | 0.24 | |
| | Accuracy (%) | 101.86 | | | 1.86 | 101.15 | | | 1.15 |
| 28 days—at −20 °C | Determined concentration (ng/mL) | 138.46 | ±2.15 | 1.55 | | 3516.93 | ±38.64 | 1.10 | |
| | Accuracy (%) | 92.30 | | | −7.70 | 93.78 | | | −6.22 |
| 28 days—at −80 °C | Determined concentration (ng/mL) | 135.19 | ±1.57 | 1.16 | | 3524.58 | ±53.28 | 1.51 | |
| | Accuracy (%) | 90.13 | | | −9.87 | 93.99 | | | −6.01 |
| 118 days—at −80 °C | Determined concentration (ng/mL) | 152.07 | ±5.64 | 3.71 | | 3219.72 | ±46.30 | 1.44 | |
| | Accuracy (%) | 101.38 | | | 1.38 | 85.86 | | | −14.14 |
| | | In CSF ($n$ = 3) | | | | | | | |
| 0 h | Determined concentration (ng/mL) | 62.77 | ±4.33 | 6.90 | | 792.67 | ±49.44 | 6.24 | |
| | Accuracy (%) | 104.61 | | | 4.61 | 105.69 | | | 5.69 |
| 6 h-at room temperature | Determined concentration (ng/mL) | 62.57 | ±0.64 | 1.03 | | 768.33 | ±2.52 | 0.33 | |
| | Accuracy (%) | 104.28 | | | 4.28 | 102.44 | | | 2.44 |
| 23.5 h-in the automatic sampler * | Determined concentration (ng/mL) | 60.45 | ±0.71 | 1.17 | | 707.28 | ±5.09 | 0.72 | |
| | Accuracy (%) | 100.75 | | | 0.75 | 94.30 | | | −5.70 |
| Repeated freeze-thaw three times from −20 °C to room temperature | Determined concentration (ng/mL) | 62.53 | ±1.07 | 1.71 | | 766.00 | ±14.42 | 1.88 | |
| | Accuracy (%) | 104.22 | | | 4.22 | 102.13 | | | 2.13 |
| Repeated freeze-thaw three times from −80 °C to room temperature | Determined concentration (ng/mL) | 61.43 | ±1.59 | 2.59 | | 781.33 | ±22.68 | 2.90 | |
| | Accuracy (%) | 102.39 | | | 2.39 | 104.18 | | | 4.18 |
| 90 days—at −20 °C | Determined concentration (ng/mL) | 68.53 | ±0.62 | 0.62 | | 734.52 | ±17.95 | 2.44 | |
| | Accuracy (%) | 114.22 | | | 14.22 | 97.94 | | | −2.06 |
| 90 days—at −80 °C | Determined concentration (ng/mL) | 65.62 | ±0.89 | 1.36 | | 753.42 | ±6.14 | 0.82 | |
| | Accuracy (%) | 109.37 | | | 9.37 | 100.46 | | | 0.46 |

* The stability of the samples after treated.

### 2.2.8. ME

As shown in Table 5, the matrix factors (MF) of contezolid at three concentrations in plasma were 97.32%, 98.33%, 98.25%, 98.00%, 94.80%, and 95.36%, respectively. Under the selected chromatographic and MS conditions, the %CVs of analytes at low, medium, and high levels were all less than 13.38%, which did not affect the quantification.

**Table 5.** ME and matrix factor of the method for determining contezolid and linezolid in six individual matrices.

| Number | LQC (150 ng/mL) | | | MQC (2000 ng/mL) | | | HQC (3750 ng/mL) | | |
|---|---|---|---|---|---|---|---|---|---|
| | Area Ratio of Analyte | Area Ratio of IS | MF | Area Ratio of Analyte | Area Ratio of IS | MF | Area Ratio of Analyte | Area Ratio of IS | MF |
| In plasma | | | | | | | | | |
| 1 | 0.8719 | 0.8896 | 0.9801 | 0.9442 | 0.9589 | 0.9847 | 0.9581 | 0.9760 | 0.9817 |
| 2 | 0.8314 | 0.8589 | 0.9680 | 0.9390 | 0.9529 | 0.9854 | 0.8791 | 0.8903 | 0.9874 |
| 3 | 0.8229 | 0.8313 | 0.9899 | 0.9110 | 0.9240 | 0.9859 | 0.8979 | 0.9153 | 0.9810 |
| 4 | 0.9893 | 1.0272 | 0.9631 | 0.9108 | 0.9319 | 0.9774 | 0.9367 | 0.9459 | 0.9903 |
| 5 | 0.7877 | 0.7994 | 0.9854 | 1.2185 | 1.2303 | 0.9904 | 1.2196 | 1.2489 | 0.9765 |
| 6 | 0.7650 | 0.8032 | 0.9524 | 0.9052 | 0.9273 | 0.9762 | 0.9204 | 0.9408 | 0.9783 |
| Mean | - | - | 0.9732 | - | - | 0.9833 | - | - | 0.9825 |
| SD | - | - | ±0.0144 | - | - | ±0.0055 | - | - | ±0.0053 |
| %CV | - | - | 1.48 | - | - | 0.56 | - | - | 0.54 |
| In CSF | | | | | | | | | |
| 1 | 0.9520 | 1.1002 | 0.8653 | 0.9000 | 0.9550 | 0.9424 | 0.8648 | 1.0114 | 0.8551 |
| 2 | 0.9370 | 0.9040 | 1.0365 | 0.9190 | 0.9743 | 0.9432 | 0.9371 | 1.0000 | 0.9371 |
| 3 | 0.9700 | 1.0200 | 0.9510 | 0.9429 | 0.9058 | 1.0410 | 0.9837 | 1.0343 | 0.9511 |
| 4 | 0.9970 | 1.0156 | 0.9817 | 0.9381 | 0.8822 | 1.0634 | 0.9907 | 1.0435 | 0.9494 |
| 5 | 0.9970 | 0.9265 | 1.0761 | 0.9333 | 0.9893 | 0.9434 | 0.9510 | 0.8902 | 1.0683 |
| 6 | 0.9760 | 1.0067 | 0.9695 | 0.9190 | 0.9850 | 0.9330 | 0.9604 | 1.0000 | 0.9604 |
| Mean | - | - | 0.9800 | - | - | 0.9777 | - | - | 0.9536 |
| SD | - | - | ±0.0729 | - | - | ±0.0582 | - | - | ±0.0681 |
| %CV | - | - | 7.44 | - | - | 5.95 | - | - | 7.14 |

## 3. Materials and Methods

### 3.1. Reference Materials

Contezolid (99.92%) was provided by MicuRx Pharmaceuticals, Inc, Shanghai, China. Linezolid (99.7%) was purchased from ApexBio Technology LLC, Houston, TX, USA.

### 3.2. Reagents

The following reagents were purchased from Thermo Fisher Scientific (China) Co., Ltd., Shanghai China: dimethyl sulfoxide (DMSO, above 99%), ultrapure water, acetonitrile and methanol of chromatographic purity, and high-performance liquid chromatography-grade ammonium.

### 3.3. Biological Matrix

Artificial CSF (R22153) was purchased from Shanghai Yuanye Bio-technology Co., Ltd. Individual or mixed human plasma samples in K2EDTA-anticoagulated tubes were taken from healthy volunteers and stored at −10 to −30 °C. Human hemolytic and hyperlipidemia plasma samples from our hospital's Respiratory Department were obtained and stored at −10 to −30 °C. Plasma and CSF samples were stored at −80 °C.

### 3.4. LC-MS/MS System

The chromatographic system used in the study was an Agilent 1260 high-performance liquid chromatography (HPLC) system (Agilent Technologies Inc., Santa Clara, CA, USA), consisting of a vacuum degasser, a binary pump, and an automatic sampler.

Agilent EclipsePlus C18 column (2.1 × 100 mm, 3.5 μm) was used for the elution of the analyte and IS. As mobile phases, we used 0.1% FA in a 10 mM aqueous ammonium acetate solution (A) and 0.1% FA in a 5 mM ammonium acetate acetonitrile: water (90:10, $v/v$) solution (B). The elution gradient started at 15% B, increased to 50% B from 0.5 min to 4.0 min, remained at 50% B until 5.0 min, and then returned to 15% B at 5.1 min. During each injection, 0.4 mL/min of flow rate was achieved, 5 μL of fluid was injected, and 6 min were spent in total.

Using an Agilent 6460A triple-quadrupole mass spectrometer (Agilent Technologies, Palo Alto, USA) connected to an electrospray ion source, positive MRM was carried out with this mass spectrometer. The nebulizer pressure was set to 30 psi. The flow of the drying gas was 11 L/min, and the drying gas temperature was held at 350 °C. The electrospray capillary voltage was optimized to 4000 V for positive and 3500 V for negative. Q1 and Q3 were both set at unit resolution. The *m/z* transition of contezolid was 409.15→269.14 (fragmentor, $F$ = 80 V, collision energy, CE = 30 eV), and that of linezolid was 338.14→195.1 ($F$ = 65 V, CE = 30 eV). We used the Agilent MassHunter Workstation to complete data acquisition and analysis.

### 3.5. Preparation of Calibration Standard Curves, Quality Control (QC) and IS Samples

Contezolid reference standards were precisely weighed and diluted completely with DMSO to prepare the stock solutions of 1.00 mg/mL as standard curve stock solution and the QC stock solution. The stock solutions were stored in a refrigerator of between −10 and −30 °C for future use.

Using a 50% acetonitrile aqueous solution as dilution solvent, the working standard solutions for plasma or CSF samples were prepared. The working solutions were diluted 20-fold into plasma or CSF. The concentrations of 5000, 4000, 2500, 1000, 500, 250, 100 and 50.0 ng/mL were for plasma, and 1000, 800, 600, 400, 200, 100, 40.0 and 20.0 ng/mL were for CSF calibration standards. The concentrations of 3750 (HQC), 2000 (MQC), 150 (LQC), and 50.0 (LLOQ) ng/mL were for plasma, and 750 (HQC), 350 (MQC), 60.0 (LQC), and 20.0 (LLOQ) for CSF QCs. The IS working solution, with a concentration of 800 ng/mL for plasma and 150 ng/mL for CSF samples, was prepared by diluting the IS stock solution with acetonitrile: methanol (1:1, $v/v$) and stored in the refrigerator at −10 to −30 °C.

### 3.6. Sample Preparation

The plasma and CSF samples (50 μL) were mixed with 250 μL of the IS working solution in 1.5 mL EP tubes. The samples were mixed for 1 min by using a vortexer, then centrifuged 10 min at 14,000 rpm (2 to 8 °C). A 100 μL plasma supernatant was added to a 200 μL acetonitrile: water (1:1) solution. After mixing, 200 μL was transferred to the HPLC vial, and 5 μL was injected into the LC-MS/MS system for analysis.

A 200 μL CSF supernatant was directly transferred to the HPLC vial, and 5 μL of the supernatant was entered into the LC-MS/MS system for analysis.

### 3.7. Method Validation and Acceptance Criteria

The method was verified according to *Bioanalytical Method Validation (M10)* of the International Council for Harmonisation of Technical Requirements for Pharmaceuticals for Human Use. The method parameters included linearity of calibration curves, selectivity and specificity, sensitivity, accuracy, precision, recovery, stability, dilution integrity, and matrix effect.

#### 3.7.1. Specificity and Selectivity

The potential interference of analyte and IS in the liquid chromatography peak region was studied by analyzing at least six normal individual blank plasma, hemolytic plasma, hyperlipidemic plasma, and six normal individual blank CSF samples, then the specificity of this method was evaluated. By analyzing a double blank (without IS or analytes) plasma and CSF, a single blank (only IS) plasma and CSF, plasma, and CSF with only analytes, as well as plasma and CSF containing analytes and IS, the specificity and selectivity of the method were determined.

The response of the interference component should not be higher than 20% of the LLOQ and not more than 5% of the IS response in each matrix LLOQ sample.

3.7.2. ME

The matrix effect of three concentrations (150 ng/mL, 2000 ng/mL and 3750 ng/mL in plasma; 60.0 ng/mL, 350 ng/mL and 750 ng/mL in CSF) of contezolid in six kinds of substrates was determined. The ME is calculated from the IS normalized MF. In the sample group prepared under "sample preparation", the peak area to mean value ratio of contezolid was recorded as A1, and the peak area to mean value ratio of IS was recorded as A2. MF = A2/A1.

The precision (the percentage of the coefficient of variation, %CV) of all individual source analytes should not be greater than 15%.

3.7.3. Calibration Curve and Range

The calibration curves were prepared by using double blank samples, single blank samples (only IS), and calibration standard samples with eight concentration levels (including LLOQ and the upper limit of quantification, ULOQ).

At the LLOQ, the accuracy of back-calculated concentrations of calibrated standards should be within ±20% of the nominal concentration, while at all other levels, should be within ±15%. It is recommended that all standard curves contain at least six effective concentrations and that at least 75% of the calibration samples meet these criteria.

3.7.4. Sensitivity

The LLOQ refers to the lowest level of concentration of analyte whose response is at least five-fold greater than the blank (SNR $\geq$ 5), whose precision is less than 20%, and whose accuracy is in the range of 80.0—120% of the nominal value.

3.7.5. Precision and Accuracy

Samples of LLOQ and QC were prepared six times for three consecutive runs on three different dates and analyzed three times in each run.

Except for LLOQ, the overall accuracy of each concentration level QC sample should be within ±15% of the nominal concentration, and the accuracy of LLOQ should be less than ±20%. Except for LLOQ, the precision of each concentration level QC sample should not exceed 15% and of LLOQ should not exceed 20%.

3.7.6. Carry-Over

Immediately after the high concentration of the sample (5000 ng/mL in plasma, 1000 ng/mL in CSF), the blank samples were analyzed to estimate the carry-over.

Blank samples after ULOQ should not carry over more than 20% of the LLOQ sample's response and 5% of the IS sample's response.

3.7.7. Dilution Integrity

Contezolid (20,000 ng/mL) was added to plasma and 1750 ng/mL to CSF. These samples were diluted to five-fold of plasma or ten-fold of CSF and determined repeatedly for six times. The mean accuracy of dilution QC should be within ±15% of the nominal concentration, and the %CV should not exceed 15%.

3.7.8. Stability

The stability of the stock solution, the stability of samples after frozen-thawed or treated, short- and long-term stability of samples were investigated. By analyzing the repetition of plasma samples (150 ng/mL and 3750 ng/mL) and CSF samples (60.0 ng/mL and 750 ng/mL) added under different conditions, the stability of contezolid in human plasma and CSF was evaluated. We evaluated the stability of the samples stored at room temperature (bench top stability) for six hours and compared it to the stability of samples stored in the injector for short-terms and repeated freeze-thaw cycles. A comparison was made with the concentration obtained after a period of at least 28 days in order to determine long-term stability. Through proper dilution, the stability of the stock solution was investigated by the response of the detector.

## 4. Conclusions

A simple, rapid, and sensitive LC-MS/MS method of the analysis of contezolid in human plasma and CSF without time-consuming and expensive extraction procedure was developed and validated. We also verified that it was feasible to use artificial CSF as the matrix of standard curve, which made up for the defect that it was not easy to obtain CSF from patients. By developing the method, the concentration of drugs in the plasma and CSF can be monitored and clinical treatment can be guided.

**Supplementary Materials:** The following supporting information can be downloaded at: https://www.mdpi.com/article/10.3390/ph16010032/s1.

**Author Contributions:** G.Z. contributed to the conception and design of study, laboratory data acquisition, data analysis and drafting of article. N.Z. contributed to the design of study and critical revision. L.D. and N.B. contributed to the laboratory data acquisition and data analysis. Y.C. contributed to the conception and design of study, analysis of data and drafting of article and critical revision. All authors have read and agreed to the published version of the manuscript.

**Funding:** This research was funded by the National Natural Science Foundations of China (81770004 and 82073894), Cultivation Project of PLA General Hospital for Distinguished Young Scientists (2020-JQPY-004) and New Medicine Clinical Research Fund (4246Z512).

**Institutional Review Board Statement:** The Medical Ethics Committee of Chinese PLA General Hospital had approved this study (S2021-609-01).

**Informed Consent Statement:** Not applicable.

**Data Availability Statement:** Data is contained within the article and supplementary material.

**Acknowledgments:** MicuRx Pharmaceuticals, Inc. provided the study drug.

**Conflicts of Interest:** The authors declare no conflict of interest.

## Abbreviation

| | |
|---|---|
| CE | collision energy |
| CNS | central nervous system |
| CSF | cerebrospinal fluid |
| CV | coefficient of variation |
| DMSO | dimethyl sulfoxide |
| ESI | electrospray ionization |
| FA | formic acid |
| HPLC | high-performance liquid chromatography |
| HQC | high quality control |
| IS | internal standard |
| LC-MS/MS | liquid chromatography-tandem mass spectrometry |
| LLOQ | lower limit of quantification |
| LOQ | low quality control |
| ME | matrix effect |
| MF | matrix factors |
| MQC | medium quality control |
| MRM | multiple reaction monitoring |
| MRSA | methicillin-resistant *Staphylococcus aureus* |
| PRSP | penicillin-resistant *Streptococcus pneumoniae* |
| QC | quality control |
| RE | relevant error |
| SNR | signal-to-noise ratio |
| ULOQ | upper limit of quantification |
| VRE | vancomycin-resistant *enterococci* |

## References


1. Gao, W.; Chen, Y.; Zhang, Y.; Zhang, Q.; Zhang, L. Nanoparticle-based local antimicrobial drug delivery. *Adv. Drug Deliv. Rev.* **2018**, *127*, 46–57. [CrossRef] [PubMed]
2. Boucher, H.W.; Talbot, G.H.; Bradley, J.S.; Edwards, J.E.; Gilbert, D.; Rice, L.B.; Scheld, M.; Spellberg, B.; Bartlett, J. Bad bugs, no drugs: No ESKAPE! An update from the Infectious Diseases Society of America. *Clin. Infect. Dis.* **2009**, *48*, 1–12. [CrossRef] [PubMed]
3. Fang, G.; Li, W.; Shen, X.; Perez-Aguilar, J.M.; Chong, Y.; Gao, X.; Chai, Z.; Chen, C.; Ge, C.; Zhou, R. Differential Pd-nanocrystal facets demonstrate distinct antibacterial activity against Gram-positive and Gram-negative bacteria. *Nat. Commun.* **2018**, *9*, 129. [CrossRef] [PubMed]
4. Kozarich, J.W. The biochemistry of disease: Desperately seeking syzygy. *Annu. Rev. Biochem.* **2009**, *78*, 55–63. [CrossRef]
5. Jean, S.S.; Tsai, M.H.; Chen, W.C.; Ho, S.J.; Wu, M.H.; Hsueh, P.R. Fatal community-acquired pneumonia caused by *Legionella pneumophila* and fluoroquinolone-resistant *Salmonella enterica* serotype Choleraesuis bacteraemia. *Int. J. Antimicrob. Agents* **2009**, *33*, 189–190. [CrossRef]
6. Shinabarger, D. Mechanism of action of the oxazolidinone antibacterial agents. *Expert Opin. Investig. Drugs* **1999**, *8*, 1195–1202. [CrossRef]
7. Jiang, J.; Hou, Y.; Duan, M.; Wang, B.; Wu, Y.; Ding, X.; Zhao, Y. Design, synthesis and antibacterial evaluation of novel oxazolidinone derivatives nitrogen-containing fused heterocyclic moiety. *Bioorg. Med. Chem. Lett.* **2021**, *32*, 127660. [CrossRef]
8. Scaiola, A.; Leibundgut, M.; Boehringer, D.; Caspers, P.; Bur, D.; Locher, H.H.; Rueedi, G.; Ritz, D. Structural basis of translation inhibition by cadazolid, a novel quinoxolidinone antibiotic. *Sci. Rep.* **2019**, *9*, 5634. [CrossRef]
9. Baddour, L.M.; Wilson, W.R.; Bayer, A.S.; Fowler, V.G., Jr.; Tleyjeh, I.M.; Rybak, M.J.; Barsic, B.; Lockhart, P.B.; Gewitz, M.H.; Levison, M.E.; et al. Infective Endocarditis in Adults: Diagnosis, Antimicrobial Therapy, and Management of Complications: A Scientific Statement for Healthcare Professionals from the American Heart Association. *Circulation* **2015**, *132*, 1435–1486. [CrossRef]
10. Grau, S.; Rubio-Terrés, C. Pharmacoeconomics of linezolid. *Expert Opin. Pharmacother.* **2008**, *9*, 987–1000. [CrossRef]
11. Rabon, A.D.; Fisher, J.P.; MacVane, S.H. Incidence and Risk Factors for Development of Thrombocytopenia in Patients Treated with Linezolid for 7 Days or Greater. *Ann. Pharmacother.* **2018**, *52*, 1162–1164. [CrossRef] [PubMed]
12. Gorchynski, J.; Rose, J.K. Complications of MRSA Treatment: Linezolid-induced Myelosuppression Presenting with Pancytopenia. *West. J. Emerg. Med.* **2008**, *9*, 177–178. [PubMed]
13. Gordeev, M.F.; Yuan, Z.Y. New Potent Antibacterial Oxazolidinone (MRX-I) with an Improved Class Safety Profile. *J. Med. Chem.* **2014**, *57*, 4487–4497. [CrossRef] [PubMed]
14. Li, C.-R.; Zhai, Q.-Q.; Wang, X.-K.; Hu, X.-X.; Li, G.-Q.; Zhang, W.-X.; Pang, J.; Lu, X.; Yuan, H.; Gordeev Mikhail, F.; et al. In Vivo Antibacterial Activity of MRX-I, a New Oxazolidinone. *Antimicrob. Agents Chemother.* **2014**, *58*, 2418–2421. [CrossRef] [PubMed]
15. Wang, S.; Cai, C.; Shen, Y.; Sun, C.; Shi, Q.; Wu, N.; Zheng, S.; Qian, J.; Zhang, R.; Zhou, H. In vitro Activity of Contezolid Against Methicillin-Resistant *Staphylococcus aureus*, Vancomycin-Resistant Enterococcus, and Strains with Linezolid Resistance Genes From China. *Front. Microbiol.* **2021**, *12*, 729900. [CrossRef]
16. Zhao, M.; Wu, J.; Huang, J.; Yu, J.; Zhang, J.; Guo, B. Development and validation of a rapid and robust LC-MS/MS method for quantification of a new oxazolidinone antibiotic MRX-I in human plasma and urine. *Clin. J. Infect. Chemother.* **2014**, *14*, 210–215. (In Chinese) [CrossRef]
17. Wang, Y.; Wu, H.; Wu, J.; Fan, Y.; Liu, X.; Li, Y.; Hu, J.; Zhang, J.; Guo, B. Development and validation of ultra-performance liquid chromatography-tandem mass spectrometric methods for simultaneous and rapid determination of contezolid and its major metabolite M2 in plasma and urine samples and its application to a study in subjects with moderate liver impairment. *J. Chromatogr. B Anal. Technol. Biomed. Life Sci.* **2022**, *1191*, 123129. [CrossRef]
18. John, C.C.; Carabin, H.; Montano, S.M.; Bangirana, P.; Zunt, J.R.; Peterson, P.K. Global research priorities for infections that affect the nervous system. *Nature* **2015**, *527*, S178–S186. [CrossRef]
19. Kielian, T.; Esen, N.; Bearden, E.D. Toll-like receptor 2 (TLR2) is pivotal for recognition of S. aureus peptidoglycan but not intact bacteria by microglia. *Glia* **2005**, *49*, 567–576. [CrossRef]
20. Lutsar, I.; McCracken, G.H., Jr.; Friedland, I.R. Antibiotic pharmacodynamics in cerebrospinal fluid. *Clin. Infect. Dis.* **1998**, *27*, 1117–1129, quiz 1128–1119. [CrossRef]
21. Dong, L.; Bai, N.; Wang, T.; Cai, Y. Development and validation of a liquid chromatography-tandem mass spectrometry (LC-MS/MS) method for the quantification of voriconazole in human cerebrospinal fluid. *Anal. Methods* **2021**, *13*, 4585–4593. [CrossRef] [PubMed]

*pharmaceuticals*

*Article*

# Insights into HPLC-MS/MS Analysis, Antioxidant and Cytotoxic Activity of *Astragalus fruticosus* against Different Types of Cancer Cell Lines

**Mohamed Fayez Dekinash [1], Tarek M. Okda [2,*], Ehab Kotb Elmahallawy [3,*], Fathy Kandil El-Fiky [4], Gamal Abd El Hay Omran [2], Emil Svajdlenka [5], Naief Dahran [6], Manal F. El-Khadragy [7], Wafa A. Al-Megrin [7] and El Moataz Bellah Ali El Naggar [1]**

1 Department of Pharmacognosy, Faculty of Pharmacy, Damanhour University, Damanhour 22511, Egypt
2 Department of Biochemistry, Faculty of Pharmacy, Damanhour University, Damanhour 22511, Egypt
3 Department of Zoonoses, Faculty of Veterinary Medicine, Sohag University, Sohag 82524, Egypt
4 Department of Pharmacognosy, Faculty of Pharmacy, Delta University for Science and Technology, Mansoura 35511, Egypt
5 Department of Natural Drugs, Faculty of Pharmacy, Masaryk University, 60300 Brno, Czech Republic
6 Department of Anatomy, Faculty of Medicine, University of Jeddah, Jeddah 21959, Saudi Arabia
7 Department of biology, College of Science, Princess Nourah bint Abdulrahman University, P.O. Box 84428, Riyadh 11671, Saudi Arabia
* Correspondence: tarekokda@pharm.dmu.edu.eg (T.M.O.); eehaa@unileon.es (E.K.E.)

**Citation:** Dekinash, M.F.; Okda, T.M.; Elmahallawy, E.K.; El-Fiky, F.K.; Omran, G.A.E.H.; Svajdlenka, E.; Dahran, N.; El-Khadragy, M.F.; Al-Megrin, W.A.; El Naggar, E.M.B.A. Insights into HPLC-MS/MS Analysis, Antioxidant and Cytotoxic Activity of *Astragalus fruticosus* against Different Types of Cancer Cell Lines. *Pharmaceuticals* **2022**, *15*, 1406. https://doi.org/10.3390/ph15111406

Academic Editors: Jan Oszmianski and Sabina Lachowicz-Wiśniewska

Received: 10 October 2022
Accepted: 7 November 2022
Published: 14 November 2022



**Abstract:** Plants from the genus *Astragalus* are gaining attention for their pharmacological importance. However, the information available regarding the HPLC–MS/MS chemical profile of *A. fruticosus* is inadequate. In this study, we performed HPLC–MS/MS analysis using electrospray ionization (ESI) and atmospheric pressure chemical ionization (APCI). We tentatively identified 11 compounds in the *A. fruticosus* methanolic extract, including five flavonoidal and six saponin glycosides. The extract showed moderate antioxidant activity with 21.05% reduction in DPPH UV absorption. The preliminary cytotoxic screening against seven human cancer cell lines using 100 µg/mL extract showed prominent cytotoxic potential against colorectal cancer HCT–116 with 3.368% cell viability. It also showed moderate cytotoxic potential against prostate (DU–145), ovarian (SKOV–3) and lung (A–549) cancer cell lines with cell viability of 14.25%, 16.02% and 27.24%, respectively. The $IC_{50}$ of the total extract against HCT–116 and DU–145 cell lines were 7.81 µg/mL and 40.79 µg/mL, respectively. The observed cytotoxicity of the total methanolic extract from the leaves against colorectal cancer might facilitate future investigations on cytotoxic agent(s) for disease management.

**Keywords:** *Astragalus fruticosus*; HPLC–MS/MS profiling; colorectal cancer; cytotoxic activity

## 1. Introduction

The genus *Astragalus* is the largest genus in the family *Leguminosae* (*Fabaceae*), with over 3000 species of annual and perennial small shrubs and herbs. It is one of the largest and most diverse genera among angiosperms [1]. *Astragalus* has been used as a medicinal plant for over 1000 years. Radix Astragali (*A. membranaceus* and *A. mongholicus*) is an ancient and well-known Chinese traditional herbal medicine that is used to enhance resistance against bacterial and viral infections, treat immunological disorders, stimulate the circulatory system, control excessive sweating, and as a diuretic, hepatoprotective and heart tonic [2]. The chemical composition of *Astragalus* uniformly consists of major active compounds, including polysaccharides, flavonoids (free or glycosidic forms) and saponins [3]. *Astragalus* contains both cycloartane and oleanane saponins with the latter being relatively rare [2]. The saponin-rich purified mixture isolated from *A. corniculatus* showed protective activity against the hamster myeloid graffi tumor [4]. The saponins extracted from *Astragalus hamosus* showed anticancer activity against breast carcinoma cell lines [2].

Using *Astragalus*-based Chinese medicine in combination with chemotherapy for treating colorectal and lung cancer increases the treatment efficacy, reduces the side effects and improves the patients' quality of life [5]. *Astragalus* extracts also have other biological activities, including antioxidant [6], hepatoprotective [7] and immunomodulatory activities [8]. The saponins isolated from *A. hamosus*, *A. kahiricus* and *A. peregrinus* showed dose-related modulation of lymphocyte proliferation [9]. There are 37 species of *Astragalus* in Egypt as described by Tackholm [10], while Boulos recorded only 32 species [11]. To our knowledge, until early 2022, previous phytochemical and biological research on members of genus *Astragalus* in Egypt included the following 15 species: *A. sieberi* [12–14], *A. hamosus* [9], *A. kahiricus* [7,9,15], *A. vogelii* [16], *A. eremophilus* [16,17], *A. bombycinus* [18,19], *A. peregrinus* [8,19,20], *A. annularis* [21], *A. trimestris* [21], *A. tomentosus* [22–24], *A. spinosus* [25–27], *A. tribuloides* [28], *A. trigonus* [29–31], *A. alexandrinus* [29,32] and *A. cremophilos* [33]. A recent study discussed potential somatic embryogenesis and micropropagation of the endangered plant *Astragalus fruticosus* to ensure its conservation. The study also identified the presence of luteolin and kaempferol flavonoids, pyrogallol, protocatechuic acid, *p*–coumaric acid, chlorogenic acid, ferulic acid and ellagic acid by comparing the high-performance liquid chromatography-ultraviolet (HPLC–UV) chromatograms of the plant's ethanolic extract and reference compounds. *Astragalus fruticosus* ethanolic extract showed α–glucosidase inhibitory activity ($IC_{50}$ = 44.8 μg/mL) and exhibited cytotoxic activity against breast (MCF–7) and leukemia (HL–60) cell lines ($IC_{50}$ of 28.3 μg/mL and 49 μg/mL, respectively) [34].

Notably, the combination of the HPLC chromatographic resolution and the structural information by MS and $MS^2$ allows tentative identification of phytochemicals in plant extracts [35–38]. The information available regarding the HPLC–MS/MS chemical profiling of *A. fruticosus* and their activity on different cell lines is limited. Here we performed a preliminary cytotoxic screening of the *A. fruticosus* Forssk methanolic extract against a panel of seven cancer cell lines and determined its $IC_{50}$ against the most sensitive cell lines. Additionally, we performed high performance liquid chromatography coupled to tandem mass spectrometry (HPLC–MS/MS) analysis of the methanolic extract, which is ideal for analyzing phytochemicals, including saponins and flavonoidal glycosides, due to its high sensitivity and selectivity.

## 2. Results and Discussion

### 2.1. HPLC–MS/MS Identification of Compounds

Using HPLC–MS/MS analysis, we tentatively identified 11 compounds from *Astragalus fruticosus*. Figure 1 shows the extracted ion chromatogram of the seven compounds identified using ESI-Ve ionization mode in the tandem mass spectrometry, and their HPLC/MS–MS data and spectra are listed in Table 1. The HPLC/MS–MS data and spectra of compounds identified using atmospheric pressure chemical ionization (APCI)+Ve ionization mode in the MS–MS are listed in Table 2. The mass spectra for tentatively identified compounds using HPLC–MS/MS showing glycon daughter peaks are provided in the Supplementary Figures S1–S11.

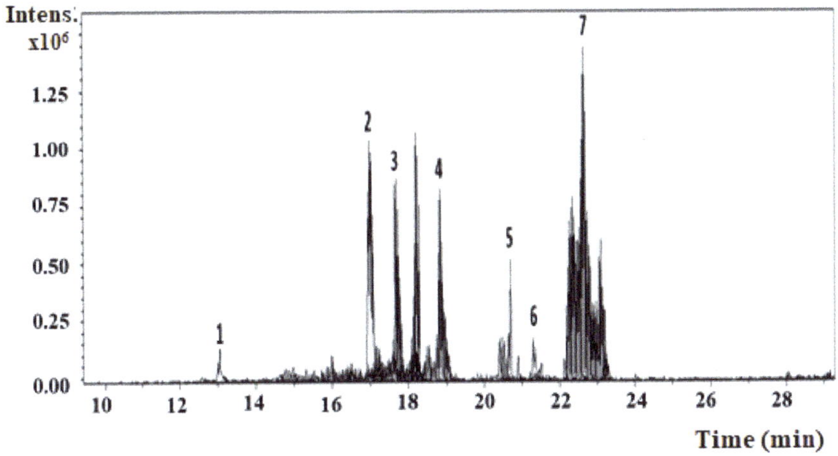


**Figure 1.** The extracted ions chromatogram of the identified compounds using ESI-Ve ionization mode in tandem mass spectrometry.

**Table 1.** Compounds identified from *A. fruticosus* and their HPLC–MS/MS$^2$ data using ESI-Ve ionization.

| Cpd No. | Rt | MS$^1$ | MS$^2$ Data | Compound Name |
|---|---|---|---|---|
| 1 | 13.0 | 640 (M-H)- | 519 (M-ring A-H)-<br>477 (M-glucopyranose-H)-<br>316 (a glycon) (M-2 x glucopyranose)- | 5-hydroxy isomucronulatol-2/,5/-di-*O*-glucoside [39] |
| 2 | 17.1 | 579 (M-H)- | 447 (M-arabinopyranose)-<br>429 (M-arabinopyranose-$H_2O$)-<br>284.8 (a glycon)<br>(M-glucopyranose-arabinopyranose)- | 2/,4/-trihydroxy-flavone-8-*C*-α-arabinopyranoside-7-*O*-β-glucopyranoside [18] |
| 3 | 17.7 | 737.5 (M-H)- | 675 (M- acetic acid)-<br>635 (M-$C_4H_7O_3$)-Mclafferty<br>593 (M-rhamnopyranose)-<br>285 (a glycon) (Kaempferol-H)- | kaempferol-3-*O*-α-L-rhamnopyranosyl-(1→2)-[6-*O*-(3-hydroxy-3-methylglutaryl)-β-D-galactopyranoside] [40] |
| 4 | 18.7 | 753.3 (M-H)- | 651.1 (M-$C_4H_7O_3$)-Mclafferty<br>609.2 (M-rhamnopyranose)-,<br>302.0 (a glycon) Quercetin- | Quercetin-3-*O*-α-L-rhamnopyranosyl-(1→2)-[6-*O*-(3-hydroxy-3-methylglutaryl)-β-D-galactopyranoside] [41] |
| 5 | 20.7 | 931 (M-H)- | 869 (M-isopropyl alcohol)-<br>786 (M-rhamnopyranose)-,<br>562<br>(M-rhamnose-glucose-isopropyl alcohol)-<br>440 (a glycon)<br>(M-rhamnose-glucose-$H_2O$)- | Trojanoside C [42] |
| 6 | 21.3 | 900 (M-H)- | 837 (M-isopropyl alcohol)-<br>797 (M-$C_6H_{14}O$)-Side chain<br>755 (M-rhamnopyranose)-<br>456 (a glycon)<br>(M-Rhamnopyranose-2 xylose)- | Astrasieversianin XV [43,44] |
| 7 | 22.5 | 751.5 (M-H)- | 689 (M-OCH3-2OH)-<br>650 (M-$C_4H_7O_3$)-<br>607 (M-rhamnopyranose)-<br>299.1 (a glycon) 7-methoxykaempferol- | 7-Methylkaempferol-3-*O*-α-L-rhamnopyranosyl -(1→2)- [6-*O*-(3-hydroxy-3-methylglutaryl)-β-D-galactopyranoside] [40] |

**Table 2.** Compounds identified in *A. fruticosus* and their HPLC–MS/MS$^2$ data using APCI + Ve ionization.

| Cpd No. | Rt | MS$^1$ | MS$^2$ Data | Compound Name |
|---|---|---|---|---|
| **8** | 20.9 | 679 (M+H)+ | 661 (M-$H_2O$)+<br>435 (M-acetylated glucose-2$H_2O$)+ | kahiricoside III [15] |
| **9** | 22.2 | 679 (M+H)+ | 661 (M-$H_2O$)+<br>643 (M-2$H_2O$)+<br>453 (a glycon-$H_2O$)+<br>(M-acetylated glucose-$H_2O$)+<br>435 (M-acetylated glucose-2$H_2O$)+ | kahiricoside IV [15] |
| **10** | 34.6 | 579.3 (M+H)+ | 533 (M-$C_2H_5O$)+<br>385 (M-xylose-$C_2H_5O$)+ | deacetyl tomentoside I [22] |
| **11** | 44.4 | 577.5 (M+H)+ | 415 (M-glucopyranose) + (a glycon)<br>387 (M-glucopyranose-$C_2H_4$)+<br>355 (M-glucopyranose-$C_3H_8O$)+ | β-sitosterol-β-D-glucoside [45–47] |

#### 2.1.1. Compounds Identified Using ESI Mode

The peak at Rt 13 and *m/z* 640 was identified as the $(M–H)^-$ of 5–hydroxy isomucronulatol-2′,5′-di-*O*-glucoside($C_{29}H_{37}O_{16}$). This was further confirmed by MS$^2$, where the loss of one glucopyranosyl unit (M–162), two sugar units and ring A resulted in the base peak at *m/z* 477, glycon signal at *m/z* 316 and the daughter peak at *m/z* 520, respectively. The 5–hydroxy isomucronulatol–2′,5′–di–*O*–glucoside was previously identified as a flavonoid component in Radix Astragali [39], and its structure and fragments are illustrated in Figure 2.

The peak at Rt 17.1 and *m/z* 579 was identified as the $(M–H)^-$ of 2′,4′–trihydroxy–flavone–8–*C*–α–arabinopyranoside–7–*O*–β–glucopyranoside $(C_{26}H_{27}O_{15})^-$ that was previously isolated and fully characterized from *A. bombycinus* [18]. The MS$^2$ spectrum showed a distinctive base peak at *m/z* 284.8 representing the genin compartment due to the loss of the sugar units (M–(162) glucopyranosyl–(132) arabinopyranosyl)$^-$. The daughter peaks at *m/z* 447 $(M–132)^-$ and *m/z* 429 $(M–132–18)^-$ were due to the loss of the arabinopyranosyl moiety. The loss of a glucose caused the daughter peak at *m/z* 417. Figure 3 shows the chemical structure and fragmentation of 2′,4′–trihydroxy–flavone–8–C–α–arabinopyranoside–7–*O*–β–glucopyranoside.

The compound eluted at Rt 17.7 and *m/z* 737.5 using ESI -Ve was identified as the $(M–H)^-$ of kaempferol–3–*O*–α–L– rhamnopyranosyl–(1→2)–[6–*O*–(3–hydroxy–3–methyl–glutaryl)–β–D–galactopyranoside] (KaeHMG), a flavonoidal glycoside that was previously identified in *A. depressus* [41], *A. monospessulanus subsp. Illyricus* [41] and *A. gombiformis* [40]. The MS$^2$ spectrum demonstrated a peak at *m/z* 593 representing the loss of the terminal rhamnopyranose unit, which was previously documented in the MS$^2$ spectrum for KaeHMG when the phenolic components of *A. gombiformis* were analyzed [40]. The peak at *m/z* 635 was due to the cleavage of the 3–hydroxy–3–methyl glutaryl moiety through McLafferty rearrangement at the carboxylic group attached to the sugar moiety. A peak at *m/z* 675 is probably due to loss of acetic acid from the 3–hydroxy–3–methyl glutaryl moiety and the formation of a stable tertiary carbocation attached to the methyl and hydroxyl groups. The aglycon kaempferol is observed at *m/z* 285. Figure 4 demonstrates the chemical structure and fragmentation of KaeHMG.

The peak eluted at Rt 18.9 and *m/z* 753.3 using the ESI - Ve mode was the $(M–H)^-$ $(C_{33}H_{37}O_{20})^-$ of quercetin–3–*O*–α–L–rhamnopyranosyl–(1→2)–[6–*O*–(3–hydroxy–3–methylglutaryl)–β–D–galactopyranoside] (QueHMG), which was confirmed by the MS$^2$ spectrum where the base peak at *m/z* 609.2 was due to the loss of the rhamnopyranosyl terminal sugar unit. The same peak at *m/z* 609.1462 was observed in the MS$^2$ spectrum of QueHMG, identified in *A. monspessulanus* using high resolution ESI–MS (HRESIMS) [41]. Another peak in the MS$^2$ spectrum at *m/z* 651.1$(M–C_4H_7O_3)^-$ was due to cleavage of the

3–hydroxy–3–methyl glutaryl moiety through McLafferty rearrangement at the carboxylic group attached to the sugar moiety. The genin was observed as a daughter peak at $m/z$ 302 (M–rhamnopyranose– gluco–pyranose–$H_2O$)$^-$. Figure 5 shows the chemical structure and fragmentation of QueHMG.

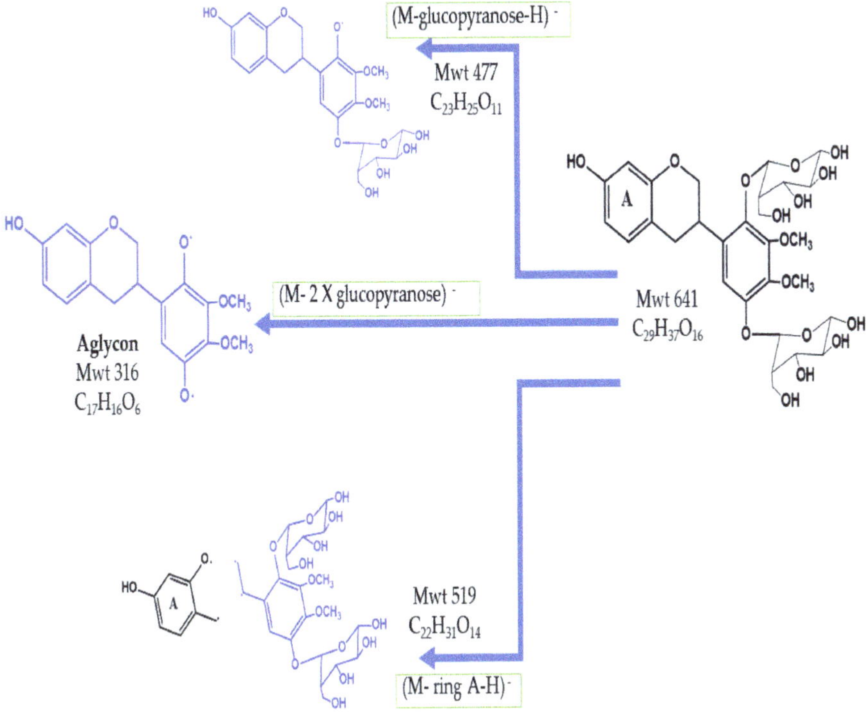


**Figure 2.** The structure and fragmentation of 5–hydroxy isomucronulatol–2′,5′–di–*O*–glucoside.

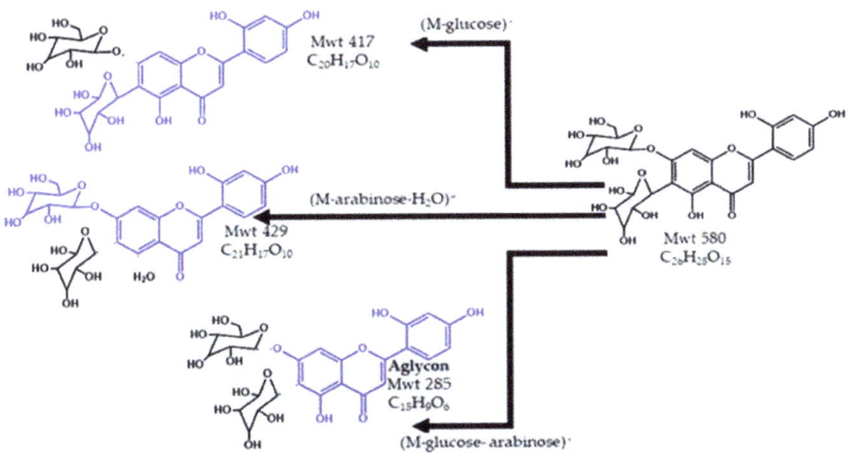


**Figure 3.** The structure and fragmentation of 2′,4′–trihydroxy–flavone–8–*C*–$\alpha$–arabinopyranoside –7–*O*–$\beta$–glucopyranoside.

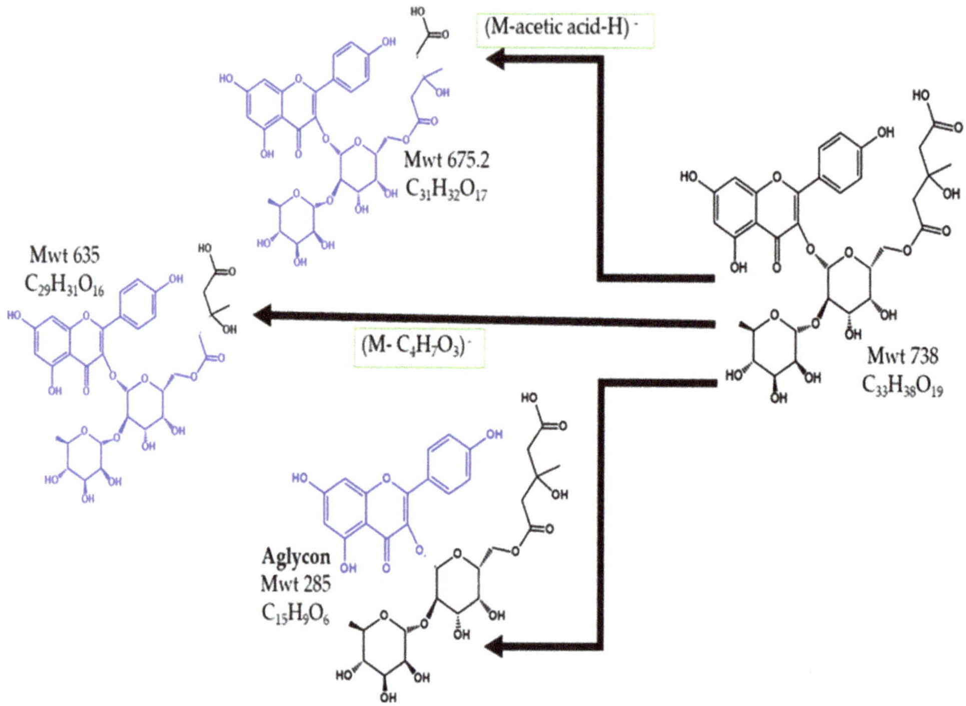


**Figure 4.** The structure and fragments of KaeHMG.

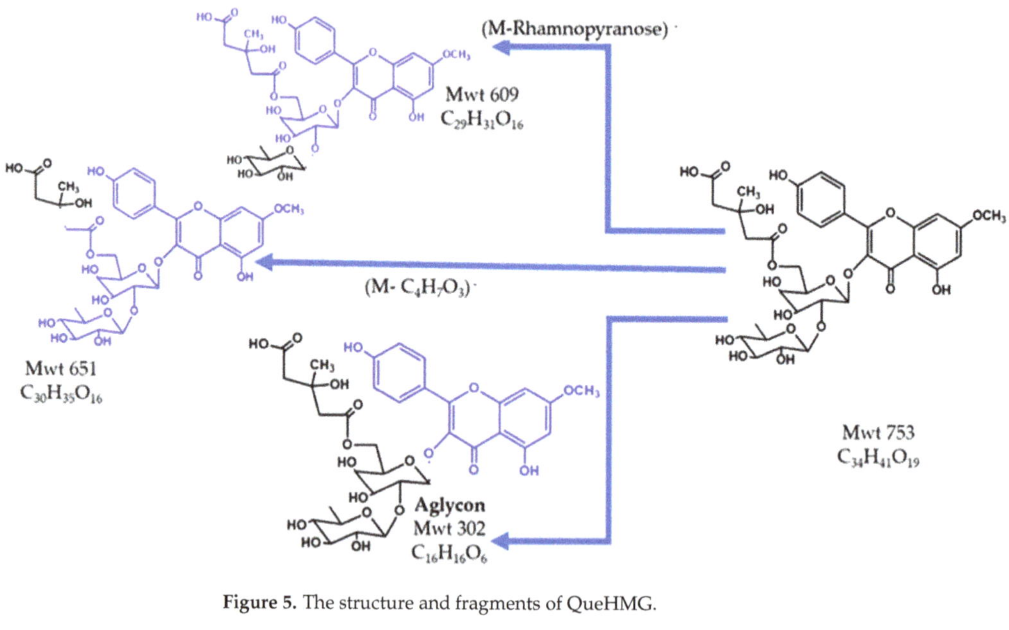


**Figure 5.** The structure and fragments of QueHMG.

The component eluted at Rt 20.7 and at $m/z$ 931 using the ESI - Ve mode was identified as the (M–H)$^-$ of 3–*O*–[α–L–rhamnopyranosyl–(1→2)–β–D–xylopyranosyl]–24–*O*–β–D–glucopyranosyl–3β,6α,16β(24*S*),25–pentahydroxycycloartane, a cycloartane–type triglycoside saponin that was isolated and fully characterized from *A. trojanus*, and named as trojanoside C [42]. The MS$^2$ spectrum showed a base peak at $m/z$ 786 due to the loss of the terminal rhamnopyranosyl unit (M–146)$^-$. The loss of isopropyl alcohol from the side chain created the daughter peak at $m/z$ 869, and while that at $m/z$ 562 was due to the loss of the terminal rhamnopyranose from $C_3$, glucopyranose and isopropyl alcohol from $C_{24}$ in the side chain (M–146–162–60)$^-$. Figure 6 illustrates the chemical structure and fragmentation of trojanoside C.

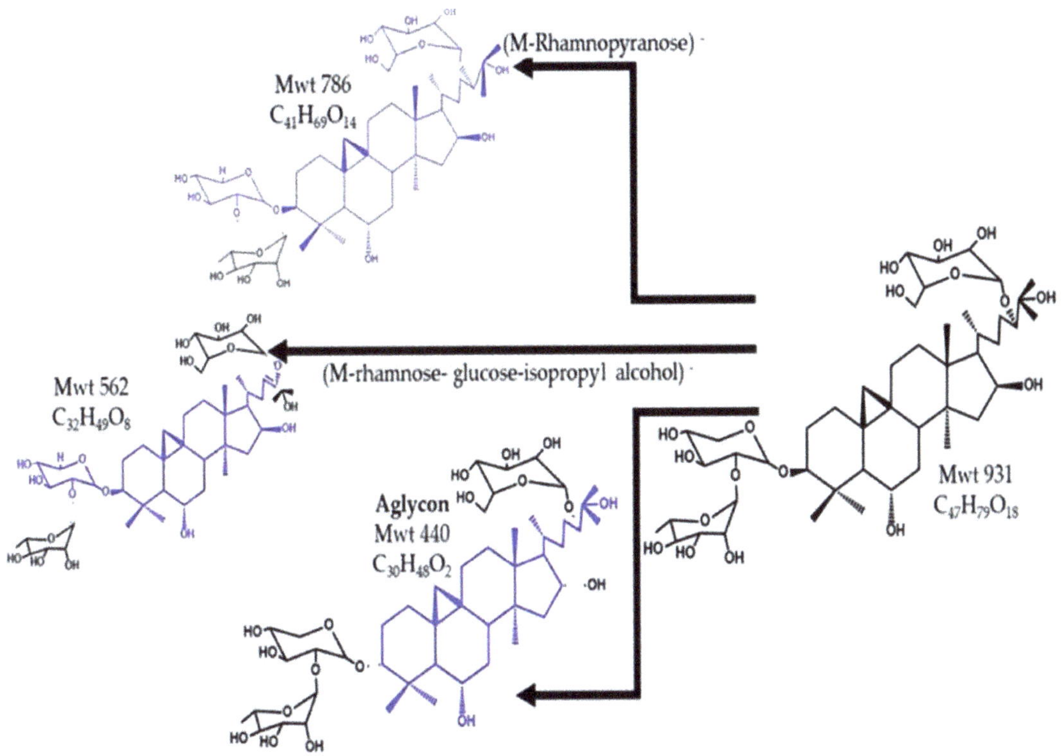


**Figure 6.** The structure and fragments of trojanoside C.

The peak at Rt 21.3 and $m/z$ 900 obtained using the ESI - Ve mode was identified as (M–H)$^-$ of 3–*O*–[α–L–Rhamnopyranosyl–(1→2)–β–D–xylopyranosyl]– 6–*O*–β–D–xylopyranosyl–20(*R*), 24(*S*)–epoxy–3β,6α,16β,25–tetrahydroxycycloartane, a cycloartane-type triglycoside saponin that was isolated and fully characterized from *A. sieversianus* [43] and *A. trojanus* [44], and named as Astrasieversianin XV. The MS$^2$ spectrum demonstrated daughter peaks at $m/z$ 755 and $m/z$ 837 due to the loss of the terminal rhamnopyranosyl unit and isopropyl alcohol from the side chain, respectively. The peak at $m/z$ 798 (M–$C_6H_{14}O$)$^-$ was created after cleavage of the side chain, resulting in a stable carbocation on the $C_{20}$ carboxylic carbon. Figure 7 shows the chemical structure and fragmentation of Astrasieversianin XV.

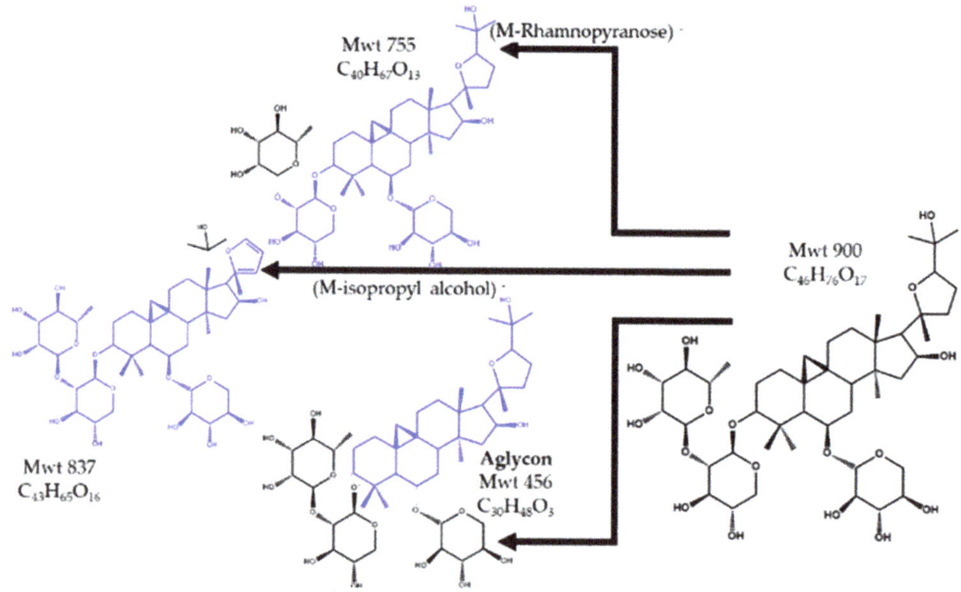


**Figure 7.** The structure and fragments of Astrasieversianin XV.

One of the main components eluted at Rt 22.5, $m/z$ 751.5 using ESI - Ve mode was identified as $(M–H)^-$ of 7–Methylkaempferol–3–O–α–L–rhamnopyranosyl–(1→2)–[6–O–(3–hydroxy–3–methylglutaryl)–β–D–galactopyranoside] (MethKaeHMG). The $MS^2$ showed a base peak at $m/z$ 607 due to the loss of the rhamnopyranosyl unit. The peak at $m/z$ 689 might arise from the loss of one methoxy and two hydroxyl groups from the flavonoidal nucleus. Another fragment at $m/z$ 650 $(M–C_4H_7O_3)^-$ emerged from cleavage of the 3–hydroxy–3–methyl glutaryl moiety through McLafferty rearrangement at the carboxylic group attached to the sugar moiety. The aglycon peak of 7–methoxy kaempferol was observed at m/z 299. The peak at $m/z$ 607 in the $MS^2$ spectrum of MethKaeHMG was previously observed while studying the flavonoidal constituents of *A. gombiformis* [40]. Figure 8 provides the chemical structure and fragmentation of MethKaeHMG.

### 2.1.2. Compounds Tentatively Identified Using APCI Mode

The two main components eluted at Rt 20.9 and 22.2 at $m/z$ 679 were identified as $(M+H)^+$ of kahiricoside III and IV, respectively. Both are cycloartane-type saponin glycosides attached to acetylated glucose, where the acetyl group is attached to $C_2$ in kahiricoside III and $C_6$ in kahiricoside IV. They were previously isolated and fully characterized from *A. kahiricus* where the $(M+H)^+$ was determined by high resolution fast atom bombardment (HRFAB)–MS at $m/z$ 679.4412 for $C_{38}H_{62}O_{10}$. Kahiricoside III was eluted first, followed by kahiricoside IV using $C_{18}$ reversed-phase HPLC and methanol–water elution [15]. The $MS^2$ spectrum indicated daughter peaks at $m/z$ 661 and $m/z$ 452 due to the loss of water and the acetylated glucose moiety with one water molecule, respectively. Those at $m/z$ 643 and $m/z$ 435 were due to the loss of two water molecules and the acetylated glucose compartment and two water molecules, respectively. Figure 9 shows the chemical structure and fragmentation of kahiricoside III and IV.

The peak eluted at Rt 34.6 and $m/z$ 579 using APCI + Ve mode was identified as $(M+H)^+$ of deacetyl tomentoside I, which was previously isolated and fully characterized from the Egyptian *A. tomentosus*. The daughter peak at $m/z$ 533 was observed earlier in the $MS^2$ spectrum of deacetyl tomentoside I isolated from *A. tomentosus*, probably caused by loss of the ethoxy group $C_2H_5O$ from the side chain. The peak at $m/z$ 385 represents the genin compartment with the dissociation of the ethoxy group. Figure 10 shows the structure and fragmentation of deacetyl tomentoside I.

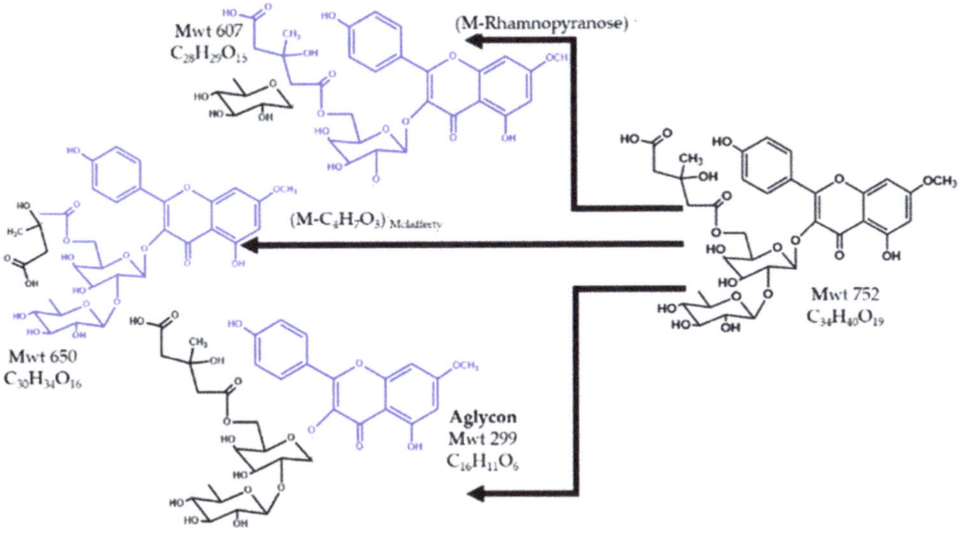


**Figure 8.** The structure and fragments of MethKaeHMG.

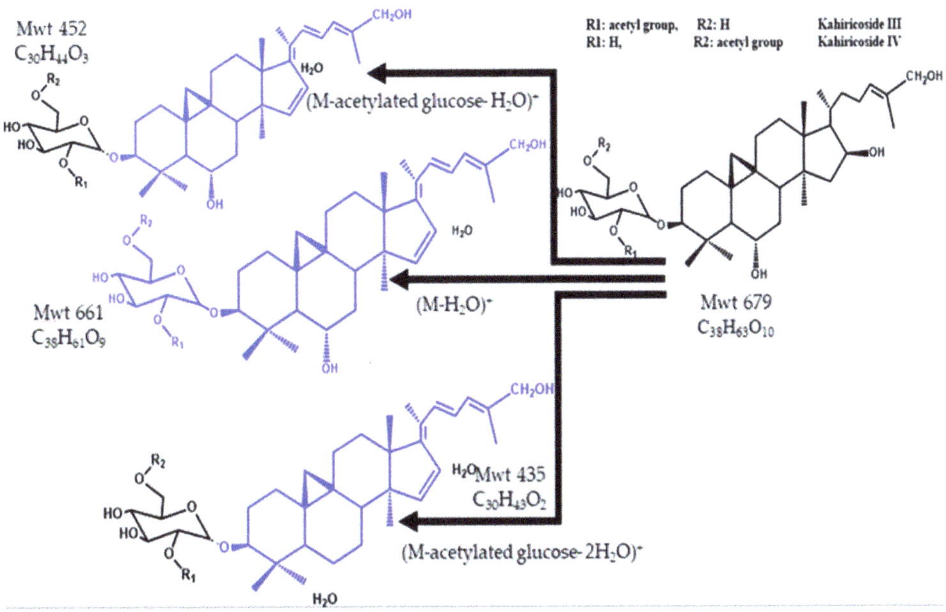


**Figure 9.** The structure and fragmentation of kahiricoside III and IV.

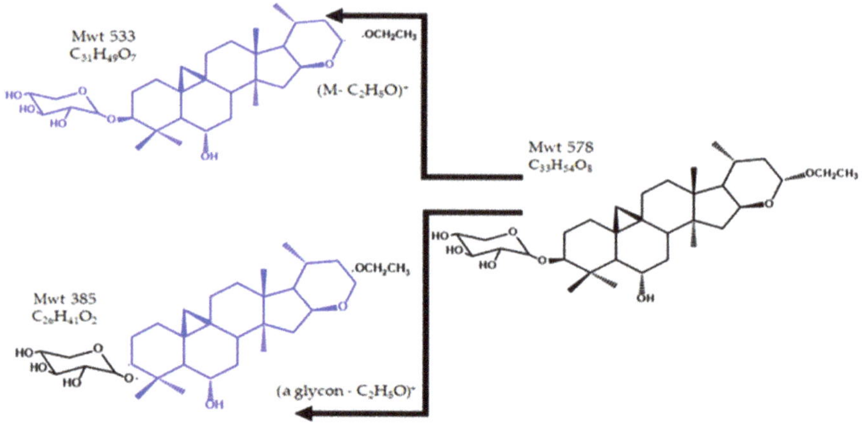


**Figure 10.** The structure and fragmentation of deacetyl tomentoside I.

The peak at Rt 44.4 and *m/z* 577 using APCI + Ve mode was identified as the $(M+H)^+$ of β–sitosterol–β–D–glucoside. In the $MS^2$ spectrum, the peak at *m/z* 415 (genin) represents the cleavage of the glycosidic linkage and loss of the glucopyranosyl unit (M–162). Loss of $C_2H_4$ from the genin moiety's side chain. The formation of a secondary carbocation in the side chain was also observed at *m/z* 387 $(M–162–28)^+$. The loss of the sugar and $C_3H_7$ from the genin side chain probably resulted in the peak at *m/z* 355. β–sitosterol–β–D–glucoside was previously isolated from *A. tomentosus* [24], *A. tanae* [45], *A. sieversianus* [47] and *A. altaicus* [46]. Figure 11 illustrates the structure and fragmentation of *β*–sitosterol–β–D–glucoside.

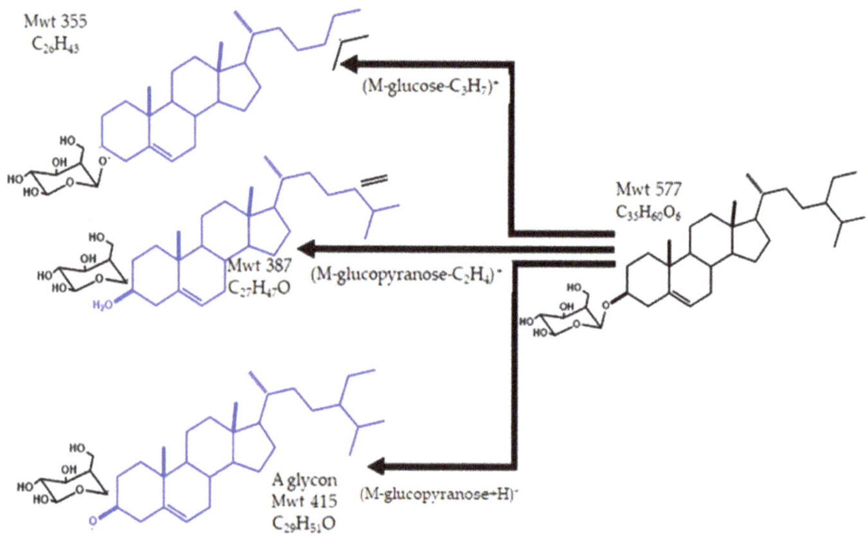


**Figure 11.** The structure and fragmentation of *β*–sitosterol–*β*–D–glucoside.

## 2.2. Cytotoxic Activity

Several natural products can prevent or even treat various cancers. Medicinal plants have been used to treat cancer as they can prevent or delay cancer onset, improve the immune system and the physiological status (35). Table 3 summarizes the cytotoxic screening of *A. fruticosus* leaf methanolic extract on seven cancer cell lines using concentrations ranging from 10 μg/mL–100 μg/mL. Following the preliminary cytotoxic screening, we generated dose response curves for different concentrations (0.02, 0.2, 2, 20 and 200 μg/mL) of the extract against colorectal (HCT–116) and prostate (DU–145) cancer cells, the most sensitive cell lines, (Figure 12) with $IC_{50}$ values of 7.81 μg/mL and 40.79 μg/mL, respectively. Meanwhile, the results for different concentrations of the methanolic extract against HCT–116 and DU–145 cells are represented in Supplementary Tables S1 and S2. At 10 μg/mL, the leaf methanolic extract was not cytotoxic against all the tested cancer cell lines, with the cell viability ranging from 96% to 102%. However, at 100 μg/mL, this extract exhibited prominent cytotoxicity against the colorectal cancer cells (HCT–116) with only 3.368% cell viability. It also showed relatively moderate cytotoxicity against prostate (DU–145), ovarian (SKOV–3) and lung (A–549) cancer cell lines with cell viability of 14.25%, 16.02% and 27.24%, respectively. The breast cancer (MCF–7) cells were the most resistant (54.1% cell viability), while the osteosarcoma (MG–63) and hepatocellular cancer (HepG2) cells exhibited cell viability of 39.09% and 37.13%, respectively.

**Table 3.** Summary for the preliminary cytotoxic screening of *A. fruticosus* leaf methanolic extract.

| Cell Line | Cell Viability % | |
|---|---|---|
| | **10 μg/mL** | **100 μg/mL** |
| Colorectal cancer HCT-116 | 76.3179 | 3.68 |
| Prostate cancer DU-145 | 95.093 | 14.253 |
| Ovarian cancer SKOV-3 | 96.0288 | 16.0264 |
| Lung cancer A-549 | 97.5202 | 27.2388 |
| Hepatocellular carcinoma HepG2 | 99.9973 | 37.1369 |
| Osteosarcoma MG-63 | 101.836 | 39.0968 |
| Breast cancer MCF-7 | 99.0308 | 54.104 |

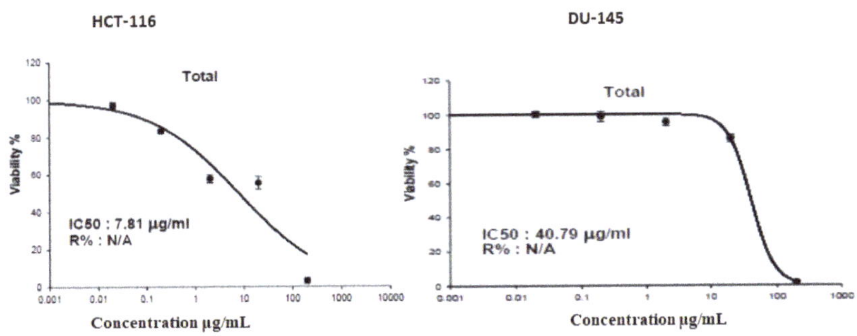


**Figure 12.** Dose response curves for the methanolic extract of *A. fruticosus* against HCT–116 and DU–145 cell lines.

These results are consistent with previous studies on the *Astragalus* species in Egypt where the ethyl acetate fraction of the *A. sieberie* whole plant methanolic extract showed higher cytotoxicity against HCT–116 cells than MCF–7 cells with $IC_{50}$ 32.2 and 69.6 μg/mL, respectively [13]. Moreover, compounds isolated from *A. spinosus* roots [48] and *A. kahiricus* aerial parts exhibited cytotoxicity against the ovarian cancer cell line (A2780) with $IC_{50}$ ranging from 16–47 μg/mL [15]. The methanolic extract of *A. fruticosus* showed dose dependent cytotoxicity against the two most sensitive cell lines, HCT–116 (colorectal) and DU–145 (prostate) cancer cells. The extract showed moderate cytotoxic activity against DU–145 cells ($IC_{50}$ = 40.79 μg/mL), while that against HCT–116 cell line was significant

($IC_{50}$ = 7.81 μg/mL). Figure 12 represents the dose response curves for the methanolic extract of *A. fruticosus* against HCT–116 and DU–145 cell lines.

### 2.3. Antioxidant Activity

DPPH is a stable radical compound with an intense violet color that fades upon reaction with antioxidants [49]. We found 21.05% reduction in the absorbance of the DPPH radical, indicating weak to intermediate antioxidant capacity. The percentage scavenging of DPPH radical was then expressed as trolox equivalent antioxidant capacity (TEAC) that was calculated as approximately 2.51 μg/mL from the calibration curve. This shows that the percentage scavenging activity of 100 μg/mL of the extract was equivalent to the antioxidant potential of 2.31 μg/mL of trolox. Combination of natural products eliminates toxic side effects of chemotherapy and decreases colon cancer incidence [50]. This is consistent with studies showing low to intermediate antioxidant potential of some members of Astragalus growing in Egypt, such as *A. sieberi* [13] and *A. bombycinus* [18].

## 3. Materials and Methods

### 3.1. Plant Materials

The aerial parts of *Astragalus fruticosus* were collected during the flowering stage from Rashid, 40 km east of Alexandria, Egypt in March 2021 [31°23′07.7″ N, 30°25′13.5″ E]. The plant was kindly identified by Professor Sherif Sharawy, Professor of Plant Taxonomy, Faculty of Science, Ain Shams University. A voucher specimen (a.f.001) was deposited in the herbarium of the Pharmacognosy Department, Faculty of Pharmacy, Damanhour University. After air–drying, 10 gm of the aerial parts were pulverized and extracted with 100 mL absolute methanol using the Soxhlet apparatus at 70 °C. The methanolic extract was evaporated to dryness using a rotary evaporator at 60 °C.

### 3.2. HPLC–ESI/APCI–MS/MS Analysis

HPLC separation was performed using the Eclipse XDB C18 column (50 × 2.1 mm, 1.8 μm) (Agilent, Santa Clara, California, USA) using an HPLC system (Agilent HP 1100, USA). The mobile phase was composed of (A) methanol with 0.05% formic acid and (B) water with 0.05% formic acid. The compounds were separated using gradient elution profile: 0 min, A:B 10:90; 36 min, A:B 100:0; 50 min, A:B 100:0. Post-run time, 16 min. Chromatography was performed at 30 °C with a flow rate of 0.3 mL/min using the Bruker daltonik mass spectrometer (Bremen, Germany) equipped with ESI and APCI interfaces. For ESI, the ionization was on negative ion mode with Turbo Spray source, scan type Q1 MS 50–1200 $m/z$, scan rate 2000 Da/s, CUR gas 25, temperature 450 °C, gas 1 50, gas 2 40 and ion spray voltage 4500 V. For APCI, the ionization was on positive ion mode; the settings for the nitrogen drying and nebulizer gas were 5 mL/min (325 °C) and 60 psi. The APCI temperature and the capillary amperage were investigated before optimizing other parameters. Data analysis was performed using LC/MSD Trap Software 5.3 (Bruker daltonik). The compounds were tentatively identified by constructing an in-house mass spectra ($MS/MS^2$) library of various *Astragalus* compounds from literature and comparing their molecular weights and fragmentation ($MS/MS^2$).

### 3.3. Cytotoxicity Assay

We used 10 and 100 μg/mL of the leaf methanolic extract in DMSO to perform the preliminary cytotoxic assay against seven cancer cell lines. The ovarian (SKOV-3) and colorectal (HCT–116) cancer cells were maintained in RPMI media, while the others (lung A–549, prostate DU–145, breast MCF–7, osteosarcoma MG–63 and hepatocellular HepG2 cancer cell lines) were cultured in DMEM media. Both were supplemented with 100 mg/mL of streptomycin, 100 units/mL penicillin and 10% of heat-inactivated fetal bovine serum in humidified, 5% ($v/v$) $CO_2$ atmosphere at 37 °C [50]. Cell viability was assessed using the sulforhodamine B (SRB) assay [51,52]. In 96-well plates, 100 μL cell suspensions ($5 \times 10^3$ cells) were incubated in complete media for 24 h. The cells were treated with

100 µL media containing various concentrations of plant extracts. After treating for 72 h, the cells were fixed by replacing the media with 150 µL of 10% TCA and incubated at 4 °C for 1 h. The TCA solution was removed, and the cells were washed five times with distilled water. Then, 70 µL SRB solution (0.4% *w*/*v*) was added and incubated at room temperature for ten min in the dark. The plates were washed thrice with 1% acetic acid and allowed to air-dry overnight. Then, 150 µL of TRIS (10 mM) was added to dissolve the protein-bound SRB stain, and the absorbance was measured in triplicates at 540 nm using a BMG LABTECH®– FLUOstar Omega microplate reader (Ortenberg, Germany). All the cell lines were obtained from Nawah Scientific Inc., (Mokatam, Cairo, Egypt). After the preliminary screening, five different concentrations (0.02, 0.2, 2, 20 and 200 µg/mL) of the extract were tested for their cytotoxic activity against colorectal (HCT–116) and prostate (DU–145) cancer cell lines, the most sensitive cell lines. Dose response curves were generated, and the cytotoxic activity of the plant extract was expressed as $IC_{50}$.

### *3.4. Antioxidant Activity of the Methanolic Extract*

The antioxidant capacity of 100 µg/mL *A. fruticosus* methanolic extract was determined using DPPH assay based on the redox potential of DDPH (Sigma–Aldrich, St. Louis, MO, USA). We prepared 1 mM DPPH solution (0.394 mg/mL) in methanol and then diluted it 1:10 to obtain a 100 µM solution (Abs at 515 nm = 0.5–0.6). Then, 500 µL each of the test sample and 100 µM DPPH solution were mixed in a cuvette. The negative control contained 500 µL each of methanol and DPPH solution. Both solutions were incubated in the dark at room temperature for 15 min, and absorbance was read at 515 nm using methanol as blank. The results were expressed as the percentage reduction in the radical absorbance [18].

$$[\text{Abs max (negative control)} - (\text{Abs sample} + \text{DPPH})/\text{Abs max}] \times 100$$

The antioxidant potential of the extract was compared to trolox as a reference antioxidant agent, and the results were expressed as TEAC. We constructed the calibration curve between inhibition percentage and trolox concentrations (12.5–0.3 µg/mL). We obtained a high correlation coefficient value of 0.9915, which indicated good linearity. We calculated the TEAC of the extract by substituting the inhibition percentage of the tested extract in the regression equation of the calibration curve.

### *3.5. Statistical Analysis*

Data were expressed as the mean ± SD. The results were calculated using one way analysis of variance followed by Tukey multiple comparisons test. GraphPad Prism software (version 5) was used for all statistical analyzes and creating graphs.

## **4. Conclusions**

To the best of our knowledge, this study is the first HPLC–MS/MS chemical profiling of *A. fruticosus* that demonstrates the predominance of flavonoidal and cycloartane-type saponin glycosides. These results agree with previous studies showing that the genus *Astragalus* is rich in these compounds. The observed cytotoxic activity of the total methanolic extract against colorectal cancer cells might facilitate further research efforts to isolate cytotoxic agent(s) for disease management. Further studies are underway to isolate, characterize and evaluate the potential efficacy of these cytotoxic phytochemicals against colorectal cancer cells. We have also initiated tissue culture experiments to enhance the production of potential compounds from the endangered plant *A. fruticosus*.

**Supplementary Materials:** The following supporting information can be downloaded at: https://www.mdpi.com/article/10.3390/ph15111406/s1: Table S1. Cytotoxic activity of different concentrations of the methanolic extract against colorectal cancer cell line HCT–116; Table S2. Cytotoxic activity of different concentrations of the methanolic extract against prostate cancer cell line DU–145; Figures S1–S11. The mass spectra for tentatively identified compounds using HPLC–MS/MS showing glycon daughter peaks are provided in the Supplementary file.

**Author Contributions:** M.F.D., T.M.O., F.K.E.-F., G.A.E.H.O. and E.M.B.A.E.N. were involved in the conception of the idea and methodology design and performed data analysis and interpretation. E.K.E., E.S., N.D., M.F.E.-K. and W.A.A.-M. participated in the methodology design, sampling and laboratory work and data analysis. T.M.O. and E.K.E. contributed their scientific advice, prepared the manuscript for publication and completed revisions. All authors have read and agreed to the published version of the manuscript.

**Funding:** This study was supported by Princess Nourah bint Abdulrahman University Researchers Supporting Project number (PNURSP2022R23), Princess Nourah bint Abdulrahman University, Riyadh, Saudi Arabia.

**Institutional Review Board Statement:** Permission to collect the plant samples for research purpose was granted from Pharmacognosy Department, Faculty of Pharmacy, Damanhour University. No other permissions were needed.

**Informed Consent Statement:** Not applicable.

**Data Availability Statement:** Data is contained within the article and the supplementary materials.

**Conflicts of Interest:** The authors declare no conflict of interest.

## Abbreviations

HPLC: High performance liquid chromatography, MS/MS$^2$: Tandem mass spectroscopy, ESI: Electrospray ionization, APCI: Atmospheric pressure chemical ionization, Cpd No.: Compound number, *m/z*: Mass to charge ratio, Rt: Retention time, M$^+$: Molecular ion, DMSO: Dimethyl sulfoxide, SRB: Sulphorhodamine B, TCA: Trichloroacetic acid, µg/mL: microgram per milliliter, TRIS: tris(hydroxymethyl)aminomethane, RPMI: Roswell Park Memorial Institute, DPPH: 1,1-diphenyl-2-picryl-hydrazyl radical, QueHMG: quercetin-3-*O*-α-L-rhamnopyranosyl-(1→2)-[6-*O*-(3-hydroxy-3-methylglutaryl)-β-D-galactopyranoside], MethKaeHMG: 7-Methylkaempferol-3-*O*-α-L-rhamnopyranosyl-(1→2)-[6-*O*-(3-hydroxy-3-methylglutaryl)-β-D-galactopyranoside], KaeHMG: kaempferol-3-*O*-α-L-rhamnopyranosyl-(1→2)-[6-*O*-(3-hydroxy-3-methylglutaryl)-β-D-galactopyranoside].

## References

1. Rather, S.A.; Wei, Y.; Wang, J.; Zhao, L.; Chang, Z. Comparative study of floral developments and its systematic importance in the genus *Astragalus* L. and *Oxytropis* DC. (*Fabaceae*, Leguminosae: Papilionoideae). *Biologia* **2021**, *76*, 865–888. [CrossRef]
2. Ionkova, I.; Shkondrov, A.; Krasteva, I.; Ionkov, T. Recent progress in phytochemistry, pharmacology and biotechnology of *Astragalus* saponins. *Phytochem. Rev.* **2014**, *13*, 343–374. [CrossRef]
3. Rios, J.; Waterman, P. A review of the pharmacology and toxicology of Astragalus. *Phytother. Res.* **1997**, *11*, 411–418. [CrossRef]
4. Krasteva, I.; Toshkova, R.; Nikolov, S. Protective effect of *Astragalus corniculatus* saponins against myeloid graffi tumour in hamsters. *Phytother. Res. Int. J. Devoted Pharmacol. Toxicol. Eval. Nat. Prod. Deriv.* **2004**, *18*, 255–257. [CrossRef]
5. Lin, S.; An, X.; Guo, Y.; Gu, J.; Xie, T.; Wu, Q.; Sui, X. Meta-analysis of astragalus-containing traditional chinese medicine combined with chemotherapy for colorectal cancer: Efficacy and safety to tumor response. *Front. Oncol.* **2019**, *9*, 749. [CrossRef]
6. Asgarpanah, J.; Motamed, S.M.; Farzaneh, A.; Ghanizadeh, B.; Tomraee, S. Antioxidant activity and total phenolic and flavonoid content of *Astragalus squarrosus* Bunge. *Afr. J. Biotechnol.* **2011**, *10*, 19176–19180. [CrossRef]
7. Allam, R.M.; Selim, D.A.; Ghoneim, A.I.; Radwan, M.M.; Nofal, S.M.; Khalifa, A.E.; Sharaf, O.A.; Toaima, S.M.; Asaad, A.M.; El-Sebakhy, N.A. Hepatoprotective effects of Astragalus kahiricus root extract against ethanol-induced liver apoptosis in rats. *Chin. J. Nat. Med.* **2013**, *11*, 354–361. [CrossRef]
8. Verotta, L.; Guerrini, M.; El-Sebakhy, N.; Asaad, A.M.; Toaima, S.M.; Abou-Sheer, M.E.; Luo, Y.-D.; Pezzuto, J.M. Cycloartane saponins from *Astragalus peregrinus* as modulators of lymphocyte proliferation. *Fitoterapia* **2001**, *72*, 894–905. [CrossRef]

9. Verotta, L.; Guerrini, M.; El-Sebakhy, N.A.; Assad, A.M.; Toaima, S.M.; Radwan, M.M.; Luo, Y.-D.; Pezzuto, J.M. Cycloartane and oleanane saponins from Egyptian Astragalus spp. as modulators of lymphocyte proliferation. *Planta Med.* **2002**, *68*, 986–994. [CrossRef]
10. Tackholm, V.; Boulos, L. *Students' Flora of Egypt*, 2nd ed.; Cairo University: Cairo, Egypt, 1974; pp. 261–271.
11. Boulos, L. *Flora of Egypt*; Al-Hadara Publishing: Cairo, Egypt, 1999; Volume 1, pp. 320–336. [CrossRef]
12. El-Mawla, A.; Attia, A. Production of flavonoids in cell cultures of *Astragalus sieberi* DC. *Bull. Pharm. Sci. Assiut* **2002**, *25*, 79–83. [CrossRef]
13. Salem, M.A.; Farid, M.M.; El-Shabrawy, M.; Mohammed, R.; Hussein, S.R.; Marzouk, M.M. Spectrometric analysis, chemical constituents and cytotoxic evaluation of *Astragalus sieberi* DC. (*Fabaceae*). *Sci. Afr.* **2020**, *7*, e00221. [CrossRef]
14. Verotta, L.; Tatò, M.; El-Sebakhy, N.A.; Toaima, S.M. Cycloartane triterpene glycosides from *Astragalus sieberi*. *Phytochemistry* **1998**, *48*, 1403–1409. [CrossRef]
15. Radwan, M.M.; El-Sebakhy, N.A.; Asaad, A.M.; Toaima, S.M.; Kingston, D.G. Kahiricosides II–V, cycloartane glycosides from an Egyptian collection of *Astragalus kahiricus*. *Phytochemistry* **2004**, *65*, 2909–2913. [CrossRef] [PubMed]
16. Perrone, A.; Masullo, M.; Plaza, A.; Hamed, A.; Piacente, S. Flavone and flavonol glycosides from *Astragalus eremophilus* and *Astragalus vogelii*. *Nat. Prod. Commun.* **2009**, *4*, 487–490. [CrossRef]
17. Perrone, A.; Masullo, M.; Bassarello, C.; Bloise, E.; Hamed, A.; Nigro, P.; Pizza, C.; Piacente, S. Unusual cycloartane glycosides from *Astragalus eremophilus*. *Tetrahedron* **2008**, *64*, 5061–5071. [CrossRef]
18. Ibrahim, L.F.; Marzouk, M.M.; Hussein, S.R.; Kawashty, S.A.; Mahmoud, K.; Saleh, N.A. Flavonoid constituents and biological screening of *Astragalus bombycinus* Boiss. *Nat. Prod. Res.* **2013**, *27*, 386–393. [CrossRef]
19. Hussein, S.R.; Marzouk, M.M.; Kawashty, S.A. Flavonoids as chemosystematic markers of *Astragalus bombycinus* Boiss. and *Astragalus peregrinus* Vahl growing in Egypt. *Egypt. Pharm. J.* **2018**, *17*, 67–70. [CrossRef]
20. Abd El-Latif, R.; Shabana, M.; El-Gandour, A.; Mansour, R.; Sharaf, M. A new isoflavone from Astragalus peregrinus. *Chem. Nat. Compd.* **2003**, *39*, 536–537. [CrossRef]
21. El-Hawiet, A.M,; Toaima, S.M.; Asaad, A.M.; Radwan, M.M.; El-Sebakhy, N.A. Chemical constituents from *Astragalus annularis* Forssk. and *A. trimestris* L., Fabaceae. *Rev. Bras. De Farmacogn.* **2010**, *20*, 860–865. [CrossRef]
22. Radwan, M.M.; Farooq, A.; El-Sebakhy, N.A.; Asaad, A.M.; Toaima, S.M.; Kingston, D.G. Acetals of Three New Cycloartane-Type Saponins from Egyptian Collections of *Astragalus tomentosus*. *J. Nat. Prod.* **2004**, *67*, 487–490. [CrossRef]
23. El-Sebakhy, N.A.; Harraz, F.M.; Abdallah, R.M.; Asaad, A.M.; Orsini, F.; Pelizzoni, F.; Sello, G.; Verotta, L. Cycloartane triterpene glycosides from Egyptian Astragalus species. *Phytochemistry* **1990**, *29*, 3271–3274. [CrossRef]
24. Abdallah, R.; Ghazy, N.; Assad, A.; El-Sebakhy, A.; Pirillo, A. Constituents of the Egyptian *Astragalus tomentosus* Lam. *Pharmazie* **1994**, *49*, 377–378.
25. Korany, M.; Moneeb, M.; Asaad, A.; El-Sebakhy, N.; El-Banna, A. Analysis of astragalosides I, II and IV in some Egyptian Astragalus species and Astragalus dietary supplements using high-performance liquid chromatography/evaporative light scattering detector and non-parametric regression. *Phytochem. Anal. PCA* **2020**, *31*, 594–605. [CrossRef] [PubMed]
26. Shabana, M. New flavonol glycoside from *Astragalus spinosus* forssk. *Egypt. J. Pharm. Sci.* **2002**, *43*, 19–27.
27. Aboul-Enein, A.M.; El-Ela, F.A.; Shalaby, E.A.; El-Shemy, H.A. Traditional medicinal plants research in Egypt: Studies of antioxidant and anticancer activities. *J. Med. Plants Res.* **2012**, *6*, 689–703. [CrossRef]
28. El-Sebakhy, N.A.; Asaad, A.M.; Abdallah, R.M.; Toaima, S.M.; Verotta, L.; Orsini, F. Constituents of Egyptian Astragalus tribuloides Del. *Nat. Prod. Sci.* **2000**, *6*, 11–15.
29. El-Sebakhy, N.A.; Asaad, A.M.; Abdallah, R.M.; Toaima, S.M.; Abdel-Kader, M.S.; Stermitz, F.R. Antimicrobial isoflavans from Astragalus species. *Phytochemistry* **1994**, *36*, 1387–1389. [CrossRef]
30. Shaker, K.H.; Bernhardt, M.; Elgamal, M.H.A.; Seifert, K. Triterpenoid saponins from *Astragalus trigonus*. *Z. Für Nat. C* **2001**, *56*, 699–702. [CrossRef]
31. Gariboldi, P.; Pelizzoni, F.; Tatò, M.; Verotta, L.; El-Sebakhy, N.; Asaad, A.M.; Abdallah, R.M.; Toaima, S.M. Cycloartane triterpene glycosides from *Astragalus trigonus*. *Phytochemistry* **1995**, *40*, 1755–1760. [CrossRef]
32. Orsini, F.; Verotta, L.; Barboni, L.; El-Sebakhy, N.A.; Asaad, A.M.; Abdallah, R.M.; Toaima, S.M. Cycloartane triterpene glycosides from *Astragalus alexandrinus*. *Phytochemistry* **1994**, *35*, 745–749. [CrossRef]
33. Makboul, M.A.; El-Shanawany, M.; Abdel-Baky, A. Study of the lipid and flavonoid contents of Astragalus cremophilos Boiss. growing in Egypt. *Bull. Pharm. Sci. Assiut* **1984**, *7*, 380–388. [CrossRef]
34. Zayed, R.A.; El-Sayed, A.S.; Ismaeil, W.M. In vitro Micropropagation, biological activities and phenolic profile of *Astragalus fruticosus* Forssk. *Asian J. Plant Sci.* **2022**, *21*, 192–202. [CrossRef]
35. Mendonça, S.C.; Simas, R.C.; Reis Simas, D.L.; Leitão, S.G.; Leitão, G.G. Mass spectrometry as a tool for the dereplication of saponins from *Ampelozizyphus amazonicus* Ducke bark and wood. *Phytochem. Anal.* **2021**, *32*, 262–282. [CrossRef] [PubMed]
36. Moqbel, H.; El Hawary, S.S.E.D.; Sokkar, N.M.; El-Naggar, E.M.B.; El Boghdady, N.; El Halawany, A.M. HPLC-ESI-MS/MS characterization of phenolics in *Prunus amygdalus*, cultivar "umm alfahm" and its antioxidant and hepatoprotective activity. *J. Food Meas. Charact.* **2018**, *12*, 808–819. [CrossRef]
37. Ibrahim, R.M.; El-Halawany, A.M.; Saleh, D.O.; Naggar, E.M.B.E.; El-Shabrawy, A.E.-R.O.; El-Hawary, S.S. HPLC-DAD-MS/MS profiling of phenolics from *Securigera securidaca* flowers and its anti-hyperglycemic and anti-hyperlipidemic activities. *Rev. Bras. Farmacogn.* **2015**, *25*, 134–141. [CrossRef]

38. El Newehy, N.M.; Abd-Alhaseeb, M.M.; Omran, G.A.; Harraz, F.M.; Shawky, E. Comparative metabolomics reveal intraspecies variability in bioactive compounds of different cultivars of pomegranate fruit (*Punica granatum* L.) and their waste by-products. *J. Sci. Food Agric.* **2022**, *102*, 5891–5902. [CrossRef]
39. Chu, C.; Qi, L.-W.; Liu, E.-H.; Li, B.; Gao, W.; Li, P. Radix Astragali (*Astragalus*): Latest advancements and trends in chemistry, analysis, pharmacology and pharmacokinetics. *Curr. Org. Chem.* **2010**, *14*, 1792–1807. [CrossRef]
40. Montoro, P.; Teyeb, H.; Masullo, M.; Mari, A.; Douki, W.; Piacente, S. LC–ESI-MS quali-quantitative determination of phenolic constituents in different parts of wild and cultivated *Astragalus gombiformis*. *J. Pharm. Biomed. Anal.* **2013**, *72*, 89–98. [CrossRef]
41. Shkondrov, A.M.; Krasteva, I.N. High resolution LC-MS/MS screening for secondary metabolites in bulgarian species of genus *Astragalus* L. *Química Nova* **2021**, *44*, 683–688. [CrossRef]
42. Bedir, E.; Çalis, I.; Aquino, R.; Piacente, S.; Pizza, C. Secondary metabolites from the roots of *Astragalus trojanus*. *J. Nat. Prod.* **1999**, *62*, 563–568. [CrossRef]
43. Li-Xian, G.; Xiano-Bing, H.; Yu-Qun, C. Astrasieversianins IX, XI and XV, cycloartane derived saponins from *Astragalus sieversianus*. *Phytochemistry* **1986**, *25*, 1437–1441. [CrossRef]
44. Bedir, E.; Tatli, I.I.; Calis, I.; Khan, I.A. Trojanosides I—K: New cycloartane-type glycosides from the aerial parts of *Astragalus trojanus*. *Chem. Pharm. Bull.* **2001**, *49*, 1482–1486. [CrossRef]
45. Kavtaradze, N.; Alaniya, M.; Masullo, M.; Cerulli, A.; Piacente, S. New Flavone Glycosides from *Astragalus tanae* Endemic to Georgia. *Chem. Nat. Compd.* **2020**, *56*, 70–74. [CrossRef]
46. Yili, A.; Aisa, H.; Isaev, M. Betulinic acid and sterols from *Astragalus altaicus*. *Chem. Nat. Compd.* **2009**, *45*, 592–594. [CrossRef]
47. Iskenderov, D.; Keneshov, B.; Isaev, M. Triterpene glycosides from Astragalus and their genins. LXXVI. Glycosides from *A. sieversianus*. *Chem. Nat. Compd.* **2008**, *44*, 319–323. [CrossRef]
48. Radwan, M.M.; El-Sebakhy, N.A.; Asaad, A.M.; Toaima, S.M.; Kingston, D.G. Spinocoumarin I, a new coumarin Derivative from *Astragalus spinosus* Forssk. *Nat. Prod. Commun.* **2007**, *2*, 919–922. [CrossRef]
49. Zhang, Y.; Wang, B.; Lu, F.; Wang, L.; Ding, Y.; Kang, X. Plant-derived antioxidants incorporated into active packaging intended for vegetables and fatty animal products: A review. *Food Addit. Contam. Part A* **2021**, *38*, 1237–1248. [CrossRef] [PubMed]
50. Tarek, M.O.; Sary, K.A.; Mohamed, A.K.; Mohamed, M.A. Chemopreventive and anticancer activities of indomethacin and vitamin D combination on colorectal cancer induced by 1,2-dimethylhydrazine in rats. *Biomed. Rep.* **2021**, *14*, 27. [CrossRef]
51. Allam, R.M.; Al-Abd, A.M.; Khedr, A.; Sharaf, O.A.; Nofal, S.M.; Khalifa, A.E.; Mosli, H.A.; Abdel-Naim, A.B. Fingolimod interrupts the cross talk between estrogen metabolism and sphingolipid metabolism within prostate cancer cells. *Toxicol. Lett.* **2018**, *291*, 77–85. [CrossRef]
52. Skehan, P.; Storeng, R.; Scudiero, D.; Monks, A.; McMahon, J.; Vistica, D.; Warren, J.T.; Bokesch, H.; Kenney, S.; Boyd, M.R. New colorimetric cytotoxicity assay for anticancer-drug screening. *JNCI J. Natl. Cancer Inst.* **1990**, *82*, 1107–1112. [CrossRef]

*pharmaceuticals*

*Article*

# *Pelargonium sidoides* Root Extract: Simultaneous HPLC Separation, Determination, and Validation of Selected Biomolecules and Evaluation of SARS-CoV-2 Inhibitory Activity

**Manal A. Alossaimi [1], May A. Alzeer [2], Fatma M. Abdel Bar [2,3,*] and Mai H. ElNaggar [4]**

1 Department of Pharmaceutical Chemistry, College of Pharmacy, Prince Sattam Bin Abdulaziz University, Al-Kharj 11942, Saudi Arabia
2 Department of Pharmacognosy, College of Pharmacy, Prince Sattam Bin Abdulaziz University, Al-Kharj 11942, Saudi Arabia
3 Department of Pharmacognosy, Faculty of Pharmacy, Mansoura University, Mansoura 35516, Egypt
4 Department of Pharmacognosy, Faculty of Pharmacy, Kafrelsheikh University, Kafrelsheikh 33516, Egypt
* Correspondence: f.abdelbar@psau.edu.sa; Tel.: +96-65-4540-3617

**Abstract:** This study aimed to establish a validated HPLC-UV analytical method for the determination of gallic acid, catechin, scopoletin, and umckalin in phytoformulations containing *P. sidoides*. Also, to assess the anti-SARS-CoV-2 effect of *P. sidoides* and these biomolecules in vitro. An HPLC-UV method was developed and verified by testing the commercial forms, Kalobin® and Umca®. It revealed low detectable scopoletin and high umckalin levels. *Pelargonium sidoides* exhibited a significant reduction of SARS-CoV-2-induced cytopathic effect in Vero E6 cells ($IC_{50}$ 13.79 μg/mL and selectivity index, SI 6.3), whereas scopoletin showed a remarkable anti-SARS-CoV-2 activity with better selectivity ($IC_{50}$ 17.79 μg/mL and SI 14.22). An in-silico prediction of the drugability indicated that the studied biomolecules are under the acceptable norms of Lipinski's rule, water-soluble, and showed high GIT absorption and bioavailability. Docking study towards the essential molecular targets for viral replication and entry of SARS-CoV-2 indicated good binding affinity of scopoletin (−6.4 Kcal/mol) towards the interface region between the SARS-CoV-2 spike protein RBD and the ACE2 surface receptor indicating the probability of interference with the viral entry to the human cells and showed H-bonding with His-41 in the active site of the main protease which may explain its high antiviral activity.

**Keywords:** COVID-19; HPLC-UV; quality control of Umckaloabo; *Pelargonium sidoides* root extract; umckalin

**Citation:** Alossaimi, M.A.; Alzeer, M.A.; Abdel Bar, F.M.; ElNaggar, M.H. *Pelargonium sidoides* Root Extract: Simultaneous HPLC Separation, Determination, and Validation of Selected Biomolecules and Evaluation of SARS-CoV-2 Inhibitory Activity. *Pharmaceuticals* **2022**, *15*, 1184. https://doi.org/10.3390/ph15101184

Academic Editors: Jan Oszmianski and Sabina Lachowicz-Wiśniewska

Received: 15 August 2022
Accepted: 19 September 2022
Published: 23 September 2022



## 1. Introduction

Medicinal plants have been used by humans for thousands of years and they are the basis for today's advanced medications. One of the commonly used medicinal plants indigenous to South Africa is *Pelargonium sidoides* DC. (Geraniaceae), also known as Umckaloabo. The root extract of the perennial flowering plant ($EP_S$ 7630) has been used by the local population for a very long time to treat a variety of symptoms, such as dysentery, diarrhea, hepatic complaints, wounds, cold, and various infections of the respiratory tract, including tuberculosis [1]. The major constituents and/or extracts of *P. sidoides* (Umckaloabo) were reported to exhibit antibacterial and antiviral activities against several bacteria and respiratory viruses [2,3]. Umckaloabo was recently suggested as a promising adjuvant treatment for the pandemic disease, COVID-19 [4]. However, more research is required to fully investigate the efficacy of this plant against SARS-CoV-2 and the compound (s) that could be responsible for this activity.

Several studies have examined the composition of the plant and found a plethora of coumarins and phenolic metabolites [1]. As reported before, some highly oxygenated coumarins are among the factors governing the pharmacological efficacy of the *P. sidoides* plant [5]. These include 7-hydroxy-5,6-di-methoxycoumarin (umckalin), 6-hydroxy-7-methoxycoumarin (scopoletin), and 6,8-dihydroxy-5,7-dimethoxycoumarin, in addition to gallic acid, and flavonoids (such as catechin) [5,6]. Recently, the chemical profile of *P. sidoides* root was characterized by high-resolution MS with the identification of 33 compounds, including phenolic acids, coumarins, flavonoids, polyphenols, vitamins, and nucleotides [7].

The pharmacological activities of herbal medications cannot be attributed to a single compound within the medication but rather to a multitude of compounds [8]. Hence, the development of systematic quality control systems and validation methods is essential to ensure that the produced herbal medicines are both safe and efficient in achieving the targeted action [9]. The use of the *Pelargonium* spp., including *P. sidoides* DC. and/or *P. reniforme* Curt. is accepted by the European Pharmacopoeia without defining specific differentiation parameters [10]. It is worth noting that umckalin was only reported in *P. sidoides* rather than *P. reniforme* [11,12]. The control of the *P. sidoides* root and its pharmaceutical preparations is usually performed through a pharmacopeial method based on the determination of tannins [10,13], which is not satisfactory for the quality control of the herbal medications and the corresponding formulations.

In this study, four main representative biomolecules, including one phenolic acid (i.e., gallic acid), one flavan derivative (i.e., catechin), and two coumarins (i.e., scopoletin, and umckalin) have been selected for the HPLC analysis and biological evaluation of this plant's phytopreparation. The chemical structures of these biomolecules represent the main phytochemical classes of *P. sidoides* which are coumarins and phenolic compounds as depicted in Figure 1. The separation and determination of these biomolecules from other plants have been extensively studied in the literature. Numerous analytical methods have been used for gallic acid [14,15], catechin [16,17], and scopoletin [18,19]. However, to the best of the authors' knowledge, only one analytical method has been reported for the determination of umckalin, through an HPLC-RP method using a C18 column, an isocratic solvent system of acetonitrile-water (45:55, *v*/*v*) flowing at a rate of 0.75 mL/min, and by UV detection at 330 nm [20]. Other analytical procedures suitable for the quantitative and qualitative determination of umckalin can be found in [7,21]. To date, no analytical method has been proposed to simultaneously separate or detect these four important biomolecules by HPLC. In addition, neither validation nor optimization has been described for the analytical methods reported in the literature.

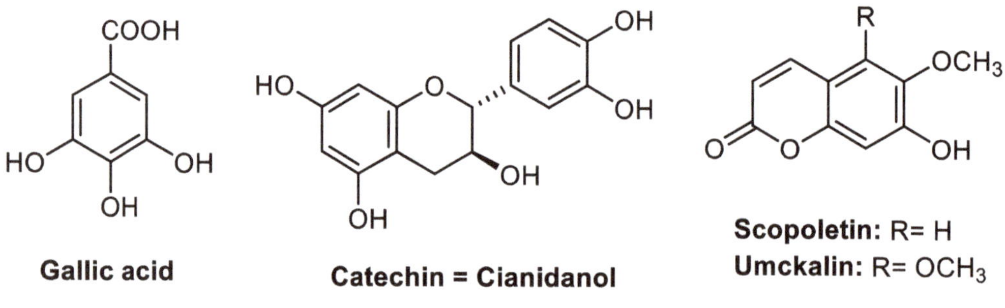


**Figure 1.** Chemical structures of the four studied biomolecules of *Pelargonium sidoides* in phytopreparations.

Hence, the objective of this work is to develop a novel validated HPLC method for the determination of the four suggested biomolecules (*viz.*, gallic acid, catechin, scopoletin, and umckalin) in commercial solutions and in tablet formulations containing *P. sidoides* (Umckaloabo) extracts. Also, to evaluate and compare the efficiency of different sample preparation procedures in extracting the studied biomolecules. As well as to investigate the antiviral activity of *P. sidoides* root extract and the investigated biomolecules against SARS-CoV-2. The physicochemical and pharmacokinetic properties of the studied biomolecules were investigated in silico to predict their potential drugability and effectiveness. Moreover, to get insight into the binding interactions of these biomolecules against the essential molecular targets for viral replication and entry of SARS-CoV-2, a computational docking study was performed. The docking study involved several targets, including the papain-like protease (PLpro), the main protease (Mpro), RNA helicase (nsp13), and RNA-dependent RNA polymerase (RdRp, nsp12) that are essential for viral replication [22–27]. Also, the ability of these compounds to inhibit viral entry to the human cells via interfering with the interaction between the SARS-CoV-2 spike protein and the human angiotensin-converting enzyme 2 (ACE2) receptors was virtually examined using the docking study [28,29].

## 2. Results and Discussion

The quality assurance of phytomedications starts from the correct and reliable identification of the raw materials. Suitable identification and detection procedures must be able to distinguish between multiple active compounds and related species. To complicate things further, a major challenge when dealing with this type of medication is the noticeable variation in the concentrations of the active compounds between different instances of the same species of plants. This variation is attributed to several important factors, including the soil type, geographical location, climate, age, time of harvesting, storage condition ... etc. Analytical methods, such as chromatography and spectroscopy can be used to achieve proper quality control (QC) in herbal medications. It is generally advisable to use combinations of chromatographic and spectroscopic methods to overcome their limitations [11].

### *2.1. HPLC Method Development and Optimization*

An extensive survey of the literature related to our studied four biomolecules, gallic acid, catechin, scopoletin, and umckalin, was carried out to identify the known physical and chemical properties of these compounds [30–33]. By varying one parameter at a time while keeping the remaining parameters constant, an optimization of simultaneous chromatographic separation conditions proceeded, including the detection wavelength, mobile phase, stationary phase, and sample preparation procedure. A series of trials were carried out with discrete values of the acetonitrile to water ratio and with pH values of 3, 8, and 9 and two brands of columns (Hypersil BDS column (250 × 4.6 mm, 5 μm particle size) (Thermo Scientific Inc., Bremen, Germany) and Waters XBridge® C18 column (200 mm × 4.6 mm, 5 μm particle size, Milford, MA, USA). The most suitable chromatographic conditions for this study were found to be acetonitrile to water ratio of 20: 80 *v*/*v* (pH 3), 1 mL/min flow rate, 20 μL injection volume, 25 min run time, using the Waters XBridge® C18 column (200 mm × 4.6 mm, 5 μm particle size, Milford, MA, USA) occupied at 30 °C, and a detection wavelength of 210 nm. When the four compounds under the study were eluted simultaneously, the chromatogram showed symmetrical peaks within a minimal analysis time. The retention times of gallic acid, catechin, scopoletin, and umckalin were found to be 3.3, 4.0, 8.4, and 19.2 min, respectively. The results are depicted in Figure 2a. An additional peak appears at 2.5 min. This peak is the solvent front which was coming from the solvent that was used to dissolve the reference standards (methanol) as shown in the blank run in Figure 2b.

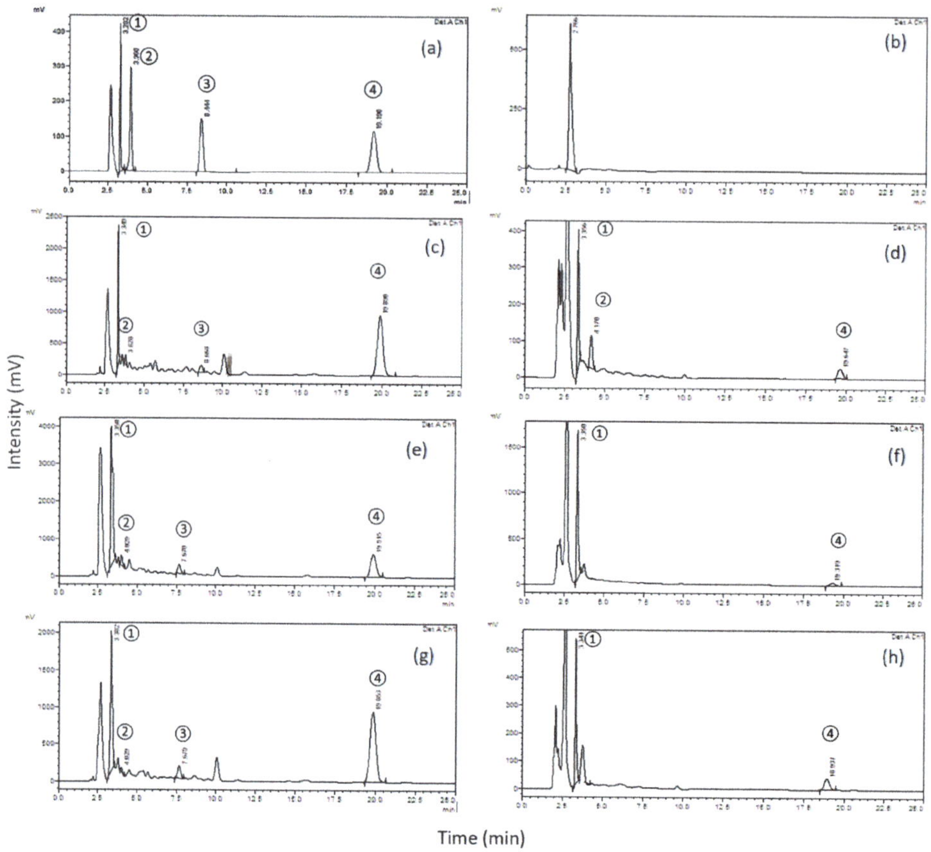


**Figure 2.** HPLC-RP chromatograms (at 210 nm) of: (**a**) 0.16 µg/mL standards mixture of (1) gallic acid, (2) catechin, (3) scopoletin, and (4) umckalin obtained at retention times of 3.3, 3.96, 8.4 and 19.2 min, respectively; (**b**) Blank (methanol); (**c**) Kalobin® solution extracted with ethyl acetate; (**d**) Kalobin® solution extracted with methanol; (**e**) Umca® solutions extracted with ethyl acetate; (**f**) Umca® solutions extracted with methanol; (**g**) Umca® tablets extracted with ethyl acetate; and (**h**) Umca® tablets extracted with methanol.

### 2.2. Method Validation

The proposed separation and detection method was validated by using the validation parameters, including specificity, linearity, detection and quantitation limits, precision, and robustness as mentioned in the material and methods section according to the guidelines of the International Council for Harmonisation (ICH) [34].

#### 2.2.1. Specificity

The tincture samples were exposed to acid or alkali hydrolysis to produce potential degradation products. The spectral purity of the biomolecules was then compared and used to evaluate the specificity and selectivity of the method [35]. It was found that the method was specific and selective.

#### 2.2.2. Linearity

The obtained calibration curves (Figure S1) were interpolated by means of the least-squares linear regression (LSLR) method [36]. To test the linearity of the proposed method, five or six distinct concentrations were considered for each standard as detailed in Table 1. The concentrations of gallic acid and catechin were taken in the range of 0.2–1 μg/mL, while the scopoletin and umckalin concentrations were set within the range of 0.1–1 μg/mL. For each concentration, three different samples were injected. The algebraic mean of the resulting peak areas was calculated and recorded. The accuracy of the regression process was identified by means of the coefficient of determination ($R^2$), which had values of 0.9995, 0.9999, 0.9998, and 0.9995 for gallic acid, catechin, scopoletin, and umckalin, respectively. Since all values were extremely close to 1, the proposed method is considered to have good linearity. The obtained regression line equations for the gallic acid, catechin, scopoletin, and umckalin biomolecules were $y = 129747x + 1655.6$, $y = 164167x + 2039.8$, $y = 162562x + 2223.7$, and $y = 188550x + 3047$, respectively.

**Table 1.** Analytical performance data for the determination of the studied compounds by the proposed method.

| Parameter * | Gallic Acid | Catechin | Scopoletin | Umckalin |
|---|---|---|---|---|
| Concentration range (μg/mL) | 0.2–1 | 0.2–1 | 0.1–1 | 0.1–1 |
| Correlation coefficient (*r*) | 0.9995 | 0.9999 | 0.9998 | 0.9995 |
| Slope | 129,746.6 | 164,166.5 | 162,561.9 | 188,550.2 |
| Intercept | 1655.6 | 2039.8 | 2223.6 | 3046.9 |
| $S_{y/x}$, S.D. of the residuals | 1059.3 | 670.3 | 827.0 | 1456.5 |
| $S_a$, S.D. of the intercept | 943.78 | 597.28 | 589.57 | 1038.3 |
| $S_b$, S.D. of the slope | 1700.6 | 1076.2 | 1159.9 | 2042.8 |
| S.D. | 1.95 | 1.08 | 1.87 | 2.12 |
| %RSD [a] | 1.95 | 1.08 | 1.86 | 2.11 |
| %Error [b] | 0.873 | 0.484 | 0.762 | 0.866 |
| LOD [c] (μg/mL) | 0.024 | 0.012 | 0.012 | 0.018 |
| LOQ [d] (μg/mL) | 0.073 | 0.036 | 0.036 | 0.055 |

* All determinations were conducted at 210 nm. [a] Percentage relative standard deviation. [b] Percentage relative error. [c] Limit of detection. [d] Limit of quantitation.

### 2.2.3. Detection and Quantitation Limits

When assessing the sensitivity of any method, it is important to quantify its limit of detection (LOD) and limit of quantitation (LOQ). The LOD refers to the smallest concentration of the measured compound that can be detected by the studied method to a certain high confidence. The LOQ is the smallest concentration that can be determined with a reasonable level of precision and accuracy [36]. Both limits were calculated from the regression equations shown in Table 1 and the resulting LOD and LOQ were: 0.024 and 0.073 μg/mL for gallic acid; 0.012 and 0.036 μg/mL for catechin; 0.012 and 0.036 μg/mL for scopoletin; and 0.018 and 0.055 μg/mL for umckalin, respectively. The equation used to calculate these values is illustrated below:

$$LOD = 3.3\ S/b$$

$$LOQ = 10\ S/b$$

S = Standard deviation of the response
b = Slope of the calibration curve

### 2.2.4. Precision and Accuracy

To identify the accuracy and precision of the proposed analytical procedure, three replicates of each QC sample concentration were considered along with the five calibrators. Recall that the concentrations of the QC samples were chosen as 0.2, 0.3, and 0.5 μg/mL (low, medium, and high). As shown in Table 2, the accuracies of the proposed method concerning gallic acid, catechin, scopoletin, and umckalin were found to be 100.89%, 99.57%,

100.03%, and 100.51%, respectively. Table 2 shows the precision of the method for all four biomolecules. The relative standard deviation for all samples was below 1%.

**Table 2.** Statistical evaluation of the precision and accuracy for the analysis of the biomolecules in their pure forms by the proposed method.

| Responses | Concentrations (µg/mL) * | | | |
|---|---|---|---|---|
| | Gallic Acid | Catechin | Scopoletin | Umckalin |
| | Concentration:0.2 µg/mL | | | |
| **Response at 0.2 µg/mL** | 28,551 | 34,214 | 34,084 | 40,077 |
| | 28,740 | 34,350 | 34,079 | 40,223 |
| | 28,340 | 34,153 | 34,090 | 40,122 |
| **X** | 28,543.67 | 34,239.00 | 34,084.33 | 40,140.67 |
| **SD** | 200.10 | 100.85 | 5.51 | 74.77 |
| **%RSD** | 0.70 | 0.29 | 0.02 | 0.19 |
| **Conc. Found *(µg/mL)** | 0.204 | 0.196 | 0.196 | 0.196 |
| **% Found** | 102.10 | 98.20 | 98.00 | 98.40 |
| | Concentration: 0.3 µg/mL | | | |
| **Response at 0.3 µg/mL** | 40,552 | 51,184 | 51,610 | 60,560 |
| | 41,019 | 51,435 | 51,633 | 60,443 |
| | 40,391 | 51,099 | 51,621 | 60,492 |
| **X** | 40,654.00 | 51,239.33 | 51,621.33 | 60,498.33 |
| **SD** | 326.19 | 174.70 | 11.50 | 58.76 |
| **%RSD** | 0.80 | 0.34 | 0.02 | 0.10 |
| **Conc. Found *(µg/mL)** | 0.307 | 0.298 | 0.304 | 0.304 |
| **% Found** | 102.33 | 99.53 | 101.30 | 101.57 |
| | Concentration: 0.5 µg/mL | | | |
| **Response at 0.5 µg/mL** | 65,527 | 75,560 | 72,072 | 83,973 |
| | 65,256 | 75,118 | 72,253 | 84,031 |
| | 64,322 | 75,233 | 72,165 | 83,992 |
| **X** | 65,035.00 | 75,303.67 | 72,163.33 | 83,998.67 |
| **SD** | 632.17 | 229.32 | 90.51 | 29.57 |
| **%RSD** | 0.97 | 0.30 | 0.13 | 0.04 |
| **Conc. Found *(µg/mL)** | 0.491 | 0.505 | 0.504 | 0.507 |
| **% Found** | 98.24 | 101.00 | 100.80 | 101.56 |
| $\bar{X} \pm SD$ | 100.89 ± 2.2 | 99.57 ± 1.4 | 100.03 ± 1.7 | 100.51 ± 1.8 |

* Each result is the average of three separate determinations. X = average SD = standard deviation %RSD = Percentage relative standard deviation Conc. Found = concentration found % Found = Percentage found. $\bar{X}$ = Mean Accuracy.

Our method showed lower LOD and LOQ for gallic acid [14,15], catechin [16], scopoletin [7,18] compared to other detection methods. However, better sensitivities were reported for catechin [17] and scopoletin [19] due to the highly sensitive detectors (i.e., UHPLC–MS/MS and fluorescence detectors, respectively). Although Panara, et al. calculated the LOD of gallic, catechin, and scopoletin, no values were calculated for umckalin [7]. In our study, umckalin showed almost comparable LOD to the previously reported method using a DAD detector [20].

2.2.5. Robustness

The robustness of the proposed method was investigated. It was found that temperature variations did not have an impact on the analytical results. However, the rest of the conditions, including the brand of the column, the flow rate of the mobile phase, and its composition all had a significant influence on the quantifications. This resulted in some small changes in the retention times of the detected peaks, and consequently, the integration was impacted. In addition, when the prepared samples were kept at a temperature of 4 °C, they remained stable for at least 4 days. To further investigate the stability of the standards, calibration solutions were injected into the HPLC system at 30-day intervals after being stored at 4 °C. The obtained peaks were almost identical and could be superimposed. This indicates that the calibration solutions were stable.

### 2.3. Determination of the Four Biomolecules in Kalobin®, Umca® Solutions and Umca® Tablets

The measured average concentration percentages of the four biomolecules in different dosage forms, namely Kalobin® and Umca® solutions and the Umca® tablets are shown in Table 3. The obtained results were satisfactory, and no interference was observed. Figure 2c–f depicts the chromatograms corresponding to the Kalobin® and Umca® solutions, while Figure 2g,h shows that of the Umca® tablets. In these figures, two extraction solvents are shown: ethyl acetate, Figure 2c,e,g, and methanol, Figure 2d,f,h. It can be observed that ethyl acetate was efficient at extracting all biomolecules, whereas methanol was only suitable for the extraction of gallic acid, catechin, and umckalin and poorly detects scopoletin. Hence, with the appropriate choice of the extraction solvent, all four biomolecules can be efficiently detected in Kalobin®, Umca® solutions and Umca® tablets using the proposed analytical method for assay. The proposed HPLC analytical method using ethyl acetate extraction in sample preparation showed high contents of gallic acid and umckalin. However, it revealed lower detectable levels of catechin and scopoletin, with the latter maximum concentration of 27.19 µg/mL in Umca® tablets, Table 3 and Figure 2.

### 2.4. In Vitro Inhibitory Activity against SARS-CoV-2

The in vitro antiviral screening assay indicated that *P. sidoides* root extract (Kalobin) has promising antiviral activity against SARS-CoV-2 (Table 3 and Figure 3) and the concentration required to cause a 50% fall in the viral-induced cytopathic effect ($IC_{50}$) was found to be 13.79 µg/mL. On the other hand, the concentration required to cause a 50% growth inhibition of the normal Vero E6 cells ($CC_{50}$) was found to be 87.25 µg/mL. Thus, the *P. sidoides* root extract showed a selectivity index (SI) for antiviral activity relative to cellular toxicity equals 6.3. The obtained results were in agreement with those published before on rhinovirus 16 (RV16), which showed that *P. sidoides* reduces the infection and improves the survival rate of the host, human bronchial epithelial cells (hBEC) [37]. Scopoletin showed the highest antiviral activity among the tested biomolecules with a lower $IC_{50}$ (17.79 µM) than the standard references chloroquine and hydroxychloroquine ($IC_{50}$ of 22.7 µM and 32.8 µM, respectively) [38]. It also showed a SI of 14.22 which was comparable to that of chloroquine and hydroxychloroquine (SI 16.64 and 10.85, respectively) [38]. Scopoletin was previously suggested as a potential antiviral agent against SARS-CoV-2 based on its in-silico binding interactions with the main viral protease (Mpro) [39]. Although many studies correlated the therapeutic applications of *P. sidoides* root extract to its main biomolecule (umckalin) [40], however, umckalin showed a poor antiviral effect against SARS-CoV-2 with an $IC_{50}$ value of 311.6 µM. These results were in harmony with those published by Trun et al. (2006) against Leishmania-induced cytopathic effect in RAW 264.7 cells, which supports our findings as umckalin appears to be inactive as an antimicrobial drug [41]. It could be concluded that the antiviral activity against SARS-CoV-2 can be attributed to scopoletin. In addition, further pharmacological studies are still required to identify the other phytochemicals responsible for the immunological and antimicrobial activities of the *P. sidoides* plant.

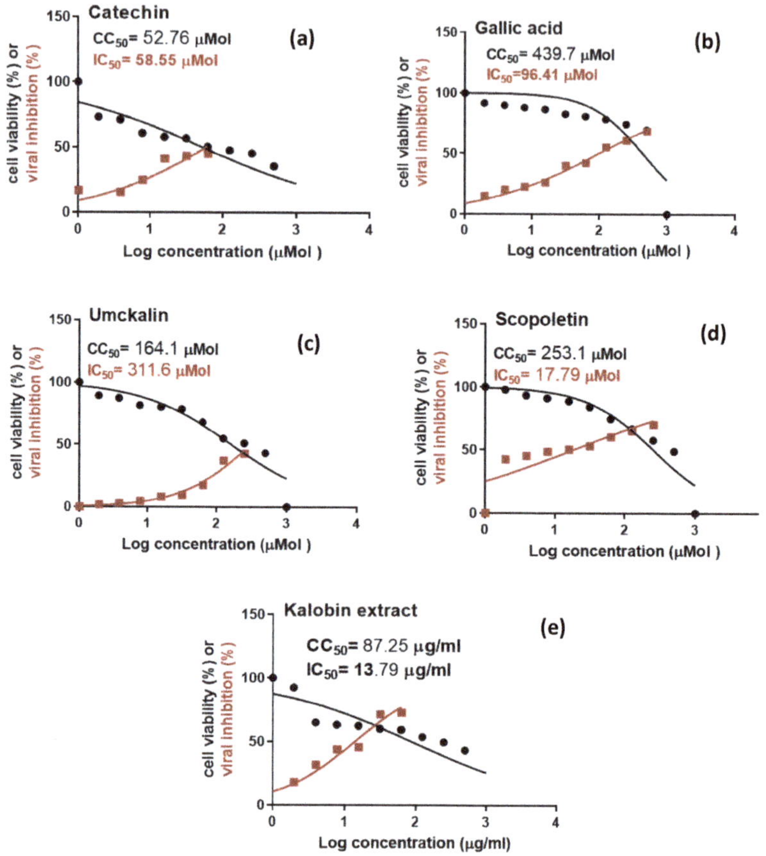


**Figure 3.** Dose-response curves of the tested compounds: (**a**) Catechin, (**b**) Gallic acid, (**c**) Umckalin, (**d**) Scopoletin, and (**e**) Kalobin extract showing their 50% Cytotoxic concentration on Vero E6 cells ($CC_{50}$) and their 50% Inhibitory concentration ($IC_{50}$) against SARS-CoV-2 on Vero E6 cells.

**Table 3.** The 50% cytotoxic concentration ($CC_{50}$) on normal Vero E6 cells, the 50% inhibitory concentration ($IC_{50}$) against SARS-CoV-2-induced cytopathic effect in Vero E6 cells by *P. sidoides* root extract ± S.E.M., and the assay results of gallic acid, catechin, scopoletin, and umckalin content in Kalobin®, Umca® solutions (S) and Umca® tablets (T).

| Test Sample | $CC_{50}$ (μM) [a] | $IC_{50}$ (μM) [b] | Selectivity Index (SI) | Concentration of Investigated Biomolecules (μg/mL) ** | | | | | |
|---|---|---|---|---|---|---|---|---|---|
| | | | | Kalobin®-S | | Umca®-S | | Umca®-T | |
| | | | | Ethyl Acetate | Methanol | Ethyl Acetate | Methanol | Ethyl Acetate | Methanol |
| Gallic acid | 439.7 ± 0.065 | 96.41 ± 0.030 | 4.6 | 89.54 | 22.00 | 188.71 | 98.91 | 146.66 | 23.72 |
| Catechin | 52.76 ± 0.079 | 58.55 ± 0.088 | 0.9 | 14.05 | 10.96 | 39.09 | 0 | 11.26 | 0 |
| Scopoletin | 253.1 ± 0.45 | 17.79 ± 0.91 | 14.1 | 14.86 | 0 | 6.31 | 0 | 27.19 | 0 |
| Umckalin | 164.1 ± 0.54 | 311.6 ± 0.043 | 0.5 | 121.07 | 3.09 | 79.46 | 2.86 | 126.33 | 4.54 |
| *P. sidoides* root extract | 87.25 ± 0.093 [c] | 13.79 ± 0.034 [c] | 6.3 | | | | | | |
| Chloroquine * | 377.7 | 22.7 | 16.64 | | | | | | |
| Hydroxychloroquine * | 356 | 32.8 | 10.85 | | | | | | |

[a] The cytotoxicity was determined using an MTT assay and values were calculated by nonlinear regression analysis using the GraphPad Prism software version 8 (San Diego, CA, USA). [b] Values were calculated by plotting the log of the inhibitor's concentrations versus normalized response (variable slope) using the GraphPad Prism software version 8 (San Diego, CA, USA). [c] Values are measured in μg/mL for the crude extract. * The $IC_{50}$ and $CC_{50}$ values of the antiviral positive controls (chloroquine and hydroxychloroquine) were reported by the same laboratory [38]. ** Concentrations are expressed as the mean of triplicate injection by the proposed HPLC method.

### 2.5. In Silico Investigation of the Physicochemical and Pharmacokinetics Properties Using SwissADME Online Platform

The physicochemical and pharmacokinetic properties affecting absorption, distribution, metabolism, and excretion were computationally evaluated to provide an understanding of the in vivo antiviral activity of the tested compounds. Results (Table S2) articulated that the four studied biomolecules are under the acceptable norms of Lipinski's rule. They are predicted to be water-soluble and to have high gastrointestinal absorption and bioavailability.

Scopoletin and umckalin showed higher lipophilicity than catechin, and gallic acid and showed the ability to cross the blood-brain barrier as indicated by their presence in the yolk of the brain or intestinal estimated permeation (BOILED-Egg) model (Figure S2) [42]. Only catechin is predicted to be effluated from the central nervous system by the P-glycoprotein.

### 2.6. Docking Study

Computational methods are increasingly used tools in drug discovery. They showed a potential role in the screening and development of several important drugs [43]. In this study, the four selected *P. sidoides* biomolecules were subjected to a computational docking study against several essential targets for SARS-CoV-2 viral replication and multiplication. These targets included the viral proteases; the papain-like protease (PLpro), and the main protease (Mpro) responsible for the proteolytic processing of the polyproteins that are translated from the viral RNA [22]. This proteolytic process is essential for the production of the functional proteins responsible for viral replication [23,24]. In addition to the viral proteases, SARS-CoV-2 RNA helicase (nsp13) plays an essential role in viral replication by unwinding duplex oligonucleotides into single strands. While, RNA-dependent RNA polymerase (RdRp, nsp12) is responsible for replicating the viral RNA genome [25]. So, they are among the most important therapeutic targets that could be used for inhibiting viral replication [26,27]. The ability of *P. sidoides* biomolecules for interfering with the interaction between the SARS-CoV-2 spike protein and the human angiotensin-converting enzyme 2 (ACE2) receptors, involved in the viral entry to the human cell [28,29] was also investigated.

Generally, the obtained results confirmed the presence of a strong relationship between the antiviral activity and the in-silico docking calculations of the studied *P. sidoides* biomolecules. The docking study demonstrated that scopoletin exhibited a good binding affinity to the tested SARS-CoV-2 targets that were higher or comparable to that of the co-crystallized ligands or standard inhibitors in most cases (Table 4).

**Table 4.** Docking scores of *Pelargonium sidoides* biomolecules against the SARS-CoV-2 main therapeutic targets using AutoDock Vina in PyRx 0.8.

| Compound | Binding Energy (Kcal/mol) | | | | |
|---|---|---|---|---|---|
| | **Mpro** | **PLpro** | **Nsp13** | **RdRp** | **The Interface of RBD of Spike Protein with Its Human ACE2 Receptor** |
| **Catechin** | −7.2 | −6.1 | −7.2 | −6.4 | −7.7 |
| **Gallic acid** | −5.1 | −4.5 | −5.7 | −5.4 | −6.4 |
| **Umckalin** | −5.3 | −4.8 | −6.6 | −5.4 | −6.3 |
| **Scopoletin** | −5.2 | −4.9 | −6.5 | −5.8 | −6.4 |
| **Reference inhibitor** | −4.9 | −6.7 | −5.7 | −8.9 | −6.6 |

The coumarin derivative, scopoletin showed good binding affinity towards RNA helicase, nsp13 (−6.5 Kcal/mol) which was greater than that of the co-crystalized ligand (−5.7 Kcal/mol). It also showed good binding affinity (−6.4 Kcal/mol) towards the interface region between the receptor-binding domain (RBD) of SARS-CoV-2 spike protein and the human cell angiotensin-converting enzyme 2 (ACE2) surface receptor indicating the probability of its interference with the viral entry to the human cells. Additionally, it demonstrated reasonable binding free energy (−5.2 Kcal/mol) with Mpro which is comparable to that of the co-crystalized ligand (−4.9 Kcal/mol), Table 4.

Although umckalin showed close binding scores to that of scopoletin, visualization of the docking results indicated that scopoletin had better interaction with the amino acids important for protein activity. This may explain the higher in vitro antiviral activity of scopoletin in comparison to umckalin. Scopoletin formed hydrogen bonding with His-41 amino acid residue (Figure 4a) in the active pocket of the main protease (Mpro), which is reported to be involved in the Cys/His catalytic dyad essential for the proteolytic activity of Mpro [44]. It also showed hydrogen bonding with some amino acids essential for the interaction between the SARS-CoV-2 spike protein RBD (Gly-496), and the ACE2 surface receptor (Lys-353 and Glu-37), Figure 4b [29]. Remarkably, it showed hydrogen bonding with several amino acids, including Ser-289, Lys-288, and Arg-567, which are important for the NTPase activity of the RNA helicase [45], Figure 4c.

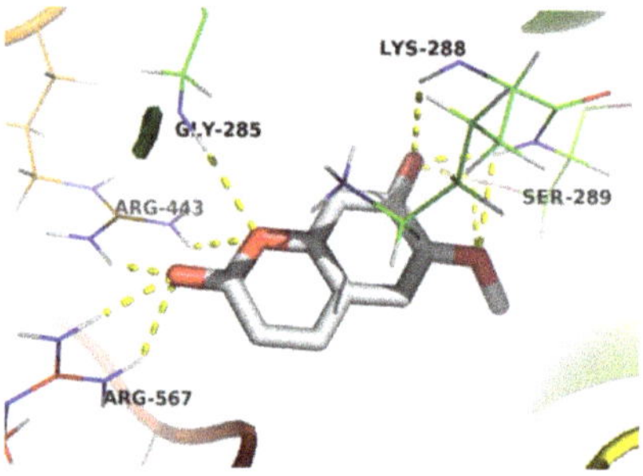


(a)

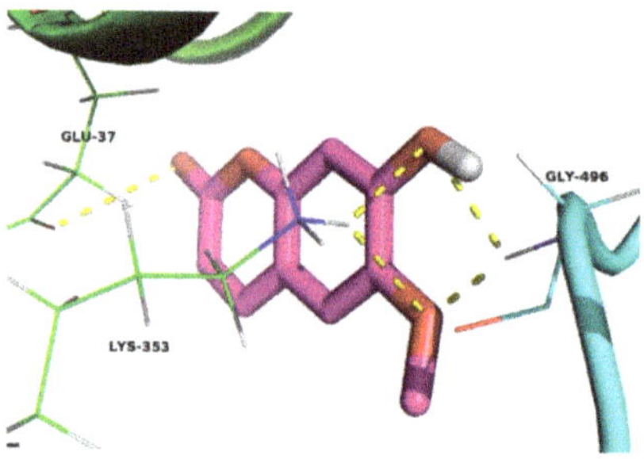


(b)

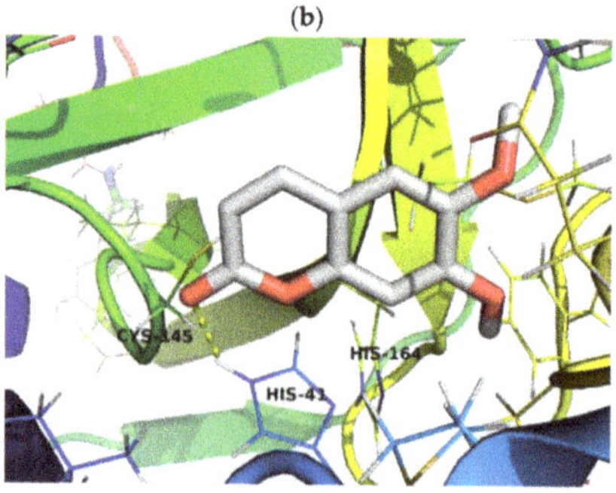


(c)

**Figure 4.** Molecular binding model of scopoletin within the active site of (**a**) the main protease (Mpro, PDB: 5R82); (**b**) within the interface region between the spike RBD (colored cyan) and human ACE2 (colored green), (PDB code: 6M0j); (**c**) within the active site of RNA helicase, nsp13 (PDB code: 5RL9) obtained by AutoDock Vina in PyRx 0.8 (Scripps Research, La Jolla, CA, USA).

Although catechin demonstrated high docking scores towards the investigated protein targets as depicted in Table 4, it showed lower in vitro antiviral activity ($IC_{50}$ 58.55 μM) than scopoletin ($IC_{50}$ 17.79 μM). This may be attributed to its predicted active efflux by the P-glycoprotein and lower lipophilicity in comparison to scopoletin.

## 3. Materials and Methods

### *3.1. Chromatographic Procedures*

#### 3.1.1. Solvents and Mobile Phases

All the reagents used in the study were of an HPLC grade. The acetonitrile for HPLC ≥ 99.9% was acquired from SIGMA-ALDRICH® (Saint Louis, MO, USA). The methanol was of HPLC grade and purchased from Fisher Chemical (Waltham, MA, USA). The ethyl acetate was acquired from SIGMA-ALDRICH® (Saint Louis, MO, USA). In addition, HPLC grade water was obtained from a Milli-Q ultrapure water system. We used orthophosphoric acid for HPLC 85–90% (Honeywell Fluka™, Seelze, Germany) to adjust the pH. Finally, a 0.45 μm Nylon membrane HNWP filter was purchased from MERCK MILLIPORE® Ltd. (Billerica, MA, USA).

#### 3.1.2. Samples

This experimental study utilized two brands of syrup phytopreparations containing *P. sidoides* tincture, including Kalobin® (Marcyrl Pharmaceutical Industries, Cairo, Egypt) Batch No: 1944426, Expiry date: November 2022 and Umca® (ABDIBRAHIM, Istanbul, Turki) Batch No: 2220420, Expiry date: March 2022, in addition to one brand of film-coated tablets (Umca®, ABDIBRAHIM, Istanbul, Turki) Batch No: 2480520, Expiry date: February 2025.

#### 3.1.3. Instrumentation and Chromatographic Conditions

In this study, the UFLC-SHIMADZU 1200 series system (Shimadzu Corporation, Santa Clara, CA, USA) was used to perform the chromatographic analysis of the samples. The system was attached to a binary pump, an online degasser, and the autosampler (SIL-20 A). The final separation was carried out using a 5 μm particle size XBridge® C18 column (Milford, MA, USA) with the dimensions of 200 mm × 4.6 mm supplied by Waters, Ireland. The elution was isocratic at 1 mL/min with a mobile phase system of acetonitrile-water (20:80, *v*/*v*) adjusted to pH 3 by dropwise addition of 10% orthophosphoric acid. Finally, the mobile phase was filtered through 0.45 μm membrane filters and degassed by sonication for 15 min prior to use. The injection volume was 20 μL, and the flow rate was maintained at 1 mL/min with a total run time of 25 min at 30 °C. The liquid chromatographic analysis system was also equipped with a photodiode array detector (SPD—20 A) with a spectral setpoint of 210 nm. The system was connected to a computer running the LabSolutions software on a Microsoft Windows 7 operating platform.

#### 3.1.4. Analytical Standards

The four biomolecules required for this study were all purchased from SIGMA-ALDRICH® (Taufkirchen, Germany) including umckalin (7-hydroxy-5,6-dimethoxycoumarin), HPLC determined purity ≥ 95.0%, CAS Number: 43053-62-9, scopoletin (7-hydroxy-6-methoxycoumarin), HPLC determined purity ≥ 97.0%, CAS Number: 92-61-5, (+)−catechin (syn. cianidanol), HPLC determined purity ≥ 99.0%, CAS Number: 154-23-4, and gallic acid (3,4,5-trihydroxybenzoic acid), HPLC determined purity ≥ 99.0%, CAS Number: 149-91-7.

#### 3.1.5. Preparation of the Test Solutions

To perform the intended analysis of this study, two different forms were considered, solution and tablet dosage forms. As for the solution form, Kalobin® and Umca® 20 mL-package were used. Each of the two solutions was processed by two different methods:

(a) Dilution of 1.1 mL solution, which contains roughly 100 mg of the dry root *P. sidoides* extract, in a 10 mL-volumetric flask. The dilution was carried out using methanol and the solution was subjected to sonication and filtration to obtain a 10 mg/mL solution.
(b) Liquid-liquid extraction (LLE) of a 20 mL tincture sample containing roughly 2000 mg of dry root *P. sidoides* extract with 3 × 20 mL of ethyl acetate. A rotary evaporator was then used to evaporate the combined ethyl acetate extracts to dryness at 45 °C. Finally, the resulting residue weighing 190.9 mg was dissolved in methanol and the volume was adjusted in a 5 mL-volumetric flask.

For the tablet dosage form of the plant extract, 10 Umca® tablets containing 20 mg of dry *P. sidoides* root extract with an average weight of 0.403 g were crushed in a glass mortar. The obtained powder was equally divided into two portions that were processed by two different methods:

(a) The first portion (2.015 g) was extracted using a 10 mL methanol solution, sonicated for 5 min, and filtered. Finally, the volume of the solution is adjusted using a 10 mL-volumetric flask to obtain a 10 mg/mL concentration.
(b) The second portion (2.015 g) was suspended in 10 mL of water, sonicated, and extracted using 3 × 20 mL of ethyl acetate. Then, a rotary evaporator was used to evaporate the combined extracts to dryness at 45 °C. The produced residue was reconstituted in methanol and the volume was adjusted in a 5 mL-volumetric flask.

The final solution obtained from each procedure was filtered through a 0.45 µm membrane and the leading few mL of the filtrate were discarded. This procedure was repeated three times and injected in triplicate. Each time the peak area was calculated and recorded. The four biomolecules' contents were calculated using the average and compared with the standards.

#### 3.1.6. Method Development and Validation

##### Preparation of the Standard Solution and the Quality Control Samples

Each of the four biomolecules' reference solutions, namely gallic acid, catechin, scopoletin, and umckalin were diluted to a concentration of 1 mg/mL by dissolving 10 mg into 10 mL of methanol. The resulting solution was diluted again with methanol to produce a standard stock solution with a concentration of 100 µg/mL. This was, then, diluted once more with methanol to different levels to achieve concentrations in the interval of 0.1–1 µg/mL. Finally, further dilution of the four 100 µg/mL stock solutions with the same solvent was used to produce three quality control samples with the concentrations: 0.2 µg/mL (low quality control, LQC), 0.3 µg/mL (medium quality control, MQC), and 0.5 µg/mL (high quality control, HQC).

##### Specificity

HPLC chromatogram of the *P. sidoides* solutions and tablets was used to compare the UV spectra of the four biomolecules under the study. This procedure was repeated twice with samples of the solutions produced after acid (0.1 M HCl) and base (0.1 M NaOH) treatment of samples of the *P. sidoides* tincture [16]. The solution (2 mL) was added to the aqueous acid or base (2 mL) and the mixture was shaken for one hour at 25 °C. The mixture was then neutralized and injected.

##### Linearity

To prepare the standard calibration curves, five calibrators were utilized within the concentration interval of 0.2–1 µg/mL for gallic acid and catechin, and 0.1–1 µg/mL for scopoletin and umckalin. Linear least square (LLS) regression was used to interpolate the peak area with respect to the drug concentration curves. The accuracy of the regression for all the four biomolecules was measured employing the coefficient of determination ($R^2$) and the results were 0.9995, 0.9999, 0.9998, and 0.9995 for gallic acid, catechin, scopoletin, and umckalin, respectively.

Accuracy and Precision

To assess the accuracy and precision of the experimental procedure, the three QC samples with concentrations of 0.2, 0.3 and 0.5 μg/mL (low, medium, and high) were reproduced at three different time instances together with the five calibrators of the four standards. The precision was measured by the percentage relative standard deviation (% RSD), whereas the accuracy percentage was calculated from Equation (1):

$$\%\ \text{Accuracy} = \frac{\text{Calculated concentration}}{\text{Nominal concentration}} \times 100 \quad (1)$$

Robustness

The robustness of the experimental chromatographic procedure was determined for different parameter values. All experiments performed in this study were repeated three times. Solutions with a concentration of 0.16 μg/mL were used to assess the stability of the procedure. The produced solutions were analyzed at four-time instances: immediately following the preparation, after 7 days, after 14 days, and after 30 days. For the latter three, the solutions were stored at a temperature of 4 °C. The obtained responses from all four experiments were plotted and compared. On the other hand, the stability of the samples was investigated by considering chromatographic analysis immediately following the preparation, and after 24 h of storage at the temperature of 25 °C.

### 3.2. *Assay Procedure for the Determination of the Four Biomolecules in Kalobin, Umca Solutions, and Umca Tablets*

Accurately weighed 10 mg portions of gallic acid, catechin, scopoletin, and umckalin reference standards were placed into 10 mL volumetric flasks and diluted with 10 mL of methanol to produce 1 mg/mL standard solutions. More diluted solutions with a concentration of 200 μg/mL were then obtained by mixing 5 mL of each reference standard with methanol in a 25 mL-volumetric flask. Then, 70 μg/mL working standard solutions were obtained by depositing 8.75 mL of each solution to a 25 mL-volumetric flask and completing the volume with acetonitrile-methanol (1:1 *v*/*v*). Finally, 20 μL portions of the working solutions were injected and the peak area was calculated each time. The corresponding chromatograms were recorded for the standard and the test solutions. Responses of the peak areas were obtained from the plots and concentration was calculated compared with the standard.

### 3.3. *Evaluation of the In Vitro Inhibitory Activity against SARS-CoV-2*

#### 3.3.1. The Used Cells and the Viral Strain

Vero E6 cells were obtained from the American Type Culture Collection (ATCC No. CRL-1586). The used SARS-CoV-2 strain, hCoV-19/Egypt/NRC-03/2020 (Accession Number on GSAID: EPI_ISL_430820) was isolated on 18 March 2020 from an oropharyngeal swab specimen obtained from a 34-year-old Egyptian woman [46]. All the experiments were executed according to the protocol approved by the ethics committee of the National Research Centre (NRC), Giza, Egypt (Approval number: NRC-20074).

#### 3.3.2. MTT Cytotoxicity Assay against Normal Cell Line (Vero E6)

The half-maximal cytotoxic concentration ($CC_{50}$) of the test compounds was evaluated using the 3-(4,5-dimethylthiazol-2-yl)-2,5-diphenyltetrazolium bromide (MTT) assay with slight modifications [47] and according to the previously published procedure [48]. The percentage of cytotoxicity was measured compared to the untreated cells from Equation (2):

$$\%\ \text{cytotoxicity} = \frac{\text{Absorbance without treatment} - \text{Absorbance with treatment}}{\text{Absorbance without treatment}} \times 100 \quad (2)$$

Plotting% cytotoxicity versus the sample concentrations was used to calculate the concentration of the test compound that is required to reduce the cell viability by 50% ($CC_{50}$).

3.3.3. Cytopathic Effect (CPE) Inhibition Assay

The inhibitory concentration 50 ($IC_{50}$) of the test compounds required to reduce the SARS-CoV-2-induced cytopathic effect on Vero E6 cells by 50% relative to the virus control was measured. The viral proliferation was assessed by visualization of the cytopathic effects on infected Vero E6 cells using crystal violet staining according to the previously published procedure [48]. The optical density of the produced color was measured at 570 nm using an Anthos Zenyth plate reader (Model: 200rt, Anthos Labtec Instruments, Heerhugowaard, The Netherlands).

3.3.4. Statistical Analysis

The data obtained were analyzed using GraphPad Prism Software version 8 (San Diego, CA, USA). Data are shown as mean ± standard deviation (SD). One-Way ANOVA followed by Dunnett's test was used to analyze the data. All concentrations of the tested samples showed statistical significance ($p < 0.05$) from the untreated cytopathic cells except for the lowest tested concentration of umckalin.

### 3.4. *In Silico Studies*

3.4.1. Evaluation of Molecular and Pharmacokinetic Properties

The physicochemical and pharmacokinetics properties were investigated using the SwissADME online platform [49].

3.4.2. Docking Studies

Molecular docking calculations were performed with AutoDock Vina [50] embedded in PyRx 0.8 software (Scripps Research, La Jolla, CA, USA). Docking was performed as described in the Supplementary Materials using crystal structures for the studied enzyme targets retrieved from RCSB-Protein Data Bank with the PDB codes displayed in Table S1. The docking results were visualized using Pymol Molecular Graphics System (Schrödinger, LLC, NY, USA) to view the docked protein-ligand interactions.

## 4. Conclusions

A new HPLC-UV analytical method has been developed and validated for separating and detecting four selected biomolecules (i.e., gallic acid, catechin, scopoletin, and umckalin) of *P. sidoides* root extract in different commercial dosage forms. The efficiency of the proposed method was verified by an assay testing procedure considering the commercial forms: Kalobin® and Umca® solutions and Umca® tablets. The developed method was accurate, precise, fast, and superior in multiple respects to other previously reported methods of the analysis of the *P. sidoides'* biomolecules. Unlike existing methods, the developed method can separate all investigated biomolecules simultaneously in their pure form and in different dosage forms. These results indicated that the proposed method can be successfully employed for the routine analysis of the four investigated biomolecules in bulk and commercial formulations containing this plant extract. Out of the four tested biomolecules, scopoletin, in addition to *P. sidoides* DC. root extract, showed remarkable in vitro antiviral activity against SARS-CoV-2 using cytopathic effect (CPE) inhibition assay on Vero E6 cells. However, the chemotaxonomic marker "umckalin" exhibited weak antiviral activity against SARS-CoV-2.

**Supplementary Materials:** The following supporting information can be downloaded at: https://www.mdpi.com/article/10.3390/ph15101184/s1, Table S1. PDB codes of the crystal structures and grid box coordinates for the enzymes used in the docking study; Table S2. The in-silico predicted physicochemical and ADME properties of *Pelargonium sidoides* biomarkers; Figure S1. Calibration curve of: (a) gallic acid, (b) catechin, (c) scopoletin, and (d) umckalin; Figure S2. BOILED-Egg representation of lipophilicity and polarity of *Pelargonium sidoides* biomarkers.

**Author Contributions:** Conceptualization, F.M.A.B.; methodology, M.A.A. (Manal A. Alossaimi), F.M.A.B. and M.H.E.; software, M.H.E.; validation, M.A.A. (Manal A. Alossaimi) and M.A.A. (May A. Alzeer); formal analysis; M.A.A. (Manal A. Alossaimi) and M.A.A. (May A. Alzeer); investigation, M.A.A. (May A. Alzeer); resources, F.M.A.B.; data curation, M.A.A. (Manal A. Alossaimi) and M.H.E.; writing—original draft preparation, M.A.A. (Manal A. Alossaimi) and M.H.E.; writing—review and editing, F.M.A.B.; visualization, M.A.A. (Manal A. Alossaimi) and M.A.A. (May A. Alzeer); supervision, F.M.A.B. All authors have read and agreed to the published version of the manuscript.

**Funding:** This research received no external funding.

**Institutional Review Board Statement:** Not applicable.

**Informed Consent Statement:** Not applicable.

**Data Availability Statement:** Data is contained within the article and Supplementary Materials.

**Acknowledgments:** The Deanships of Graduate Studies and Scientific Research at Prince Sattam bin Abdulaziz University are acknowledged for supporting this research. We are thankful to the Vaccine Development and Virological Tests Unit, Center of Scientific Excellence for Influenza Viruses, National Research Centre, 12622, Giza, Egypt for carrying out the in vitro SARS-CoV-2 antiviral assay.

**Conflicts of Interest:** The authors declare no conflict of interest.

## References

1. Kolodziej, H. Traditionally used *Pelargonium* species: Chemistry and biological activity of umckaloabo extracts and their constituents. *Curr. Top. Phytochem.* **2000**, *3*, 77–93.
2. Kayser, O.; Kolodziej, H. Antibacterial activity of extracts and constituents of *Pelargonium sidoides* and *Pelargonium reniforme*. *Planta Med.* **1997**, *63*, 508–510. [CrossRef]
3. Careddu, D.; Pettenazzo, A. *Pelargonium sidoides* extract EPs 7630: A review of its clinical efficacy and safety for treating acute respiratory tract infections in children. *Int. J. Gen. Med.* **2018**, *11*, 91–98. [CrossRef] [PubMed]
4. Ahmad, A.; Rehman, M.; Alkharfy, K. An alternative approach to minimize the risk of coronavirus (COVID-19) and similar infections. *Eur. Rev. Med. Pharmacol. Sci.* **2020**, *24*, 4030–4034. [PubMed]
5. Colling, J.; Groenewald, J.H.; Makunga, N.P. Genetic alterations for increased coumarin production lead to metabolic changes in the medicinally important *Pelargonium sidoides* DC (Geraniaceae). *Metab. Eng.* **2010**, *12*, 561–572. [CrossRef]
6. Latté, K.P.; Kayser, O.; Tan, N.; Kaloga, M.; Kolodziej, H. Unusual coumarin patterns of *Pelargonium* Species forming the origin of the traditional herbal medicine Umckaloabo. *Z. Nat. C* **2000**, *55*, 528–533. [CrossRef]
7. Panara, A.; Aalizadeh, R.; Thomaidis, N.S. Chemical characterisation of *Pelargonium sidoides* root based on LC-QToF-MS non-target screening strategies. *Phytochem. Anal.* **2022**, *33*, 40–56. [CrossRef]
8. Kolodziej, H.; Kiderlen, A.F. In vitro evaluation of antibacterial and immunomodulatory activities of *Pelargonium reniforme, Pelargonium sidoides* and the related herbal drug preparation EPs 7630. *Phytomedicine* **2007**, *14* (Suppl. 6), 18–26. [CrossRef]
9. Luypaert, J.; Zhang, M.; Massart, D.L. Feasibility study for the use of near infrared spectroscopy in the qualitative and quantitative analysis of green tea, *Camellia Sinensis* (L.). *Anal. Chim. Acta* **2003**, *478*, 303–312. [CrossRef]
10. HMPC-Committee-on-Herbal-Medicinal-Products. *Assessment Report on Pelargonium sidoides DC and/or Pelargonium reniforme Curt., radix*; EMA/HMPC/444251/2015; European Medicines Agency: London, UK, 2018.
11. Maree, J.E.; Viljoen, A.M. Phytochemical distinction between *Pelargonium sidoides* and *Pelargonium reniforme*—A quality control perspective. *S. Afr. J. Bot.* **2012**, *82*, 83–91. [CrossRef]
12. Kolodziej, H. Fascinating metabolic pools of *Pelargonium sidoides* and *Pelargonium reniforme*, traditional and phytomedicinal sources of the herbal medicine Umckaloabo. *Phytomedicine* **2007**, *14* (Suppl. 6), 9–17. [CrossRef] [PubMed]
13. European-Pharmacopoeia. *European Pharmacopoeia*, 9th ed.; The Council of Europe: Strasbourg, France, 2016.
14. Canas, S.; Trindade, C.; Sun, B.; Naves, P. Phenolic compounds involved in pine wilt disease: HPLC-based method development and validation for their quantification. *J. Plant Biochem. Biotechnol.* **2020**, *30*, 343–353. [CrossRef]
15. Dhanani, T.; Shah, S.; Kumar, S. A validated high-performance liquid chromatography method for determination of tannin-related marker constituents gallic acid, corilagin, chebulagic acid, ellagic acid and chebulinic Acid in four *Terminalia* species from India. *J. Chromatogr. Sci.* **2015**, *53*, 625–632. [CrossRef]
16. Fiori, J.; Pasquini, B.; Caprini, C.; Orlandini, S.; Furlanetto, S.; Gotti, R. Chiral analysis of theanine and catechin in characterization of green tea by cyclodextrin-modified micellar electrokinetic chromatography and high performance liquid chromatography. *J. Chromatogr. A* **2018**, *1562*, 115–122. [CrossRef]
17. Svoboda, P.; Vlčková, H.; Nováková, L. Development and validation of UHPLC–MS/MS method for determination of eight naturally occurring catechin derivatives in various tea samples and the role of matrix effects. *J. Pharm. Biomed. Anal.* **2015**, *114*, 62–70. [CrossRef]

18. Choi, S.-I.; Kwon, H.-Y.; La, I.-J.; Jo, Y.-H.; Han, X.; Men, X.; Lee, S.-J.; Kim, Y.-D.; Seong, G.-S.; Lee, O.-H. Development and validation of an analytical method for deacetylasperulosidic acid, asperulosidic acid, scopolin, asperuloside and scopoletin in fermented *Morinda citrifolia* L. (Noni). *Separations* **2021**, *8*, 80. [CrossRef]
19. Wang, L.; Wu, H.L.; Yin, X.L.; Hu, Y.; Gu, H.W.; Yu, R.Q. Simultaneous determination of umbelliferone and scopoletin in Tibetan medicine *Saussurea laniceps* and traditional Chinese medicine Radix *Angelicae pubescentis* using excitation-emission matrix fluorescence coupled with second-order calibration method. *Spectrochim. Acta Part A Mol. Biomol. Spectrosc.* **2017**, *170*, 104–110. [CrossRef]
20. Franco, L.; de Oliveira, B.H. Determination of umckalin in commercial tincture and phytopreparations containing *Pelargonium sidoides* by HPLC: Comparison of sample preparation procedures. *Talanta* **2010**, *81*, 1368–1372. [CrossRef] [PubMed]
21. Masondo, N.A.; Makunga, N.P. Advancement of analytical techniques in some South African commercialized medicinal plants: Current and future perspectives. *S. Afr. J. Bot.* **2019**, *126*, 40–57. [CrossRef]
22. Dömling, A.; Gao, L. Chemistry and Biology of SARS-CoV-2. *Chem* **2020**, *6*, 1283–1295. [CrossRef] [PubMed]
23. Zhang, L.; Lin, D.; Sun, X.; Curth, U.; Drosten, C.; Sauerhering, L.; Becker, S.; Rox, K.; Hilgenfeld, R. Crystal structure of SARS-CoV-2 main protease provides a basis for design of improved α-ketoamide inhibitors. *Science* **2020**, *368*, 409–412. [CrossRef]
24. Liu, C.; Zhou, Q.; Li, Y.; Garner, L.V.; Watkins, S.P.; Carter, L.J.; Smoot, J.; Gregg, A.C.; Daniels, A.D.; Jervey, S. Research and development on therapeutic agents and vaccines for COVID-19 and related human coronavirus diseases. *ACS Cent. Sci.* **2020**, *6*, 315–331. [CrossRef]
25. Jia, Z.; Yan, L.; Ren, Z.; Wu, L.; Wang, J.; Guo, J.; Zheng, L.; Ming, Z.; Zhang, L.; Lou, Z. Delicate structural coordination of the Severe Acute Respiratory Syndrome coronavirus Nsp13 upon ATP hydrolysis. *Nucleic Acids Res.* **2019**, *47*, 6538–6550. [CrossRef]
26. Liu, W.; Morse, J.S.; Lalonde, T.; Xu, S. Learning from the past: Possible urgent prevention and treatment options for severe acute respiratory infections caused by 2019-nCoV. *Chembiochem* **2020**, *21*, 730–738.
27. Lung, J.; Lin, Y.S.; Yang, Y.H.; Chou, Y.L.; Shu, L.H.; Cheng, Y.C.; Liu, H.T.; Wu, C.Y. The potential chemical structure of anti-SARS-CoV-2 RNA-dependent RNA polymerase. *J. Med. Virol.* **2020**, *92*, 693–697. [CrossRef]
28. Wu, C.; Liu, Y.; Yang, Y.; Zhang, P.; Zhong, W.; Wang, Y.; Wang, Q.; Xu, Y.; Li, M.; Li, X. Analysis of therapeutic targets for SARS-CoV-2 and discovery of potential drugs by computational methods. *Acta Pharm. Sin. B* **2020**, *10*, 766–788. [CrossRef]
29. Gurung, A.B.; Ali, M.A.; Lee, J.; Farah, M.A.; Al-Anazi, K.M. Identification of potential SARS-CoV-2 entry inhibitors by targeting the interface region between the spike RBD and human ACE2. *J. Infect. Public Health* **2021**, *14*, 227–237. [CrossRef]
30. NCBI. *PubChem Compound Summary for CID 5316862, Umckalin*; National Center for Biotechnology Information: Bethesda, MD, USA, 2022.
31. NCBI. *PubChem Compound Summary for CID 370, Gallic Acid*; National Center for Biotechnology Information: Bethesda, MD, USA, 2022.
32. NCBI. *PubChem Compound Summary for CID 9064, Cianidanol*; National Center for Biotechnology Information: Bethesda, MD, USA, 2022.
33. NCBI. *PubChem Compound Summary for CID 5280460, Scopoletin*; National Center for Biotechnology Information: Bethesda, MD, USA, 2022.
34. ICH-guideline. *Validation of Analytical Procedures: Text and Methodology*; International Council for Harmonisation: Geneva, Switzerland, 1995.
35. Shabir, G.A. Validation of high-performance liquid chromatography methods for pharmaceutical analysis. Understanding the differences and similarities between validation requirements of the US Food and Drug Administration, the US Pharmacopeia and the International Conference on Harmonization. *J. Chromatogr. A* **2003**, *987*, 57–66. [CrossRef]
36. Miller, J.N. Basic statistical methods for analytical chemistry Part 2: Calibration and regression methods. A review. *Analyst* **1991**, *116*, 3–14. [CrossRef]
37. Roth, M.; Fang, L.; Stolz, D.; Tamm, M. *Pelargonium sidoides* radix extract EPs 7630 reduces rhinovirus infection through modulation of viral binding proteins on human bronchial epithelial cells. *PLoS ONE* **2019**, *14*, e0210702. [CrossRef]
38. Seliem, I.A.; Panda, S.S.; Girgis, A.S.; Moatasim, Y.; Kandeil, A.; Mostafa, A.; Ali, M.A.; Nossier, E.S.; Rasslan, F.; Srour, A.M. New quinoline-triazole conjugates: Synthesis, and antiviral properties against SARS-CoV-2. *Bioorg. Chem.* **2021**, *114*, 105117. [CrossRef] [PubMed]
39. Ikanovic, T.; Sehercehajic, E.; Saric, B.; Tomic, N.; Hadziselimovic, R. In silico analysis of scopoletin interaction with potential SARS-CoV-2 target. In *New Technologies, Development and Application IV*; Karabegović, I., Ed.; Springer: Berlin/Heidelberg, Germany, 2021; Volume 233.
40. White, A.G.; Davies-Coleman, M.T.; Ripley, B.S. Measuring and optimising umckalin concentration in wild-harvested and cultivated *Pelargonium sidoides* (Geraniaceae). *S. Afr. J. Bot.* **2008**, *74*, 260–267. [CrossRef]
41. Trun, W.; Kiderlen, A.F.; Kolodziej, H. Nitric oxide synthase and cytokines gene expression analyses in Leishmania-infected RAW 264.7 cells treated with an extract of *Pelargonium sidoides* (Eps 7630). *Phytomedicine* **2006**, *13*, 570–575. [CrossRef] [PubMed]
42. Daina, A.; Zoete, V. A boiled-egg to predict gastrointestinal absorption and brain penetration of small molecules. *ChemMedChem* **2016**, *11*, 1117. [CrossRef] [PubMed]
43. Sliwoski, G.; Kothiwale, S.; Meiler, J.; Lowe, E.W. Computational methods in drug discovery. *Pharmacol. Rev.* **2014**, *66*, 334–395. [CrossRef] [PubMed]

44. Ton, A.T.; Gentile, F.; Hsing, M.; Ban, F.; Cherkasov, A. Rapid identification of potential inhibitors of SARS-CoV-2 main protease by deep docking of 1.3 billion compounds. *Mol. Inform.* **2020**, *39*, 2000028. [CrossRef]
45. Gurung, A.B. In silico structure modelling of SARS-CoV-2 Nsp13 helicase and Nsp14 and repurposing of FDA approved antiviral drugs as dual inhibitors. *Gene Rep.* **2020**, *21*, 100860.
46. Kandeil, A.; Mostafa, A.; Hegazy, R.R.; El-Shesheny, R.; El Taweel, A.; Gomaa, M.R.; Shehata, M.; Elbaset, M.A.; Kayed, A.E.; Mahmoud, S.H.; et al. Immunogenicity and Safety of an Inactivated SARS-CoV-2 Vaccine: Preclinical Studies. *Vaccines* **2021**, *9*, 214. [CrossRef]
47. Mosmann, T. Rapid colorimetric assay for cellular growth and survival: Application to proliferation and cytotoxicity assays. *J. Immunol. Methods* **1983**, *65*, 55–63. [CrossRef]
48. Mostafa, A.; Kandeil, A.; Elshaier, Y.A.M.M.; Kutkat, O.; Moatasim, Y.; Rashad, A.A.; Shehata, M.; Gomaa, M.R.; Mahrous, N.; Mahmoud, S.H.; et al. FDA-Approved Drugs with Potent In Vitro Antiviral Activity against Severe Acute Respiratory Syndrome Coronavirus 2. *Pharmaceuticals* **2020**, *13*, 443. [CrossRef]
49. Daina, A.; Michielin, O.; Zoete, V. SwissADME: A free web tool to evaluate pharmacokinetics, drug-likeness and medicinal chemistry friendliness of small molecules. *Sci. Rep.* **2017**, *7*, 42717. [CrossRef] [PubMed]
50. Trott, O.; Olson, A.J. AutoDock Vina: Improving the speed and accuracy of docking with a new scoring function, efficient optimization, and multithreading. *J. Comput. Chem.* **2010**, *31*, 455–461. [CrossRef] [PubMed]

*pharmaceuticals*

MDPI

*Article*

# In the Search for Novel, Isoflavone-Rich Functional Foods—Comparative Studies of Four Clover Species Sprouts and Their Chemopreventive Potential for Breast and Prostate Cancer

**Agnieszka Galanty [1,*], Monika Niepsuj [1], Marta Grudzińska [1], Paweł Zagrodzki [2], Irma Podolak [1] and Paweł Paśko [2]**

1 Department of Pharmacognosy, Faculty of Pharmacy, Jagiellonian University Medical College, Medyczna 9, 30-688 Cracow, Poland; monika.niepsuj13@gmail.com (M.N.); grrudzinska@gmail.com (M.G.); irma.podolak@uj.edu.pl (I.P.)

2 Department of Food Chemistry and Nutrition, Faculty of Pharmacy, Jagiellonian University Medical College, Medyczna 9, 30-688 Cracow, Poland; pawel.zagrodzki@uj.edu.pl (P.Z.); p.pasko@uj.edu.pl (P.P.)

* Correspondence: agnieszka.galanty@uj.edu.pl

**Citation:** Galanty, A.; Niepsuj, M.; Grudzińska, M.; Zagrodzki, P.; Podolak, I.; Paśko, P. In the Search for Novel, Isoflavone-Rich Functional Foods—Comparative Studies of Four Clover Species Sprouts and Their Chemopreventive Potential for Breast and Prostate Cancer. *Pharmaceuticals* **2022**, *15*, 806. https://doi.org/10.3390/ph15070806

Academic Editors: Jan Oszmianski and Sabina Lachowicz-Wiśniewska

Received: 10 June 2022
Accepted: 24 June 2022
Published: 28 June 2022

**Publisher's Note:** MDPI stays neutral with regard to jurisdictional claims in published maps and institutional affiliations.



**Abstract:** Despite a significant amount of research, the relationship between a diet rich in isoflavones and breast and prostate cancer risk is still ambiguous. The purpose of the current study was to pre-select the potential candidate for functional foods among red, white, crimson, and Persian clover sprouts, cultured for different periods of time (up to 10 days), with respect to the isoflavone content (determined by HPLC-UV-VIS), and to verify their impact on hormone-dependent cancers in vitro. The red clover sprouts were the richest in isoflavones (up to 426.2 mg/100 g dw), whereas the lowest content was observed for the crimson clover. Each species produced isoflavones in different patterns, which refer to the germination time. Hormone-insensitive MDA-MB-231 breast cancer cells were more resistant to the tested extracts than estrogen-dependent MCF7 breast cancer cells. Regarding prostate cancer, androgen-dependent LNCap cells were most susceptible to the tested sprouts, followed by androgen-insensitive, high metastatic PC3, and low metastatic DU145 cells. The observed cytotoxic impact of the tested sprouts is not associated with isoflavone content, as confirmed by chemometric analysis. Furthermore, the sprouts tested revealed a high antioxidant potential, and were characterized by high safety for normal breast and prostate cells.

**Keywords:** clover sprouts; *Trifolium*; isoflavones; prostate cancer; breast cancer; cytotoxic

## 1. Introduction

Plants from the *Fabaceae* family, as well as their sprouts and by-products, are an attractive object of research which are currently gaining more and more scientific interest. This is mainly due to their nutritional value, but also their health-promoting potential, associated with the presence of polyphenolic compounds [1], among which, isoflavones, a class of so-called phytoestrogens, are the most intriguing. As they are structurally similar to 17-β-estradiol, isoflavones reveal estrogenic activity, both by affecting estrogen receptors and aromatase activity (which regulates transformation of androstenedione into estrone). Furthermore, isoflavones can also affect 5-α-reductase activity, which is responsible for the transformation of testosterone to dihydrotestosterone. This may be of great importance in terms of the development of hormone-sensitive cancers, including breast and prostate cancers, as these compounds can compete with 17-β-estradiol for the binding site of estrogen receptors. Literature data indicate that isoflavones trigger epigenomic effects that could be beneficial in breast cancer prevention and/or treatment [2], and they can reduce the risk of inducing estrogen-dependent tumorigenesis [3]. However, some studies point out that at low concentrations, isoflavones may stimulate the growth of breast cancer cells [4].

Regarding prostate cancer, isoflavones, in addition to affecting the estrogen receptor, also modulate the androgen receptor, resulting in a decreased risk of the carcinogenesis process, but the detailed signalling pathways are still unknown [5]. Despite a significant amount of research, the relationship between a diet rich in isoflavones and breast and prostate cancers is still ambiguous and controversial [6–8].

Functional food, in addition to its nutritional effect, reduces the risk of various diseases, including cancer. Sprouts are an attractive and dynamically developing example of functional food, with benefits such as high nutritional and low caloric values. Additionally, during the germination process, the level of active compounds can increase significantly, as was demonstrated especially for phenolic acids and flavonoids [9], but also for isoflavones [10]. Sprouts rich in phytoestrogens (including isoflavones) are an example of functional food that may reveal chemopreventive potential against the development of hormone-dependent cancers. However, as the isoflavones provided with the diet can influence the hormonal balance in the human body, it is crucial to know what their content in a given product is. This is especially important for people with the risk of the development or already-existing hormone-dependent cancers. During the search for novel dietary sources of isoflavones, we focused on clover (*Trifolium*) species, belonging to the *Fabaceae* family, of which, red clover (*T. pratense* L.) is the most well-known. In addition to soy, red clover is one of the richest sources of isoflavones, and its aerial parts are used as dietary supplements to reduce menopausal symptoms. Only a few studies described the level of these compounds in red clover sprouts, and the results are very promising [10–12], although significant differences were also observed between particular studies. The results also indicate the impact of red clover and its sprouts on breast cancer cells and their estrogenic activity in vitro [13–15]. Therefore, we decided to include red clover sprouts in our study as an example of the most-known clover species, and to compare them with the sprouts of three other clovers, namely, white (*T. repens* L.), crimson (*T. incarnatum* L.), and Persian (*T. resupinatum* L.), in terms of the isoflavone profile and the impact on breast and prostate cells. Interestingly, except for the latter, these clover species have never been used to grow sprouts. As far as Persian clover sprouts are concerned, the results on their antioxidant properties have been published [16,17], but no information on the content of isoflavones can be found.

The purpose of this study was to preselect the potential candidate for functional foods from among the four clover sprouts mentioned above, cultured for different periods of time, with respect to the isoflavone content and the importance in the chemoprevention of hormone-dependent cancers. To achieve the assumed goal, except from HPLC analysis, we used a specially designed in vitro cellular model, comprising breast and prostate cancer cells, differing in their hormone sensitivity. Normal breast and prostate cells were also included in the study to determine the safety of the examined clover sprouts. To better justify our results and reveal potential relationships between the studied factors, chemometric analysis was applied.

## 2. Results and Discussion

This study aimed at the comparative analysis of four clover species sprouts, in the context of the search for new candidates for functional food with chemopreventive potential in hormone-dependent breast and prostate cancers. We used three clover species and compared them to red clover sprouts, which are a popular food product. It is also worth noting that this is the first study to describe the sprouts of white and crimson clover, and also to define the isoflavone content of Persian clover.

### *2.1. Four Clover Species Sprouts Differ in Their Isoflavones Profile*

In the first step of the experiment, the qualitative profile of isoflavones and the quantitative accumulation dynamics in the sprouts grown for 3, 5, 7, and 10 days were determined by HPLC, and the results are presented in Table 1. Two isoflavone glycosides, namely, ononin and daidzin, together with three free isoflavones: genistein, daidzein, and formononetin, were identified in the examined sprout extracts. Most of the seed extracts contained only trace amounts of isoflavones, apart from PCS and CCS, where genistein, daidzein, and formononetin were detected in small amounts. None of the tested species revealed the presence of biochanin A or genistin. Notable qualitative differences were observed between the four clover species. The PC and RC sprouts contained five (genistein, daidzein, ononin, formononetin, daidzin) and four (genistein, ononin, formononetin, daidzin) of the examined isoflavones, respectively, whereas only two isoflavones were observed in the WC (ononin, formononetin) and CC (genistein, formononetin) sprouts. The isoflavone profile for RC sprouts is consistent with the observations of other authors [11]; however, some of them also indicated the presence of biochanin A, genistin, sissotrin, or glycytein [10,12]. No information has been published so far on the isoflavone profile in WC, PC, and CC sprouts; thus, our analysis is the first one concerning their presence.

In terms of quantitative analysis, the absolute winner in isoflavone content among the tested species is RC, with the sprouts containing between 204.1 and 426.2 mg/100 g dw of isoflavones sum (Figure 1). The WC and PC sprouts contained 12.92–48.12 and 0.26–47.06 mg/100 g dw of isoflavone sum, respectively, whereas their sum in the CC sprouts was the lowest (0.07–23.31 mg/100 g dw). Ononin and formononetin were present in the highest amount in the examined sprouts. Interestingly, ononin was the predominant isoflavone in RC and PC sprouts, whereas formononetin was predominant in WC and CC sprouts (Table 1). Similar observations were made by [10] for red clover sprouts; however, the total isoflavone content was twice as much as our results, due to the higher amount of formononetin (up to 215.55 mg/100 g dw for 10-day sprouts). The contents of ononin, daidzin, and genistein were comparable to those described by [10] for ten-day sprouts, whereas [12] noted a significantly lower amount of formononetin in 5-day red clover sprouts compared to our study. Such a difference may result from the details in the growth conditions, or the variety used for the experiment.

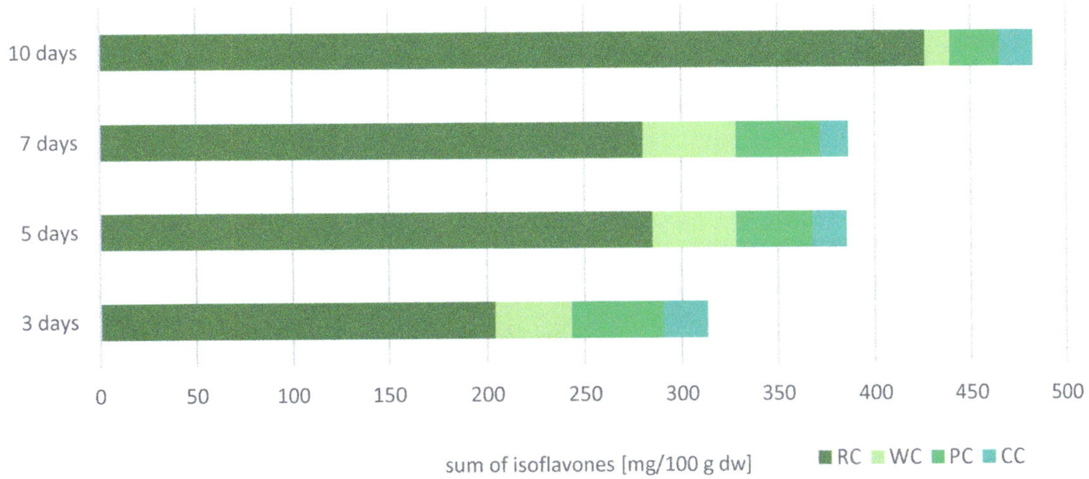


**Figure 1.** Cumulation dynamics of isoflavones sum in red (RC), white (WC), Persian (PC), and crimson (CC) clover sprouts harvested for 3, 5, 7, and 10 days.

**Table 1.** Concentration of isoflavones and antioxidant activity of red (RC), white (WC), Persian (PC), and crimson (CC) clover seeds (S) and sprouts harvested for 3, 5, 7, and 10 days.

| RCS | RC3 | RC5 | RC7 | RC10 | WCS | WC3 | WC5 | WC7 | WC10 | PCS | PC3 | PC5 | PC7 | PC10 | CCS | CC3 | CC5 | CC7 | CC10 |
|---|---|---|---|---|---|---|---|---|---|---|---|---|---|---|---|---|---|---|---|
| **ISOFLAVONE GLYCOSIDES [mg/100 g dw, $n$ = 3]** | | | | | | | | | | | | | | | | | | | |
| **daidzin** | | | | | | | | | | | | | | | | | | | |
| Tr | 2.81 ± 0.08 | 8.46 ± 0.12 | 5.52 ± 0.40 | 18.58 ± 1.40 | Tr | Tr | Tr | Tr | Tr | Tr | 1.47 ± 0.42 | 1.22 ± 0.05 | 0.99 ± 0.13 | 0.60 ± 0.08 | Tr | Tr | Tr | Tr | Tr |
| **ononin** | | | | | | | | | | | | | | | | | | | |
| Tr | 193 ± 6 | 191 ± 3 | 229 ± 5 | 332 ± 10 | Tr | 11.87 ± 0.32 | 15.24 ± 0.35 | 19.93 ± 0.20 | Tr | Tr | 37.91 ± 4.48 | 32.18 ± 5.68 | 37.22 ± 0.64 | 21.95 ± 0.38 | Tr | Tr | Tr | Tr | Tr |
| **ISOFLAVONE AGLYCONES [mg/100 g dw, $n$ = 3]** | | | | | | | | | | | | | | | | | | | |
| **daidzein** | | | | | | | | | | | | | | | | | | | |
| Tr | Tr | Tr | Tr | Tr | Tr | Tr | Tr | Tr | Tr | 0.13 ± 0.00 | 1.62 ± 0.20 | 1.16 ± 0.05 | 1.03 ± 0.06 | 0.68 ± 0.03 | Tr | Tr | Tr | Tr | Tr |
| **formononetin** | | | | | | | | | | | | | | | | | | | |
| Tr | 5.97 ± 0.04 | 77.69 ± 2.39 | 34.96 ± 0.76 | 60.45 ± 1.39 | Tr | 27.12 ± 1.60 | 27.68 ± 0.68 | 28.19 ± 0.86 | 12.92 ± 0.39 | Tr | 3.79 ± 0.36 | 3.37 ± 0.35 | 3.14 ± 0.64 | 1.56 ± 0.04 | 0.07 ± 0.03 | 19.80 ± 1.98 | 14.82 ± 2.48 | 11.57 ± 0.44 | 7.98 ± 0.42 |
| **genistein** | | | | | | | | | | | | | | | | | | | |
| Tr | 2.10 ± 0.14 | 6.00 ± 0.15 | 7.98 ± 0.20 | 15.12 ± 0.32 | Tr | Tr | Tr | Tr | Tr | 0.13 ± 0.01 | 2.42 ± 0.10 | 1.28 ± 0.08 | 0.94 ± 0.07 | 0.57 ± 0.04 | 0.07 ± 0.03 | 3.51 ± 0.62 | 2.90 ± 0.73 | 3.75 ± 0.89 | 9.75 ± 1.25 |
| **ANTIOXIDANT ACTIVITY FRAP [μM/ $Fe^{2+}$ /g dw, $n$ = 3]** | | | | | | | | | | | | | | | | | | | |
| 1.23 ± 0.16 | 44.60 ± 2.71 | 63.50 ± 1.30 | 33.87 ± 1.60 | 65.08 ± 3.16 | 1.17 ± 0.18 | 13.55 ± 0.92 | 29.53 ± 1.12 | 17.45 ± 0.78 | 31.54 ± 1.36 | 1.40 ± 0.17 | 26.00 ± 2.83 | 59.35 ± 2.06 | 28.44 ± 2.20 | 51.56 ± 1.62 | 1.03 ± 0.25 | 9.48 ± 1.02 | 23.47 ± 1.04 | 20.33 ± 0.96 | 29.10 ± 1.28 |
| **ANTIOXIDANT ACTIVITY DPPH [μM Trolox/g dw, $n$ = 3]** | | | | | | | | | | | | | | | | | | | |
| 0.29 ± 0.02 | 5.85 ± 0.24 | 9.65 ± 0.99 | 6.98 ± 0.31 | 13.88 ± 1.30 | 0.54 ± 0.10 | 17.88 ± 1.02 | 24.92 ± 1.26 | 11.65 ± 0.96 | 25.74 ± 1.57 | 0.32 ± 0.05 | 3.05 ± 0.22 | 5.05 ± 0.22 | 3.60 ± 0.29 | 7.28 ± 0.32 | 0.24 ± 0.09 | 2.04 ± 0.09 | 3.09 ± 0.16 | 2.78 ± 0.17 | 4.11 ± 0.14 |

Tr—traces; significant differences are shown in Supplementary Materials, Table S1 (quantitative analysis) and Table S4 (antioxidant activity).

### 2.2. Different Pattern in Isoflavones Accumulation Dynamics among the Tested Sprouts

The accumulation dynamics of isoflavones revealed different patterns in their synthesis among the tested sprouts. In RC sprouts, the content of ononin and genistein was constantly increasing with the sprouting time, whereas formononetin and daidzin were produced in an increase and fall mode, but with different maximum content (in RC5 and RC10 sprouts, respectively). In WC sprouts, the content of ononin and formononetin was increasing up to the seventh day, followed by a drastic decrease on the last day of sprouting. The PC3 sprouts had the highest content of all isoflavones examined, with a further decrease, especially observed on the last day of culturing. In the case of CC sprouts, two different patterns of isoflavone accumulation were observed: the rise and fall mode for genistein, with its highest content in PC10, and the drastic increase followed by a constant decrease for formononetin, with its highest amount noted for PC3 sprouts. Our results are just opposite to the accumulation dynamics patterns in red clover sprouts described by [10], with maximum content of formononetin, daidzin, and genistein in 10-day sprouts, and ononin in 7 day-sprouts. Another study noted a constant decrease in daidzin synthesis in red clover sprouts cultured for 5 days, whereas the amount of formononetin and genistein increased with the time of culture, with the maximum in the fourth and third day, respectively, followed by a decrease [12]. The results obtained, compared to those obtained by other authors, suggest considerable fluctuations of isoflavone amounts and accumulation in clover sprouts, as a result of the details in the culturing and analytical conditions.

### 2.3. Comparison of Cytotoxic Potential of Four Clover Species Sprouts

In the next step of the experiment, we focused on the cytotoxic potential of the tested sprouts which varied in isoflavone content. Taking into account the estrogenic properties of isoflavones, as a cellular model, we have chosen hormone-dependent and independent cancer cells, namely, breast and prostate. The model was also completed with appropriate normal breast and prostate cells to verify the safety of the tested samples.

#### 2.3.1. Impact of the Tested Sprouts on Viability of Breast Cancer and Normal Cells

To evaluate the cytotoxic impact of the sprouts tested, we used two breast cancer cell lines, differing not only in metastatic potential, but also in the expression of hormone receptors: low invasive, estrogen- and progesterone-receptor positive MCF7, and high metastatic, estrogen- and progesterone-receptor negative MDA-MB-231 cells. Normal epithelial breast MCF10A cells were also included in the experimental model. The results are presented in Table 2 as $IC_{50}$ values, and Figure 2 (for the highest concentration tested of 500 μg/mL). The highly invasive MDA-MB-231 cells were found to be less susceptible to the tested extracts compared to MCF7 cells, with the exception of the CC10 sprouts ($IC_{50}$ 56.7 and 61.1 μg/mL, respectively). However, for most samples, the $IC_{50}$ values exceeded the maximum concentration used in the assay. The highest cytotoxic effect against MCF7 cells was observed for CC7 and CC10, but also PC5 and PC10 sprouts, and the differences between the $IC_{50}$ values for these samples were not statistically significant. The most important thing is that at the concentrations cytotoxic to cancer cells, the sprouts tested were characterized by high safety for normal breast cells, and the decrease in their viability was observed only at the higher concentrations tested (300–500 μg/mL). The most interesting observation from our study comes from the relationship between the isoflavone content and the cytotoxic potential of the examined sprouts. Sprouts rich in isoflavones were almost inactive in breast cancer cells, whereas CC sprouts, with the lowest amount of isoflavones, expressed a high cytotoxic impact and were safe for normal cells at the same time.

**Table 2.** Cytotoxic activity of the extracts of red (RC), white (WC), Persian (PC), and crimson (CC) clover seeds (S) and sprouts harvested for 3, 5, 7, and 10 days to breast (MCF10A) and prostate (PNT2) normal and cancer (MCF7, MDA-MB-231, and DU145, PC3, LNCaP, respectively) cells, expressed as $IC_{50}$ values (µg/mL).

| RCS | RC3 | RC5 | RC7 | RC10 | WCS | WC3 | WC5 | WC7 | WC10 | PCS | PC3 | PC5 | PC7 | PC10 | CCS | CC3 | CC5 | CC7 | CC10 |
|---|---|---|---|---|---|---|---|---|---|---|---|---|---|---|---|---|---|---|---|
| **PANEL OF BREAST CELLS** | | | | | | | | | | | | | | | | | | | |
| **MCF10A** | | | | | | | | | | | | | | | | | | | |
| ↑ | ↑ | ↑ | ↑ | ↑ | ↑ | ↑ | ↑ | ↑ | ↑ | ↑ | ↑ | ↑ | ↑ | 440.6 | ↑ | ↑ | ↑ | 331.4 | 315.7 |
| **MCF7** | | | | | | | | | | | | | | | | | | | |
| ↑ | ↑ | ↑ | ↑ | ↑ | ↑ | 352.2 | ↑ | ↑ | ↑ | 352.2 | 153.4 | 61.1 | 361.3 | 71.3 | ↑ | ↑ | ↑ | 58.9 | 61.1 |
| **MDA-MB-231** | | | | | | | | | | | | | | | | | | | |
| ↑ | ↑ | ↑ | ↑ | ↑ | ↑ | 457.5 | ↑ | ↑ | ↑ | ↑ | ↑ | ↑ | ↑ | 412.9 | ↑ | ↑ | ↑ | 224.9 | 56.7 |
| **PANEL OF PROSTATE CELLS** | | | | | | | | | | | | | | | | | | | |
| **PNT2** | | | | | | | | | | | | | | | | | | | |
| ↑ | ↑ | ↑ | ↑ | ↑ | ↑ | ↑ | ↑ | ↑ | ↑ | ↑ | ↑ | 315 | ↑ | 429.1 | ↑ | ↑ | 481.4 | 237.9 | 119.5 |
| **DU145** | | | | | | | | | | | | | | | | | | | |
| 430.5 | ↑ | 482.3 | ↑ | 488.6 | ↑ | ↑ | ↑ | 401.9 | 355 | ↑ | 117.9 | ↑ | ↑ | 261.1 | ↑ | ↑ | ↑ | 190.3 | 176.4 |
| **PC3** | | | | | | | | | | | | | | | | | | | |
| 226.7 | 269.8 | 248.9 | 243.6 | 206.3 | 297.6 | 293.4 | 351.7 | 298.1 | 286.5 | ↑ | 203.1 | 324.9 | 415.9 | 187.6 | ↑ | 272.2 | ↑ | 72.2 | 32.9 |
| **LNCaP** | | | | | | | | | | | | | | | | | | | |
| ↑ | 192.7 | 263.5 | 165.3 | 190.2 | ↑ | 354.5 | 260.7 | 189.3 | 239.4 | ↑ | 70.6 | 83.4 | 213.7 | 188.1 | ↑ | 37.6 | 95.8 | 234.0 | 153.8 |

↑ >Cmax.

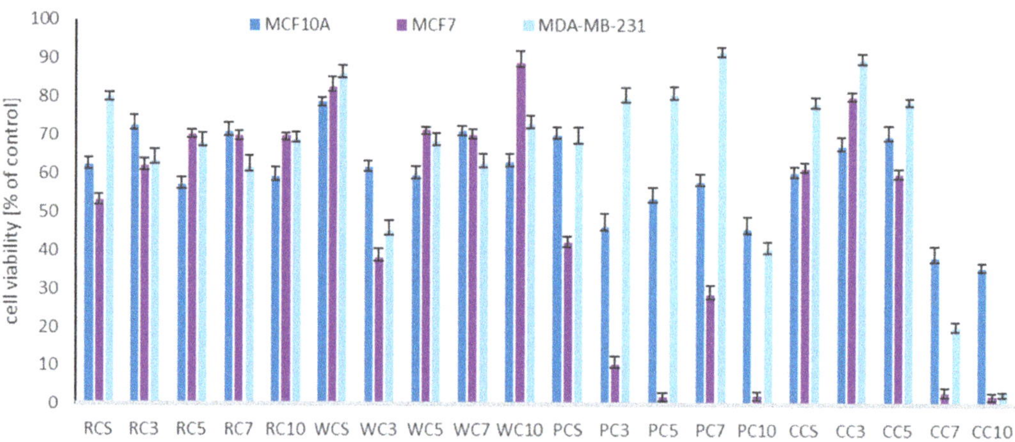


**Figure 2.** Cytotoxic effect of the extracts of red (RC), white (WC), Persian (PC), and crimson (CC) clover seeds (S) and sprouts harvested for 3, 5, 7, and 10 days to breast normal (MCF10A) and cancer (MCF7, MDA-MB-231) cells. Cells were treated with 500 µg/mL of sprout extracts ($n$ = 3) for 24 h. Values are presented as the mean ± SD (standard deviation). Significant differences are shown in Table S2.

Recently, red clover sprouts were reported to influence MCF7 and MDA-MB-231 cells by [15], and the $IC_{50}$ values were 15 mg of isoflavones in dw of sprouts/mL, which is a much weaker effect than that obtained in our study. The estrogenic potential of commercial red clover sprouts was also observed, as a stimulation of MCF7 cell proliferation [13]. Our results for WC, CC, and PC sprouts described their cytotoxic activity to breast cancer cells for the first time. The observed cytotoxic impact of the tested sprouts, not associated with isoflavones, may be due to the presence of other compounds, such as chlorogenic or gallic acid, which were also present in noticeable amounts in the extracts (data not shown). Some recent studies demonstrated the interesting cytotoxic potential of the two compounds in breast cancer cells [18–21], but further analysis is needed to prove this relationship.

2.3.2. Impact of the Tested Sprouts on Viability of Prostate Cancer and Normal Cells

As information on the impact of isoflavones on prostate cancer cells is scarce, and its results are ambiguous, we decided to verify the effect of the examined clover sprouts, differing in isoflavone content, on androgen-dependent LNCap, and androgen-insensitive prostate cancer cells, DU145 and PC3, with low and high metastatic potential, respectively. The model was also completed with normal prostate epithelial cells, PNT2. The results are presented in Table 2 as $IC_{50}$ values, and Figure 3 (for the highest concentration tested of 500 µg/mL). Interestingly, androgen-dependent LNCap cells were most susceptible to the tested sprouts, followed by high metastatic but androgen-insensitive PC3 cells, whereas low metastatic DU145 cells were most resistant. Similar to the effect obtained for breast cancer cells, CC and PC sprouts, with low amounts of isoflavones, revealed the highest cytotoxic impact to the examined cancer cells. Interestingly, the effect observed for normal prostate cells was very weak, with the lowest $IC_{50}$ of 119.5 and 237.9 µg/mL only for CC10 and CC7 sprouts, respectively, which suggests that at the doses cytotoxic to cancer cells, the tested extracts (with the exception of CC10 and CC7 samples) were safe to normal cells. The observed cytotoxicity of four clover species sprouts in prostate cancer and normal cells is probably the first attempt to describe such an effect for any clover species. However, again, the cytotoxic impact of the extracts tested did not correlate with the isoflavone content. Based on some published studies so far [22–24], the observed effect may also be associated with the presence of chlorogenic and gallic acid, but this speculation needs further examination.

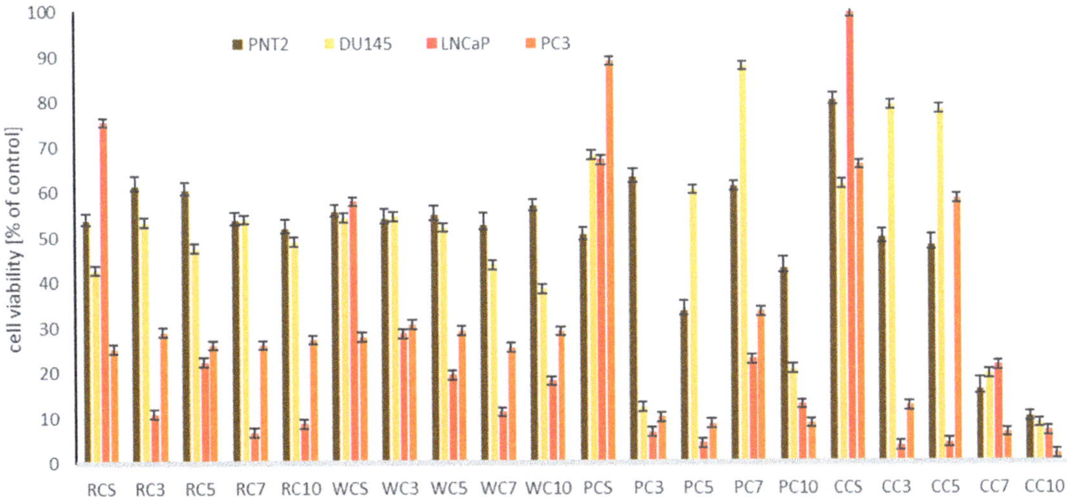


**Figure 3.** Cytotoxic effect of the extracts of red (RC), white (WC), Persian (PC), and crimson (CC) clover seeds (S) and sprouts harvested for 3, 5, 7, and 10 days to prostate normal (PNT2) and cancer (DU145, PC3, LNCaP) cells. Cells were treated with 500 µg/mL of sprout extracts ($n$ = 3) for 24 h. Values are presented as the mean ± SD (standard deviation). Significant differences are shown in Table S3.

### 2.4. Antioxidant Potential of Clover Sprouts Not Always Corresponds with Isoflavones Amount

Isoflavones are not only known for their estrogenic activity, but because they belong to the class of polyphenolics, they also reveal antioxidant properties [25]. Oxidative damage is often related to the initiation and progression of carcinogenesis [26]. Moreover, some antioxidants can also protect normal tissues from chemo- or radiotherapy side effects [27]. Therefore, in the last step of the experiment, we decided to verify and compare the antioxidant potential of clover sprouts tested by the FRAP and DPPH assay, and the results are presented in Table 1. RC and PC sprouts revealed the strongest antioxidant potency measured by FRAP, whereas in the DPPH assay, WC sprouts achieved the highest results. Interestingly, for each species examined, the highest results were observed for both the FRAP and DPPH assays for 5- and 10-day sprouts. The activity was not dependent on the content of isoflavones in the sprouts, apart from the RC sprouts, where daidzin (CW = 0.154), ononin (CW = 0.144), and formononetin (CW = 0.132) were positively correlated with the antioxidant activity evaluated by the FRAP method. In a similar study, 2-day red clover sprouts revealed the highest antioxidant activity, measured by DPPH, followed by a constant decrease in activity up to the last, fifth day of culturing. In contrast to our results for RC sprouts, the antioxidant potential did not correspond to total isoflavones [28]. Two studies described the antioxidant potential of PC sprouts; however, the results are presented as % of activity, and thus, it is hard to compare with our observations [16,17]. In the latter study, the sprouts were grown for 4 days without light, and the antioxidant activity increased during the culturing time, which is opposite to the pattern observed in our study, probably due to the different light conditions. In the study of [17], the examination was performed on commercial sprouts of unknown age. It is worth highlighting that the antioxidant potential of WC and CC sprouts was demonstrated for the first time in our study.

### 2.5. Chemometric Analysis Reveals Some Relationships between the Studied Factors

The statistically significant PCA model of two significant components was derived, with eigenvalues of 4.05 and 3.23, respectively. The model explained 72.8% of the variance of the original parameters and the results are presented in Figures 4 and 5. The first principal component in this model had predominantly negative weights for the original variables. Among them were: ononin, formononetin, daidzin, genistein, and FRAP. Therefore, all the compounds mentioned had high correlation weights with the analyzed antioxidant status index. They were divided into two clusters containing three first, and two last parameters, respectively. The second principal component was loaded, mainly positively, by MCF7 and three other mutually correlated parameters, PNT2, MCF10A, and MDA-MB-231, which were in one tight cluster (Figure 4). In particular, the highest positive correlation weights based on this component, with similar values, were disclosed between PNT2, MCF7, and MCF10A. This confirmed that both breast cancer cell lines, MDA-MB-231 and MCF7, were almost unaffected by the sprouts tested, to a similar degree as normal breast and prostate cells. There were no negative correlations between these parameters and any other parameters (Table S5). The parameter clearly different from the others, in both principal components, was LNCaP. This can be explained due to its androgen sensitivity, which is a distinguishing feature from the other two prostate cancer cell lines used in the study.

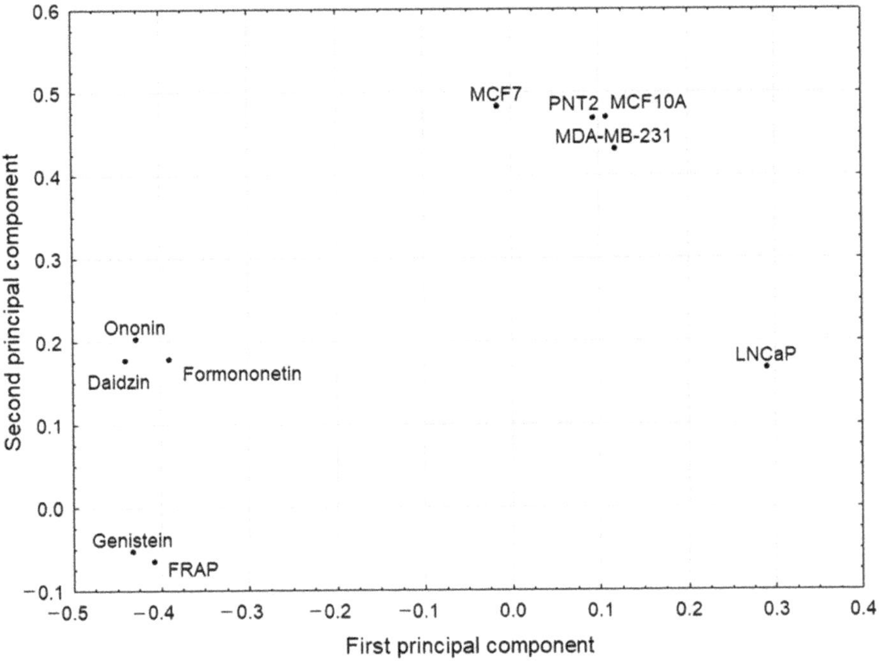


**Figure 4.** The variable loadings on the first and second principal components in the PCA model.

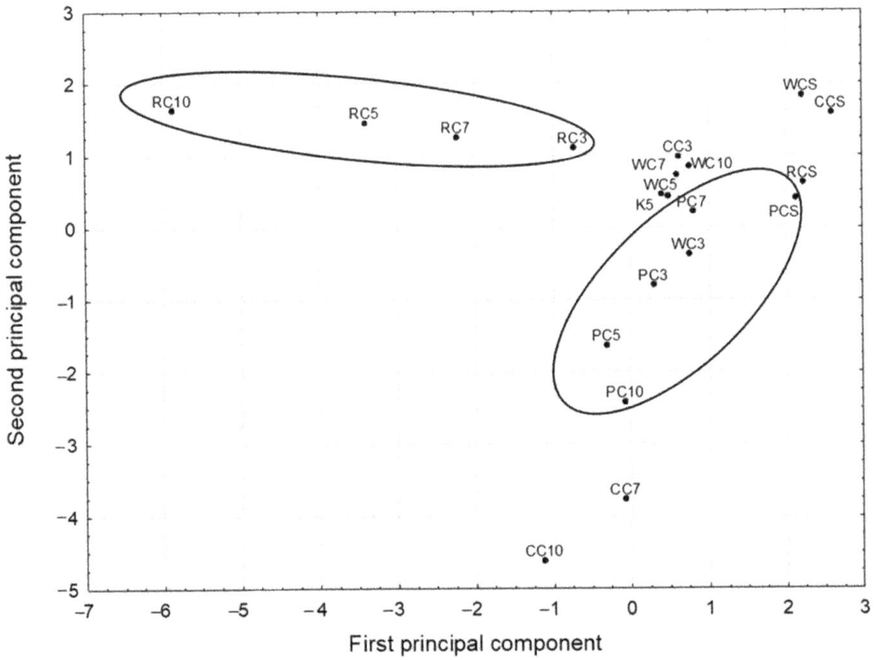


**Figure 5.** The projection of various species of clover sprouts into the space defined by the first two principal components of the PCA model.

## 3. Materials and Methods

### 3.1. Reagents

Dimethyl sulfoxide (DMSO), chloroform, HPLC grade acetonitrile, water, and formic acid were purchased from Sigma-Aldrich (Seelze, Germany). Reference standards for HPLC analysis of isoflavones: genistein, genistin, glycytein, daidzein, daidzin, ononin, biochanin A, and formononetin were purchased from Fluka Chemie. Trolox (6-hydroxy-2,5,7,8,-tetramethyl-chroman-2-carboxylic acid); and $FeCl_3{\cdot}6H_2O$; 1,1-diphenyl-2-picrylhydrazyl (DPPH) were from Sigma Chemical Co. (St. Louis, MO, USA). 2,4,6-Trispyridyl-s-triazine (TPTZ) was purchased from Fluka Chemie (Buchs, Switzerland). Methanol, acetic acid, ammonium hydroxide solution, hydrochloric acid, sodium acetate, and sodium carbonate were from Avantor Performance Materials Poland S.A. (Gliwice, Poland). All reagents were of analytical grade. Distilled water was purchased from Sigma-Aldrich.

### 3.2. Plant Material

The seeds of four clover species, namely, red clover (*Trifolium pratense* L., Nike variety), white clover (*T. repens* L., Grasslands variety), crimson clover (*T. incarnatum* L., Opolska variety), and Persian clover (*T. resupinatum* L., Celtico variety), obtained from plants cultivated in Poland, were purchased in Małopolska Hodowla Roślin (Kraków, Poland). Voucher specimens (KFg/2021/T1p; KFg/2021/T2r; KFg/2021/T3i; KFg/2021/T4rs) were placed in the Department of Pharmacognosy Jagiellonian University Medical College. The seeds were transferred to EQMM Easy Green Microfarm, and grown for 3, 5, 7, and 10 days after seeding, at 25 ± 2 °C, 70% humidity, in sunlight exposure (10 h/day), being watered 3 times a day. For the purpose of this manuscript, the obtained samples are denoted RC for red clover, WC for white clover, CC for crimson clover, and PC for Persian clover, with appropriate numbers indicating the cultivation period. The samples prepared from the seeds are denoted with the letter S, added to the acronym of the clover species.

### 3.3. Preparation of Extracts and Quantitative Analysis

The plant material was Soxhlet extracted with methanol, as previously described [29]. Extracts obtained were decanted, centrifuged, and stored in darkness in a freezer at −20 °C prior to analysis for the quantity and antioxidant capacity of isoflavones. The methanol extracts were further evaporated, and the dry residues were dissolved in DMSO and used for the determination of cytotoxicity.

The quantitative analysis of isoflavones in clover sprouts was performed, as previously described [29], on the Dionex HPLC system, equipped with a PDA 100 UV-VIS detector and a Hypersil Gold (C-18) column (5 μm, 250 × 4.6 mm, Thermo EC). Analysis was carried out in gradient mode, with 1% formic acid in water (A) and acetonitrile (B), 5–60% B in 60 min, at a flow rate of 1 mL/min. The detection wavelengths used were 254 and 285 nm. The compounds mentioned above were identified by comparing their retention times with those of the reference standards. The isoflavone content was calculated by measuring the peak area with respect to the appropriate standard curve (concentration range 0.0625–1 mg/mL). All analyses were performed in three independent experiments, and the mean value was expressed in mg/100 g of dw.

### 3.4. Determination of the Antioxidant Capacities

The analysis was performed using the DPPH and FRAP methods, as previously described [29]. Briefly, DPPH methanolic solution (3.9 mL, 25 mg/L) was mixed with clover sprout extracts (0.1 mL). The reaction was monitored at 515 nm until the absorbance was constant. Each sample was measured in three replicates. The mean capacity was expressed as μM Trolox/g dw. Fresh, working FRAP solution (900 μL; 2.5 mL 10 mM ferric-tripiridyltriazine in 40 mM HCl, 2.5 mL 20 mM $FeCl_3{\cdot}H_2O$ and 25 mL 0.3 mol/L acetate buffer, pH 3.6) was mixed with 90 μL of distilled water and 30 μL of clover sprout extracts, and measured at 593 nm. All analyses were performed in three independent experiments. The mean capacity was expressed as μM $Fe^{2+}$/g dw. The absorbance of both antioxidant assays was measured using a Biotek Synergy microplate reader (BioTek Instruments Inc., Winooski, VT, USA).

### 3.5. Cell Cultures and Viability Assay

Cytotoxic activity was tested on two panels of human cancer and normal cells: prostate panel (androgen-insensitive prostate carcinoma DU-145, derived from the metastatic site: brain, ATCC HTB-81; androgen-insensitive, grade IV prostate carcinoma, PC-3, derived from the metastatic site: bone, ATCC CRL-1435; androgen-sensitive prostate adenocarcinoma LNCaP, derived from the metastatic site: lymph node, ATCC CRL-1740; prostate epithelial cells, PNT2, ECACC 95012613), and breast panel (ER-positive breast adenocarcinoma MCF7, ATCC HTB-22; ER-negative breast adenocarcinoma MDA-MB-231, ATCC HTB-26; breast epithelial MCF10A, ATCC CRL-10317). Cells were grown under standard conditions (37 °C, 5% CO2, relative humidity) and culture media (DMEM/F12 for PNT2, PC3, MDA-MB-231; DMEM Low Glucose for DU145; RPMI1640 with sodium pyruvate for LNCaP; MEM with NEAA for MCF7; DMEM/F12 with 20 ng/mL epidermal growth factor (EGF), 10 μg/mL insulin, 0.5 μg/mL hydrocortisone, 100 ng/mL cholera toxin), supplemented with 10% fetal bovine serum (FBS) or 5% donor horse serum for MCF10A, and 1% antibiotics solution (10, 000 U penicillin and 10 mg streptomycin/mL). The stock solutions of the examined extracts, prepared in DMSO, were then diluted in the culture medium to the working concentrations (from 0 to 500 μg/mL). Cell viability was determined after 24 h of incubation by MTT assay, as previously described [30]. The absorbance was measured at 570 nm using a Biotek Synergy microplate reader (BioTek Instruments Inc., Winooski, VT, USA). All analyses were performed in three independent experiments, and the results are expressed as % of cell viability (mean ± SD) and $IC_{50}$ values (concentration at which viability is inhibited by 50 percent).

### 3.6. Statistical Analysis

Statistical analysis was performed using Statistica v.13 (Statsoft, Tulsa, OK, USA). The results obtained were analyzed using one-way analysis of variance (ANOVA) followed by a post hoc Tukey's test. All experiments were carried out in triplicate, and the data were reported as the mean ± standard deviation (SD). Furthermore, the differences between the groups were considered statistically significant when the $p$-values were 0.05 or less.

The principal component analysis (PCA) model was used to describe the correlation structure between the parameters. Parameters with large loadings on the first two principal components (>0.3) were assumed to be correlated with each other. To express the strength of bivariate associations, for pairs of correlated parameters, the algebraic products of their corresponding loadings and the cosine of the corresponding angle were calculated (these coefficients are called the correlation weights). The "corresponding angle" means the angle determined by the two lines connecting the origin with the coordinates of both parameters on the PCA loadings plot. The PCA approach was also applied to check whether the clusters of various species of clover sprouts appear in the PCA score plot. Statistical analyses were performed using STATISTICA v. 13.3. package (TIBCO Software Inc., Palo Alto, CA, USA) and SIMCA-P v.9 (Umetrics, Umeå, Sweden). The software provided by MP System Co. (Chrzanów, Poland) was used to calculate correlation weights for the pairs of parameters in the PCA model.

## 4. Conclusions

The *Trifolium* genus can be an interesting source for new candidates for functional food, as the sum of isoflavones in red clover sprouts (up to 426.2 mg/100 g dw) is comparable to what is described for soy (150–450 mg/100 g dw), and significantly higher than commercially available alfalfa sprouts (180 mg/100 g dw) [31,32]. The sum of isoflavones calculated for fresh red clover sprouts is approximately 20 mg/100 g, and this amount is comparable to some dietary supplements used for menopause. The time of sprouting had various effects on the concentration of isoflavones, depending on the species. The observed cytotoxic effect on breast and prostate hormone sensitive cancer cells was associated with a low amount of isoflavones, as it was especially observed for crimson and Persian clover sprouts. Importantly, all of the sprouts tested were safe for normal breast and prostate cells.

Our results indicate the need to implement some optimization and/or standardization procedures in the culture of clover sprouts to obtain batches with a more stable/defined level of these compounds, and a more probable prediction of their biological impact. Thus, special caution should be undertaken before including clover sprouts in the diet of consumers with the risk of hormone-dependent cancers. Moreover, women taking isoflavone-rich dietary supplements for reducing their menopausal symptoms should also be aware that some clover sprouts can deliver an additional amount of isoflavones. These observations may be also important for functional food producers. More in-depth studies are needed on the influence of isoflavone-rich sprouts on other aspects of cell functioning.

**Supplementary Materials:** The following are available online at https://www.mdpi.com/article/10.3390/ph15070806/s1, Table S1: Statistical differences for quantitative analysis of isoflavones in four clover species sprouts, Table S2: Statistical differences for cytotoxic activity of four clover species sprouts (500 µg/mL) to breast cancer and normal cells, Table S3: Statistical differences for cytotoxic activity of four clover species sprouts (500 µg/mL) to prostate cancer and normal cells, Table S4: Statistical differences for antioxidant activity of four clover species sprouts, Table S5: Correlation weights for the pairs of parameters based on PCA model.

**Author Contributions:** Conceptualization, A.G. and P.P.; methodology, A.G., P.P. and P.Z.; software, A.G., P.P. and P.Z.; validation, P.Z.; formal analysis, A.G., P.P., M.G., M.N., P.Z.; investigation, A.G., P.P., M.G., M.N., P.Z.; resources, A.G., P.P.; data curation, A.G., P.P., P.Z.; writing—original draft preparation, A.G., P.P.; writing—review and editing, A.G., P.P., I.P., M.G., M.N., P.Z.; visualization, P.P.; supervision, A.G., P.P., I.P.; project administration, A.G., P.P.; funding acquisition, A.G., I.P., P.P. All authors have read and agreed to the published version of the manuscript.

**Funding:** This research received no external funding.

**Institutional Review Board Statement:** Not applicable.

**Informed Consent Statement:** Not applicable.

**Data Availability Statement:** Data is contained within the article and supplementary material.

**Acknowledgments:** The publication was created with the use of equipment (Biotek Synergy microplate reader and Dionex HPLC system) co-financed by the qLIFE Priority Research Area under the program "Excellence Initiative—Research University" (No. 06/IDUB/2019/94) at Jagiellonian University.

**Conflicts of Interest:** The authors declare no conflict of interest.

## References

1. Grela, E.R.; Kiczorowska, B.; Samolińska, W.; Matras, J.; Kiczorowski, P.; Rybiński, W.; Hanczakowska, E. Chemical composition of leguminous seeds: Part I—content of basic nutrients, amino acids, phytochemical compounds, and antioxidant activity. *Eur. Food Res. Technol.* **2017**, *243*, 1385–1395. [CrossRef]
2. Basu, P.; Maier, C. Phytoestrogens and breast cancer: In vitro anticancer activities of isoflavones, lignans, coumestans, stilbenes and their analogs and derivatives. *Biomed. Pharmacother.* **2018**, *107*, 1648–1666. [CrossRef] [PubMed]
3. Hajirahimkhan, A.; Dietz, B.M.; Bolton, J.L. Botanical Modulation of Menopausal Symptoms: Mechanisms of Action? *Planta Med.* **2013**, *79*, 538–553. [CrossRef] [PubMed]
4. Lecomte, S.; Demay, F.; Ferrière, F.; Pakdel, F. Phytochemicals Targeting Estrogen Receptors: Beneficial Rather Than Adverse Effects? *Int. J. Mol. Sci.* **2017**, *18*, 1381. [CrossRef] [PubMed]
5. Sivoňová, M.K.; Kaplán, P.; Tatarková, Z.; Lichardusová, L.; Dušenka, R.; Jurečeková, J. Androgen receptor and soy isoflavones in prostate cancer. *Mol. Clin. Oncol.* **2019**, *10*, 191–204. [CrossRef]
6. Applegate, C.C.; Rowles, J.L.; Ranard, K.M.; Jeon, S.; Erdman, J.W. Soy consumption and the risk of prostate cancer: An updated systematic review and meta-analysis. *Nutrients* **2018**, *10*, 40. [CrossRef]
7. Perez-Cornago, A.; Appleby, P.N.; Boeing, H.; Gil, L.; Kyrø, C.; Ricceri, F.; Murphy, N.; Trichopoulou, A.; Tsilidis, K.K.; Khaw, K.-T.; et al. Circulating isoflavone and lignan concentrations and prostate cancer risk: A meta-analysis of individual participant data from seven prospective studies including 2,828 cases and 5,593 controls. *Int. J. Cancer* **2018**, *143*, 2677–2686. [CrossRef]

8. Zhao, T.-T.; Jin, F.; Li, J.-G.; Xu, Y.; Dong, H.-T.; Liu, Q.; Xing, P.; Zhu, G.-L.; Xu, H.; Miao, Z.-F. Dietary isoflavones or isoflavone-rich food intake and breast cancer risk: A meta-analysis of prospective cohort studies. *Clin. Nutr.* **2019**, *38*, 136–145. [CrossRef]
9. Paśko, P.; Galanty, A.; Tyszka-Czochara, M.; Żmudzki, P.; Zagrodzki, P.; Gdula-Argasińska, J.; Prochownik, E.; Gorinstein, S. Health Promoting vs Anti-nutritive Aspects of Kohlrabi Sprouts, a Promising Candidate for Novel Functional Food. *Plant Foods Hum. Nutr.* **2021**, *76*, 76–82. [CrossRef]
10. Budryn, G.; Gałązka-Czarnecka, I.; Brzozowska, E.; Grzelczyk, J.; Mostowski, R.; Żyżelewicz, D.; Cerón-Carrasco, J.P.; Pérez-Sánchez, H. Evaluation of estrogenic activity of red clover (*Trifolium pratense* L.) sprouts cultivated under different conditions by content of isoflavones, calorimetric study and molecular modelling. *Food Chem.* **2018**, *245*, 324–336. [CrossRef]
11. Butkutė, B.; Taujenis, L.; Norkevičienė, E. Small-Seeded Legumes as a Novel Food Source. Variation of Nutritional, Mineral and Phytochemical Profiles in the Chain: Raw Seeds-Sprouted Seeds-Microgreens. *Molecules* **2018**, *24*, 133. [CrossRef] [PubMed]
12. Chiriac, E.R.; Chiţescu, C.L.; Borda, D.; Lupoae, M.; Gird, C.E.; Geană, E.-I.; Blaga, G.-V.; Boscencu, R. Comparison of the Polyphenolic Profile of Medicago sativa L. and Trifolium pratense L. Sprouts in Different Germination Stages Using the UHPLC-Q Exactive Hybrid Quadrupole Orbitrap High-Resolution Mass Spectrometry. *Molecules* **2020**, *25*, 2321. [CrossRef] [PubMed]
13. Boué, S.M.; Wiese, T.E.; Nehls, S.; Burow, M.E.; Elliott, S.; Carter-Wientjes, C.H.; Shih, B.Y.; McLachlan, J.A.; Cleveland, T.E. Evaluation of the Estrogenic Effects of Legume Extracts Containing Phytoestrogens. *J. Agric. Food Chem.* **2003**, *51*, 2193–2199. [CrossRef] [PubMed]
14. Spagnuolo, P.; Rasini, E.; Luini, A.; Legnaro, M.; Luzzani, M.; Casareto, E.; Carreri, M.; Paracchini, S.; Marino, F.; Cosentino, M. Isoflavone content and estrogenic activity of different batches of red clover (*Trifolium pratense* L.) extracts: An in vitro study in MCF-7 cells. *Fitoterapia* **2014**, *94*, 62–69. [CrossRef] [PubMed]
15. Zakłos-Szyda, M.; Budryn, G. The Effects of Trifolium pratense L. Sprouts' Phenolic Compounds on Cell Growth and Migration of MDA-MB-231, MCF-7 and HUVEC Cells. *Nutrients* **2020**, *12*, 257. [CrossRef]
16. Shirvani, A.; Goli, S.A.H.; Shahedi, M.; Soleimanian-Zad, S. Changes in nutritional value and application of thyme (Thymus vulgaris) essential oil on microbial and organoleptic markers of Persian clover (*Trifolium resupinatum*) sprouts. *LWT* **2016**, *67*, 14–21. [CrossRef]
17. Sayyed-Alangi, S.Z.; Nematzadeh, M. Formulation, development and evaluation of bifunctionalized nanoliposomes containing Trifolium resupinatum sprout methanolic extract: As effective natural antioxidants on the oxidative stability of soybean oil. *BMC Chem.* **2019**, *13*, 77. [CrossRef]
18. Suberu, J.O.; Romero-Canelón, I.; Sullivan, N.; Lapkin, A.A.; Barker, G.C. Comparative Cytotoxicity of Artemisinin and Cisplatin and Their Interactions with Chlorogenic Acids in MCF7 Breast Cancer Cells. *ChemMedChem* **2014**, *9*, 2791–2797. [CrossRef]
19. Lee, H.L.; Lin, C.S.; Kao, S.H.; Chou, M.C. Gallic acid induces G1 phase arrest and apoptosis of triple-negative breast cancer cell MDA-MB-231 via p38 mitogen-activated protein kinase/p21/p27 axis. *Anti-Cancer Drugs* **2017**, *28*, 1150–1156. [CrossRef]
20. Wang, K.; Zhu, X.; Zhang, K.; Zhu, L.; Zhou, F. Investigation of Gallic Acid Induced Anticancer Effect in Human Breast Carcinoma MCF-7 Cells. *J. Biochem. Mol. Toxicol.* **2014**, *28*, 387–393. [CrossRef]
21. Zeng, A.; Liang, X.; Zhu, S.; Liu, C.; Wang, S.; Zhang, Q.; Zhao, J.; Song, L. Chlorogenic acid induces apoptosis, inhibits metastasis and improves antitumor immunity in breast cancer via the NF-κB signaling pathway. *Oncol. Rep.* **2021**, *45*, 717–727. [CrossRef] [PubMed]
22. Thomas-Charles, C.; Fennell, H. Anti-Prostate Cancer Activity of Plant-Derived Bioactive Compounds: A Review. *Curr. Mol. Biol. Rep.* **2019**, *5*, 140–151. [CrossRef]
23. Souza, L.D.S.D.; Horta, I.P.C.; Rosa, L.D.S.; Lima, L.G.B.; da Rosa, J.S.; Montenegro, J.; Santos, L.D.S.; de Castro, R.B.N.; Freitas-Silva, O.; Teodoro, A.J. Effect of the roasting levels of *Coffea arabica* L. extracts on their potential antioxidant capacity and antiproliferative activity in human prostate cancer cells. *RSC Adv.* **2020**, *10*, 30115–30126. [CrossRef]
24. Tuli, H.S.; Mistry, H.; Kaur, G.; Aggarwal, D.; Garg, V.K.; Mittal, S.; Yerer, M.B.; Sak, K.; Khan, A. Gallic acid: A dietary polyphenol that exhibits anti-neoplastic activities by modulating multiple oncogenic targets. *Anti-Cancer Agents Med. Chem.* **2022**, *22*, 499–514. [CrossRef]
25. Křížová, L.; Dadáková, K.; Kasparovska, J.; Kašparovský, T. Isoflavones. *Molecules* **2019**, *24*, 1076. [CrossRef] [PubMed]
26. Liguori, I.; Russo, G.; Curcio, F.; Bulli, G.; Aran, L.; DELLA-Morte, D.; Gargiulo, G.; Testa, G.; Cacciatore, F.; Bonaduce, D.; et al. Oxidative stress, aging, and diseases. *Clin. Interv. Aging* **2018**, *13*, 757–772. [CrossRef]
27. Nakayama, A.; Alladin, K.P.; Igbokwe, O.; White, J.D. Systematic Review: Generating Evidence-Based Guidelines on the Concurrent Use of Dietary Antioxidants and Chemotherapy or Radiotherapy. *Cancer Investig.* **2011**, *29*, 655–667. [CrossRef]
28. Chiriac, E.R.; Chiţescu, C.L.; Sandru, C.; Geană, E.-I.; Lupoae, M.; Dobre, M.; Borda, D.; Gird, C.E.; Boscencu, R. Comparative Study of the Bioactive Properties and Elemental Composition of Red Clover (*Trifolium pratense*) and Alfalfa (*Medicago sativa*) Sprouts during Germination. *Appl. Sci.* **2020**, *10*, 7249. [CrossRef]
29. Paśko, P.; Galanty, A.; Zagrodzki, P.; Żmudzki, P.; Bieniek, U.; Prochownik, E.; Domínguez-Álvarez, E.; Bierła, K.; Łobiński, R.; Szpunar, J.; et al. Varied effect of fortification of kale sprouts with novel organic selenium compounds on the synthesis of sulphur and phenolic compounds in relation to cytotoxic, antioxidant and anti-inflammatory activity. *Microchem. J.* **2022**, *179*, 107509. [CrossRef]
30. Grudzińska, M.; Paśko, P.; Wróbel-Biedrawa, D.; Podolak, I.; Galanty, A. Antimelanoma Potential of *Cladonia mitis* Acetone Extracts—Comparative in Vitro Studies in Relation to Usnic Acid Content. *Chem. Biodivers.* **2022**, e202200408. [CrossRef]

31. Wang, S.Y.; Zhang, Y.J.; Zhu, G.Y.; Shi, X.C.; Chen, X.; Herrera-Balandrano, D.D.; Liu, F.Q.; Laborda, P. Occurrence of isoflavones in soybean sprouts and strategies to enhance their content: A review. *J. Food Sci.* **2022**, *87*, 1961–1982. [CrossRef] [PubMed]
32. Plaza, L.; de Ancos, B.; Cano, P.M. Nutritional and health-related compounds in sprouts and seeds of soybean (Glycine max), wheat (*Triticum aestivum*. L) and alfalfa (*Medicago sativa*) treated by a new drying method. *Eur. Food Res. Technol.* **2003**, *216*, 138–144. [CrossRef]

pharmaceuticals

Article

# Flowers and Leaves Extracts of *Stachys palustris* L. Exhibit Stronger Anti-Proliferative, Antioxidant, Anti-Diabetic, and Anti-Obesity Potencies than Stems and Roots Due to More Phenolic Compounds as Revealed by UPLC-PDA-ESI-TQD-MS/MS

**Sabina Lachowicz-Wiśniewska [1,2,3,*], Anubhav Pratap-Singh [3], Ireneusz Kapusta [4], Angelika Kruszyńska [5], Andrzej Rapak [5], Ireneusz Ochmian [2], Tomasz Cebulak [4], Wioletta Żukiewicz-Sobczak [1] and Paweł Rubiński [1]**

[1] Department of Food and Nutrition, Calisia University, 4 Nowy Świat Street, 62-800 Kalisz, Poland; wiola.zukiewiczsobczak@gmail.com (W.Ż.-S.); p.rubinski@akademiakaliska.edu.pl (P.R.)
[2] Department of Horticulture, West Pomeranian University of Technology in Szczecin, 71-434 Szczecin, Poland; ireneusz.ochmian@zut.edu.pl
[3] Faculty of Land and Food Systems (LFS), The University of British Columbia, Vancouver Campus 213-2205 East Mall, Vancouver, BC V6T 1Z4, Canada; anubhav.singh@ubc.ca
[4] Department of Food Technology and Human Nutrition, College of Natural Science, Rzeszow University, 4 Zelwerowicza Street, 35-601 Rzeszow, Poland; ikapusta@ur.edu.pl (I.K.); tomcebulak@gmail.com (T.C.)
[5] Laboratory of Tumor Molecular Immunobiology, Ludwik Hirszfeld Institute of Immunology and Experimental Therapy, Polish Academy of Sciences, 53-114 Wroclaw, Poland; angelika.kruszynska@hirszfeld.pl (A.K.); andrzej.rapak@hirszfeld.pl (A.R.)
* Correspondence: s.lachowicz-wisniewska@akademiakaliska.edu.pl

**Citation:** Lachowicz-Wiśniewska, S.; Pratap-Singh, A.; Kapusta, I.; Kruszyńska, A.; Rapak, A.; Ochmian, I.; Cebulak, T.; Żukiewicz-Sobczak, W.; Rubiński, P. Flowers and Leaves Extracts of *Stachys palustris* L. Exhibit Stronger Anti-Proliferative, Antioxidant, Anti-Diabetic, and Anti-Obesity Potencies than Stems and Roots Due to More Phenolic Compounds as Revealed by UPLC-PDA-ESI-TQD-MS/MS. *Pharmaceuticals* **2022**, *15*, 785. https://doi.org/10.3390/ph15070785

Academic Editor: Célia Cabral

Received: 25 April 2022
Accepted: 20 June 2022
Published: 23 June 2022



**Abstract:** The present work aims to assess the biological potential of polyphenolic compounds in different parts (flowers, leaves, stems, and roots) of *Stachys palustris* L. Towards secondary metabolites profile, 89 polyphenolic compounds (PCs) were identified by UPLC-PDA-ESI-TQD-MS/MS, with a total average content of 6089 mg/100 g of dry matter (d.m.). In terms of biological activity, antioxidant activity (radical activity, reducing power), digestive enzyme inhibitory ($\alpha$-glucosidase, $\alpha$-amylase, pancreatic lipase) effect, and antiproliferative activity (inhibition of cell viability and induction of apoptosis in different human cancer cell lines) were explored. Leaves, flowers, stems, and roots of *S. palustris* L. have not been studied in this regard until now. Vescalagin and cocciferin d2, isoverbascoside (verbascoside), luteolin 6-C-glucoside, luteolin 6-C-galactoside, apigenin 6-C-glucoside, (−)-epicatechin, ellagic acid, and malvidin 3-*O*-diglucoside were detected as main ingredients in the studied parts. Methanolic extract of *S. palustris* L. leaves and flowers revealed the highest amount of PCs with the strongest antiradical (18.5 and 15.6 mmol Trolox equivalent (TE)/g d.m., respectively) and reducing power effects (7.3 and 5.6 mmol TE/g d.m.). Leaf extracts exhibited better $\alpha$-amylase and pancreatic lipase inhibition effects, while flower extracts exhibited better $\alpha$-glucosidase inhibition effect. Regarding antiproliferative activity, extracts of the leaves and flowers significantly reduced cell viability and induced a high level of apoptosis in human lung, pancreatic, bladder, and colon cancer cell lines, as well as in human acute myeloid leukemia; whereas the extracts from stems and roots revealed the weaker effects. The results of this work showed anti-proliferative and potentially anti-diabetic, anti-obesity properties of *S. palustris* L., especially for flowers and leaves, which may have wide potential applications in the functional food, special food, pharmaceutical, cosmetics industries, and/or in medicine.

**Keywords:** bioactive compounds; in vitro biological potency; medical plant; marsh woundwort

## 1. Introduction

In recent years, edible plants, forgotten plants, wild plants, as well as medical plants have emerged as the potential sources of secondary metabolites for therapeutic interventions [1], which has opened doors for their use as active ingredients in food, pharmaceutical, cosmetics, and/or medical products.

*Stachys palustris* L. (sect. Stachys; family Lamiaceae, subfamily Lamioideae), otherwise known as marsh woundwort, is one such edible plant found in both lowlands and mountains, which has been used in traditional medicines, such as emetic, antiseptic, nervine, sedative, antispasmodic, emmenagogue, hemostatic, vulnerable, expectorant, and tonic agent [2–5]. However, it is most effective in treating internal and external hemorrhages, cramps, and joint pains, as well as the treatment of gout [2]. This plant can grow up to 1 m tall, has square stems, purplish-red flowers arranged in whorls, opposite leaves, and dry four-chambered schizocarp fruit [2].

The biosynthesis of polyphenolic compounds is an intensive process, largely conditional on abundant factors linked with the plant and their environment. The distribution and concentration of bioactive compounds in different underground and above-ground parts of the plant can be strongly varied. This is mainly tied to the role of polyphenols in the growth phase and in plants' life cycle [6]. Previous research has shown that *S. palustris* L. plants exhibit a high total polyphenol content of about 213 mg gallic acid equivalent (GAE)/g dry matter (d.m.), high antiradical activity against DPPH radical, and promising anti-proliferative properties against cervix adenocarcinoma cells (HeLa) [2,7]. On the other hand, the chromatographic analysis showed the presence of only eight compounds belonging to the group of phenylethanoid glycosides, isoscutellarein derivatives, phenolic acid, and iridoids [2].

Despite the above data and traditional knowledge about *S. palustris* L. being a high nutraceutical potential, little is known about the secondary metabolites profile and biological activity of edible leaves, flowers, stems, and roots of *S. palustris* L. [6,7]. Therefore, this study aimed to assess the profile and amount of polyphenolic compounds by UPLC-PDA-ESI-TQD-MS/MS and their in vitro biological activity (antioxidant and anti-proliferative activity). Results, on the polyphenol profile and anti-proliferative activity, as well as inhibitory activity against digestive enzymes, could be interesting for potential applications in functional foods, natural health products, pharmaceuticals, cosmetics industries, and/or medicine.

## 2. Results and Discussion

### 2.1. Profile of Polyphenolic Compounds

The UPLC–PDA-ESI–TQD–MS/MS fingerprint of *Stachys palustris* L. flowers, leaves, stems and roots (Table 1, and Figure S1) discovered the presence of 89 polyphenolic ingredients based on their UV maxima, $m/z$, $[M-H]^-$ and $[M-H]^+$, retention times, peak areas, available data, and standards. Amongst identified compounds, 40 hydrolysable tannins, 16 phenylethanoid glycosides (PhGs), 4 anthocyanins, 1 flavan-3-ol, 6 phenolic acids, 1 flavonols, and 21 flavones were found. However, the qualitative composition of the tested bioactive ingredients was strongly dependent on the extracted part of the plant; with 56, 55, 50, and 49 compounds identified, respectively, in flowers, roots, leaves, and stems of *S. palustris* L (Tables 1 and 2). It is worth emphasizing that thus far only eight compounds had been identified in *S. palustris* L. belonging to secondary metabolites, such as verbascoside, echinacoside, two isoscutellarein derivatives, monomelittoside, chlorogenic acid, harpagide, and its derivative 8-*O*-acetyl-harpagide [2]. However, in our study, compounds, such as monomelittoside, chlorogenic acid, harpagide, and its derivative 8-*O*-acetyl-harpagide [2], were not observed. Thus, apart from the compounds confirmed by Venditti et al. [2], the remaining compounds were detected in the flowers, leaves, stems, and roots extracts of *S. palustris* L. for the first time.

**Table 1.** Characterization of polyphenolic compounds in *Stachys palustris* L. by LC–PDA-ESI–TQD–MS/MS.

| Identified Compounds | Rr [min] | Δd [nm] | [M-H]⁻/MS-MS | Ref. |
|---|---|---|---|---|
| Hydrolysable tannins | | | | |
| Grandinin | 2.49 | 203 | 1065/975/931/301 | [8–19] |
| Grandinin isomer | 2.71 | 227 | 1065/975/931/301 | [8–19] |
| Grandinin isomer | 2.77 | 203 | 1065/975/931/301 | [8–19] |
| Castalagin/vescalagin isomer | 2.83 | 224 | 933/915/631/301 | [8–19] |
| Vescalagin | 2.93 | 224 | 933/915/631/613/569/467/301 | [8–19] |
| Castalagin/vescalagin isomer | 3.01 | 224 | 933/915/613/569/467/301 | [8–19] |
| Castalagin/vescalagin (HHDP–NHTP–glucose) isomer | 3.25 | 224 | 933/915/889/871/631/613/467/301 | [8–19] |
| Pedunculagin isomer (diHHDP-glucose) | 3.36 | 232 | 783/481/301 | [8–19] |
| Castalagin/vescalagin isomer | 3.43 | 204 | 933/915/889/631/301 | [8–19] |
| Cocciferin d2 isomer (HHDP-NHTP-glucose-galloyldiHHDP-glucose) | 3.59 | 224 | 933/915/631/301 | [8] |
| Castalagin/vescalagin isomer | 3.69 | 204 | 933/631/461/301 | [8–19] |
| Pedunculagin isomer (diHHDP-glucose) | 3.95 | 204 | 783/481/301 | [8–19] |
| Castalagin/vescalagin isomer | 4.02 | 208 | 933/915/631/301 | [8–19] |
| Pedunculagin isomer (diHHDP-glucose) | 4.24 | 230, 275 | 783/481/301 | [8–19] |
| Castalagin/Vescalagin isomer | 4.29 | 245 | 933/915/871/613/569/301 | [8–19] |
| Casuarictin (galloyl-DiHHDP-glucose) | 4.45 | 205 | 935/783/633/301 | [8–19] |
| Cocciferin d2 isomer (HHDP-NHTP-glucose-galloyldiHHDP-glucose) | 4.51 | 223 | 933/915/631/390/301 | [8] |
| Chebulanin | 4.54 | 205 | 651/481/463/337/319/275/169 | [8–19] |
| Castalagin/vescalagin isomer | 4.58 | 212/270 | 933/631/569/301 | [8–19] |
| Chebulanin | 4.60 | 206 | 651/481/463/337/319/275/169 | [9,11–19] |
| Sanguiin H-10 isome (digalloyltriHHDPdiglucose) | 4.87 | 315 | 1567/1265/1103/933/631/481/301 | [20,21] |
| Castalagin/vescalagin isomer | 4.93 | 216 | 933/633/481/301 | [8] |
| Castalagin/Vescalagin isomer | 5.05 | 276/353 | 933/915/871/631/613/467/301 | [8] |
| Casuarinin (diHHDP-galloyl-glucose) | 5.18 | 217 | 935/783/633/481/301 | [8–19] |
| Castalagin/vescalagin isomer | 5.33 | 220 | 933/783/633/434/301 | [8] |
| Pedunculagin (diHHDP-glucose) | 5.43 | 313 | 783/707/633/481/301 | [8–19] |
| Sanguiin H-10 isomer (digalloyltriHHDPdiglucose) | 5.51 | 240 | 1567/783/631/481/301 | [20,21] |
| Roburin E | 5.63 | 230 | 1064/301 | [8–19] |
| Vescalagin isomer | 5.91 | 218 | 933/631/467/301 | [8–19] |
| Castalagin isomer | 6.16 | 222 | 933/631/467/301 | [8–19] |
| Geraniin isomer | 6.25 | 209 | 951/933/633/301/257 | [8–19] |
| Casuarinin/potentilin (galloyl-diHHDP-glucose) | 6.51 | 279 | 935/783/633/301 | [8–19] |
| Tellimagrandin I (digalloyl-HHDP-glucose) | 6.53 | 218/277 | 785/615/483/301 | [8–19] |
| Pentagalloyl-glucose | 6.95 | 280 | 939/787/769/617/599/447 | [8–19] |
| Tellimagrandin I isomer (digalloyl-HHDP-glucose) | 6.99 | 218/277 | 785/615/483/301 | [8–19] |
| Trigalloyl-HHDP-glucose | 7.07 | 280 | 937/767/635/465/301 | [8,14] |
| Trigalloyl-HHDP-glucose | 7.74 | 280 | 937/767/635/465/301 | [8,14] |
| Trigalloyl-HHDP-glucose | 7.80 | 281 | 937/767/635/465/301 | [8,14] |
| Trigalloyl-HHDP-glucose | 7.93 | 281 | 937/767/635/465/301 | [8,14] |
| Chebulagic acid (galloyl-chebuloyl-HHDP-glucose) | 8.00 | 220/272 | 953/785/633/463/337/301/169 | [14] |

**Table 1.** *Cont.*

| Identified Compounds | Rr [min] | Δd [nm] | [M-H]⁻/MS-MS | Ref. |
| --- | --- | --- | --- | --- |
| Phenylethanoid glycosides | | | | |
| Echinacoside | 6.27 | 329 | 785/623/461/161 | [22,23] |
| Betonyoside E | 6.32 | 326 | 785/639//621/609/193/161 | [22,23] |
| Stachysoside A | 6.38 | 272 | 755/623/461/593/179/161 | [24] |
| B-OH-Forsythoside B methylether | 7.29 | 330 | 785/755/623/347/161 | [22,23] |
| Stachysoside A | 7.41 | 330 | 755/623/593/461/179/161 | [24] |
| Isoacteoside (isoverbascoside) | 7.60 | 330 | 623/461/161 | [2,25–27] |
| B-OH-Forsythoside B methylether | 7.80 | 323 | 785/755/623/347/161 | [22,23] |
| Forsythoside B isomer | 7.91 | 330 | 755/623/607/593/461/161 | [22] |
| Forsythoside B isomer | 7.97 | 325 | 755/623/593/461/161 | [22] |
| Forsythoside B isomer | 8.07 | 326 | 755/623/461/447/161 | [24] |
| Cistanoside A | 8.22 | 328 | 799/637/623/475/315 | [22,23] |
| Alyssonoside | 8.34 | 329 | 769/593/575/447/315/161 | [24] |
| Alyssonoside isomer | 8.59 | 329 | 769/593/575/447/315/161 | [24] |
| Martynoside | 9.34 | 283 | 651/475/457/328/161 | [24] |
| Samioside | 10.02 | 329 | 755/593/461/315/161 | [24] |
| Leucoseptoside A | 10.29 | 313 | 637/461/315/193/175/161 | [24,28] |
| Stachysoside E | 10.43 | 319 | 669/625/583/380/264 | [24] |
| Anthocyanins | | | | |
| Delphinidin 3-*O*-glucoside | 3.65 | 520 | 465/303 | [29] |
| Malvidin 3-*O*-diglucoside | 4.51 | 525 | 665/493/331 | [29] |
| Cyanidin 3-*O*-glucoside | 5.06 | 517 | 449/287 | [29] |
| Malvidin 3-*O*-acetylglucoside | 6.00 | 525 | 535/331 | [29] |
| Flavan-3-ols | | | | |
| (-)Epicatchin | 4.83 | 280 | 289 | [8] |
| Phenolic acid | | | | |
| Ellagic acid glucoside | 5.74 | 370 | 463/301 | [25] |
| Ellagic acid pentoside | 6.05 | 254/362 | 433/301 | [25] |
| Ellagic acid | 7.52 | 255/365 | 301 | [25] |
| 3,4-dicaffeoyl quinic acid | 8.91 | 324 | 515/353/191/179 | [25] |
| 3,5-dicaffeoyl quinic acid | 9.04 | 324 | 515/353/191/179 | [25] |
| 4,5-dicaffeoyl quinic acid | 11.07 | 324 | 515/353/191/179 | [25] |
| Flavonols | | | | |
| Kaempferol hexose glucuronide | 10.85 | 343 | 623/285 | [30] |
| Flavones | | | | |
| Chrysoeriol acetyl-allopyranosyl-glucopyranoside | 5.82 | 269/332 | 677/299 | [31,32] |
| Luteolin-6-C-galactoside | 6.59 | 269/349 | 447/357/327/297/285 | [33,34] |
| Luteolin-6-C-glucoside | 6.74 | 267/347 | 447/357/327/299/285 | [33,34] |
| Apigenin-6-C-galactoside | 7.29 | 268/336 | 431/341/311/283/269 | [35] |
| Apigenin-6-C-glucoside | 7.43 | 269/336 | 431/341/311/283/269 | [35] |
| Apigenin 7-*O*-*β*-*D*-(6-*p*-coumaroyl)-glucopyranoside | 8.26 | 325 | 577/432/407/269 | [31,36] |
| Apigenin acetyl-allosyl-glucoside | 9.01 | 332 | 635/269 | [26] |
| Chrysoeriol 7-*O*-acetylallosylglucoside | 10.84 | 311 | 667/299 | [25] |
| 4′-O-methylisoscutellarein-diacetyl-allosyl-glucopyranoside | 8.62 | 346 | 707/299 | [25,26] |
| 4′-O-methylisoscutellarein-acetyl-allosyl-glucopyranoside | 8.87 | 329 | 665/485/299 | [25,26] |
| Isoscutellarein-acetylallosyl-(glucopyranoside)apiose | 9.19 | 276/330 | 651/429/285 | [25,26] |

**Table 1.** *Cont.*

| Identified Compounds | Rr [min] | Δd [nm] | [M-H]$^{-}$/MS-MS | Ref. |
|---|---|---|---|---|
| Isoscutellarein-acetylallosyl-(glucopyranoside)apiose isommer | 9.32 | 326 | 651/637/429/285 | [25,26] |
| Isoscutellarein-acetylallosyl-(glucopyranoside)apiose isommer | 9.44 | 329 | 651/607/429/285 | [25,26] |
| Isoscutellarein-acetylallosyl-(glucopyranoside)apiose isommer | 9.49 | 329 | 651/607/429/285 | [25,26] |
| Isoscutellarein-acetylallosyl-(glucopyranoside)apiose isommer | 9.63 | 329 | 651/285 | [25,26] |
| Isoscutellarein-acetylallosyl-(glucopyranoside)apiose isommer | 9.72 | 329 | 651/285 | [25,26] |
| Isoscutellarein-acetylallosyl-(glucopyranoside)apiose isommer | 9.79 | 314 | 651/285 | [25,26] |
| Isoscutellarein-acetylallosyl-glucopyranoside | 9.85 | 328 | 651/285 | [25,26] |
| Isoscutellarein-acetylallosyl-glucopyranoside isommer | 10.10 | 329 | 651/285 | [25,26] |
| 4′-*O*-methylisoscutellarein-acetyl-allosyl-glucopyranoside isomer | 11.23 | 277/305 | 665/299 | [25,26] |

Rt, retention time.

**Table 2.** Total polyphenol content in different parts of *Stachys palustris* L. [mg/100 g d.m.].

| Polyphenolic Compounds | Flowers | Leaves | Stems | Roots |
|---|---|---|---|---|
| Hydrolysable Tannins (HT) | | | | |
| Grandinin | 172.25 ± 2.07a [a] | 76.69 ± 0.92b | 62.67 ± 0.75c | 32.34 ± 0.39d |
| Grandinin isommer | 399.62 ± 4.80b | 533.26 ± 6.40a | 230.27 ± 2.76c | nd |
| Grandinin isommer | 265.51 ± 3.19a | nd [b] | 71.98 ± 0.86c | 120.43 ± 1.45b |
| Castalagin/vescalagin isomer | 214.78 ± 2.58b | 494.02 ± 5.93a | 212.20 ± 2.55b | nd |
| Vescalagin | 2370.75 ± 18.97a | 1528.04 ± 12.22b | 1236.86 ± 9.89c | 403.8 ± 3.23d |
| Castalagin/vescalagin isomer | 36.41 ± 0.44c | 224.02 ± 2.69a | 95.17 ± 1.14b | nd |
| Castalagin/vescalagin (HHDP–NHTP–glucose) isomer | 500.80 ± 6.01a | 129.82 ± 1.56c | 220.30 ± 2.64b | nd |
| Pedunculagin isomer (diHHDP-glucose) | nd | nd | nd | 78.36 ± 0.94a |
| Castalagin/vescalagin isomer | 945.11 ± 1.89b | 1041.14 ± 2.08a | 617.74 ± 1.24c | 51.56 ± 0.62d |
| Cocciferin d2 isomer (HHDP-NHTP-glucose-galloyldiHHDP-glucose) | 1552.66 ± 18.63a | 1388.25 ± 16.66b | 867.36 ± 10.41c | 230.08 ± 2.76d |
| Castalagin/vescalagin isomer | 636.42 ± 7.64b | 990.62 ± 11.89a | 572.37 ± 6.87c | 37.70 ± 0.45d |
| Pedunculagin isomer (diHHDP-glucose) | 63.94 ± 0.77b | 80.22 ± 0.96a | 27.38 ± 0.33c | 2.59 ± 0.03d |
| Castalagin/vescalagin isomer | 162.72 ± 1.95c | 298.50 ± 3.58a | 137.30 ± 1.65b | 16.72 ± 0.20c |
| Pedunculagin isomer (diHHDP-glucose) | 110.36 ± 1.32b | 143.92 ± 1.73a | 47.51 ± 0.57c | 6.05 ± 0.07d |
| Castalagin/Vescalagin isomer | 100.25 ± 1.20b | 117.42 ± 1.41a | 57.71 ± 0.69c | 13.61 ± 0.16d |
| Casuarictin (galloyl-diHHDP-glucose) | 33.64 ± 0.40b | 85.27 ± 1.02a | 22.76 ± 0.27c | 34.51 ± 0.41b |
| Cocciferin d2 isomer (HHDP-NHTP-glucose-galloyldiHHDP-glucose) | 21.33 ± 0.26b | 29.19 ± 0.35a | 8.57 ± 0.10c | nd |
| Chebulanin | nd | nd | nd | 2.62 ± 0.03a |
| Castalagin/vescalagin isomer | 40.36 ± 0.48a | nd | nd | nd |
| Chebulanin | nd | nd | nd | 12.15 ± 0.15a |

**Table 2.** *Cont.*

| Polyphenolic Compounds | Flowers | Leaves | Stems | Roots |
|---|---|---|---|---|
| Sanguiin H-10 isomer (digalloyltriHHDPdiglucose) | 4.73 ± 0.06c | 9.11 ± 0.11b | 3.84 ± 0.05cd | 18.68 ± 0.22a |
| Castalagin/vescalagin isomer | 27.11 ± 0.33a | 28.71 ± 0.34a | 17.43 ± 0.21b | 4.79 ± 0.06c |
| Castalagin/Vescalagin isomer | 18.71 ± 0.22a | 1.32 ± 0.02b | 0.95 ± 0.01b | nd |
| Casuarinin (diHHDP-galloyl-glucose) | 38.79 ± 0.47a | 18.53 ± 0.22b | 7.82 ± 0.09c | 3.42 ± 0.04d |
| Castalagin/vescalagin isomer | 18.55 ± 0.22a | 15.14 ± 0.18b | 8.00 ± 0.10c | 5.39 ± 0.06d |
| Pedunculagin (diHHDP-glucose) | 5.76 ± 0.07b | 11.64 ± 0.14a | 5.43 ± 0.07b | 1.47 ± 0.02c |
| Sanguiin H-10 isomer (digalloyltriHHDPdiglucose) | 8.20 ± 0.10b | 45.22 ± 0.54a | 3.68 ± 0.04c | nd |
| Roburin E | 27.43 ± 0.33a | 20.53 ± 0.25b | 4.11 ± 0.05c | nd |
| Vescalagin isomer | 5.48 ± 0.07c | 8.96 ± 0.11a | 6.31 ± 0.08b | nd |
| Castalagin isomer | 26.95 ± 0.32a | 14.71 ± 0.18b | 4.85 ± 0.06c | 2.54 ± 0.03d |
| Geraniin isomer | 1.63 ± 0.02b | 5.28 ± 0.06a | 4.57 ± 0.05a | nd |
| Casuarinin/potentilin (galloyl-diHHDP-glucose) | nd | 1.39 ± 0.02b | 3.85 ± 0.05a | 1.59 ± 0.02b |
| Tellimagrandin I (digalloyl-HHDP-glucose) | 5.38 ± 0.06a | nd | nd | 2.32 ± 0.03b |
| Pentagalloyl-glucose | 1.83 ± 0.02a | nd | nd | nd |
| Tellimagrandin I isomer (digalloyl-HHDP-glucose) | 5.28 ± 0.06a | nd | nd | nd |
| Trigalloyl-HHDP-glucose | 8.55 ± 0.10a | 4.79 ± 0.06b | 1.94 ± 0.02c | nd |
| Trigalloyl-HHDP-glucose | 6.40 ± 0.08a | 1.30 ± 0.02b | 0.75 ± 0.01c | nd |
| Trigalloyl-HHDP-glucose | 6.23 ± 0.07a | 3.23 ± 0.04b | 0.61 ± 0.01c | nd |
| Trigalloyl-HHDP-glucose | 25.87 ± 0.31a | nd | nd | nd |
| Chebulagic acid (galloyl-chebuloyl-HHDP-glucose) | 75.04 ± 0.90a | 3.33 ± 0.04b | 1.64 ± 0.02c | nd |
| Phenylethanoid glycosides (PhG) | | | | |
| Echinacoside | nd | nd | nd | 1.11 ± 0.01a |
| Betonyoside E | 1.53 ± 0.02a | 1.44 ± 0.02a | 0.57 ± 0.01bc | 1.10 ± 0.01b |
| Stachysoside A | 6.62 ± 0.08a | nd | nd | nd |
| B-OH-Forsythoside B methylether | nd | nd | nd | 15.01 ± 0.18a |
| Stachysoside A | nd | nd | nd | 30.72 ± 0.37a |
| Isoacteoside (isoverbascoside) | nd | nd | nd | 114.26 ± 1.37a |
| B-OH-Forsythoside B methylether | nd | nd | nd | 1.04 ± 0.01a |
| Forsythoside B | nd | nd | 5.76 ± 0.07b | 10.84 ± 0.13a |
| Forsythoside B | nd | nd | nd | 1.75 ± 0.02a |
| Forsythoside B isomer | nd | nd | nd | 2.69 ± 0.03a |
| Cistanoside A | nd | nd | nd | 1.33 ± 0.02a |
| Alyssonoside | nd | nd | nd | 13.05 ± 0.16a |
| Alyssonoside isomer | nd | nd | nd | 12.45 ± 0.15a |
| Martynoside | nd | nd | nd | 4.17 ± 0.05a |
| Samioside | nd | nd | nd | 0.92 ± 0.01a |
| Leucoseptoside A | nd | 1.76 ± 0.02a | nd | nd |
| Stachysoside E | nd | 1.70 ± 0.02a | nd | nd |
| Anthocyanins (ANT) | | | | |
| Delphinidin 3-*O*-glucoside | 3.36 ± 0.04a | nd | nd | nd |
| Malvidin 3-*O*-diglucoside | 9.99 ± 0.12a | nd | nd | nd |
| Cyanidin 3-*O*-glucoside | 3.07 ± 0.04a | nd | nd | nd |
| Malvidin 3-*O*-acetylglucoside | 3.57 ± 0.04a | nd | nd | nd |
| Flavan-3-ols (F3O) | | | | |
| (-)-Epicatchin | 31.95 ± 0.38a | 4.81 ± 0.06c | 13.69 ± 0.16b | nd |

**Table 2.** *Cont.*

| Polyphenolic Compounds | Flowers | Leaves | Stems | Roots |
|---|---|---|---|---|
| Phenolic acid (PA) | | | | |
| Ellagic acid glucoside | nd | 4.24 ± 0.05a | 0.71 ± 0.01b | nd |
| Ellagic acid pentoside | 5.35 ± 0.06a | 3.56 ± 0.04b | 0.83 ± 0.01c | 0.84 ± 0.01c |
| Ellagic acid | 48.45 ± 0.58a | 26.35 ± 0.32b | 7.18 ± 0.09c | 3.16 ± 0.04d |
| 3,4-dicaffeoyl quinic acid | 0.64 ± 0.01c | 1.41 ± 0.02b | 0.38 ± 0.00c | 21.85 ± 0.26a |
| 3,5-dicaffeoyl quinic acid | 1.98 ± 0.02b | 2.21 ± 0.03a | 1.16 ± 0.01c | 2.03 ± 0.02b |
| 4,5-dicaffeoyl quinic acid | 0.89 ± 0.01a | 0.25 ± 0.00a | 0.30 ± 0.00a | 0.18 ± 0.00a |
| Flavonols (FL) | | | | |
| Kaempferol hexose glucuronide | 7.36 ± 0.09a | nd | nd | nd |
| Flavones (FLN) | | | | |
| Chrysoeriol acetyl-allopyranosyl-glucopyranoside | nd | 3.86 ± 0.05a | 3.29 ± 0.04a | 1.02 ± 0.01b |
| Luteolin-6-C-galactoside | 22.57 ± 0.27c | 462.91 ± 5.55a | 72.75 ± 0.87b | 0.46 ± 0.01d |
| Luteolin-6-C-glucoside | 45.72 ± 0.55c | 267.68 ± 3.21a | 40.74 ± 0.49b | nd |
| Apigenin-6-C-galactoside | 17.59 ± 0.21c | 237.74 ± 2.85a | 39.72 ± 0.48b | nd |
| Apigenin-6-C-glucoside | 14.16 ± 0.17c | 394.48 ± 4.73a | 65.50 ± 0.79b | nd |
| Apigenin 7-*O*-*β*-*D*-(6-*p*-coumaroyl)-glucopyranoside | 1.39 ± 0.02c | 2.70 ± 0.03a | 1.52 ± 0.02b | nd |
| Apigenin acetyl-allosyl-glucoside | nd | nd | nd | 3.72 ± 0.04a |
| Chrysoeriol 7-*O*-acetylallosylglucoside | nd | 2.01 ± 0.02a | nd | nd |
| 4′-O-methylisoscutellarein-diacetyl-allosyl-glucopyranoside | 36.59 ± 0.44a | 31.61 ± 0.38b | 17.67 ± 0.21c | nd |
| 4′-O-methylisoscutellarein-acetyl-allosyl-glucopyranoside | nd | nd | nd | 5.97 ± 0.07a |
| Isoscutellarein-acetylallosyl-(glucopyranoside)apiose | 0.37 ± 0.00c | 2.96 ± 0.04b | 0.80 ± 0.01c | 93.21 ± 1.12a |
| Isoscutellarein-acetylallosyl-(glucopyranoside)apiose isommer | nd | nd | nd | 2.91 ± 0.03a |
| Isoscutellarein-acetylallosyl-(glucopyranoside)apiose isommer | nd | nd | nd | 18.45 ± 0.22a |
| Isoscutellarein-acetylallosyl-(glucopyranoside)apiose isommer | nd | nd | nd | 11.22 ± 0.13a |
| Isoscutellarein-acetylallosyl-(glucopyranoside)apiose isommer | nd | nd | nd | 3.58 ± 0.04a |
| Isoscutellarein-acetylallosyl-(glucopyranoside)apiose isommer | nd | nd | nd | 4.42 ± 0.05a |
| Isoscutellarein-acetylallosyl-(glucopyranoside)apiose isommer | nd | nd | nd | 1.13 ± 0.01a |
| Isoscutellarein-acetylallosyl-glucopyranoside | nd | nd | nd | 5.18 ± 0.06a |
| Isoscutellarein-acetylallosyl-glucopyranoside isommer | nd | nd | nd | 2.69 ± 0.03a |
| 4′-*O*-methylisoscutellarein-acetyl-allosyl-glucopyranoside isomer | nd | nd | nd | 14.21 ± 0.17a |

**Table 2.** *Cont.*

| Polyphenolic Compounds | Flowers | Leaves | Stems | Roots |
|---|---|---|---|---|
| Procyjanidyny polimery (PP) | 336.61 ± 4.04b | 444.87 ± 5.34a | 102.29 ± 1.23d | 133.55 ± 1.6c |
| Degree of polymerization (DP) | 2.96c | 2.07d | 3.72b | 4.74a |
| The total sum of phenolic compounds | 8544.63b | 9252.11a | 4938.76c | 1622.94d |

[a] Values that are expressed as the mean (n = 3) ± standard deviation and different letters (between morphological parts) within the same row indicate statistically significant differences by Duncan's test ($p < 0.05$); [b] nd, not identified.

#### 2.1.1. Hydrolysable Tannins

The main class of polyphenolic compounds identified in the flowers, leaves, stems, and roots of *S. palustris* L. was hydrolysable tannins. In this study, *S. palustris* L. was found to contain 40 compounds of this class, including 36 hydrolysable tannins in the flowers, 31 in the leaves, 32 in the stems, and 22 in the roots. This group contains the derivatives of gallic acid, ellagitannins, and gallotannins [8]. Gallic acid is known to dimerize into ellagic acid formed by intramolecular dilactonization of HHDP acid (hexahydroxydiphenoyl acid). Ellagitanins, constituting the most abundant fraction of hydrolysable tannins, are formed by the oxidative coupling of adjacent galloyl fractions from gallotannins [8,10]. The group of HHDP is further divided into two groups: a NHTP (nonahydroxytriphenoyl) group by coupling to the next galloyl group, or a Chebuloyl group by double oxidation [8,10]. The structure of thirty-four compounds were confirmed, based on identification of their fragmentation pattern on the basis of fragmentation earlier described in the Chestnut of *Castanea sativa* by Miller [8], the different plant parts of *Terminalia arjuna* [9], the fraction of *Sanguisorba officinalis* L. [11], the different parts of myrtle berry from Italy [12], *Fragaria vesca* L. berries [13], *Terminalia chebula* fruits [14], plant medical [15,16], Myrtaceae family [17], *Genista tinctoria* L. [18], and the leaves of *Phyllagathis rotundifolia* [19]. These compounds were identified as grandinin, two grandinin derivatives, roburin E (*m*/*z* 1065), fourteen castalagin/vescalagin isomers (*m*/*z* 933), four pedunculagin isomer [diHHDP-glucose] (*m*/*z* 783), two casuarinin [diHHDP-galloyl-glucose] (*m*/*z* 935), casuarictin [galloyl-DiHHDP-glucose] (*m*/*z* 935), two chebulanin (*m*/*z* 651), geraniin isomer (*m*/*z* 951), pentagalloyl-glucose (*m*/*z* 939), two tellimagrandin I [digalloyl-HHDP-glucose] (*m*/*z* 785), and four trigalloyl-HHDP-glucose (*m*/*z* 937).

The fragmentation of the above hydrolysable tannins resulted in losses of typical residues such as gallic acid ($[M-H-170]^-$), galloyl ($[M-H-152]^-$), HHDP ($[M-H-302]^-$), HDDP glucose ($[M-H-482]^-$), galloyl-HDDP-glucose ($[M-H-634]^-$), or galloyl-glucose ($[M-H-332]^-$) residues [8,10,11]. According to Esposito et al. [8], the compound of chebulagic acid (galloylo-chebuloyl-HHDP-glucose) contains chebuloyl group generated by the oxidation of HHDP residues and further supported by the loss of carboxylic and chebuloyl groups. In addition, the compound cocciferin d2 isomer (HHDP-NHTP-glucose-galloyl-diHHDP-glucose), which is a dimer ellagitannin, was noted as $[M-2H]^{2-}$ (double charged pseudomolecular ion). Sanguiin H-10 isomers $[M-H]^-$ at *m*/*z* 1567 were identified on the basis of data published by Kool et al. [20] in Boysenberry (*Rubus loganbaccus* × *baileyanus* Britt.), and by Mullen et al. [21] during the fragmentation MS/MS loss of two HHDP (302 mass unit (m.u.), glucosyl (162 m.u.), and galloyl (170 m.u.) moieties.

#### 2.1.2. Flavanones

The compounds belonging the group of flavones usually subsist as glycosides and seldom as free aglycones [37]. In the full-scan LC-MS/MS, the deprotonated pseudomolecular ions $[M-H]^-$ of luteolin, apigenin, isoscutellarein, and chrysoeriol (*m*/*z* 285.0, *m*/*z* 269.0, *m*/*z* 285.0, and *m*/*z* 299, respectively) were observed [33]. A total of 21 compounds are reported in different parts of *S. plasturis* L., including 7 flavones in the flowers, 9 in the leaves, 8 in the stems where apigenins and luteolins dominated, and 15 in the roots where isoscutellarein and chrysoeriol dominated. The pathway for flavones C-glycosides showed the decomposition of glycan with a loss of neutral residues with the decomposition of a flavonoid moiety associated to the residual part of glycan [38]. These compounds for the genus *Stachys* are considered important chemosystematics markers [25].

The first class of flavones was luteolin ($[M\text{-}H]^-$ at $m/z$ 447) determined based on characteristic UV absorption maxima at 269 and 349 nm. According to Lin et al. [37], these compounds are characterized by two UV spectral maximums: 250–300 nm (band I), and 342–350 nm (band II). Based on the fragmentation of the peak that indicated the presence of a *C*-glycosylated derivative, these compounds were confirmed as luteolin 6-*O*-galactoside and luteolin 6-*O*-glucoside (previously identified in the *Elaeis guineensis* Jacq.) [33,34].

Apart from luteolins, four apigenin compounds ($m/z$ 431, 577, and 635) were also identified based on the UV spectrum at 260 and 336 nm. Two compounds with molecular ion at $m/z$ 431 and the fragmentation of MS/MS indicative of a *C*-glycosylated derivative and loss of glucosyl (162 m.u.) moiety were identified as apigenin 6-*C*-galactoside and apigenin 6-*C*-glucoside, previously noted in *Cyclanthera pedata* leaves, fruits, and dietary supplements [35]. The compound with pseudomolecular ion at $m/z$ 635 was tentatively identified on the basis of fragmentation earlier described in *Stachys parviflora* L. [26] as apigenin acetylallosyl-glucoside. Mass spectra of apigenin with *p*-coumaric acid derivative showed molecular ion at $m/z$ 577 and was confirmed based on previous results for *Stachys byzantina* [36] and *Stachys bombycina* [31] as apigenin 7-*O*-*β*-*D*-(6-*p*-coumaroyl)-glucopyranoside.

Two compounds belonging to chrysoeriol derivatives with pseudomolecular ion at $m/z$ 667 and $m/z$ 677 were confirmed based on the fragmentation pathway described in the *Stachys subgenus* [32] and *Stachys bombycina* [31] and were identified as chrysoeriol 7-*O*-acetylallosylglucoside and chrysoeriol-acetyl-allopyranosyl-glucopyranoside.

The last class of flavones were twelve isoscutellarein derivatives with the pseudomolecular ion $[M\text{-}H]^-$ $m/z$ 667, $m/z$ 707, $m/z$ 665, and $m/z$ 651. The presence of Aglycon $[A\text{-}H]^-$ fragment and *D*-allose, which is characteristic of a large group of *Stachys* plants, was noted [25]. As constituents acetylated on the internal or external allose unit showing the intermediate ions $[M\text{-}180]^-$ or $[M\text{-}180\text{-}OAc]^-$ [25], these compounds were identified as isoscutellarein-acetylallosyl-(glucopyranoside)apiose isomers (nine compounds), 4′-*O*-methylisoscutellarein-acetyl-allosyl-glucopyranoside, and their isomers similar to *Stachys recta* L. [25] and *Stachys parviflora* L. [26].

### 2.1.3. Phenylethanoid Glycosides (PhGs)

Phenylethanoid glycosides (PhGs) are appreciated for a strong biological potency [39,40]. The typical spectra exhibited UV-Vis maxima between 320–340 nm [39]. 16 PhG compounds, including 2 in the flowers, 3 in the leaves, 2 in the stems, and 13 in the roots, were detected in *S. palustris* L. The detected PhGs showed deprotonated molecular ions $[M\text{-}H]^-$ at $m/z$ 785, $m/z$ 755, $m/z$ 623, $m/z$ 799, $m/z$ 769, $m/z$ 651, $m/z$ 755, and $m/z$ 669. During the fragmentation of the above, PhGs showed losses of typical residues, such as the neutral loss of the caffeoyl residue (162 m.u.), *p*-coumaric moiety (176 m.u.), the rhamnose residue (146 m.u.), the glucose residue (162 m.u.), the $COCH_2$ group (42 m.u.), a $CH_2$ radical (14 m.u.), and the methoxy group (30 m.u.) [25,26,39,40].

Compounds with pseudomolecular ions at $m/z$ 785, $m/z$ 623, $m/z$ 755, and $m/z$ 669 were confirmed as betonyoside E, two B-OH-Forsythoside B methylether isomers, isoacteoside, two forsythoside B, and their isomer based on stachysoside E, previously isolated from *Stachys recta* L. [25], *Stachys parviflora* L. [26], *Stachys officinalis* L. [27], and *Stachys alopecuros* L. [2].

Mass spectra for two compounds were tentatively assigned as echinacoside ($m/z$ 799) and cistanoside A ($m/z$ 785), previously described in *Cistanche deserticola* [22] and *Cistanche armena* [23].

Three compounds with pseudomolecular ions at $m/z$ 756 and 637 was tentatively detected as two stachysoside A and leucoseptoside A, with a characteristic fragmentation pathway presented in *Lagopsis supina* [28].

The last three compounds belonging to PhGs were tentatively specified on the basis [24] noted in *Phlomis* species. Two compounds having pseudomolecular ions at $m/z$ 769, and on the basis of fragmentation MS/MS the losses of *p*-coumaric residue, pentose, and glucose moiety, were tentatively assigned as alyssonoside and alyssonoside isomer [24]. Compounds, such as samioside ($m/z$ 755) and martynoside ($m/z$ 651), were assigned on the losses of glucose, pentose, rhamnose moiety, and *p*-coumaric and rhamnose moiety, according to the fragmentation pattern determined by Kirmizibekmes et al. [24].

2.1.4. Other Phenolic Acids, Anthocyanins, Flavonol, and Flavan-3-ol

The last groups of polyphenolic compounds identified in the different parts of *S. palustris* L. were phenolic acids, anthocyanins, flavonol, and flavan-3-ol. Among detected phenolic acids, three compounds presented similar maximum absorbance at 325 nm, which is typical for derivatives of dicaffeoylquinic acid [25]. All of them had common pseudomolecular ions at $m/z$ 515. The first compound was detected as 3,5-di-affeoylquinic acid compared with the standard. The next two compounds noted the characteristic ions at $m/z$ 173, which indicate the presence of quinic acid in position four. Thus, these compounds were determined as 3,4-di-caffeoylquinic and 4,5-di-caffeoylquinic acids, respectively, on the basis of the order of elution [25]. The next three phenolic acid tentatively assigned as ellagic acid ($m/z$ 301), ellagic acid glucoside ($[M\text{-}H\text{-}463]^-$), and ellagic acid pentoside ($[M\text{-}H\text{-}461]^-$) were previously determined in *Stachys officinalis* [33].

All anthocyanins and flavonol assigned in *S. palustris* L. was noted in just flowers for the first time. In the full-scan LC-MS/MS, the deprotonated molecular ions $[M\text{-}H]^-$ of delphinidin, malvidin, and cyanidin ($m/z$ 303, $m/z$ 331, and $m/z$ 287, respectively) [29] were detected. These compounds were detected as delphinidin 3-*O*-glucoside, malvidin 3-*O*-diglucoside, cyanidin 3-*O*-glucoside, and malvidin 3-*O*-acetylglucoside on the basis of their fragmentation pathways [29]. One flavonol (deprotoned molecular ion at $m/z$ 285) was specified as kaempferol hexose glucuronide from the losses of hexose moiety (162 m.u.) and glucuronide moiety (176 m.u.) [30]. The existence of (−)-epicatechin was confirmed via the standard.

*2.2. Content of Polyphenolic Compounds and Polymeric Procyanidins*

Polyphenols are a very important group of secondary metabolites because they exhibit a wide range of health benefits and biological activities [41]. A statistical test on one-factor analysis indicated a significant effect ($p < 0.05$) of research plant parts on the total polyphenol content (TPC). The average TPC in the *S. palustris* L. was 6090 mg/100 g d.m. (Table 2). The highest amount of secondary metabolites were measured in the leaves (9252 mg/100 g d.m.), and this value was 1.9 and 5.7 times higher than in the stems and roots.

The main group of the analyzed polyphenols detected in the flowers, leaves, stems, and roots of *S. palustris* were hydrolysable tannins (constituting an average of 82.9% of all polyphenolic compounds) > flavones (8%) > flavan-3-ols (monomers and polymers; 5%) > phenylethanoid glycosides (3.3%) > phenolic acids, anthocyanins, and flavanols (<0.8%). Compared with other *Stachys* species, hydrolysable tannins were also a predominant group detected in *S. cretica* ssp. *anatolica* [42]. Nonetheless, higher TPC detected in the flowers and leaves is because these organs actively metabolize these compounds during photosynthesis [43,44].

A similar trend for TPC was noted in the studies on different parts of *Astragalus macrocephalus* subsp. *finitimus* [45]. In the *S. palustris* L. isolated from Hungary and France, the TPC was 17,630 and 24,980 mg GAE/100 g d.m., respectively [2], and was 1.8 and 2.5 times higher than in the leaves of *S. palustris* L. from Poland. In turn, the TPC of our extracts were significantly higher than those reported for other *Stachys* species (ranging from 430 mg GAE/100 g d.m. in *S. trinervis* to 4450 mg GAE/100 g d.m. in *S. fruticulosa*) from Iran by Khanavi et al. [46]. According to Oracz et al. [47], TPC may be influenced by many factors, e.g., origin, soil, weather conditions, analytical method, and the preparation of test samples. Carev et al. [42] reported that methanol extracts of *S. cretica* ssp. *anatolica* from Turkey showed 4330 mg/100 g d.m. TPC assessed by LC–ESI–MS/MS and 1.6 times lower TPC by spectrophotometric method [42]; the results were significantly lower than those reported in the flowers and leaves noted in our work. Likewise, during the analysis of *S. cretica ssp. Vacillans,* the record showed 3604 mg/100 g d.m. TPC using the HPLC system, and this value was 2.3 times lower compared to the spectrophotometric method [41]. On the other hand, the total concentration of TPC isolated from stems of *S. officinalis* L. were 6120 mg GAE/100 g dry extract [33], 1610–3330 mg GAE/100 g [48], which were 1.2 times higher and 2 times lower compared to stems of *S. palustris* L. The total amount of bioactive compounds extracted from leaves and roots of *Stachys officinalis* L. from Czech Republic were, respectively, 7495–8050 mg GAE/100 g d.m. and 2286–3164 mg caffeic acid/100 g d.m. (depending on the cultivar tested) [49]. TPC in the *S. recta* was 2050 mg/100 g d.m. and was around 4.8 times lower than in the leaves of *S. palustris* L. [25].

The most abundant group of tested parts of *S. palustris* L. contained a range of 1083 to 7945 mg/100 g d.m. for roots and flowers, respectively. The total tannins content identified in the different organs of *Calluna vulgaris* L. Hull in flowering time were 5, 4.3, and 3 times lower than our data for flowers, leaves, and stems [43]. In addition, the predominant compounds among 40 hydrolysable tannins were vescalagin and cocciferin d2 constituting 21% and 19%, respectively, in the leaves and 37% and 21% in the roots. This was also confirmed in the study by Esposito et al. [8] and Singh et al. [9].

The concentration of the second numerous group [42] ranged from 138.9 in the flowers to 1406.0 mg/100 g d.m. in the leaves. The most abundant compound in the flowers were luteolin 6-C-galactoside and 4′-*O*-methylisoscutellarein-glucoside-rhamnoside (constituting 33 and 26% of all flavones); in the leaves it was luteolin 6-C-galactoside and apigenin 6-C-glucoside (constituting 33 and 28%); in the stems it was luteolin 6-C-glucoside and apigenin 6-C-glucoside (constituting 33 and 27%); in the roots it was isoscutellarein-acetylallosyl-(glucopyranoside)apiose (constituting 55%). In turn, in studies by Carev et al. [42] the major compound in *S. cretica* ssp. *anatolica* was apigenin-7-glucoside and this was 2540 mg/100 g d.m. The total of flavonoids as quercetin equivalent in *S. tmolea* [50] and in *S. cretica ssp. vacillans* [41] were 500 and 5010 mg/100 g; in *Stachys cretica subsp. kutahyensis* there was 4020 mg/100 g extract [51]. In addition, apigenin and luteolin were reported to reveal a high anxiolytic potency in rats [52] and have high anticancer, antioxidant, and anti-inflammatory activities [50,53].

The next quantitatively important group was flavan-3-ols, including (−)-epicatechin as a monomer (constituting 5% of all flavan-3-ols) and polymeric procyanidins (PP; constituting 95%). The average amount of (−)-epicatechin in measured parts of the plant was 12.6 mg/100 g d.m. and PP—254.3 mg/100 g d.m. The (−)-epicatechin of measured parts may be organized in the following sequence: flowers > stems > leaves, and for PP: leaves > flowers > roots > stems. A similar trend for PP was noted in the different tested organs of *Rumex crispus* L. and *Rumex obtusifolius* L. [6]. The total of catechins as catechin equivalent in *Stachys marrubiifolia viv.* the leaf was 40 mg/100g extract [54]. The lowest amount of flavan-3-ols was detected in the stems and also confirmed by Feduraev et al. [6], probably by the inside metabolic action of the tissues and cells and the molecular constitution of the exudate carried by the phloem channels [6]. In addition, the alkaline solution of the central cavity of stems is exposed to oxidation PCs, including flavan-3-ols [6].

Moreover, the most opulent in PhGs were the roots, and the content was 209.5 mg/100 g d.m., and this amount was an average of 34 times higher than in the rest parts of *S. palustris* L. The main compound detected in the roots was isoverbascoside (verbascoside). This was also confirmed in the research on *S. cretica ssp. anatolica* by Carev et al. [47], on *S. cretica* subsp. *mersinaea* by Bahadori et al. [41], on *S. tmolea* by Elfalleh et al. [50], and on *Stachys cretica subsp. kutahyensis* [51]. During the measurement of *S. recta,* the content of PhGs was 607 mg/100 g d.m. and the concentration was around three times higher than in the stems of *S. palustris* L. [25]. In addition, the biological activity of verbascoside was confirmed, such as, for example, anti-tumor, anti-inflammatory, anti-radical, and anti-thrombotic effects [50].

In turn, the phenolic acids were the most abundant in the flowers compared to the rest parts of *S. palustris* L. tested, and their amount was 2.2, 3, and 8 times higher compared to the rest fractions. The total content of phenolic acids measured in the *S. recta* was 709 mg/100 g d.m. [25] and was significantly higher than our results. The anthocyanins, and flavonols were noted in the flowers. These compounds present less than 1% of all PCs in the measured extract of *S. palustris* L. The total of anthocyanins and flavonols as quercetin and cyanidin 3-glucoside equivalent in the *Stachys marrubiifolia viv.* leaves were 537 and 70 mg/100 g extract [54]. According to Bahadori et al. [41], the phenolic acid content measured in *S. cretica* subsp. *mersinaea* was 401 mg/100 g extract; in *S. tmolea* was 118 mg/100 g [50]. In turn, anthocyanin compounds, in addition to the health benefits, are responsible for the color of the raw material. According to the analysis, the flowers are the darkest and red with the addition of dark yellow color (Figure S1). In addition, the NAI index indicates an intense dark red pigment located in the flowers. On the other hand, the type of anthocyanins identified indicates a reddish-purple color. Of course, depending on the pH, it can change [2,29]. This is consistent with the botanical color of the flowers of the plant under study [55]. The root also seems interesting because it showed a dark red color with the addition of dark yellow [55].

### 2.3. *In Vitro Biological Activity*

#### 2.3.1. Antiradical and Reducing Potential

The antioxidant properties of *S. palustris* L. was assessed as reducing power (FRAP assay) and radical scavenging activity (ABTS assay). Results indicate that significant differences were noted between parts of the research plant. Table 3 suggests that the flowers and leaves had the highest radical scavenging activities (18.5 and 15.6 mmol TE/g d.m. respectively) and the highest FRAP reduction potential (5.6 and 7.3 mmol TE/g d.m., respectively). The ability to scavenge synthetic ABTS radicals determined for the roots and stems was about 2 and 3.8 times and 4.4 and 4.5 times weaker compared to the flowers and the leaves. Similar conclusions were found for the iron (III) reduction capacity, which was 5.3 and 18.5 times and 4 and 14.3 times weaker compared to the flowers and the leaves.

**Table 3.** In vitro biological activity.

| Parts of Plant | α-Amylase [$IC_{50}$ (mg/mL)] | α-Glucosidase [$IC_{50}$ (mg/mL)] | Pancreatic Lipase [$IC_{50}$ (mg/mL)] | ABTS [mmol TE/g d.m.] | FRAP [mmol TE/g d.m.] |
|---|---|---|---|---|---|
| Leaves | 6.85 ± 0.11a [a] | 12.71 ± 0.20b | 27.46 ± 0.44a | 15.55 ± 0.25b | 7.25 ± 0.12a |
| Flowers | 8.14 ± 0.13b | 11.20 ± 0.18a | 46.23 ± 0.74c | 18.49 ± 0.30a | 5.57 ± 0.09b |
| Stems | 16.43 ± 0.26c | 19.09 ± 0.31c | 38.90 ± 0.62b | 7.81 ± 0.12c | 1.37 ± 0.02c |
| Roots | 26.34 ± 0.42d | 34.81 ± 0.56d | 47.94 ± 0.77c | 4.10 ± 0.07d | 0.39 ± 0.01d |

[a] Values that are expressed as the mean (n = 3) ± standard deviation and different letters (between morphological parts) within the same row indicate statistically significant differences ($p < 0.05$).

The results indicate that the flowers and the leaves are more effective at eliminating excess reactive oxygen species (ROS) that cause oxidative stress in the body compared to the roots and the stems [41,45]. When assessing the reducing power in *S. anisochila*, *S. beckeana*, *S. plumosa*, and *S. alpina* spp. *Dinarica*, a value of 1.9, 1.8, 0.5, and 1.4 mmol $Fe^{2+}$/g d.m. has been reported [56]. Likewise, significantly lower values of antioxidant activity for the FRAP test were noted for the areal part of *S. trinervis*, *S. byzantina*, *S. setifera*, *S. subaphylla*, *S. turcomanica*, *S. inflata*, *S. laxa*, *S. persica*, and *S. fruticulosa* [45]. In *S. tmolea* (the areal part), the antiradical activity and reducing power was 32.3 and 41.9 mg TE/g d.m., respectively [50]. In studies conducted by Benabderrahim et al. [51], the antioxidant activity measured by ABTS and FRAP assays in *S. cretica* subsp. *Kutahyensis* were 175.8 and 239.1 mg TE/g of extract. The results obtained by Bahadori et al. [41] in *Stachys cretica* subsp. *Mersinaea* were 292.7 and 236.4 mg TE/g of extract for ABTS and FRAP tests, respectively. Whereas, in the extracts of *S. cretica* ssp. *Anatolica* for antiradical activity and reducing power were 112.2 and 127.2 mg TE/g of extract [42]. Moreover, our obtained results were not directly comparable. Besides, the research by Saravanakumar et al. [57] and Yadav et al. [58] shows that the differences in the value of antioxidant activity may significantly depend on the purity of solvents and their polarity, extraction procedure, and fractionation methods.

#### 2.3.2. In Vitro Enzyme Inhibition

Digestive enzymes, such as pancreatic lipase, are involved in fat metabolism and support the digestion of dietary fats into fatty acids. A-glucosidase and α-amylase are engaged in the carbohydrate metabolism and break down complex sugars into monosaccharides and oligosaccharides [57,59]. We analyze different parts of the plant with the digestive enzyme inhibitory properties to evaluate potential anti-obesity and anti-diabetic properties (Table 3).

The higher pancreatic lipase enzyme inhibitory activity was reported for leaves of *S. palustris* L. In addition, the flowers and leaves noted the higher α-amylase and α-glucosidase enzyme inhibitory properties. The least ability to inhibit the activity of digestive enzymes was noted for the roots; the ability to inhibit the activity of α-amylase was on average 3.5 times weaker compared to the leaves and the flowers, while the α-glucosidase activity was 3 times weaker. The ability to inhibit pancreatic lipase activity was noted for both the roots and the flowers.

The higher enzyme (α-amylase and α-glucosidase) inhibitory activity was evaluated in *Stachys japonica* of methanol extract, and the result was 7.43 ug extract/ug acarbose equivalent (ACE) [57]. Whereas the methanol extract from *S. cretica* subsp. *Smyrnaea* [41,60], *S. cretica* subsp. *Mersinaea* [41,60], and *S. cretica* subsp. *Kutahyensis* [51] had α-amylase inhibitory activity at the level 61.5, 418.6, and 315.5 mg ACE/g extract, respectively [50]. Furthermore, the inhibitory activity against α-glucosidase for *Stachys cretica* subsp. *Mersinaea* was 734.5 mg ACE/g of methanol extract [41]. Another study indicated that the α-amylase inhibition activity for *S. iberica* subsp. *Iberica var. densipilosa* and *S. byzantina* was 219.5 and 200.1 mg ACE/g, respectively [61].

We conclude that the leaves and flowers of *S. palustris* L. have PP and PA compounds that show potentially high antidiabetic activities and may be used for the inhibition of these enzyme activities. To our knowledge, the inhibition of α-amylase and α-glucosidase activity had not been reported before for *S. palustris* L. extracts.

#### 2.3.3. Anti-Proliferative Activity

Anti-proliferative activity in the leaves, flowers, stems, and roots of *S. palustris* L. was tested in the A549 (lung adenocarcinoma), BxPC3 (pancreatic ductal adenocarcinoma), HT-29 and CACO-2 (colorectal adenocarcinoma), HCV29T (bladder cancer), and AML-NEV007 (acute myeloid leukemia) cell lines (Figures 1 and 2). We chose tumors that are particularly resistant to chemotherapy and difficult to treat. Cells were treated with ethanol extract for 48 h, after which the MTS viability test and the analysis of the induction of apoptosis were performed.

The most significant results were obtained for leaves and flower extracts of *S. palustris* L. In particular, leaves extract markedly decreased the metabolic activity of all tested cell lines to 8–22%. Extract from the flower showed weaker inhibition effects (10–52% remaining activity). In addition, the roots and stems extracts were found weakest and reduced the viability to 30–90%. Diluted ethanol (1%) had no effect on cell lines.

We then tested the induction of apoptosis using Annexin V double staining and propidium iodide. As in the MTS test, the leaf and flower extracts significantly induced apoptosis in all tested cell lines in the range of 69–86%. The CACO-2 line was the most resistant, with the apoptosis level around 45%. A study by Kokhdan et al. [62] evaluated a methanol extract of *S. pilifera* against HT-29 cell line (colon adenocarcinoma) viability and demonstrated favorable inhibitory properties and significant anti-proliferative effects. It was also reported that *S. laxa* chloroform extract significantly prevented the proliferation of HT-29 and T47D (ductal carcinoma) cell lines, and the total extract of *S. subaphylla* also showed anti-proliferative properties against T47D cell line [63]. Furthermore, Haznagy-Radnai et al. [7], noted that some *Stachys* species, such as *S. palustris and S. recta,* stems in methanolic extracts displayed significant antiproliferative activity against cervix adenocarcinoma cells (HeLa); *S. germanica* flowers against breast adenocarcinoma (MCF-7) cells. In addition, Lachowicz et al. [11] noted that the methanolic extract of *Sanguisorba officinalis* L. leaves and flowers exhibited significant antiproliferative activity versus bladder cancer (HCV29T), colorectal adenocarcinoma (DLD-1), pancreatic ductal adenocarcinoma (BxPC3), and Jurkat cell lines. The obtained extracts contain a number of different substances that may have an additive or synergistic anti-proliferative effect on cancer cells. Many of these substances affect various intracellular pathways that depend on the type of cancer cells [64]. The externalization of Annexin V indicates the activation of caspases. Most likely, the process of apoptosis follows the classical mitochondrial pathway. Polyphenolic compounds are abundant in the leaves and flowers of *S. palustris* L. Luteolin, ellagic acid, and apigenin derivatives are known for their activating effects on the p53 transcription factor. They also affect the cell cycle and reduce the expression of the proapoptotic proteins from the Bcl-2 family [65]. Polyphenols have anti-inflammatory properties by inhibiting the activity of COX2 and NFKB, which is important in cancer [66]. Some natural compounds have exhibited synergism with established anticancer agents, and thus may reduce the side effects of chemotherapy. Further research is needed to identify active compounds and test their anti-cancer properties in appropriate in vivo models. The above results indicate that the anti-proliferative activity is mainly influenced by the part of the plant studied and material extraction, as well as the profile of PCs.

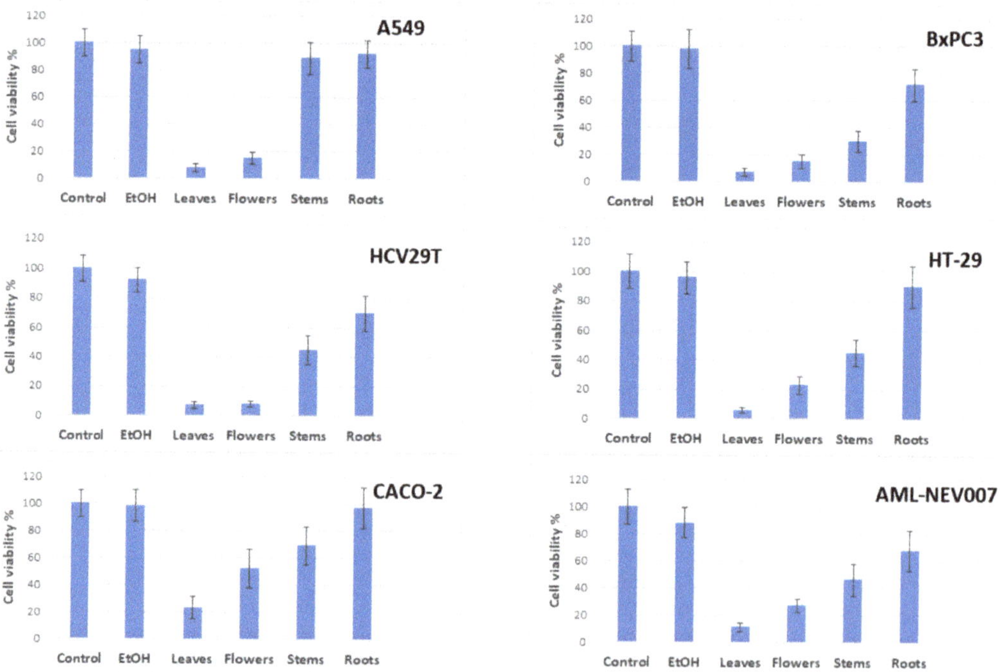


**Figure 1.** The viability of different cancer cells after treatment with 30% ethanolic extracts for 48 h. One 1:10 solid-to-liquid (S/L) ratio was used for all parts of the plant. Obtained extracts were diluted 30 times to final ethanol concentration 1%.

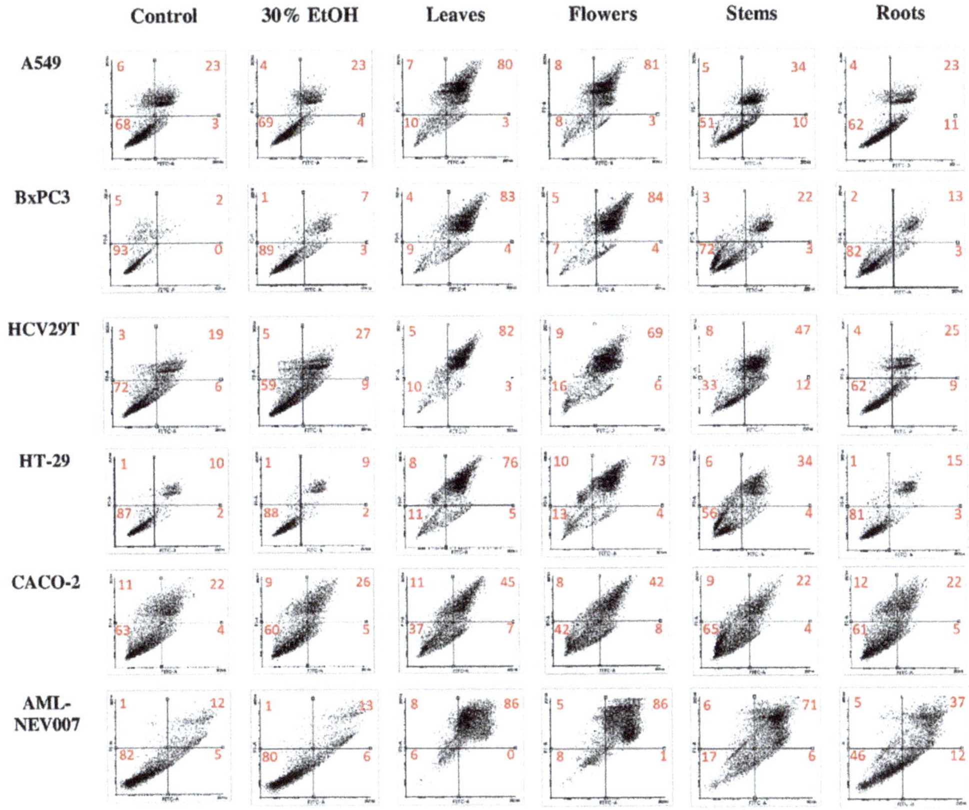

**Figure 2.** Induction of apoptosis after treatment for 48 h.

## 2.4. Multivariate Analysis

The results show a significant relationship between the high value of particular groups of the PCs and enzyme inhibitory activities, antioxidant, and antiproliferative activity, which was confirmed by the Pearson correlation coefficient. It is well known that PCs indicate numerous properties, including redox properties, and they, therefore, act as singlet oxygen quenchers and hydrogen donors, as well as exhibiting antioxidant activity [50]. Moreover, it is the synergistic effect between bioactive compounds in the tested material that provides antioxidant activity depending on, among others, their concentration [50]. Thus, a very strong correlation was noted between the antioxidant potential (for ABTS and FRAP assay, respectively) and the overall value of PCs ($R^2$ 0.953 and 0.956), as well as individual groups of substances, such as hydrolysable tannins ($R^2$ 0.967 and 0.901), polymeric procyanidins ($R^2$ 0.853 and 0.978), and phenolic acids ($R^2$ 0.792 and 0.685).

In addition, anthocyanins, flavan-3-ols, and flavonols had a stronger correlation with antiradical activity than FRAP assay ($R^2$ 0.699, 0.684, 0.699, respectively), but flavones had a stronger correlation with reducing power than ABTS assay ($R^2$ 0.700). A strong correlation between PCs and antioxidant assays was also reported by Bahadori et al. [41, 60] and Khanaki et al. [45] in research concerning the analysis of *Stachys cretica* subsp. *Mersinaea*. On the other hand, the anthocyanins, flavan-3-ols, and flavonols showed negative correlation with BxPC3 cell line, and flavones noted negative correlation with $\alpha$-glucosidase inhibitory activities.

Interestingly, PhGs, especially compounds identified in the roots, were the only group of PCs to show a negative correlation with enzyme inhibitory activities, a negative strong correlation with antioxidant activity, regardless of the method used, and antiproliferative activity in the all cells line.

The strongest relationships to inhibition of α-amylase activity were noted for hydrolysable tannins, flavones, and PP against the inhibition of α-glucosidase activity—hydrolysable tannins, anthocyanins, flavan-3-ols, phenolic acid, and flavonols; whereas the strongest correlations in inhibition of pancreatic lipase activity were noted for phenolic acid and PP. In turn, α-amylase inhibitory activities indicated a positive correlation with individual PCs, apart from *p*-coumaric acid and TPC ($R^2$ −0.999) [41]. Furthermore, a strongly positive correlation was noted between the anti-proliferative potential (of all analyzed cancer cells line) and hydrolysable tannins, flavones, PP, and TPC; while the remaining groups of the polyphenolic compounds noted a positive correlation with cancer cells inhibitory activities, apart from PhGs.

Overall, the PCA results concerning different parts of plant extract data indicated a clear correlation with all PCs and anti-proliferative, anti-diabetic, anti-obesity, and antioxidant tests (Figure 3). The PCA detected two essential components that were responsible for 92.20% of data variance, including PC1 for 70.25%; while the second PC2 was only responsible for 21.94%. Figure 3 clearly shows the tested parts of *S. palustris* L. based on phytochemicals and antioxidant, anti-proliferative, anti-obesity, and anti-diabetic activities: stems and roots are found on the left of the plot, flowers and leaves are distributed on the right part. The obtained data indicates that the metabolites that distinguish the studied parts of the plant are the PCs and mainly hydrolyzed tannins and PP, which are found in high amounts in the leaves and flowers (PC1), and PhGs, which are especially present in the roots (PC2). Consequently, their type and concentration affect health-promoting properties. Therefore, the biological activities demonstrated a difference between parts of the plant. Besides, the difference in the profile and concentration of PCs in plant fractions is attributed to the difference in the morphological and anatomical structures and ongoing physiological processes [44,67]. The results of the PCA method showed similar results as the Pearson correlation.

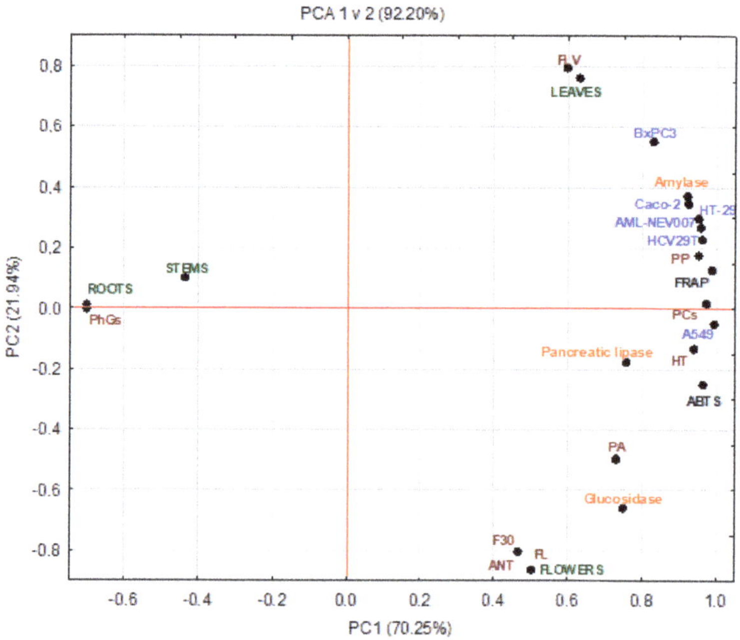


**Figure 3.** PCA of bioactive compounds and biological activities for all parts of *Stachys palustris* L.

## 3. Materials and Methods

### 3.1. Chemicals, Material and Instruments

Chemicals: Acetonitrile UHPLC, methanol, ascorbic acid, formic acid, methanol, ABTS (2,2′-azinobis(3-ethylbenzothiazoline-6-sulfonic acid), 6-hydroxy-2,5,7,8-tetramethylchroman-2-carboxylic acid (Trolox), 2,4,6-tri(2-pyridyl)-s-triazine (TPTZ), 2,2-Di(4-tert-octylphenyl)-1-picrylhydrazyl (DPPH), methanol, acetic acid, α-amylase from porcine pancreas, α-glucosidase from Rhizopus sp., lipase from porcine pancreas, 3,5-dini-trosalicylic acid, Antibiotic-Antimycotic Solution, and RPMI 1640 culture medium were purchased from Sigma-Aldrich (Steinheim, Germany). (−)-Epicatechin, ellagic acid, dicaffeic acid, kampferol-3-*O*-galactoside, malvidin-3-*O*-glucoside, delphinidin 3-*O*-glucoside, cyanidin-3-*O*-glucoside, apigenin, apigenin 6-C-glucoside, and luteolin were purchased from Extrasynthese (Lyon, France). DMEM culture medium with 10% FBS were purchased from Gibco (Thermo Fisher Scientific, Waltham, MA, USA), and MTS solution was purchased from Promega (Madison, WI, USA).

Material: Flowers, leaves, stems, and roots of *Stachys palustris* L. (~3 kg) were obtained from a private garden in Szczytna (53°33′46″ N 20°59′07″ E), Lower Silesia, Poland. The plant was collected randomly in August 2019 from different parts of the field (total area of cultivation is 1 ha). The root of the *Stachys palustris* L. plant by Professor Ireneusz Ochmian. The fresh flowers, leaves, stems, and roots were directly frozen at −25 °C, and then freeze-dried and crashed. The powders were kept frozen (−25 °C) until planned analysis around 2 weeks.

Instruments: Freeze-dryer (FreeZone 2,5, benchtop, A.G.A. Analytical, Warsaw, Poland), laboratory mill (RCMZ-800, Warsaw, Poland), Konica Minolta CM-700d spectrophotometer, ultra-performance reverse-phase liquid chromatography (UPLC-ESI-TQD-MS/MS) with Waters ACQUITY system (Waters, Milford, MA, USA), high-performance reverse-phase liquid chromatography with fluorescent detector (HPLC-FL) (Waters, Milford, MA, USA), centrifuge MPW-251 (MPW med. Instruments, Warsaw, Poland), Nicolet Evolution 300 spectrophotometer (Thermo, Watham, USA), Wallac 1420 VICTOR2 Plate Reader (PerkinElmer, Waltham, MA, USA).

### 3.2. Color Parameter

Color and shine of material were measured in a transmitted mode through Konica Minolta CM-700d spectrophotometer in 1 cm-thick glass trays. Measurements were conducted in CIE L*a*b* system, through a 10° observer type and D65 illuminant [68].

### 3.3. Polyphenolic Compounds (PCs) by UPLC-ESI-TQD-MS/MS and Procyanidin Polymers (PP) by the Phloroglucinolysis Method

The method procedure was applied according to the protocol shared by Kapusta et al. [69]. Profiles of polyphenolic compounds were analyzed using UPLC-PDA-ESI-TQD-MS/MS. Briefly, the separation was carried out using a BEH C18 column (100 mm × 2.1 mm i.d., 1.7 μm, Waters, Warsaw, Poland) that was kept at 50 °C. For the anthocyanins investigation, the following solvent system was applied: mobile phase A (2% formic acid in water, *v*/*v*) and mobile phase B (2% formic acid in 40% ACN in water, *v*/*v*). For other polyphenolic compounds, a lower concentration of formic acid was used (0.1%, *v*/*v*). The gradient program was set, as follows: 0 min 5% B, from 0 to 8 min linear to 100% B, and from 8 to 9.5 min for washing and back to initial conditions. The injection volume of samples was 5 μL (partial loop with needle overfill), and the flow rate was 0.35 mL/min. The following parameters were used for TQD: capillary voltage 3.5 kV, cone voltage, 30 V in positive and negative mode; the source was kept at 120 °C and the desolation temperature was 350 °C, con gas flow 100 L/h, and desolation gas flow 800 L/h. Argon was used as the collision gas at a flow rate of 0.3 mL/min. The profile compounds identification was based on specific PDA spectra, mass-to-charge ratio, and fragment ions obtained after collision-induced dissociation (CID). Quantification of compounds was achieved by the injection of solutions of known concentrations that ranged from 0.05 to 5 mg/mL ($R^2 \leq 0.9998$) as standards ((+)-catechin, 3,4-dicaffeoylquinic acid, luteolin 7-*O*-glucoside, apigenin 7-*O*-glucoside, kampferol-3-O-galactoside, ellagic acid, delphinidin 3-O-glucoside, cyanidin 3-O-glucoside, and malvidin 3-O-glucoside). All of the determinations were performed in duplicate and expressed as mg/L. Waters MassLynx software v.4.1 was used for data acquisition and processing. The analysis of polyphenolic compounds were assayed in triplicate, and described as mg per 100 g d.m.

The procyanidin polymers by the phloroglucinolysis method were applied according to the protocol shared by Lachowicz et al. [11]. The fractions of polymeric procyanidin were analyzed using HPLC-FL. Briefly, the separation was carried out using a BEH C18 RP column (2.1 × 5 mm, 1.7 µm; Waters Corporation, Milford, MA, USA) that was kept at 15 °C with gradient elution of solvent A as 2.5% acetic acid and solvent B as acetonitrile at a flow rate of 0.42 mL/min for a duration of 10 min (100 mm × 2.1 mm i.d., 1.7 µm, Waters) that was kept at 30 °C, and the fluorescence was recorded at the excitation and emission wavelengths at 278 and 360 nm. Quantification of compounds was made from procyanidin B2, (+)-catechin and (−)-epicatechin ($R^2 \leq 0.9995$) as standards. The degree of polymerization was calculated as the molar ratio of all flavan-3-ol units (phloroglucinol adducts + terminal units) to (−)-epicatechin and (+)-catechin, which correspond to terminal units. The analysis of polymeric procyanidins were assayed in triplicate, and described as mg per 100 g d.m.

### 3.4. *In Vitro Biological Activity*

#### 3.4.1. Extraction Procedure

The extraction was applied according to the protocol shared by Lachowicz et al. [11]. Briefly, 0.2 g of dry material was mixed with 7 mL of 80% of methanol in water and sonicated at 20 °C for 20 min, and then incubated at 4 °C for 24 h, and the next step was centrifuged. The supernatant was analyzed. The extractions method and all biological activities were assayed in triplicate.

#### 3.4.2. Antioxidant Activity

Antiradical Activity

The antiradical activity (ABTS) method was applied according to the protocol shared by Re et al. [70]. Briefly, 0.03 mL of extracted material mixed with 2.97 mL of ABTS solution and measured after 6 min at 734 nm using spectrophotometer. The result is described as mmol of Trolox equivalents (TE) per g d.m. ($y = 33.64x + 2.68$, $R^2 = 0.998$).

Reducing Potency

The reducing potency (FRAP) method was applied according to the protocol shared by Benzie and Strain [71]. Briefly, 0.1 mL of extracted material with 0.9 mL of distilled water and 3 mL of ferric complex measured after 10 min of incubation at 593 nm using spectrophotometer. The result is described as mmol of Trolox equivalents (TE) per g d.m ($y = 19.82x - 1.85$, $R^2 = 0.999$).

#### 3.4.3. Ability of α-Amylase, α-Glucosidase, Pancreatic Lipase Inhibitors

Ability of α-amylase, α-glucosidase inhibitors (anti-diabetic activity), and ability of lipase activity inhibitors (anti-obesity activity) effect of the extracted material were applied according to the protocol shared by Nakai et al. [72], Podsędek et al. [73], and Nickavar et al. [74]. Briefly, potato starch solution (0.2% ($v/v$)), the material extract, or phosphate buffer (0.1 M; pH 6.9; control) was mixed with α-amylase, and after incubation (37 °C, 10 min) the enzymatic reaction was stopped by the addition of HCl (0.4 M), the absorbance was read at 600 nm. For the α-glucosidase assay, the material extract was mixed with enzyme solution and incubated for 10 min. After that the reaction was initiated by *p*-nitrophenyl-α-D-glucopyranoside solution (5 mM), and incubated (37 °C, 20 min), and read the absorbance at 405 nm. For the pancreatic lipase assay, the material extracts were mixed with enzyme solution, and incubated (37 °C, 5 min). Then, methylumbelliferone solution (0.1 mM) was added, and incubated (37 °C, 20 min) and read at an excitation wavelength of 360 nm and at an emission wavelength of 460 nm. The value of the inhibitor, required to inhibit 50% of the enzyme activity, was defined as the $IC_{50}$ value. The $IC_{50}$ of the fruits tested was obtained from the line of the plot of the fruit concentration in 1 mL of reaction mixture versus the % inhibition.

#### 3.4.4. Antiproliferative Potency

Cell Lines and Cell Culture

The human cancer cell lines A549 (lung adenocarcinoma), BxPC3 (pancreatic ductal adenocarcinoma), HT-29, and CACO-2 (colorectal adenocarcinoma) and HCV29T (bladder cancer) were cultured in DMEM culture medium with 10% FBS (Gibco, Thermo Fisher Scientific, Walham, MA, USA) and Antibiotic Antimycotic Solution (Sigma-Aldrich, St. Louis, MO, USA). AML-NEV007 cell line (acute myeloid leukemia) was maintained in RPMI 1640 culture medium supplemented with 2 mM L-glutamine, 100 U/mL penicillin and 100 μg/mL streptomycin (Sigma-Aldrich, St. Louis, MO, USA), and 10% fetal bovine serum (FBS). All cell lines were cultured at 37 °C in a humidified atmosphere of 5% $CO_2$. The cells were seeded at densities of $5 \times 10^3$ cells/0.1 mL (0.32 $cm^2$) for cell viability assay. All cell lines were obtained from the collection of the Institute of Immunology and Experimental Therapy, Polish Academy of Sciences, Wroclaw, Poland.

Determination of Cell Viability

The plant extract was applied according to the protocol shared by Lachowicz et al. [11]. Cell viability was assessed by the CellTiter 96 Aqueous One Solution Cell Proliferation Assay (Promega), according to the manufacturer's protocol. Each treatment within a single experiment was performed in triplicate. Absorbance at 490 nm was recorded using a Wallac 1420 VICTOR2 plate reader (PerkinElmer, Waltham, MA, USA). Data were normalized to the untreated control.

Apoptosis Assay

Apoptosis was assessed by the Annexin V Apoptosis Detection Kit (Sigma-Aldrich, St. Louis, MO, USA), according to the manufacturer's protocol. Briefly, the cells were incubated with all compounds for 48 h, next were stained with Annexin V-FITC (8 μg/mL) and PI (5 μg/mL) for 15 min at RT in the dark. The cells were washed with cold PBS (with $Ca^{2+}$ and $Mg^{2+}$) containing 2.5% FBS between the steps. Data were acquired using a FACSCalibur flow cytometer (Becton Dickinson, Franklin Lakes, NJ, USA) and analyzed using Flowing Software 2.5.1 (Perttu Terho, Turku, Finland). Apoptosis was quantified as a percentage of both Annexin V-positive and Annexin V/PI-double-positive cells.

### *3.5. Statistics*

Statistical analysis included post-hoc Duncan's multiple range test ($p < 0.05$) as one-way analysis. Principal component analysis (PCA) as the multivariate analysis were performed with Statistica 13.3 (StatSoft, Kraków, Poland).

## 4. Conclusions

Overall, the studies of PCs by the LC–PDA-ESI–TQD–MS/MS technique of different parts of *Stachys palustris* L. detected 89 polyphenolic compounds, including 40 hydrolysable tannins, 16 phenylethanoid glycosides, 4 anthocyanins, 1 flavan-3-ol, 6 phenolic acids, 1 flavonol, and 21 flavones. The profile and levels of these ingredients were conditional on the parts of *S. palustris* L. used; thus, in flowers, roots, leaves, and stems, there were 56, 55, 50, 49 compounds identified, respectively. The flavonols and anthocyanins were detected only in the flowers, while PhGs dominated the roots. In addition, the main compounds evaluated in the research were vescalagin, cocciferin d2, isoverbascoside (verbascoside), luteolin 6-C-glucoside, luteolin 6-C-galactoside, apigenin 6-C-glucoside, (−)-epicatechin, ellagic acid, and malvidin 3-*O*-diglucoside.

The highest amount of PCs was detected in the leaves, followed by the flowers, stems, and roots. The strongest antioxidant activity of the ABTS and FRAP assays for the leaves and flowers extracts were exhibited. In addition, the best digestive enzyme inhibition effect as a potential antidiabetic and anti-obesity activity. One of the most important features of leaf and flower extracts is their ability to induce cell death in various tumor cell lines. In turn, in this study, the roots and stems were statistically the weakest in terms of medicinal potential.

For these reasons, *S. palustris* L. leaves and flowers rich in natural antioxidants with high biological activity should be further examined as health-beneficial ingredients for functional food, special food, cosmetics, and/or medical and pharmaceutical industries. Further investigations are required to isolate and identify active compounds from leaves and flowers with anti-microbiological and antiproliferative effects and a wider range of antiproliferative effects, as well as the analysis of the bioavailability of compounds.

**Supplementary Materials:** The following supporting information can be downloaded at: https://www.mdpi.com/article/10.3390/ph15070785/s1, Figure S1: LC-PDA-ESI-TQD-MS/MS chromatogram fragile of the *Stachys palustris* L. roots extract at 280 and 320 nm; Table S1: Parameter of color.

**Author Contributions:** Conceptualization, S.L.-W. and A.P.-S.; Data curation, S.L.-W., A.R., I.K., T.C., I.O. and A.P.-S.; Formal analysis, S.L.-W., A.R., I.K., T.C., I.O. and A.P.-S.; Funding acquisition, S.L.-W., A.P.-S., P.R., W.Ż.-S.; Investigation, S.L.-W., A.R., A.K., I.K., I.O. and A.P.-S.; Methodology, S.L.-W., A.R., I.K., T.C. and I.O.; Software, S.L.-W., A.R., I.K., T.C. and I.O.; Supervision, S.L.-W. and A.P.-S.; Writing—original draft, S.L.-W., A.R. and A.P.-S. All authors have read and agreed to the published version of the manuscript.

**Funding:** This research was partially supported by the National Science and Engineering Research Council of Canada (NSERC) Discovery grant (grant No. RGPIN-2018-04735).

**Institutional Review Board Statement:** Not applicable.

**Informed Consent Statement:** Not applicable.

**Data Availability Statement:** Not applicable.

**Acknowledgments:** Authors acknowledge the financial support provided by the A. S. Dekaban Foundation for supporting the stay of Sabina Lachowicz-Wiśniewska as Visiting Assistant Professor at the UBC Food Process Engineering Laboratory, Vancouver, Canada (01.2020-05.2020). Supported by the Foundation for Polish Science (FNP), and by the scholarship for young scientists of the Ministry of Education and Science (MEIN) for Sabina Lachowicz-Wiśniewska.

**Conflicts of Interest:** The authors declare no conflict of interest.

## References

1. Mohammed, M.J.; Anand, U.; Altemimi, A.B.; Tripathi, V.; Guo, Y.; Pratap-Singh, A. Phenolic Composition, Antioxidant Capacity and Antibacterial Activity of White Wormwood (*Artemisia herbaalba*). *Plants* **2021**, *10*, 164. [CrossRef] [PubMed]
2. Venditti, A.; Frezza, C.; Bianco, A.; Serafini, M.; Cianfaglione, K.; Nagy, D.U.; Iannarelli, R.; Caprioli, G.; Maggi, F. Polar constituents, essential oil and antioxidant activity of marsh woundwort (*Stachys palustris* L.). *Chem. Biodivers.* **2017**, *14*, e1600401. [CrossRef] [PubMed]
3. Facciola, S. *Cornocupia II. A Source Book of Edible Plants*; Kampong: Vista, CA, USA, 1998.
4. Łuczaj, Ł.J.; Svanberg, I.; Köhler, P. Marsh woundwort, *Stachys palustris* L.(Lamiaceae): An overlooked food plant. *Genet. Resour. Crop Evol.* **2011**, *58*, 783–793. [CrossRef]
5. Usher, G. *Dictionary of Plants Used by Man*; Constable and Company Ltd.: London, UK, 1974.
6. Feduraev, P.; Chupakhina, G.; Maslennikov, P.; Tacenko, N.; Skrypnik, L. Variation in phenolic compounds content and antioxidant activity of different plant organs from *Rumex crispus* L. and *Rumex obtusifolius* L. at different growth stages. *Antioxidants* **2019**, *8*, 237. [CrossRef] [PubMed]
7. Háznagy-Radnai, E.; Réthy, B.; Czigle, S.; Zupkó, I.; Wéber, E.; Martinek, T.; Falkay, G.; Máthé, I. Cytotoxic activities of Stachys species. *Fitoterapia* **2008**, *79*, 595–597. [CrossRef] [PubMed]
8. Esposito, T.; Celano, R.; Pane, C.; Piccinelli, A.L.; Sansone, F.; Picerno, P.; Zaccardelli, M.; Aquino, R.P.; Mencherini, T. Chestnut (*Castanea sativa* Miller.) burs extracts and functional compounds: UHPLC-UV-HRMS profiling, antioxidant activity, and inhibitory effects on phytopathogenic fungi. *Molecules* **2019**, *24*, 302. [CrossRef] [PubMed]

9. Singh, A.; Bajpai, V.; Kumar, S.; Sharma, K.R.; Kumar, B. Profiling of gallic and ellagic acid derivatives in different plant parts of Terminalia arjuna by HPLC-ESI-QTOF-MS/MS. *Nat. Prod. Commun.* **2016**, *11*, 239–244. [CrossRef]
10. Moilanen, J.; Sinkkonen, J.; Salminen, J.P. Characterization of bioactive plant ellagitannins by chromatographic, spectroscopic and mass spectrometric methods. *Chemoecology* **2013**, *23*, 165–179. [CrossRef]
11. Lachowicz, S.; Oszmiański, J.; Rapak, A.; Ochmian, I. Profile and content of phenolic compounds in leaves, flowers, roots, and stems of *Sanguisorba officinalis* L. determined with the LC-DAD-ESI-QTOF-MS/MS analysis and their in vitro antioxidant, antidiabetic, antiproliferative potency. *Pharmaceuticals* **2020**, *13*, 191. [CrossRef]
12. D'Urso, G.; Sarais, G.; Lai, C.; Pizza, C.; Montoro, P. LC-MS based metabolomics study of different parts of myrtle berry from Sardinia (Italy). *J. Berry Res.* **2017**, *7*, 217–229. [CrossRef]
13. Del Bubba, M.; Checchini, L.; Chiuminatto, U.; Doumett, S.; Fibbi, D.; Giordani, E. Liquid chromatographic/electrospray ionization tandem mass spectrometric study of polyphenolic composition of four cultivars of *Fragaria vesca* L. berries and their comparative evaluation. *J. Mass Spectrom.* **2012**, *47*, 1207–1220. [CrossRef]
14. Li, Y.; Li, Y.; Chen, J. Screening and identification of acetylcholinesterase inhibitors from *Terminalia chebula* fruits based on ultrafiltration and ultrα-performance liquid chromatography-quadrupole time-of-flight mass spectrometry. *Microchem. J.* **2021**, *168*, 106438. [CrossRef]
15. Ito, H. Metabolites of the ellagitannin geraniin and their antioxidant activities. *Planta Med.* **2011**, *77*, 1110–1115. [CrossRef]
16. Muccilli, V.; Cardullo, N.; Spatafora, C.; Cunsolo, V.; Tringali, C. α-Glucosidase inhibition and antioxidant activity of an oenological commercial tannin. Extraction, fractionation and analysis by HPLC/ESI-MS/MS and 1H NMR. *Food Chem.* **2017**, *215*, 50–60. [CrossRef]
17. Santos, L.S.; Alves Filho, E.G.; Ribeiro, P.R.; Zocolo, G.J.; Silva, S.M.; de Lucena, E.M.; Alves, R.E.; de Brito, E.S. Chemotaxonomic evaluation of different species from the Myrtaceae family by UPLC-qToF/MS-MS coupled to supervised classification based on genus. *Biochem. Syst. Ecol.* **2020**, *90*, 104028. [CrossRef]
18. Troalen, L.G.; Phillips, A.S.; Peggie, D.A.; Barran, P.E.; Hulme, A.N. Historical textile dyeing with *Genista tinctoria* L.: A comprehensive study by UPLC-MS/MS analysis. *Anal. Methods* **2014**, *6*, 8915–8923. [CrossRef]
19. Hooi Poay, T.; Sui Kiong, L.; Cheng Hock, C. Characterisation of galloylated cyanogenic glucosides and hydrolysable tannins from leaves of *Phyllagathis rotundifolia* by LC-ESI-MS/MS. *Phytochem. Anal.* **2011**, *22*, 516–525. [CrossRef]
20. Kool, M.M.; Comeskey, D.J.; Cooney, J.M.; McGhie, T.K. Structural identification of the main ellagitannins of a boysenberry (Rubus loganbaccus× baileyanus Britt.) extract by LC–ESI-MS/MS, MALDI-TOF-MS and NMR spectroscopy. *Food Chem.* **2010**, *119*, 1535–1543. [CrossRef]
21. Mullen, W.; Yokota, T.; Lean, M.E.; Crozier, A. Analysis of ellagitannins and conjugates of ellagic acid and quercetin in raspberry fruits by LC–MSn. *Phytochemistry* **2003**, *64*, 617–624. [CrossRef]
22. Li, L.; Tsao, R.; Yang, R.; Liu, C.; Young, J.C.; Zhu, H. Isolation and purification of phenylethanoid glycosides from *Cistanche deserticola* by high-speed counter-current chromatography. *Food Chem.* **2008**, *108*, 702–710. [CrossRef]
23. Piwowarczyk, R.; Ochmian, I.; Lachowicz, S.; Kapusta, I.; Sotek, Z.; Błaszak, M. Phytochemical parasite-host relations and interactions: A *Cistanche armena* case study. *Sci. Total Environ.* **2020**, *716*, 137071. [CrossRef]
24. Kırmızıbekmez, H.; Montoro, P.; Piacente, S.; Pizza, C.; Dönmez, A.; Çalış, İ. Identification by HPLC-PAD-MS and quantification by HPLC-PAD of phenylethanoid glycosides of five *Phlomis* species. *Phytochem. Anal. Int. J. Plant Chem. Biochem. Tech.* **2005**, *16*, 1–6. [CrossRef]
25. Karioti, A.; Bolognesi, L.; Vincieri, F.F.; Bilia, A.R. Analysis of the constituents of aqueous preparations of *Stachys recta* by HPLC–DAD and HPLC–ESI-MS. *J. Pharm. Biomed. Anal.* **2010**, *53*, 15–23. [CrossRef]
26. Shakeri, A.; D'Urso, G.; Taghizadeh, S.F.; Piacente, S.; Norouzi, S.; Soheili, V.; Asili, J.; Salarbashi, D. LC-ESI/LTQOrbitrap/MS/MS and GC–MS profiling of *Stachys parviflora* L. and evaluation of its biological activities. *J. Pharm. Biomed. Anal.* **2019**, *168*, 209–216. [CrossRef]
27. Miyase, T.; Yamamoto, R.; Ueno, A. Phenylethanoid glycosides from *Stachys officinalis*. *Phytochemistry* **1996**, *43*, 475–479. [CrossRef]
28. Yang, L.; He, J. Lagopsis supina extract and its fractions exert prophylactic effects against blood stasis in rats via anti-coagulation, anti-platelet activation and anti-fibrinolysis and chemical characterization by UHPLC-qTOF-MS/MS. *Biomed. Pharmacother.* **2020**, *132*, 110899. [CrossRef]
29. Nicoue, E.E.; Savard, S.; Belkacemi, K. Anthocyanins in wild blueberries of Quebec: Extraction and identification. *J. Agric. Food Chem.* **2007**, *55*, 5626–5635. [CrossRef]
30. Li, X.; Zhang, Y.; Zeng, X.; Yang, L.; Deng, Y. Chemical profiling of bioactive constituents in *Sarcandra glabra* and its preparations using ultra-high-pressure liquid chromatography coupled with LTQ Orbitrap mass spectrometry. *Rapid Commun. Mass Spectrom.* **2011**, *25*, 2439–2447. [CrossRef]
31. Delazar, A.; Celik, S.; Göktürk, R.S.; Unal, N.; Nahar, L.; Sarker, S.D. Two acylated flavonoid glycosides from *Stachys bombycina*, and their free radical scavenging activity. *Die Pharm.-Int. J. Pharm. Sci.* **2005**, *60*, 878–880. [CrossRef]
32. Marin, P.D.; Grayer, R.J.; Grujic-Jovanovic, S.; Kite, G.C.; Veitch, N.C. Glycosides of tricetin methyl ethers as chemosystematic markers in *Stachys subgenus* Betonica. *Phytochemistry* **2004**, *65*, 1247–1253. [CrossRef]
33. Šliumpaitė, I.; Venskutonis, P.R.; Murkovic, M.; Ragažinskienė, O. Antioxidant properties and phenolic composition of wood betony (*Betonica officinalis* L., syn. *Stachys officinalis* L.). *Ind. Crops Prod.* **2013**, *50*, 715–722. [CrossRef]

34. Tahir, N.I.; Shaari, K.; Abas, F.; Parveez, G.K.A.; Ishak, Z.; Ramli, U.S. Characterization of apigenin and luteolin derivatives from oil palm (*Elaeis guineensis* Jacq.) leaf using LC–ESI-MS/MS. *J. Agric. Food Chem.* **2012**, *60*, 11201–11210. [CrossRef] [PubMed]
35. Orsini, F.; Vovk, I.; Glavnik, V.; Jug, U.; Corradini, D. HPTLC, HPTLC-MS/MS and HPTLC-DPPH methods for analyses of flavonoids and their antioxidant activity in *Cyclanthera pedata* leaves, fruits and dietary supplement. *J. Liq. Chromatogr. Relat. Technol.* **2019**, *42*, 290–301. [CrossRef]
36. Demirtas, I.; Gecibesler, I.H.; Yaglioglu, A.S. Antiproliferative activities of isolated flavone glycosides and fatty acids from *Stachys byzantina*. *Phytochem. Lett.* **2013**, *6*, 209–214. [CrossRef]
37. Lin, L.Z.; Harnly, J.; Zhang, R.W.; Fan, X.E.; Chen, H.J. Quantitation of the hydroxycinnamic acid derivatives and the glycosides of flavonols and flavones by UV absorbance after identification by LC-MS. *J. Agric. Food Chem.* **2012**, *60*, 544–553. [CrossRef]
38. Sun, D.; Dong, L.; Guo, P.; Yan, W.; Wang, C.; Zhang, Z. Simultaneous determination of four flavonoids and one phenolic acid in rat plasma by LC–MS/MS and its application to a pharmacokinetic study after oral administration of the Herba Desmodii Styracifolii extract. *J. Chromatogr. B* **2013**, *932*, 66–73. [CrossRef]
39. Arthur, H.; Joubert, E.; De Beer, D.; Malherbe, C.J.; Witthuhn, R.C. Phenylethanoid glycosides as major antioxidants in Lippia multiflora herbal infusion and their stability during steam pasteurisation of plant material. *Food Chem.* **2011**, *127*, 581–588. [CrossRef]
40. Amessis-Ouchemoukh, N.; Abu-Reidah, I.M.; Quirantes-Piné, R.; Rodríguez-Pérez, C.; Madani, K.; Fernández-Gutiérrez, A.; Segura-Carretero, A. Tentative characterisation of iridoids, phenylethanoid glycosides and flavonoid derivatives from *Globularia alypum* L. (Globulariaceae) leaves by LC-ESI-QTOF-MS. *Phytochem. Anal.* **2014**, *25*, 389–398. [CrossRef]
41. Bahadori, M.B.; Kirkan, B.; Sarikurkcu, C.; Ceylan, O. Metabolite profiling and health benefits of *Stachys cretica* subsp. *mersinaea* as a medicinal food. *Ind. Crops Prod.* **2019**, *131*, 85–89. [CrossRef]
42. Carev, I.; Sarikurkcu, C. LC-MS/MS Profiles and in vitro biological activities of extracts of an endemic species from Turkey: *Stachys cretica* ssp. *anatolica*. *Plants* **2021**, *10*, 1054. [CrossRef]
43. Saltveit, M.E. Synthesis and metabolism of phenolic compounds. *Fruit Veg. Phytochem. Chem. Hum. Health* **2017**, *2*, 115.
44. Chepel, V.; Lisun, V.; Skrypnik, L. Changes in the content of some groups of phenolic compounds and biological activity of extracts of various parts of heather (*Calluna vulgaris* (L.) Hull) at different growth stages. *Plants* **2020**, *9*, 926. [CrossRef]
45. Sarikurkcu, C.; Zengin, G. Polyphenol profile and biological activity comparisons of different parts of *Astragalus macrocephalus* subsp. *finitimus* from Turkey. *Biology* **2020**, *9*, 231. [CrossRef]
46. Khanavi, M.; Hajimahmoodi, M.; Cheraghi-Niroomand, M.; Kargar, Z.; Ajani, Y.; Hadjiakhoondi, A.; Oveisi, M.R. Comparison of the antioxidant activity and total phenolic contents in some Stachys species. *Afr. J. Biotechnol.* **2009**, *8*, 1143–1147.
47. Oracz, J.; Zyzelewicz, D.; Nebesny, E. The content of polyphenolic compounds in cocoa beans (*Theobroma cacao* L.), depending on variety, growing region, and processing operations: A review. *Crit. Rev. Food Sci. Nutr.* **2015**, *55*, 1176–1192. [CrossRef]
48. Dušek, K.; Dušková, E.; Smékalová, K. Variability of morphological characteristic and content of active substances in *Betonica officinalis* L. in the czech republic. *Agriculture/Pol'nohospodárstvo* **2009**, *55*, 102–110.
49. Franz, C.H.; Novak, J.; Hajdari, A.; Mustafa, B. Total flavonoids, total phenolics and antioxidant activity of *Betonica officinalis* L. from Kosovo. In Proceedings of the IV International Symposium on Breeding Research on Medicinal and Aromatic Plants-ISBMAP2009, Ljubljana, Slovenia, 17–21 June 2009; Volume 860, pp. 75–80.
50. Elfalleh, W.; Kirkan, B.; Sarikurkcu, C. Antioxidant potential and phenolic composition of extracts from *Stachys tmolea*: An endemic plant from Turkey. *Ind. Crops Prod.* **2019**, *127*, 212–216. [CrossRef]
51. Benabderrahim, M.A.; Sarikurkcu, C.; Elfalleh, W.; Ozer, M.S.; Ceylan, O. Phenolic composition and biological activities of Turkish endemic plant: *Stachys cretica* subsp. *kutahyensis*. *S. Afr. J. Bot.* **2021**, *138*, 124–128. [CrossRef]
52. Kumar, D.; Bhat, Z.A. Apigenin 7-glucoside from *Stachys tibetica* Vatke and its anxiolytic effect in rats. *Phytomedicine* **2014**, *21*, 1010–1014. [CrossRef]
53. Madunić, J.; Madunić, I.V.; Gajski, G.; Popić, J.; Garaj-Vrhovac, V. Apigenin: A dietary flavonoid with diverse anticancer properties. *Cancer Lett.* **2018**, *413*, 11–22. [CrossRef]
54. Bouasla, I.; Hamel, T.; Barour, C.; Bouasla, A.; Hachouf, M.; Bouguerra, O.M.; Messarah, M. Evaluation of solvent influence on phytochemical content and antioxidant activities of two Algerian endemic taxa: *Stachys marrubiifolia* Viv. and *Lamium flexuosum* Ten. (Lamiaceae). *Eur. J. Integr. Med.* **2021**, *42*, 101267. [CrossRef]
55. Taylor, K.; Rowland, P. Biological flora of the British Isles: *Stachys palustris* L. *J. Ecol.* **2011**, *99*, 1081–1090. [CrossRef]
56. Kukić, J.; Petrović, S.; Niketić, M. Antioxidant activity of four endemic *Stachys taxa*. *Biol. Pharm. Bull.* **2006**, *29*, 725–729. [CrossRef] [PubMed]
57. Saravanakumar, K.; Park, S.; Mariadoss, A.V.A.; Sathiyaseelan, A.; Veeraraghavan, V.P.; Kim, S.; Wang, M.H. Chemical composition, antioxidant, and anti-diabetic activities of ethyl acetate fraction of *Stachys riederi* var. japonica (Miq.) in streptozotocin-induced type 2 diabetic mice. *Food Chem. Toxicol.* **2021**, *155*, 112374. [CrossRef] [PubMed]
58. Yadav, A.; Yadav, M.; Kumar, S.; Sharma, D.; Yadav, J.P. In vitro antioxidant activities and GC-MS analysis of different solvent extracts of *Acacia nilotica* leaves. *Indian J. Pharm. Sci.* **2018**, *80*, 892–902. [CrossRef]
59. Khan, S.; Nazir, M.; Saleem, H.; Raiz, N.; Saleem, M.; Anjum, S.M.M.; Zengin, G.; Mukhtar, M.; Tiusif, M.I.; Mahomoodally, F.M.; et al. Valorization of the antioxidant, enzyme inhibition and phytochemical propensities of *Berberis calliobotrys* Bien. ex Koehne: A multifunctional approach to probe for bioactive natural products. *Ind. Crops Prod.* **2019**, *141*, 111693. [CrossRef]

60. Bahadori, M.B.; Kirkan, B.; Sarikurkcu, C. Phenolic ingredients and therapeutic potential of *Stachys cretica* subsp. *smyrnaea* for the management of oxidative stress, Alzheimer's disease, hyperglycemia, and melasma. *Ind. Crops Prod.* **2019**, *127*, 82–87. [CrossRef]
61. Sarikurkcu, C.; Kocak, M.S.; Uren, M.C.; Calapoglu, M.; Tepe, A.S. Potential sources for the management global health problems and oxidative stress: *Stachys byzantina* and *S. iberica* subsp. *iberica* var. densipilosa. *Eur. J. Integr. Med.* **2016**, *8*, 631–637. [CrossRef]
62. Kokhdan, E.P.; Sadeghi, H.; Ghafoori, H.; Sadeghi, H.; Danaei, N.; Javadian, H.; Aghamaali, M.R. Cytotoxic effect of methanolic extract, alkaloid and terpenoid fractions of *Stachys pilifera* against HT-29 cell line. *Res. Pharm. Sci.* **2018**, *13*, 404.
63. Khanavi, M.; Manayi, A.; Lotfi, M.; Abbasi, R.; Majdzadeh, M.; Ostad, S.N. Investigation of cytotoxic activity in four *Stachys species* from Iran. *Iran. J. Pharm. Res. IJPR* **2012**, *11*, 589.
64. Gullett, N.P.; Amin, A.R.; Bayraktar, S.; Pezzuto, J.M.; Shin, D.M.; Khuri, F.R.; Aggarwal, B.B.; Surh, Y.-U.; Kucuk, O. Cancer prevention with natural compounds. *Semin. Oncol.* **2010**, *37*, 258–281. [CrossRef]
65. Kopustinskiene, D.M.; Jakstas, V.; Savickas, A.; Bernatoniene, J. Flavonoids as anticancer agents. *Nutrients* **2020**, *12*, 457. [CrossRef]
66. Aiello, P.; Sharghi, M.; Mansourkhani, S.M.; Ardekan, A.P.; Jouybari, L.; Daraei, N.; Peiro, K.; Mohamadian, S.; Rezaei, M.; Heidari, M.; et al. Medicinal plants in the prevention and treatment of colon cancer. *Oxid. Med. Cell. Longev.* **2019**, *2019*, 2075614. [CrossRef]
67. Bystrická, J.; Vollmannová, A.; Margitanová, E. Dynamics of polyphenolics formation in different plant parts and different growth phases of selected buckwheat cultivars. *Acta Agric. Slov.* **2010**, *95*, 225. [CrossRef]
68. Ochmian, I.; Błaszak, M.; Lachowicz, S.; Piwowarczyk, R. The impact of cultivation systems on the nutritional and phytochemical content, and microbiological contamination of highbush blueberry. *Sci. Rep.* **2020**, *10*, 16696. [CrossRef]
69. Kapusta, I.; Cebulak, T.; Oszmiański, J. Characterization of polish wines produced from the interspecific hybrid grapes grown in south-east Poland. *Eur. Food Res. Technol.* **2018**, *244*, 441–455. [CrossRef]
70. Re, R.; Pellegrini, N.; Proteggente, A.; Pannala, A.; Yang, M.; Rice-Evans, C. Antioxidant activity applying an improved ABTS radical cation decolorization assay. *Free. Radic. Biol. Med.* **1999**, *26*, 1231–1237. [CrossRef]
71. Benzie, I.F.; Strain, J.J. The ferric reducing ability of plasma (FRAP) as a measure of "antioxidant power": The FRAP assay. *Anal. Biochem.* **1996**, *239*, 70–76. [CrossRef]
72. Nakai, M.; Fukui, Y.; Asami, S.; Toyoda-Ono, Y.; Iwashita, T.; Shibata, H.; Mitsunaga, T.; Hashimoto, F.; Kiso, Y. Inhibitory effects of oolong tea polyphenols on pancreatic lipase in vitro. *J. Agric. Food Chem.* **2005**, *53*, 4593–4598. [CrossRef]
73. Podsedek, A.; Majewska, I.; Redzynia, M.; Sosnowska, D.; Koziołkiewicz, M. In vitro inhibitory effect on digestive enzymes and antioxidant potential of commonly consumed fruits. *J. Agric. Food Chem.* **2014**, *62*, 4610–4617. [CrossRef]
74. Nickavar, B.; Yousefian, N. Evaluation of α-amylase inhibitory activities of selected antidiabetic medicinal plants. *J. Verbrauch. Lebensm.* **2011**, *6*, 191–195. [CrossRef]

*pharmaceuticals*

*Article*

# Supercritical Fluid Chromatography—Tandem Mass Spectrometry for Rapid Quantification of Pentacyclic Triterpenoids in Plant Extracts

**Danil I. Falev *, Denis V. Ovchinnikov, Ilya S. Voronov, Anna V. Faleva, Nikolay V. Ul'yanovskii and Dmitry S. Kosyakov ***

Laboratory of Natural Compounds Chemistry and Bioanalytics, Core Facility Center "Arktika", M.V. Lomonosov Northern (Arctic) Federal University, Northern Dvina Emb. 17, 163002 Arkhangelsk, Russia; d.ovchinnikov@narfu.ru (D.V.O.); i.voronov@narfu.ru (I.S.V.); a.bezumova@narfu.ru (A.V.F.); n.ulyanovsky@narfu.ru (N.V.U.)

* Correspondence: d.falev@narfu.ru (D.I.F.); d.kosyakov@narfu.ru (D.S.K.)

**Abstract:** Pentacyclic triterpenoids (PCTs) are a widely distributed class of plant secondary metabolites. These compounds have high bioactive properties, primarily antitumor and antioxidant activity. In this study, a method was developed for the quantitative analysis of pentacyclic triterpenoids in plants using supercritical fluid chromatography–tandem mass spectrometry (SFC-MS/MS). Separation of ten major PCTs (friedelin, lupeol, β-amyrin, α-amyrin, betulin, erythrodiol, uvaol, betulinic, oleanolic and ursolic acids) was studied on six silica-based reversed stationary phases. The best results (7 min analysis time in isocratic elution mode) were achieved on an HSS C18 SB stationary phase using carbon dioxide—isopropanol (8%) mobile phase providing decisive contribution of polar interactions to the retention of analytes. It was shown that the use of atmospheric pressure chemical ionization (APCI) is preferred over atmospheric pressure photoionization (APPI). The combination of SFC with APCI-MS/MS mass spectrometry made it possible to achieve the limits of quantification in plant extracts in the range of 2.3–20 $\mu g \cdot L^{-1}$. The developed method was validated and tested in the analyses of birch outer layer (*Betula pendula*) bark, and licorice (*Glycyrrhiza glabra*) root, as well as lingonberry (*Vaccinium vitis-idaea*), cranberry (*Vaccinium oxycoccos*), apple (*Malus domestica "Golden Delicious"* and *Malus domestica "Red Delicious"*) peels.

**Keywords:** pentacyclic triterpenoids; plant feedstock; supercritical fluid chromatography; tandem mass spectrometry

**Citation:** Falev, D.I.; Ovchinnikov, D.V.; Voronov, I.S.; Faleva, A.V.; Ul'yanovskii, N.V.; Kosyakov, D.S. Supercritical Fluid Chromatography—Tandem Mass Spectrometry for Rapid Quantification of Pentacyclic Triterpenoids in Plant Extracts. *Pharmaceuticals* **2022**, *15*, 629. https://doi.org/10.3390/ph15050629

Academic Editors: Sabina Lachowicz-Wiśniewska and Jan Oszmianski

Received: 19 April 2022
Accepted: 17 May 2022
Published: 20 May 2022



## 1. Introduction

Pentacyclic triterpenoids (PCTs) are a widely distributed class of important plant secondary metabolites. The most abundant triterpenoids in higher plants are derivatives of pentacyclic hydrocarbons lupane, oleanane and ursane, which contain 30 carbon atoms and belong to the monool, diol, ketone and triterpenic acid types [1,2], differing significantly in polarity and solubility. PCTs possess various types of bioactivities, having anti-inflammatory, antimicrobial, antihyperlipidemic, hepatoprotective, gastroprotective, antidiabetic and hemolytic properties, as well as cardiovascular, antipyretic and wound healing effects [3–10]. Many researchers emphasize the antitumor [3,4,10–14] and antioxidant activities of PCTs [6,7,15,16], which have also been observed in triterpenoid-rich plant extracts [17–23]. Therefore, PCTs are of exceptional interest for the pharmaceutical, cosmetic and food industries. Thus, the development of new rapid, selective, and highly sensitive methods for screening and determination of triterpenoids in plant materials, drugs, and biological fluids is an important task.

Mass spectrometry techniques, especially MALDI MS, have been successfully used for the rapid screening of triterpenoids in plant materials [24], while the quantitative analysis of PCTs requires chromatographic separation of analytes. High-performance liquid chromatography–tandem mass spectrometry (HPLC-MS/MS) in multiple reaction monitoring (MRM) mode is widely used for PCT quantitative analysis [25–28]. The authors have found that it is preferable to use atmospheric pressure chemical ionization (APCI) with positive ion mode, whereas using electrospray ionization (ESI) causes a decrease in PCT ionization efficiency due to the low polarity of the analytes (except triterpenic acids in negative ion mode).

Since different types of triterpenoids strongly differ from each other in terms of polarity, and each type includes structurally similar compounds with close retention on the common HPLC stationary phases, the chromatographic separation of complex mixtures of PCTs is a nontrivial task. It requires gradient elution and long analysis time when using reversed stationary phases which, however, do not provide a good chromatographic resolution between similar compounds with the same monoisotopic mass in some critical pairs of PCTs, such as erythrodiol/uvaol, β-amyrin/α-amyrin, and oleanolic/ursolic acids. Using the stationary phase with cross-linked octadecyl groups (Nucleodur C18 Isis) with higher resolution towards positional isomers made it possible to separate nine PCTs in gradient elution mode in only 25 min and achieve the limits of quantitation (LOQ) in HPLC-MS/MS analysis in the range of 10–22 $\mu g \cdot L^{-1}$ [27]. Stationary phases with a mixed retention mechanism can be used to reduce analysis time and enhance selectivity. In our recent study [28], the application of the Acclaim Mixed-Mode WAX-1 stationary phase combining ion exchange, hydrophilic interactions, and reversed phase retention in isocratic mode allowed rapid separation of ten PCTs (including mentioned above critical pairs) in 7 min with LOQs of 8.3–150 $\mu g \cdot L^{-1}$. The advantage of the method is the ability to fine tune the separation selectivity by changing the contributions of various types of interactions to the retention mechanism with a change in the composition and pH of the mobile phase. On the other hand, due to the many factors affecting separation, the development and optimization of an analytical method for such stationary phases requires considerable effort.

A modern and "greener" alternative to HPLC in the analysis of complex mixtures is supercritical fluid chromatography (SFC) using sub- or supercritical carbon dioxide as a main component of the mobile phase. Due to the low viscosity and thus high diffusion coefficients of such media, SFC features high mass transfer rate, higher efficiency, and exceptional separation speed. It has been successfully used for rapid analysis of various natural compounds [29–32]. There are few published works on the use of SFC for the analysis of PCTs; the most recent one [33] uses a combination of SFC with light scattering detection (ELSD). Although separation of eight PCTs was achieved in isocratic mode within 16 min, the drawback of this approach is low sensitivity and selectivity of ELSD. Considering the possibility of rapid separation of PCTs by the SFC, it is promising to hyphenate it with sensitive tandem mass spectrometric detection, combining the advantages of these separation and detection techniques. Thus, the aim of the present study is to develop a novel rapid and sensitive method for the simultaneous determination of pentacyclic triterpenoids in plant biomass by SFC-MS/MS.

## 2. Results and Discussion

### 2.1. Tandem Mass Spectrometry

The ten most common representatives of monools (lupeol, α- and β-amyrins), diols (erythrodiol, betulin, uvaol), ketones (friedelin) and triterpenic acids (betulinic, oleanolic, ursolic) with friedelane, lupane, oleanane and ursane backbones (Figure 1) were chosen as target analytes. As can be seen from their structures, PCTs are characterized by rather low polarity, which makes APCI or atmospheric pressure photoionization (APPI) preferable for the mass spectrometric detection. Even though triterpenic acids can form deprotonated molecules under SFC-APCI/APPI conditions, the positive ion mode demonstrated an undoubted advantage, allowing efficient ionization of all selected analytes, including those containing no acidic groups. It has been established that all hydroxyl-containing analytes undergo dehydration in the ion source during APCI/APPI. Thus, they form

$[M + H - H_2O]^+$ species, whereas friedelin gives protonated molecule $[M + H]^+$. These ions were chosen as precursor ions in tandem mass spectrometry.

| № | Compounds | Family | R |
|---|---|---|---|
| I | Friedelin | Friedelane | =O |
| II | Betulinic acid | Lupane | -COOH |
| III | Oleanolic acid | Oleanane | -COOH |
| IV | Ursolic acid | Ursane | -COOH |
| V | Lupeol | Lupane | $-CH_3$ |
| VI | β-Amyrin | Oleanane | $-CH_3$ |
| VII | α-Amyrin | Ursane | $-CH_3$ |
| VIII | Erythrodiol | Oleanane | $-CH_2OH$ |
| IX | Betulin | Lupane | $-CH_2OH$ |
| X | Uvaol | Ursane | $-CH_2OH$ |

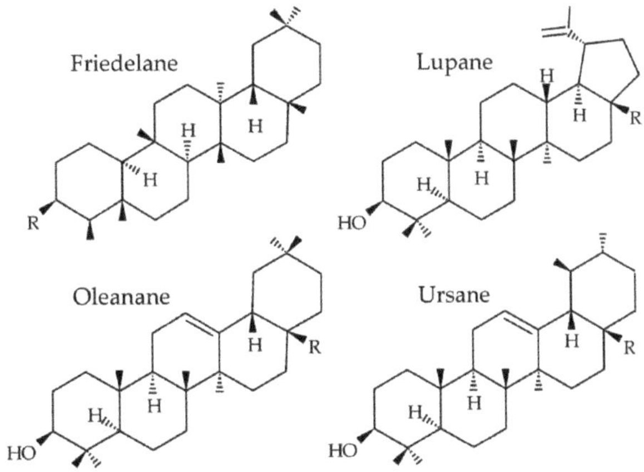


**Figure 1.** Chemical structure of the studied pentacyclic triterpenoids.

The optimization of APCI and APPI parameters was carried out in selected ion monitoring mode by flow injection of analyte standard solutions into the ion source with a flow of SFC mobile phase to achieve maximum intensities of the signals of precursor ions. The obtained optimal conditions were as follows: APCI: ion source temperature—350 °C; curtain gas, nebulizer gas and drying gas pressure—25, 20 and 30 psi, respectively; nebulizer current—3.5 μA; APPI: ion source temperature—350 °C; curtain gas, nebulizer gas and drying gas pressure—10, 20 and 30 psi, respectively; ion spray voltage—750 V. After obtaining tandem mass spectra, the most intense ion transitions were chosen and the optimal parameters (declustering and entrance potentials, collision energy) for detecting PCTs in MRM mode were determined (Table 1).

**Table 1.** Optimized conditions of PCT mass spectrometric detection in MRM mode.

| Analyte | Monoisotopic Mass, Da | Precursor Ion, *m/z* | Product Ion, *m/z* | Declustering Potential, V | Entrance Potential, V | Collision Energy, eV |
|---|---|---|---|---|---|---|
| I | 426 | 427 | 95 | 41 | 4.5 | 45 |
| II | 456 | 439 | 95 | 61 | 5.5 | 45 |
| III | 456 | 439 | 191 | 43 | 5.0 | 19 |
| IV | 456 | 439 | 191 | 43 | 6.0 | 49 |
| V | 426 | 409 | 95 | 55 | 6.5 | 47 |
| VI | 426 | 409 | 95 | 55 | 8.0 | 49 |
| VII | 426 | 409 | 95 | 55 | 7.5 | 59 |
| VIII | 442 | 425 | 95 | 50 | 5.5 | 47 |
| IX | 442 | 425 | 95 | 50 | 6.0 | 19 |
| X | 442 | 425 | 95 | 50 | 7.5 | 21 |

### *2.2. Chromatographic Separation and Column Screening*

Preliminary tests with BEH Silica and 2-ethylpyridine polar stationary phases showed high retention of most polar analytes, especially triterpenic acids, along with unacceptable chromatographic peak shapes. For this reason, the present study was focused on the use of six silica-based reversed stationary phases—Accucore C30, Nucleodur C18 Isis, Luna C18, Zorbax SB-Aq, HSS C18 SB and Nucleodur PolarTec. The latter is distinguished by the presence of amide groups embedded into alkyl chain and contributing to the retention mechanism through polar interactions. SFC column screening was carried out using the model mixture of analytes under the following basic conditions: temperature of 25 °C, backpressure 150 bar, and eluent flow rate 1.0 mL·min$^{-1}$. The mobile phase was 90% carbon dioxide and 10% methanol.

Despite the similarity of the tested stationary phases, the obtained chromatograms (Supplementary Materials, Figure S1) demonstrated significant differences in retention, selectivity and elution order of the analytes. As expected, triacontyl (C30) and cross-linked octadecyl (C18 Isis) phases are characterized by similar patterns of retention, with elution order corresponding to the decrease in polarity of analytes: triterpenic acids–diols–monools. This indicates the decisive contribution of the reversed phase (hydrophobic) interactions into the mechanism of analyte retention. The only difference between two phases is the position of the peak of friedelin, which has a higher retention on the C30 column. Unacceptably low retention factors (*k*) and thus the impossibility of separating the most polar compounds (diols and triterpenic acids) make such stationary phases unsuitable for solving the problem of PCTs analysis. This also applies to the Zorbax SB-Aq stationary phase, in which the polar surface of the silica particles apparently contributes to the retention (higher *k* values for diols) but is difficult to access for the analytes due to steric hindrances caused by the presence of isopropyl side chains in the grafted alkyl groups.

A completely different pattern was observed for non-endcapped stationary phases with residual silanol groups capable of interactions with analytes (Luna C18, HSS C18 SB), as well as for a Nucleodur PolarTec sorbent. They provide much higher *k* values for all studied PCTs due to mixed-mode retention involving both hydrophobic and polar interactions. Despite this, at a relatively high concentration of methanol (moderately polar mobile phase) the contribution of the reversed phase mechanism is decisive, and the order of analytes elution on both octadecyl phases does not differ from that of the triacontyl sorbent. Due to the greater availability and different nature of the polar groups in the Nucleodur PolarTec phase, the retention of the most polar analytes on this sorbent increases significantly and the separation selectivity is unacceptable. In this regard, based on the results of column screening, the stationary phase HSS C18 SB, which was especially designed to increase the selectivity of the separation of substances with polar groups in SFC and showed the best separation of the investigated PCTs, was chosen for further optimization of the chromatographic method.

Variations of temperature (20–55 °C) and backpressure (110–190 bar) did not significantly affect the separation of triterpenoids on an HSS C18 SB column, which is in good agreement with the published data on the minority of these parameters when optimizing separation in SFC on polar stationary phases [31]. To ensure the maximum chromatographic column lifetime and the stability of maintaining the temperature, we chose 25 °C as the optimal value. The backpressure of 150 bar typical for SFC separations was used in our further work.

The main parameter affecting the retention and selectivity in SFC is organic modifier (co-solvent) content in the carbon dioxide-based mobile phase. A decrease in the methanol concentration from 10 to 6% led to a sharp (2–4-fold) increase in the *k* values (Figure 2) for all analytes, except for friedelin, which does not possess hydroxyl or carboxyl groups. The most important feature of the obtained dependences of the retention factors on the methanol content is the inversion of the elution order of PCTs belonging to the monools and triterpenic acids (in the range of 6–8% MeOH), as well as monools and diols (in the range of 8–10% MeOH). Thus, at methanol concentrations $\leq$ 6%, this leads to the occurring the elution order (friedelin–monools–acids–diols) characteristic of retention by the mechanism, the main contribution to which is made by polar interactions of analytes with silanol groups of the sorbent. Varying the methanol concentration leads to a significant change in the contributions of hydrophobic and polar interactions to the mixed retention mechanism of PCTs with the possibility of transition from reversed-phase separation to the normal-phase one, and vice versa. The reason for this phenomenon is the competition of analytes and polar mobile phase modifier for sorption centers (silanol groups) of the stationary phase. As can be seen, achieving complete separation of all analytes requires a methanol concentration of <6%, which is irrational both from the point of view of an unacceptably high retention of diols ($k > 40$) and a decrease in the solubility of PCTs in the mobile phase with the risk of precipitation and clogging the chromatographic tract. To overcome this problem, methanol as a mobile phase modifier was replaced by the less polar isopropanol.

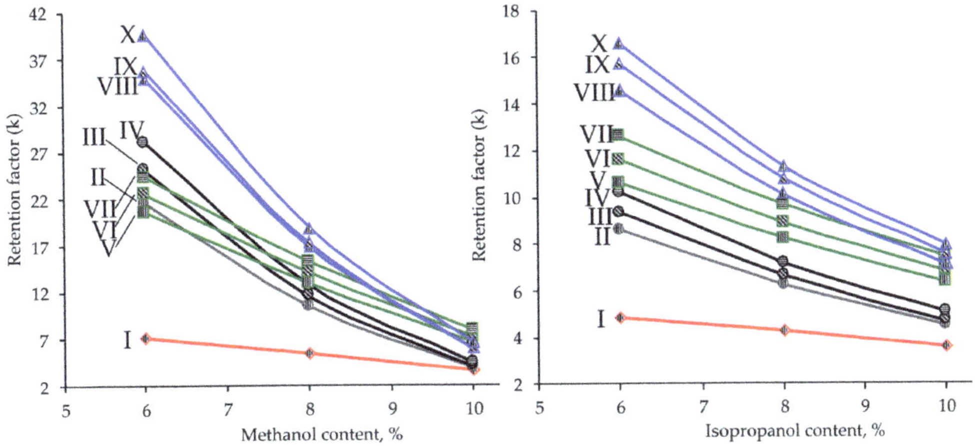


**Figure 2.** Effect of methanol (**left**) and isopropanol (**right**) content in the mobile phase on the retention factors of PCTs on HSS C18 SB stationary phase (flow rate 1.5 mL·min$^{-1}$, T = 25 °C, backpressure 150 bar).

The obtained results (Figure 2) show that the lower affinity of isopropanol with respect to the silica surface makes it possible to implement a predominantly normal-phase separation mechanism in the entire studied concentration range. Sufficiently complete separation of analytes was achieved at a modifier content of 6–8%, while providing optimal values of retention factors in the range of 4–17. To ensure the maximum separation rate, the working concentration of isopropanol in the mobile phase of 8% and increased eluent flow rate of 1.5 mL·min$^{-1}$ were chosen for analytical method development.

Thus, the optimal separation parameters of ten PCTs are: HSS C18 SB stationary phase, 92% carbon dioxide and 8% isopropanol mobile phase, elution rate—1.5 mL·min$^{-1}$, a temperature—25 °C, and backpressure—150 bar. The SFC-APCI-MS/MS chromatogram obtained under these conditions is presented in Figure 3.

This demonstrates that the analysis time (7 min) is 2–3 times lower than those reported in the literature for reversed phase HPLC [27] or even SFC on polar sorbent [33] methods and is comparable to the value attained in our recent work using HPLC on a stationary phase with mixed retention mechanisms [28]. It is worth noting that SFC, unlike mixed mode HPLC, does not require complex and time-consuming method optimization procedures due to there being fewer parameters affecting the separation. Despite the short retention times, a baseline separation was observed for most PCTs (Table S1). The minimum chromatographic resolution (*Rs*) was 0.83 for the betulinic/oleanolic acid critical pair. However, these compounds are distinguished with different ion transitions in MRM detection. The *Rs* values for other pairs of analytes (oleanolic/urosolic acids and betulin/uvaol) were close to 1.0; however, this does not create significant problems for the quantitative determination of these PCTs.

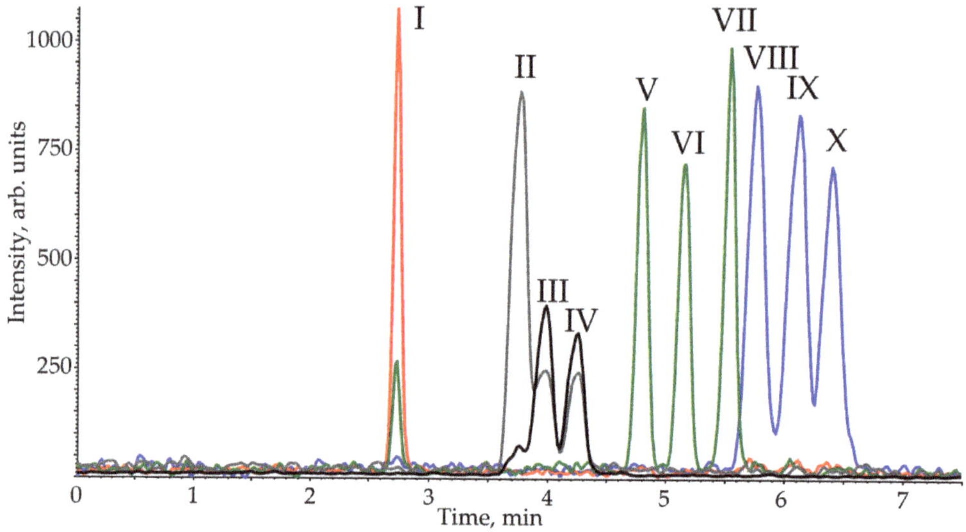


**Figure 3.** SFC-APCI-MS/MS chromatogram of analytes model mixture (I—200 μg·L$^{-1}$; II, VIII and X—100 μg·L$^{-1}$; III, IV and IX—50 μg·L$^{-1}$; V, VI and VII—25 μg·L$^{-1}$) obtained on HSS C18 SB stationary phase in the optimized conditions.

### 2.3. *Quantification and Method Validation*

Since the LOQs of triterpenoids obtained using APCI (2.3–20 μg·L$^{-1}$) were somewhat lower than those attained in APPI-MS/MS (3.5–33 μg·L$^{-1}$) (Table 2), the APCI method was used for further experiments. The reliability of the obtained LODs and LOQs was proved by the analysis of a model mixture of analytes with concentration close to the limit of quantitation (Figure S2). The developed method demonstrated a high sensitivity comparable to that reported in previous works using HPLC-MS/MS (8.3–150 μg·L$^{-1}$) [27,28], even exceeding it. The calibration dependences were linear ($R^2 > 0.999$) for all ten triterpenoids (from LOQ to the maximum used concentrations), spanning at least two orders of magnitude (Table 2).

**Table 2.** Calibration dependences ($y = a \times x$) for the area of chromatographic peak versus analyte concentration, limits of quantification of analytes by SFC–APCI–MS/MS and SFC–APPI–MS/MS methods on HSS C18 SB stationary phase.

| Analyte | APPI | | | | APCI | | | |
|---|---|---|---|---|---|---|---|---|
| | Linear Concentration Range, μg·L$^{-1}$ | *a* | $R^2$ | LOQ, μg·L$^{-1}$ | Linear Concentration Range, μg·L$^{-1}$ | *a* | $R^2$ | LOQ, μg·L$^{-1}$ |
| I | 33–2000 | 6.163 | 0.99288 | 33 | 20–2000 | 26.076 | 0.99945 | 20 |
| II | 7.0–2000 | 50.235 | 0.99278 | 7.0 | 11–2000 | 87.018 | 0.99997 | 11 |
| III | 3.5–1000 | 42.627 | 0.99422 | 3.5 | 4.0–1000 | 75.104 | 0.99997 | 4.0 |
| IV | 4.6–1000 | 38.455 | 0.99033 | 4.6 | 4.7–1000 | 68.122 | 0.99994 | 4.7 |
| V | 3.8–1000 | 139.33 | 0.99283 | 3.8 | 2.6–1000 | 219.15 | 0.99998 | 2.6 |
| VI | 4.5–1000 | 119.78 | 0.99833 | 4.5 | 2.7–1000 | 211.91 | 0.99997 | 2.7 |
| VII | 3.8–1000 | 147.89 | 0.99243 | 3.8 | 2.3–1000 | 253.00 | 0.99996 | 2.3 |
| VIII | 20–2000 | 35.314 | 0.99206 | 20 | 9.8–2000 | 92.152 | 0.99995 | 9.8 |
| IX | 11–1000 | 68.244 | 0.99116 | 11 | 5.5–1000 | 175.25 | 0.99997 | 5.5 |
| X | 27–2000 | 27.677 | 0.99410 | 27 | 13–2000 | 75.618 | 0.99993 | 13 |

The attained accuracy was close to 100%, and the standard deviation did not exceed 10% in intra-day and inter-day assays (Table S2) at levels close to LOQ.

The matrix effects assessment was carried out by the spike recovery test using the licorice root PLE extract as a real matrix. The recovery values were in the range of 88–118%, which proved the absence of significant interferences from the matrix for all analytes (Table S3). The efficient elimination of the matrix effect is due to the high sensitivity of the method, allowing sample dilution, the use of APCI not highly susceptible to interference from other components, and chromatographic separation of the triterpenoids from the matrix.

The use of isocratic elution and optimal conditions of the chromatographic separation of analytes ensure the robustness of the developed approach. Variations in pressure (145–155 bar) and temperature (23–27 °C) did not lead to a significant (>2%) shift in $t_R$ values and loss of chromatographic resolution *Rs*. The chromatographic separation did not deteriorate after repeated analysis of plant extracts (about 100 injections).

### 2.4. Plant Biomass Analyses

To test the developed method, licorice root, birch bark, and berry and apple peels were selected as real objects. These objects are characterized by a complex chemical composition and are considered an important source of biologically active substances for pharmaceutical, cosmetic and food purposes. PLE with methanol, providing near quantitative extraction of PCTs, was used as a sample preparation method [26]. The resulting chromatograms (Figure 4) demonstrate the presence of all analytes in the samples (Table 3) over a wide range of concentrations (0.0041–250 mg·g$^{-1}$).

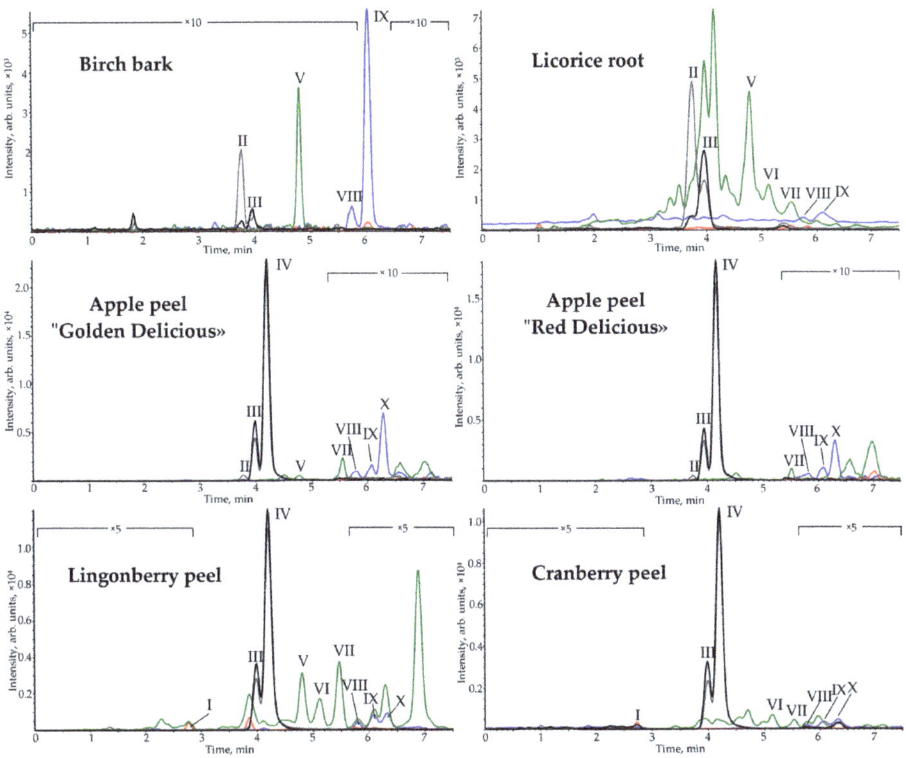


**Figure 4.** SFC-APCI-MS/MS chromatograms of plant methanolic PLE extracts on HSS C18 SB stationary phase.

**Table 3.** The content of PCTs (mg·g$^{-1}$, recalculated for the oven-dried plant material) in plant tissues ($n$ = 3, $p$ = 0.95).

| Analyte | Birch Bark | Licorice Root | Apple Peel "Golden Delicious" | Apple Peel "Red Delicious" | Lingonberry Peel | Cranberry Peel |
|---|---|---|---|---|---|---|
| I | - | - | - | - | 0.17 ± 0.01 | 0.14 ± 0.01 |
| II | 13 ± 1 | 0.16 ± 0.01 | 0.79 ± 0.01 | 0.47 ± 0.02 | - | - |
| III | 3.6 ± 0.5 | 0.10 ± 0.01 | 10 ± 1 | 6.3 ± 0.5 | 3.5 ± 0.3 | 3.2 ± 0.2 |
| IV | - | - | 49 ± 5 | 32 ± 1 | 15 ± 2 | 14 ± 1 |
| V | 4.6 ± 0.3 | 0.043 ± 0.001 | 0.26 ± 0.02 | - | 0.80 ± 0.01 | - |
| VI | - | 0.0076 ± 0.0008 | - | - | 0.58 ± 0.01 | 0.23 ± 0.01 |
| VII | - | 0.0062 ± 0.0001 | 0.088 ± 0.006 | 0.027 ± 0.001 | 0.84 ± 0.01 | 0.12 ± 0.01 |
| VIII | 2.9 ± 0.1 | 0.0041 ± 0.0001 | 0.16 ± 0.01 | 0.090 ± 0.008 | 0.058 ± 0.003 | 0.030 ± 0.003 |
| IX | 250 ± 10 | 0.0073 ± 0.0002 | 0.12 ± 0.02 | 0.077 ± 0.002 | 0.072 ± 0.004 | 0.022 ± 0.002 |
| X | - | - | 1.1 ± 0.1 | 0.48 ± 0.05 | 0.17 ± 0.01 | 0.096 ± 0.007 |

The outer layer of birch bark is dominated by betulin, erythrodiol, uvaol, betulinic and oleanolic acids (3.6–250 mg·g$^{-1}$). The content of the components is consistent with the literature data [27,28].

Betulinic acid predominates in licorice root extract. Minor substances (0.0041–0.1 mg·g$^{-1}$) are oleanolic acid, lupeol, β-amyrin, α-amyrin, erythrodiol and betulin. Most triterpenoids in licorice root have been previously reported [27,34,35].

Major triterpenoids of apple peel are betulinic, oleanolic and ursolic acids, as well as uvaol. At the same time, the content of triterpene acids in the peel of "Golden Delicious" apples is higher compared to "Red Delicious". Minor components (0.027–0.26 mg·g$^{-1}$) are lupeol, α-amyrin, betulin and erythrodiol. The presence of most components in apples has been previously reported [27,36–38]. However, the papers [37,38] reported only the sum content of the erythrodiol and uvaol, since it was not possible to separate this pair of substances.

Berry peel extracts differ from apple peel in the presence of the ketone friedelin. The main difference between lingonberry peel and cranberry peel is the increased content of monools. Their total content in the lingonberry peel is 6-fold higher than in the cranberry peel. The major components of berry peels are ursolic and oleanolic acids (3.2–15 mg·g$^{-1}$). The minor components of berries peel are diols. The presence of main triterpenoids has been reported previously [27,28,39–41].

## 3. Materials and Methods

### 3.1. Reagents and Materials

Commercially available standards of ten studied pentacyclic triterpenoids—friedelin (tech. grade), betulinic acid (≥97.0%), oleanolic acid (≥97.0%), ursolic acid (≥90.0%), lupeol (≥90.0%), β-amyrin (≥98.5%), α-amyrin (≥98.5%), erythrodiol (≥97.0%), betulin (≥98.0%), and uvaol (≥95.0%)—were purchased from Sigma-Aldrich (Steinheim, Germany).

HPLC gradient grade methanol and isopropanol (Chimmed, Moscow, Russia), and carbon dioxide (≥99.99%, Kriogen, Moscow, Russia) were used for the preparation of mobile phase. Methanol was also used for analyte solution preparation and PLE extraction of plant materials.

The stock solutions of triterpenoids in methanol (250 mg·L$^{-1}$) were prepared from an accurate sample. Calibration solutions of analytes were prepared by mixing and successive dilutions of the stock solutions with methanol. The solutions were stored in the dark at 4 °C for no more than one week. Stability was checked once a day.

### 3.2. Plant Materials and Extraction

The silver birch (*Betula pendula*) bark, cranberry (*Vaccinium oxycoccos*) and lingonberry (*Vaccinium vitis-idaea*) fruits were collected in the forests of the Arkhangelsk region of Russia in August 2021. Identification of botanical raw materials was carried out according to the herbarium of Northern (Arctic) Federal University. The apple fruits (*Malus domestica* var. *Golden Delicious* and *Malus domestica* var. *Red Delicious*) and licorice (*Glycyrrhiza glabra*) roots were purchased in the retail chain in August 2021. The outer layers of bark, root, and berry and apple peels were separated manually and dried in oven at 50 °C overnight. Dry plant material was grinded (0.5–1 mm) and stored in desiccator over silica gel in dark at room temperature.

Pressurized liquid extraction (PLE) was performed on an ASE-350 accelerated solvent extraction system (Dionex, Sunnyvale, CA, USA) according to a previously developed method [26]. A sample of dry plant material (1.0 g) was extracted with methanol (two extraction cycles of 10 min each) at 100 °C and 100 bar under nitrogen. The resulting extract was dried. The dry extract (1.00 mg) was dissolved in 1000 μL of methanol, dilute with methanol, filtered through a nylon membrane filter (0.22 μm) and subjected to chromatographic analysis.

### 3.3. Supercritical Fluid Chromatography and Mass Spectrometry

The SFC-MS/MS system was used, which consisted of a 3200 QTrap triple quadrupole mass spectrometer (ABSciex, Vaughan, ON, Canada) equipped with APCI (Turbo-V) and APPI (Photospray) ion sources, and an Acquity UPC$^2$ SFC system (Waters, Milford, MA, USA), including four pumps for supplying carbon dioxide and co-solvent, autosampler, column thermostat, and backpressure regulator. Make-up solvent and dopant (in APPI) were pumped using an additional Ultimate 3000 RS HPLC system (Thermo Scientific, Waltham, MA, USA).

Separation was carried out on the following columns: Accucore C30 150 × 2.1 mm, 2.6 μm (Thermo Scientific, Waltham, MA, USA), Luna C18 250 × 4.6 mm, 5.0 μm (Phenomenex, Torrance, CA, USA), Nucleodur C18 Isis 150 × 2.0 mm, 1.8 μm (Macherey-Nagel, Duren, Germany), Nucleodur PolarTec 150 × 3.0 mm, 3.0 μm (Macherey-Nagel, Duren, Germany), Zorbax SB-Aq 150 × 3.0 mm, 3.5 μm (Agilent, Santa Clara, CA, USA) and Acquity UPC$^2$ HSS C18 SB 150 × 3.0 mm, 1.8 μm (Waters, Milford, MA, USA). The volume of the injected sample was 5 μL. Flow rate of make-up solvent (isopropanol) was 0.1 mL·min$^{-1}$. The SFC-MS/MS system was controlled, and data were collected and processed using the Empower 3.0 software package (Waters, Milford, MA, USA).

The parameters of the APCI and APPI ionization sources varied in the ranges 250–450 °C for the temperature of the ion source; 10–20, 10–30, and 10–40 psi for curtain gas, nebulizer gas, and dryer gas, respectively; 2.5–5 mA for nebulizer current (APCI); and 650–1000 V for ion spray voltage (APPI). The parameters of the MRM mode varied in the ranges 10–100 V for declustering potential and collision energy and 1–12 V for entrance potential. Toluene (0.1 mL·min$^{-1}$) was used as a dopant in APPI.

Optimization of PCT separation was carried out by APCI-MS using the SIM (single ion monitoring) mode: I—427 *m/z*; II, III and IV—439 *m/z*; V, VI and VII—409 *m/z*; VIII, IX and X—425 *m/z*. Optimization of the ionization source parameters was also carried out using the SIM mode.

### 3.4. Method Validation

The values of the lower limit of quantification (LOQ) of ten triterpenoids were determined using a signal-to-noise ratio (S/N) criterion of 10. The intra-day precision was estimated at the lowest concentration level (close to LOQ). A series of consecutive chromatographic analyses of the analytes ($n$ = 7) was used. The inter-day precision was determined in the same manner within 48 h ($n$ = 14). The matrix effect and accuracy of PCTs quantification in plant extracts were estimated by spike-recovery test. Three concentration levels of analytes were introduced into licorice root PLE extracts and analyzed ($n$ = 3).

## 4. Conclusions

A fast, accurate, highly sensitive and green method for the analysis of pentacyclic triterpenoids of different types by supercritical fluid chromatography–tandem mass spectrometry with atmospheric pressure chemical ionization was developed and validated. The use of the HSS C18 SB stationary phase and isopropanol as mobile phase modifier allowed the rapid chromatographic separation of ten pentacyclic triterpenoids (friedelin, lupeol, β-amyrin, α-amyrin, betulin, erythrodiol, uvaol, betulinic, oleanolic and ursolic acids) with a mixed retention mechanism with prevailing polar interactions of analytes with silanol groups of the stationary phase. With an analysis time of 7 min, the developed method ensures LOQs in plant biomass extracts of 2.3–20 $\mu g \cdot L^{-1}$. The application of the developed method for the analysis of real objects—outer layer of birch (*Betula pendula*) bark, licorice (*Glycyrrhiza glabra*) root, and lingonberry (*Vaccinium vitis-idaea*), cranberry (*Vaccinium oxycoccos*), apple (*Malus domestica "Golden Delicious"* and *Malus domestica "Red Delicious"*) peels—allowed new data to be obtained on the contents of ten PCTs. The developed method can be used both for the analysis of plant materials, drugs, and biological fluids, as well as for pharmacodynamic and pharmacokinetic studies.

**Supplementary Materials:** The following supporting information can be downloaded at: https://www.mdpi.com/article/10.3390/ph15050629/s1, Table S1: Chromatographic separation parameters of ten PCTs on the stationary phase HSS C18 SB under optimal conditions. Table S2: The results of the evaluation of intraday and inter-day reproducibility analysis of the PCTs on HSS C18 SB stationary phase. Table S3: Matrix effect on the determination of PCTs on HSS C18 SB stationary phase. Figure S1: SFC-APCI-MS chromatogram (SIM mode) of analytes model mixture (250 $\mu g \cdot L^{-1}$ of each compound) obtained on various stationary phases. Figure S2: SFC-APCI-MS/MS chromatogram of analytes model mixture (I—25 $\mu g \cdot L^{-1}$; II, VIII and X—12.5 $\mu g \cdot L^{-1}$; III, IV and IX—6.25 $\mu g \cdot L^{-1}$; V, VI and VII—3.125 $\mu g \cdot L^{-1}$) on HSS C18 SB stationary phase.

**Author Contributions:** Conceptualization, D.I.F. and D.V.O.; methodology, D.I.F., D.V.O., N.V.U. and D.S.K.; validation, D.I.F., D.V.O., I.S.V. and A.V.F.; formal analysis, D.I.F., D.V.O., I.S.V. and A.V.F.; investigation, D.I.F., D.V.O., I.S.V. and A.V.F.; resources, D.I.F. and N.V.U.; writing—original draft preparation, D.I.F. and D.V.O.; writing—review and editing, D.S.K.; visualization, I.S.V., A.V.F. and N.V.U.; supervision, D.S.K.; funding acquisition, D.I.F. All authors have read and agreed to the published version of the manuscript.

**Funding:** This research was funded by Russian Science Foundation, grant number 21-73-00291.

**Institutional Review Board Statement:** Not applicable.

**Informed Consent Statement:** Not applicable.

**Data Availability Statement:** The data is contained in the article and Supplementary Materials.

**Acknowledgments:** This study was performed using an instrumentation of the Core Facility Center "Arktika" of the Lomonosov Northern (Arctic) Federal University.

**Conflicts of Interest:** The authors declare no conflict of interest.

## References

1. Cháirez-Ramírez, M.; Moreno-Jiménez, M.; González-Laredo, R.; Gallegos-Infante, J.; Rocha-Guzmán, N. Lupane-type triterpenes and their anti-cancer activities against most common malignant tumors: A review. *EXCLI J.* **2016**, *15*, 758–771. [CrossRef] [PubMed]
2. Jager, S.; Trojan, H.; Kopp, T.; Laszczyk, M.N.; Scheffler, A. Pentacyclic triterpene distribution in various plants–Rich sources for a new group of multi-potent plant extracts. *Molecules* **2009**, *14*, 2016–2031. [CrossRef] [PubMed]
3. Liu, J. Pharmacology of oleanolic acid and ursolic acid. *J. Ethnopharmacol.* **1995**, *49*, 57–68. [CrossRef]
4. Saleem, M. Lupeol, a novel anti-inflammatory and anti-cancer dietary triterpene. *Cancer Lett.* **2009**, *285*, 109–115. [CrossRef]
5. Alakurtti, S.; Makela, T.; Koskimies, S.; Yli-Kauhaluoma, J. Pharmacological properties of the ubiquitous natural product betulin. *Eur. J. Pharm. Sci.* **2006**, *29*, 1–13. [CrossRef]
6. Somova, L.O.; Nadar, A.; Rammanan, P.; Shode, F.O. Cardiovascular, antihyperlipidemic and antioxidant effects of oleanolic and ursolic acids in experimental hypertension. *Phytomedicine* **2003**, *10*, 115–121. [CrossRef]
7. Mlala, S.; Oyedeji, A.O.; Gondwe, M.; Oyedeji, O.O. Ursolic Acid and Its Derivatives as Bioactive Agents. *Molecules* **2019**, *24*, 2751. [CrossRef]
8. Jesus, J.A.; Lago, J.H.G.; Laurenti, M.D.; Yamamoto, E.S.; Passero, L.F.D. Antimicrobial activity of oleanolic and ursolic acids: An update. *Evid. Based. Complement. Alternat. Med.* **2015**, *2015*, 620472. [CrossRef]

9. Yogeeswari, P.; Sriram, D. Betulinic acid and its derivatives: A review on their biological properties. *Curr. Med. Chem.* **2005**, *12*, 657–666. [CrossRef]
10. Garcia-Oliveira, P.; Otero, P.; Pereira, A.G.; Chamorro, F.; Carpena, M.; Echave, J.; Fraga-Corral, M.; Simal-Gandara, J.; Prieto, M.A. Status and challenges of plant-anticancer compounds in cancer treatment. *Pharmaceuticals* **2021**, *14*, 157. [CrossRef]
11. Yadav, V.R.; Prasad, S.; Sung, B.; Kannappan, R.; Aggarwal, B.B. Targeting inflammatory pathways by triterpenoids for prevention and treatment of cancer. *Toxins* **2010**, *2*, 2428–2466. [CrossRef] [PubMed]
12. Parikh, N.R.; Mandal, A.; Bhatia, D.; Siveen, K.S.; Sethi, G.; Bishayee, A. Oleanane triterpenoids in the prevention and therapy of breast cancer: Current evidence and future perspectives. *Phytochem. Rev.* **2014**, *13*, 793–810. [CrossRef] [PubMed]
13. Cichewicz, R.H.; Kouzi, S.A. Chemistry, biological activity, and chemotherapeutic potential of betulinic acid for the prevention and treatment of cancer and HIV infection. *Med. Res. Rev.* **2004**, *24*, 90–114. [CrossRef]
14. Laszczyk, M.N. Pentacyclic triterpenes of the lupane, oleanane and ursane group as tools in cancer therapy. *Planta Med.* **2009**, *75*, 1549–1560. [CrossRef]
15. Nguyen, H.N.; Ullevig, S.L.; Short, J.D.; Wang, L.; Ahn, Y.J.; Asmis, R. Ursolic acid and related analogues: Triterpenoids with broad health benefits. *Antioxidants* **2021**, *10*, 1161. [CrossRef]
16. Do Nascimento, P.G.G.; Lemos, T.L.G.; Bizerra, A.M.C.; Arriaga, A.M.C.; Ferreira, D.A.; Santiago, G.M.P.; Braz-Filho, R.; Costa, J.G.M. Antibacterial and antioxidant activities of ursolic acid and derivatives. *Molecules* **2014**, *19*, 1317–1327. [CrossRef] [PubMed]
17. Neto, C.C. Cranberry and blueberry: Evidence for protective effects against cancer and vascular diseases. *Mol. Nutr. Food Res.* **2007**, *51*, 652–664. [CrossRef]
18. Neto, C.C. Cranberry and its phytochemicals: A review of in vitro anticancer studies. *J. Nutr.* **2007**, *137*, 186S–193S. [CrossRef]
19. Gerhauser, C. Cancer chemopreventive potential of apples, apple juice, and apple components. *Planta Med.* **2008**, *74*, 1608–1624. [CrossRef]
20. Huang, M.; Lu, J.-J.; Huang, M.-Q.; Bao, J.-L.; Chen, X.-P.; Wang, Y.-T. Terpenoids: Natural products for cancer therapy. *Expert Opin. Investig. Drugs* **2012**, *21*, 1801–1818. [CrossRef]
21. He, X.; Liu, R.H. Phytochemicals of apple peels: Isolation, structure elucidation, and their antiproliferative and antioxidant activities. *J. Agric. Food Chem.* **2008**, *56*, 9905–9910. [CrossRef] [PubMed]
22. Olennikov, D.N.; Vasilieva, A.G.; Chirikova, N.K. Fragaria viridis fruit metabolites: Variation of LC-MS profile and antioxidant potential during ripening and storage. *Pharmaceuticals* **2020**, *13*, 262. [CrossRef] [PubMed]
23. Lachowicz, S.; Oszmiański, J.; Wiśniewski, R. Determination of triterpenoids, carotenoids, chlorophylls, and antioxidant capacity in Allium ursinum L. at different times of harvesting and anatomical parts. *Eur. Food Res. Technol.* **2018**, *244*, 1269–1280. [CrossRef]
24. Anikeenko, E.A.; Rakhmatullina, E.N.; Falev, D.I.; Khoroshev, O.Y.; Ul'yanovskii, N.V.; Kosyakov, D.S. Application of Carbon Matrices to Screening Pentacylic Triterpenoids in Plant Feedstock by MALDI Mass Spectrometry. *J. Anal. Chem.* **2020**, *75*, 1749–1757. [CrossRef]
25. Kosyakov, D.S.; Ul'yanovskii, N.V.; Falev, D.I. Determination of Triterpenoids from Birch Bark by Liquid Chromatography-Tandem Mass Spectrometry. *J. Anal. Chem.* **2014**, *69*, 50–55. [CrossRef]
26. Falev, D.I.; Kosyakov, D.S.; Ul'yanovskii, N.V.; Ovchinnikov, D.V.; Shestakov, S.L. Subcritical extraction of birch bark pentacyclic triterpenes. *Russ. Chem. Bull.* **2017**, *66*, 875–881. [CrossRef]
27. Falev, D.I.; Ul'yanovskii, N.V.; Ovchinnikov, D.V.; Faleva, A.V.; Kosyakov, D.S. Screening and semi-quantitative determination of pentacyclic triterpenoids in plants by liquid chromatography–tandem mass spectrometry in precursor ion scan mode. *Phytochem. Anal.* **2021**, *32*, 252–261. [CrossRef]
28. Falev, D.I.; Kosyakov, D.S.; Ul'yanovskii, N.V.; Ovchinnikov, D.V. Rapid simultaneous determination of pentacyclic triterpenoids by mixed-mode liquid chromatography–tandem mass spectrometry. *J. Chromatogr. A* **2020**, *1609*, 460458. [CrossRef]
29. Laboureur, L.; Ollero, M.; Touboul, D. Lipidomics by supercritical fluid chromatography. *Int. J. Mol. Sci.* **2015**, *16*, 13868–13884. [CrossRef]
30. Tyśkiewicz, K.; Gieysztor, R.; Maziarczyk, I.; Rój, E.; Skalicka-Woźniak, K. Supercritical fluid chromatography with photodiode array detection in the determination of fat-soluble vitamins in hemp seed oil and waste fish oil. *Molecules* **2018**, *23*, 1131. [CrossRef]
31. Ovchinnikov, D.V.; Pokrovskiy, O.I.; Kosyakov, D.S.; Bogolitsyn, K.G.; Ul'yanovskii, N.V.; Falev, D.I. Evaluation of temperature and pressure effects on retention in supercritical fluid chromatography on polar stationary phases. *J. Chromatogr. A* **2020**, *1610*, 460600. [CrossRef] [PubMed]
32. Fanali, C.; D'Orazio, G.; Gentili, A.; Fanali, S. Analysis of enantiomers in products of food interest. *Molecules* **2019**, *24*, 1119. [CrossRef] [PubMed]
33. Lesellier, E.; Destandau, E.; Grigoras, C.; Fougère, L.; Elfakir, C. Fast separation of triterpenoids by supercritical fluid chromatography/evaporative light scattering detector. *J. Chromatogr. A* **2012**, *1268*, 157–165. [CrossRef] [PubMed]
34. Hayashi, H.; Fukui, H.; Tabata, M. Examination of triterpenoids produced by callus and cell suspension cultures of Glycyrrhiza glabra. *Plant Cell Rep.* **1988**, *7*, 508–511. [CrossRef]
35. Xu, J.; Wang, X.; Zhang, H.; Yue, J.; Sun, Y.; Zhang, X.; Zhao, Y. Synthesis of triterpenoid derivatives and their anti-tumor and anti-hepatic fibrosis activities. *Nat. Prod. Res.* **2020**, *34*, 766–772. [CrossRef]
36. Jemmali, Z.; Chartier, A.; Dufresne, C.; Elfakir, C. Optimization of the derivatization protocol of pentacyclic triterpenes prior to their gas chromatography-mass spectrometry analysis in plant extracts. *Talanta* **2016**, *147*, 35–43. [CrossRef]

37. De la Peña Armada, R.; Bronze, M.R.; Matias, A.; Mateos-Aparicio, I. Triterpene-Rich Supercritical $CO_2$ Extracts from Apple By-product Protect Human Keratinocytes Against ROS. *Food Bioprocess Technol.* **2021**, *14*, 909–919. [CrossRef]
38. Poirier, B.C.; Buchanan, D.A.; Rudell, D.R.; Mattheis, J.P. Differential Partitioning of Triterpenes and Triterpene Esters in Apple Peel. *J. Agric. Food Chem.* **2018**, *66*, 1800–1806. [CrossRef]
39. Szakiel, A.; Pa̧czkowski, C.; Koivuniemi, H.; Huttunen, S. Comparison of the triterpenoid content of berries and leaves of lingonberry Vaccinium vitis-idaea from Finland and Poland. *J. Agric. Food Chem.* **2012**, *60*, 4994–5002. [CrossRef]
40. Szakiel, A.; Pączkowski, C.; Pensec, F.; Bertsch, C. Fruit cuticular waxes as a source of biologically active triterpenoids. *Phytochem. Rev.* **2012**, *11*, 263–284. [CrossRef]
41. Klavins, L.; Klavins, M. Cuticular wax composition of wild and cultivated northern berries. *Foods* **2020**, *9*, 587. [CrossRef] [PubMed]

*pharmaceuticals*

*Article*

# Application of Liquid Chromatography Coupled to Mass Spectrometry in Quality Assessment of Dietary Supplements—A Case Study of Tryptophan Supplements: Release Assay, Targeted and Untargeted Studies

**Krzysztof Adam Stępień and Joanna Giebułtowicz ***

Department of Bioanalysis and Drugs Analysis, Faculty of Pharmacy, Medical University of Warsaw, 1 Banacha, 02-097 Warsaw, Poland; krzysztof.stepien@wum.edu.pl
* Correspondence: joanna.giebultowicz@wum.edu.pl; Tel.: +48-22-572-0949

**Abstract:** Dietary supplements are widely consumed in the EU and the USA. Based on their similarity to pharmaceuticals, consumers mistakenly believe that dietary supplements have also been approved for safety and efficacy. However, in the absence of mandatory testing, data on supplement quality is scarce. Thus, we applied liquid chromatography coupled with tandem mass spectrometry to analyse the quality of dietary supplements containing tryptophan (Trp). We examined 22 supplements in tablets or capsules, produced in the USA, Great Britain, Germany, France, Czech Republic, and Poland. Trp release, crucial for bioavailability and efficiency, was assessed. Additionally, we performed a qualitative analysis of the main ingredient and screened for contaminants. Among the contaminants, we detected Trp's metabolites, condensation products of Trp and carbonyl compounds, Trp degradation products, degradation products of kynurenine, and other contaminants such as glucosamine and melatonin. The main ingredient content was in the range of 55–100% in capsules and 69–87% in tablets. Surprisingly, almost no Trp release was noted from some supplements. Our study confirms the need to advance research on supplements. We believe that the high-quality analysis of supplements based on reliable analytical techniques will be an important contribution to the discussion on the regulatory framework of these products.

**Keywords:** dietary supplement; food supplement analysis; LC-MS/MS; release test; quality control; food composition

**Citation:** Stępień, K.A.; Giebułtowicz, J. Application of Liquid Chromatography Coupled to Mass Spectrometry in Quality Assessment of Dietary Supplements—A Case Study of Tryptophan Supplements: Release Assay, Targeted and Untargeted Studies. *Pharmaceuticals* **2022**, *15*, 448. https://doi.org/10.3390/ph15040448

Academic Editor: Daniela De Vita

Received: 17 February 2022
Accepted: 31 March 2022
Published: 4 April 2022



## 1. Introduction

The consumption of dietary supplements is increasing globally, due to low prices, broad prescription-free distribution, and a common belief in efficacy and safety [1]. In the EU, they are classified as food and contain ingredients with nutritional or physiological effects [2]. Similarly, in the USA, dietary supplements are also classified as food and contain amino acids, herbal substances, vitamins, minerals, and enzymes [3]. Although they are foods, they are sold in typically pharmaceutical dosage forms such as tablets, capsules, sachets, and others designed to be taken in small and defined unit quantities. Based on their similarity to pharmaceuticals, consumers mistakenly believe that dietary supplements have also been approved for safety and efficacy before marketing. However, in the absence of mandatory quality testing, data on supplement quality is scarce. When the data do appear, they indicate some issues such as the presence of contaminants [4,5], the content of the main ingredient lower than the declared one [6,7], or low release from the formulation [8]. The most commonly described contaminants are heavy metals [9], anabolic steroids in preparations for athletes [10], and dioxins in dietary supplements containing fish oil [11]. The amount of the main ingredient in the dietary supplements was only examined for melatonin supplements [1] and supplements containing eicosapentaenoic acid (EPA) and docosahexaenoic acid (DHA) [12]. Release of the main constituent from the formulation was conducted for formulations with calcium carbonate [13], melatonin [14], folic acid [15,16], iron [17], triiodothyronine [18], trans-resveratrol [19] and lutein [8]. The paucity of data on supplement quality is disadvantageous [20]. Therefore, the need for international collaboration to advance the knowledge on supplements has recently been emphasized. Not only the efficacy but also the quality of dietary supplements should be evaluated, which is essential to improve the regulatory framework [21].

Depression, the most common mental illness, is one of the most common disorders for which supplementation is also used. This disorder affects over 300 million people around the world, of different ages, and in all communities [22]. The prevalence of the illness increases with age and is more common in women and people with higher education [23]. Depression is treated pharmacologically, often with moderate efficacy [24]. This is why some people also use supplementation. There are numerous mood-enhancing supplements on the market, mainly containing the neutral amino acid tryptophan (Trp) [25]. It is one of the 20 L-amino acids incorporated into proteins during mRNA translation [26] and a precursor of serotonin (5-hydroxytryptamine), niacin (niacinamide), and melatonin. Trp enters the kynurenine pathway and is a precursor of the coenzyme NAD(P)+ [27]. It is an exogenous amino acid whose [28] deficiency leads to the insufficient synthesis of the neurotransmitter serotonin, which worsens mood. Lower levels of Trp in peripheral blood have been confirmed in patients suffering from depression. Trp supplementation significantly improved the symptoms of the disease [29]. Trp has a positive effect on mood, cognitive functions [30], sleep [31], and a decrease or maintenance of a healthy weight [32]. Response to supplementation is individual and may be influenced by genetic factors [33]. The potential efficacy of Trp in depression patients led us to select supplements with this ingredient to evaluate their quality.

Western diet usually contains about 0.5 g of Trp per day. However, only 2–3% of this amount enters the brain for conversion, via 5-hydroxytryptophan, to serotonin. It is due to extensive metabolism and competition with other long-chain neutral amino acids, e.g., histidine, isoleucine, leucine, methionine, phenylalanine [34]. Trp is an ingredient of dietary supplements [35]. Sometimes, during Trp supplementation, dose-independent side effects occur, i.e., tremors, dry mouth, mild nausea, dizziness. Contaminants, present in commercially available Trp for nutritional use (feed-grade Trp in raw materials of different manufacturers), were investigated [36]. The contaminants detected and identified in commercially available Trp sources were the known metabolites of this amino acid, oxidation products of Trp, condensation products of Trp with carbonyl compounds [37]. So far, there are little data on the quality of Trp preparations. Additionally, no data have been published on the content and release of Trp from supplements, and most studies on supplements rely on simple analytical techniques.

This study aimed to apply liquid chromatography coupled with mass spectrometry as a highly reliable analytical technique to evaluate the quality of dietary supplements containing Trp in tablets or capsules ($n$ = 22) produced in the USA, UK, Germany, France, Czech Republic, and Poland. This evaluation was performed by (i) assessment of Trp release, a key parameter for bioavailability and efficacy, (ii) qualitative analysis of the main ingredient, and (iii) screening for contaminants. We believe that a high-quality analysis of supplements will be an important contribution to the discussion of the regulatory framework for these products and that the new analytical approach will have broad applicability in the assessment of supplement quality.

## 2. Results and Discussion

### 2.1. Tentative Contaminants Present in Trp Supplements

In addition to Trp, twenty-two compounds were detected in the analysed supplements in the range of 0.02% to 43.89% of the main ingredient area (Table 1, Figure A1). Their molecular formula, retention time, experimental and theoretical mass, fragmentation, and tentative names are presented in Table 1. None of these compounds was listed on the package as a component of the supplement. Detected compounds were classified into five groups: (A) Trp's metabolites, (B) condensation products of Trp and carbonyl compounds, (C) Trp degradation products, (D) degradation products of kynurenine and (E) other contaminants.

The first group (group A) includes products of the main metabolic pathways of Trp: anthranilic acid (**I4**), indole-3-acetaldehyde (**I7**), indole acetic acid (**I9**), 5-hydroxyTrp (**I16**), formylkynurenine (**I19**) (Figure 1).

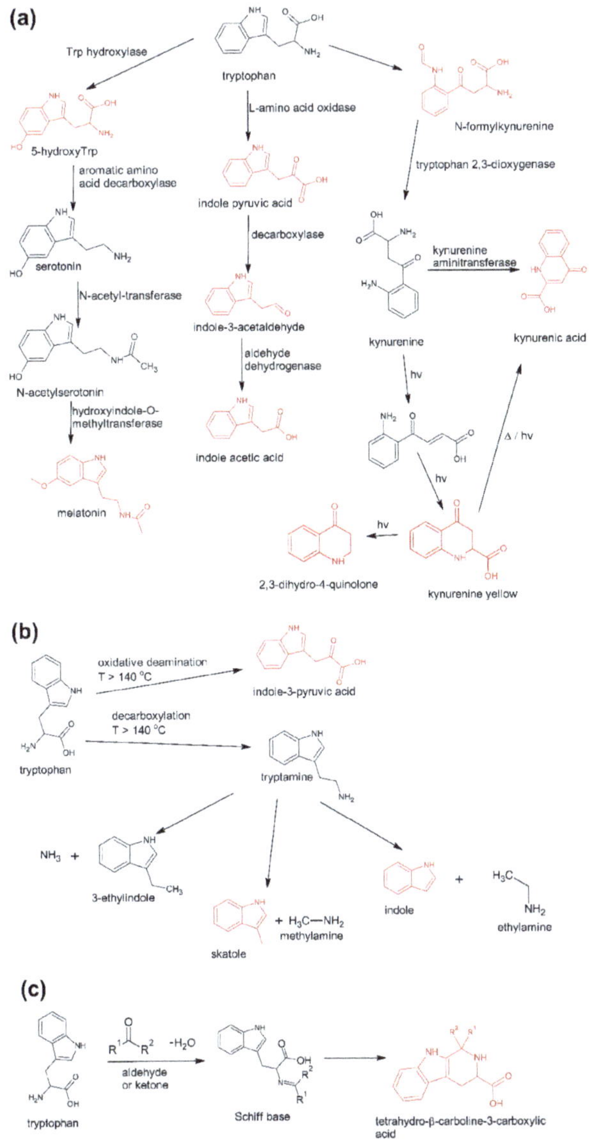


**Figure 1.** Pathways reasoning for the presence of specific contaminants in Trp supplements. (**a**) Major metabolic pathways downstream of Trp (**b**), Trp degradation products formed after exclusive exposure to heat (**c**), and reaction products of Trp with aldehydes and ketones. Compounds marked in red were detected in this study [37].

**Table 1.** Tentative identification of contaminants found in Trp supplements and their MS parameters.

| Code | Formula | Neutral Mass Calculated from the Formula [Da] | Neutral Mass Calculated from the Measured *m/z* [Da] | ΔMass [ppm] | RT [min] | Identification Confidence Level | Fragments [*m/z*] | Dietary Supplements Containing Contaminant (% of the Analysed) | Tentative Name | % of the Main Ingredient Area |
|---|---|---|---|---|---|---|---|---|---|---|
| **I1** | C8H7N | 117.05785 | 117.05792 | 0.6 | 5.0 | 3 | 91.05414 | All, (100%) | Indole | 23.19 |
| **I2** | C9H9N | 131.07350 | 131.07350 | 0.0 | 5.0 | 2 | 117.06720; 130.06493; 131.07260 | All (100%) | Skatole | 6.68 |
| **I3** | C8H7NO | 133.05276 | 133.05243 | 2.5 | 3.1 | 2 | 79.05412; 106.06493 | C1; C2; C8; C9; T1; T3; T4; T6; T7; T8; T9; T10 (55%) | Oxindole | 2.41 |
| **I4** | C7H7NO2 | 137.04768 | 137.04766 | 0.2 | 3.1 | 2 | 92.04936; 94.06511; 110.06001 | C8; C9; T1; T3; T5; T6; T7; T8; T9; T10 (45%) | Anthranilic acid | 0.42 |
| **I5** | C9H7NO | 145.05276 | 145.05284 | 0.6 | 5.0 | 2 | 91.05412; 117.05762; 118.06503 | All (100%) | 3-formy-lindole | 43.89 |
| **I6** | C9H9NO | 147.06841 | 147.06827 | 1.0 | 3.9 | 3 | 120.04422; 130.03930; 130.06487 | T6 (4.5%) | 2,3-dihydro-4-quinolone | 0.02 |
| **I7** | C10H9NO | 159.06841 | 159.06845 | 0.3 | 5.0 | 2 | 130.06490; 132.080610; 142.06616 | All (100%) | Indole-3-acetaldehyde | 2.50 |
| **I8** | C10H10N2O | 174.07931 | 174.07924 | 0.4 | 3.9 | 3 | 132.04401; 147.09129; 157.07565 | C1, C2, C3, C4, C5, C6, C7, C8, C10, C11, C12, T1, T2, T3, T4, T5, T6, T7, T8, T9, T10 (95%) | 1-Phenyl-3-methyl-5-pyrazolone | 0.20 |
| **I9** | C10H9NO2 | 175.06333 | 175.06326 | 0.4 | 3.9 | 2 | 130.06479; 146.05980; 158.05960 | C7, C11, T1, T2, T3, T4, T6, T7, T8, T8 (45%) | Indole acetic acid | 0.15 |
| **I10** | C6H13NO5 | 179.07937 | 179.07938 | 0.1 | 3.1 | 1 | 127.03854; 144.06540; 145.04945 | C8, T3 (9%) | Glucosamine | 0.07 |

**Table 1.** *Cont.*

| Code | Formula | Neutral Mass Calculated from the Formula [Da] | Neutral Mass Calculated from the Measured *m/z* [Da] | ΔMass [ppm] | RT [min] | Identification Confidence Level | Fragments [*m/z*] | Dietary Supplements Containing Contaminant (% of the Analysed) | Tentative Name | % of the Main Ingredient Area |
|---|---|---|---|---|---|---|---|---|---|---|
| **I11** | C10H7NO3 | 189.04259 | 189.04239 | 1.1 | 8.5 | 2 | 162.05463; 172.03886; 173.04672 | T4, T6 (9%) | Kynurenic acid | 2.33 |
| **I12** | C10H9NO3 | 191.05824 | 191.05811 | 0.7 | 4.0 | 3 | 150.05463; 164.07000; 174.05472 | C4, C5, C7, C8, C9, C10, C11, C12, T1, T2, T3, T4, T6, T7, T8, T9, T10 (77%) | Kynurenine yellow | 0.03 |
| **I13** | C11H10N2O2 | 202.07423 | 202.07414 | 0.4 | 3.9 | 3 | 130.06480; 157.07570; 185.07106 | C1, C2, C3, C5, C7, C8, C9, C10, C12, T1, T2, T3, T4, T5, T6, T7, T10 (77%) | Unsaturated Trp | 0.38 |
| **I14** | C11H9NO3 | 203.05824 | 203.05849 | 1.2 | 3.9 | 3 | 160.07555; 176.07080; 186.05499 | C1, C2, C3, C4, C5, C6, C7, C8, C9, C10, C11, C12, T2, T3, T5, T6, T7, T8, T9, T10 (91%) | Indole pyruvic acid | 0.03 |
| **I15** | C12H12N2O2 | 216.08988 | 216.08986 | 0.1 | 8.0 | 3 | 171.09120; 173.10748; 188.07051 | C2, C4, C6, C7, C8, C9, C11, C12, T1, T2, T3, T4, T5, T6, T7, T8, T9, T10 (82%) | Tetrahydro-β-carboline-3-carboxylic acid | 0.23 |
| **I16** | C11H12N2O3 | 220.08479 | 220.08467 | 0.5 | 3.9 | 2 | 130.06488; 158.05981; 175.08636 | All (100%) | 5-hydroxyTrp | 0.18 |
| **I17** | C13H14N2O2 | 230.10553 | 230.10533 | 0.9 | 8.2 | 3 | 168.08034; 188.07034; 214.08580 | C2, C8, C9, C11, T1, T2, T3, T4, T5, T6, T7, T8, T9 (59%) | 1-methyl-tetrahydro-β-carboline-3-carboxylic acid | 0.02 |
| **I18** | C13H16N2O2 | 232.12118 | 232.12099 | 0.8 | 9.1 | 1 | 174.09070; 204.10060; 216.10116 | C9 (4.5%) | Melatonin | 1.62 |

**Table 1.** *Cont.*

| Code | Formula | Neutral Mass Calculated from the Formula [Da] | Neutral Mass Calculated from the Measured *m/z* [Da] | ΔMass [ppm] | RT [min] | Identification Confidence Level | Fragments [*m/z*] | Dietary Supplements Containing Contaminant (% of the Analysed) | Tentative Name | % of the Main Ingredient Area |
|---|---|---|---|---|---|---|---|---|---|---|
| **I19** | C11H12N2O4 | 236.07971 | 236.07932 | 1.6 | 3.9 | 3 | 146.05975; 173.06980; 203.08099 | All (100%) | n-formylky-nurenine | 0.49 |
| **I20** | C20H19N3O2 | 333.14773 | 333.14739 | 1.0 | 8.8 | 3 | 188.07037; 205.09702; 217.09743 | C1, C2, C3, C4, C5, C6, C7, C8, C9, C10, C11, T1, T2, T3, T4, T5, T6, T7, T8, T9, T10 (95%) | 2-(3-Methyle-neindole)Trp | 0.02 |
| **I21** | C21H19N3O2 | 345.14773 | 345.14745 | 0.8 | 9.0 | 3 | 283.12204; 285.13794; 329.12610 | C1, C2, C3, C4, C5, C6, C7, C8, C9, C10, C11, T1, T2, T3, T4, T5, T6, T7, T8, T9, T10 (95%) | 1-(3-methylene-indole)-tetrahydro-β-carboline -3-carboxylic acid | 0.10 |
| **I22** | C22H23N3O4 | 393.16886 | 393.16878 | 0.2 | 8.5 | 3 | 251.31799; 277.11810; 358.15448 | C1, C2, C3, C4, C5, C6, C7, C8, C9, C10, C11, T1, T2, T3, T4, T5, T6, T7, T8, T9, T10 (95%) | 1-(2-Trp)-1-(3-indole) propane diol | 0.03 |

These compounds are formed during the fermentation of Trp in biotechnological production through the activity of Trp-degrading enzymes. The second group (group B) of contaminants corresponds to the condensation of Trp with carbonyl compounds: tetrahydro-β-carboline-3-carboxylic acid (**I15**) (condensation with formaldehyde), 1-methyl-tetrahydro-β-carboline-3-carboxylic acid (**I17**) (condensation with acetaldehyde), 1-(3-methyleneindole)-tetrahydro-β-carboline-3-carboxylic acid (**I21**) (condensation with indole-3-acetaldehyde (**I7**)). Reactions of tryptophan with aldehydes/ketones to form tetrahydro-beta-carbolines (tetHβCs), known as the Pictet–Spengler reaction, is one of the most common reactions of tryptophan with organic compounds. The transformation is usually acid-catalysed and takes place at low pH and high temperatures [38]. Highly reactive aldehydes are generated during the fermentation processes, thus tetHβCs may be detected in any biotechnologically derived Trp [39]. In the summary, the first and second groups of contaminants are associated with the process of Trp production using fermentation. The third group (group C) contains indole (**I1**), skatole (**I2**), oxindole (**I3**), 3-formylindole (**I5**), unsaturated Trp (**I13**), indole pyruvic acid (**I14**), 2-(3-Methyleneindole)Trp (**I20**), 1-(2-Trp)-1-(3-indole)propanediol (**I22**), and originate from Trp degradation [37]. Indole (**I1**), skatole (**I2**), and indole pyruvic acid (**I14**) are Trp degradation products formed after exclusive exposure to heat. At temperatures above 140 °C, decarboxylation and oxidative deamination of Trp occurs, forming tryptamine and indole pyruvic acid (**I14**). Tryptamine can degrade further, to form a possible product: indole (**I1**) or skatole (**I2**) [39]. Additionally, contaminants from this group may be the precursors in the Trp production process as chemical synthesis (indole (**I1**), 3-formylindole (**I5**)), enzymatic synthesis (indole (**I1**)), biotechnological synthesis (indole (**I1**), anthranilic acid (**I4**)). The next group of contaminants (group D) is degradation products of kynurenine following irradiation and heat: 2,3-dihydro-4-quinolone (**I6**), kynurenic acid (**I11**), kynurenine yellow (**I12**). Thermal and UV radiation cause a cascade of reactions of kynurenine, and it transforms to yield kynurenine yellow (**I12**) and 4-quinolone. Kynurenine yellow (**I12**) can react further, undergoing either oxidative decarboxylation to also afford 4-quinolone or oxidation to kynurenic acid (**I11**) [40]. Compounds from the last group (group E) are probably accidental contaminants related to production conditions, packaging method or quality, transport conditions. The contaminants include 1-phenyl-3-methyl-5-pyrazolone (**I8**), glucosamine (**I10**), melatonin (**I18**). The properties of some of them can be found in the literature. Glucosamine (**I10**) is used in the treatment of osteoarthritis [41]. Melatonin (**I18**) is centrally produced by the pineal gland and directly released in the blood, acting as a hormone. In mammals, yeast, and bacteria, melatonin (**I18**) is synthesized from tryptophan. Melatonin (**I18**) has a lot of functions: circadian and seasonal timing of organism; sleep and wakefulness cycle; endocrine functions, such as energy metabolism, glycaemic control, blood lipid profile and reproduction [42]. Glucosamine (**I10**) and melatonin (**I18**) are ingredients of many dietary supplements. Manufacturers of C8 and T3 (where we detected glucosamine (**I10**)) produce also dietary supplements containing glucosamine (**I10**). Thus, its presence (**I10**) in C8 and T3 may be the result of the insufficient purification (e.g., washing) before the manufacturing process. The same conclusion can be made in case of melatonin contaminant (**I18**).

To better visualize the results the heat map was prepared (Figure 2). We can observe the following:

a) Dietary supplements in capsules contained mainly contaminants from group C (Trp degradation products), which may indicate that Trp was obtained by chemical synthesis;
b) Dietary supplements in tablets contained mainly contaminants belonging to groups A (Trp's metabolites) and B (condensation products of Trp and carbonyls), which may indicate that Trp was obtained by biotechnology;
c) Trp from C1 and C3 dietary supplements might be produced by the same manufacturer. The supplements contained the same contaminants (difference in **I3**—Trp degradation product, which may be related to different storage conditions);

d) Trp from C6 and C11 dietary supplements were produced by the same manufacturer, supplements contained the same contaminants (difference **I17**—condensation product of Trp and carbonyls);

e) Trp from C5 and C10 dietary supplements were produced by the same manufacturer, supplements contained similar contaminants, the differentiating contaminants were classified as Trp degradation products and can be generated during supplement storage.

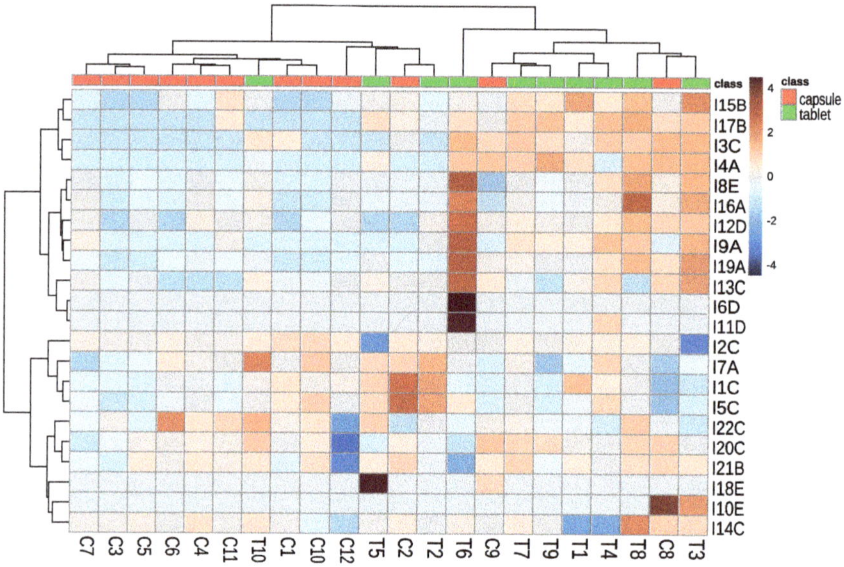


**Figure 2.** Clustering result of the tested supplements and detected contaminants (using Euclidean distance and clustering algorithm using Ward's method). The level of contaminant is presented as a heatmap (red colour indicates higher concentration and blue colour indicates lower concentration than the average) T—tablet, C—capsule, I—contaminant (with the name of a group of contaminants, i.e., A—Trp's metabolites, B—condensation products of Trp and carbonyls, C—Trp degradation products, D—degradation products of kynurenine, E—other contaminants).

To summarize, twenty-two compounds were detected in the analysed supplements in the range of 0.02% to 43.89% of the main ingredient area. Among the contaminants, there were Trp's metabolites, condensation products of Trp and carbonyl compounds, Trp degradation products, degradation products of kynurenine, and other contaminants. Some of Trp's contaminants have been already described in Trp raw material of different manufacturers [39], and melatonin supplements [1]. Melatonin can be synthesized from tryptophan by yeast and bacteria, so the occurrence of the contaminants was expected. The biological effect of Trp-related contaminants is unknown. Some Trp degradation products can impact cellular metabolism. **I12** a degradation product of Trp was shown to induce apoptosis in a human natural killer cell line. **I15** and **I17** act as antioxidants and free radical scavengers. However, the dose of **I12**, **I15**, **I17** needed to have a specific effect on cellular metabolism is unknown. The contaminants were present rather in small amounts, so they may not cause significant side effects [37].

Contamination can occur accidentally, due to poor manufacturing practices or contaminants originating from the supplement ingredients, or intentionally being added by manufacturers. The first group covers heavy metals [9] or substances found in raw materials, e.g., herbicides [43], insecticides [44], mycotoxins [45], and dioxins [11]. All detected contaminants in our study were from this group. Most of them were generated during manufacturing, under storage or transport of supplements/Trp, but some were found in the preparation by accident. Heavy metal analysis was not performed because it requires other analytical techniques such as ICP (inductively coupled plasma) or ASA (atomic absorption analysis). Moreover, these contaminants are mainly detected in herbal-based dietary supplements. Similarly, targeted screening for pesticides and mycotoxins (which were detected in herbal formulations), dioxins (detected in fish oil formulations), cyanobacterial neurotoxins (detected in shark cartilage) and microcystins (detected in algae) because they were not warranted, was not conducted in the study. For instance, pesticides were previously detected in supplements with Ginkgo [44,46] and Ginseng [47,48], whereas mycotoxins were in supplements with Ginkgo and grapes (Table A1). Many of these compounds require targeted screening as well as isolation and enrichment from the complex matrix to obtain a reliable signal [49]. The isolation methods include solid-phase extraction [50], dispersive solid-phase extraction [51], liquid-phase microextraction [52], microwave-assisted extraction [53], microwave-assisted saponification combined with simultaneous unsaponifiable extraction [54]. In our study, a simple extraction was performed without enrichment.

The second group of contaminants includes substances that are prohibited in dietary supplements and are intentionally used by the manufacturer to enhance the observed effect (anabolic steroids [10], hypoglycemic drugs [55], drugs used in potency disorders [56], weight loss products [57]). For Trp supplements, we screened for antidepressants because Trp is often used for depression [28], but no such substances were detected.

The liquid chromatography with mass spectrometry used in this experiment in the untargeted analysis is one of the most frequently used for that purpose. The increased bioavailability of high-resolution instruments improved the detection and identification of compounds in food including dietary supplements. However, confidence in these identifications varies between studies and substances, since it is not always possible or even meaningful to synthesize each substance or confirm them via complementary methods [58]. Thus, we applied the confidence identification level for our data (Table 1). To minimize the risk of false-positive identification it is recommended to search dedicated "small size" MS databases including compounds with a realistic probability to be observed [59]. In our case, the database consisted of degradation products of Trp was used. To decrease further the risk of false-positive identification, all detected compounds were fragmented to achieve a confidence level of at least 3. However, for unexpected compounds such as melatonin or glucosamine, we confirmed the structure with the reference standards. The differences in retention times of these compounds in the samples and reference standard were 0.01 min for glucosamine and 0.03 min for melatonin. The isotopic and fragmentation patterns were similar. The fragmentations are shown in Figure 3.

### 2.2. Determination of Trp in Dietary Supplements

Following the Polish Pharmacopoeia VI, the content of an active substance in tablets or capsules should not exceed the following: (1) ±10% for units with the declared active substance content below 100 mg or (2) ±5% for units with the declared content of the active substance of 100 mg and above. These requirements apply to pharmaceuticals. Due to the lack of specific guidelines for dietary supplements and the fact that dietary supplements appear in the same form as drugs, the same criteria for Trp content were adopted in this study. Therefore, none of the formulations contained the amount of Trp declared by the manufacturer (Table 2), i.e., it was not within 90–110% in each tablet or capsule. The lowest (55% of the declared content) Trp content was in supplement C6, followed by C8 (60%) and T5 (69%). The amount of Trp was within the range of 70–79% in nine supplements and 80–90% in the other nine supplements. The amount of Trp ranged from 70–79% in nine supplements and 80–90% in the other nine supplements. The low Trp content may be due to the lower amount of active ingredients used in production. The highest average Trp content (i.e., 100.45% of the claimed content) was observed in supplement C2. However, the amount of Trp in each capsule varied significantly and ranged from 174 to 251 mg/unit (CV = 19%), indicating improper mixing of the capsule mass. The concentration of the main ingredient in C6 and T2 supplements also had a high coefficient of variation: C6

(CV = 32%), T2 (CV = 35%), but the average amount of Trp was 55% and 87% of the claimed content, respectively. In these cases, both the wrong amount of active ingredient used and improper mixing of the tablet or capsule mass during the manufacturing process may be the reason for inadequate quality.

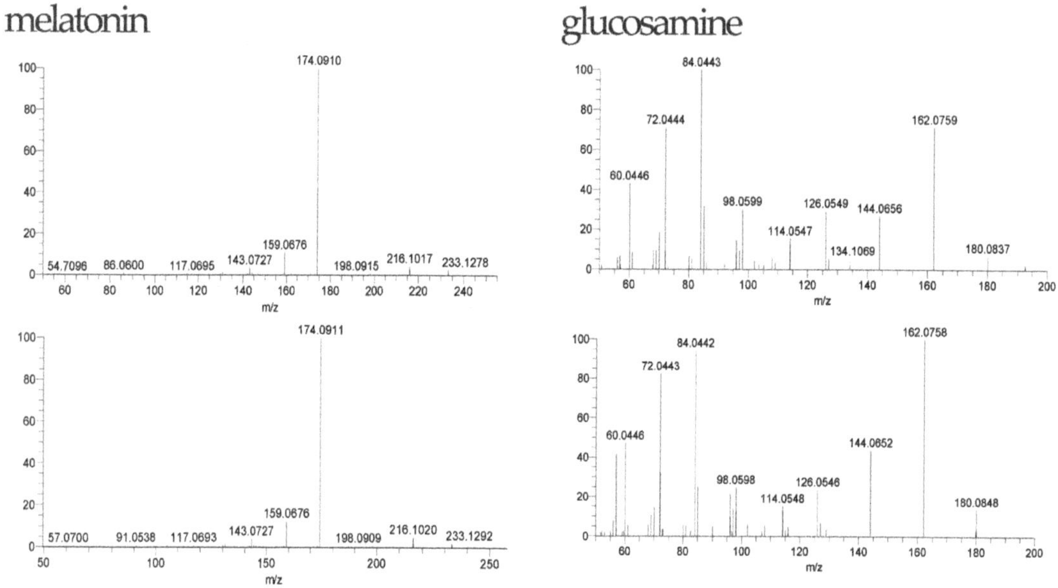


**Figure 3.** Glucosamine and melatonin fragmentation pattern in samples (**top**) and reference standards (**down**).

**Table 2.** Content of Trp in the dietary supplements (maximum error value above 40 was bolded).

| Code | Dosage Form | Source | Declared Trp Content [mg/unit] | Determined Trp Content [mg/unit] [a] | Maximum Error [%] |
|---|---|---|---|---|---|
| C1 | capsule | United Kingdom | 250 | 205 (CV = 5.0%) | −21 |
| C2 [b] | capsule | France | 220 | 221 (CV = 19%) | −21 |
| C3 | capsule | United States | 500 | 398 (CV = 4.1%) | −24 |
| C4 | capsule | Poland | 500 | 368 (CV = 4.1%) | −29 |
| C5 | capsule | Poland | 500 | 415 (CV = 13%) | −27 |
| C6 | capsule | No label | 500 | 277 (CV = 32%) | **−74** |
| C7 | capsule | Czech Republic | 160 | 143 (CV = 16%) | −29 |
| C8 | capsule | Germany | 50 | 29.8 (CV = 6.6%) | **−47** |
| C9 | capsule | Poland | 100 | 77.3 (CV = 8.4%) | −32 |
| C10 | capsule | United Kingdom | 500 | 432 (CV = 5.4%) | −19 |
| C11 | capsule | United States | 500 | 350 (CV = 14%) | **−44** |
| C12 | capsule | United States | 500 | 443 (CV = 8.7%) | −20 |
| T1 | tablet | Poland | 100 | 72.6 (CV = 10%) | −32 |
| T2 | tablet | United States | 1000 | 870 (CV = 35%) | **−48** |
| T3 | tablet | Poland | 40 | 32.6 (CV = 14%) | −27 |

**Table 2.** *Cont.*

| Code | Dosage Form | Source | Declared Trp Content [mg/unit] | Determined Trp Content [mg/unit] [a] | Maximum Error [%] |
|---|---|---|---|---|---|
| **T4** | tablet | Poland | 100 | 71.4 (CV = 7.0%) | −33 |
| **T5** | tablet | Poland | 167 | 115.5 (CV = 7.1%) | −36 |
| **T6** | tablet | Poland | 50 | 37.0 (CV = 9.0%) | −33 |
| **T7** | tablet | Poland | 50 | 41.4 (CV = 8.1%) | −24 |
| **T8** | tablet | Poland | 50 | 41.1 (CV = 16%) | −29 |
| **T9** | tablet | Poland | 50 | 37.1 (CV = 3.6%) | −29 |
| **T10** | tablet | No label | 200 | 155 (CV = 15%) | −35 |

CV—coefficient of variation; [a]—mean (standard deviation $n = 3$); [b]—three capsules were analysed, results (251 mg, 239 mg; 174 mg).

Inconsistency between the declared and determined content of the main ingredient has been previously reported for melatonin supplements [1] and lutein [8]. None of the lutein supplements ($n = 10$) and 41% of the melatonin supplements ($n = 17$) met our criteria. However, it is not clear whether the melatonin or lutein content was evenly distributed among the units. Therefore, no conclusions could be drawn regarding quality.

Content uniformity is an important critical quality attribute. High variability in active ingredient content can be caused by the following: improper particle distribution (e.g., agglomeration); poor macro- and microblending at the powder mixing stage; loss of a component (e.g., due to adsorption to the equipment surface); thief sampling and analytical errors; segregation of well-mixed blends during powder transfer, handling or further operations [60]. The controlling of all this process is required. In the case of pharmaceuticals available on the market, content uniformity is not likely to occur [61,62]. The only study showing the problem with this attribute concern tablets splitting [63]. However, a large number of articles on content uniformity and the ways of continuous monitoring tablet content uniformity [64] suggest that it is a difficult task to achieve.

For supplements with Trp, we observed only a slightly higher level of the active ingredient for capsules (79%) than for tablets (77%). Similar results were noted for melatonin (capsules—91%, tablets—87%) [1] and lutein (capsules—122%, tablets—42%). The supplements with lutein from Brazil (e.g., 0.12% or 135%) had lower quality than those from the USA (112%, 113%) [8].

### 2.3. Dissolution Test for Trp Tablets and Capsules

The Food and Drug Administration provides guidelines for drug testing. According to the dissolution test requirements, the active ingredient should release from the immediate-release oral solid drug at least 80% of its claimed content after 30 min of the release test [65].

Dietary supplements do not have dedicated guidelines for dissolution testing. Therefore, the same criteria for the dissolution test were used in this study. Trp release higher than 80% was determined for supplements C6, C7, C10 at pH 1.2 and supplement T9, T10 at pH 6.8. (Table 3, Figure 4). Supplement C11 had the lowest release (1.22%) at gastric pH (pH 1.2), while at pH 6.8 the release reached 60.2%. Supplements C2 and C3 were characterized by a release of no more than 5% Trp at both pHs (1.2 and 6.8). Thus, only up to 5% of the claimed dose of Trp could be absorbed from these supplements across biological membranes to produce a physiological effect (Figure 4a). Trp release between 10 and 20% regardless of dissolution medium was determined for C8 and between 20% and 30% for T3 and T7. In summary, Trp release for 10 of the 22 supplements was determined between 1.22 and 59.9% at both pHs (Figure 4).

**Table 3.** Comparison of the amount of Trp determined and released from tablets and capsules in two pH (gastric, pH = 1.2 and intestinal, pH = 6.8) with the expanded uncertainty.

| Code | The Average Percentage of Trp Amount Released from a Dosage Form (Standard Deviation $n$ = 6) | | Expanded Uncertainty Parameters | | | | | |
|---|---|---|---|---|---|---|---|---|
| | | | pH 1.2 | | | pH 6.8 | | |
| | pH 1.2 | pH 6.8 | $\|x_1 - x_2\|$ | $U(x_1 - x_2)$ | Equal [a] | $\|x_1 - x_2\|$ | $U(x_1 - x_2)$ | Equal [a] |
| C1 | 60 (11) | 66.0 (8.1) | 55.38 | 25.29 | No | 39.99 | 136.07 | Yes |
| C2 | 2.65 (0.55) | 2.3 (2.2) | 215.43 | 48.12 | No | 216.31 | 8.53 | No |
| C3 | 3.08 (0.76) | 4.8 (2.6) | 382.17 | 19.07 | No | 373.66 | 57.22 | No |
| C4 | 53 (13) | 41.2 (3.8) | 100.87 | 52.04 | No | 162.59 | 117.62 | No |
| C5 | 36.2 (7.5) | 17.9 (3.7) | 234.17 | 68.74 | No | 325.37 | 113.25 | No |
| C6 | 84.3 (8.3) | 76.4 (2.2) | 144.78 | 187.05 | Yes | 105.26 | 233.65 | Yes |
| C7 | 81 (18) | 40.6 (3.3) | 13.18 | 37.53 | Yes | 77.73 | 9.87 | No |
| C8 | 17.9 (1.8) | 15.58 (0.70) | 20.81 | 3.89 | No | 21.98 | 0.29 | No |
| C9 | 75.5 (3.1) | 46.6 (6.6) | 1.75 | 10.06 | Yes | 30.65 | 15.13 | No |
| C10 | 90.4 (9.4) | 68.9 (7.3) | 19.93 | 49.18 | Yes | 87.62 | 435.58 | Yes |
| C11 | 1.22 (0.31) | 60.2 (8.1) | 344.27 | 80.11 | No | 49.64 | 541.99 | Yes |
| C12 | 47.7 (5.8) | 18.0 (1.5) | 204.17 | 55.25 | No | 352.75 | 17.95 | No |
| T1 | 71.9 (9.1) | 78.0 (5.1) | 0.7 | 11.42 | Yes | 5.41 | 9.64 | Yes |
| T2 | 77.9 (5.4) | 76.0 (5.7) | 91.2 | 355.46 | Yes | 109.37 | 1071.86 | Yes |
| T3 | 27 (16) | 21.3 (6.9) | 21.9 | 7.22 | No | 24.12 | 3.37 | No |
| T4 | 54.1 (12.2) | 79.3 (8.8) | 17.3 | 11.55 | No | 7.93 | 26.61 | Yes |
| T5 | 22.8 (2.9) | 39.3 (6.2) | 77.3 | 10.29 | No | 49.75 | 36.69 | No |
| T6 | 32.9 (4.9) | 59.9 (6.1) | 20.5 | 4.31 | No | 7.01 | 3.94 | No |
| T7 | 12.1 (1.0) | 26.1 (2.4) | 35.3 | 3.87 | No | 28.37 | 1.06 | No |
| T8 | 47.7 (3.6) | 68.3 (5.0) | 17.3 | 7.84 | No | 6.80 | 2.85 | No |
| T9 | 67.4 (7.1) | 80.0 (5.9) | 3.4 | 3.26 | No | 2.76 | 3.74 | Yes |
| T10 | 71.0 (7.0) | 81.4 (2.9) | 13.3 | 28.43 | Yes | 7.49 | 11.42 | Yes |

[a] amount of Trp in the formulation and amount of Trp released are equal (yes) or not (no) within the uncertainty.

Trp was completely released from C6, C10, T1, T2, T11 supplements at both pHs (Table 3, Figure 4). The other five supplements released Trp at only one pH: gastric (supplement C7, C9) or intestinal (supplement C1, C11, T4, T9). Thus, in these supplements, the amount of Trp released was limited only by the content of the main ingredient (Table 2). No negative effect of technological parameters and excipients was observed. One of these supplements, i.e., C11 was probably designed by the manufacturer as an enteral form, which was not even mentioned in the packaging. For this supplement, a release of less than 10% was observed at pH 1.2 (as recommended by the guidelines) and a complete release was observed at pH 6.8. However, due to the lower Trp content, the complete release did not reach 80% of the claimed content as recommended.

The low release of Trp from C2 (pH = 1.2, release 2.65%; pH = 6.8, release 2.3%) and C3 (pH = 1.2, release 3.08%; pH = 6.8, release 4.8%) was mainly due to improperly selected process parameters and/or improperly selected excipients. This is because the content of Trp in the dosage form was much higher than the amount of Trp released (Figure 4a). In the remaining formulations (i.e., C4, C6, C8, C9, C11, T1, T4, T5, T6, T9), the low Trp release was due to both low compound content in the formulation and inappropriate preparation technology (poorly selected technological parameters or excipients) (Figure 4). Referring to in vivo conditions, units characterized by low release will enter the gastric juice but will not release the substance. Thus, no physiological effect will be observed.

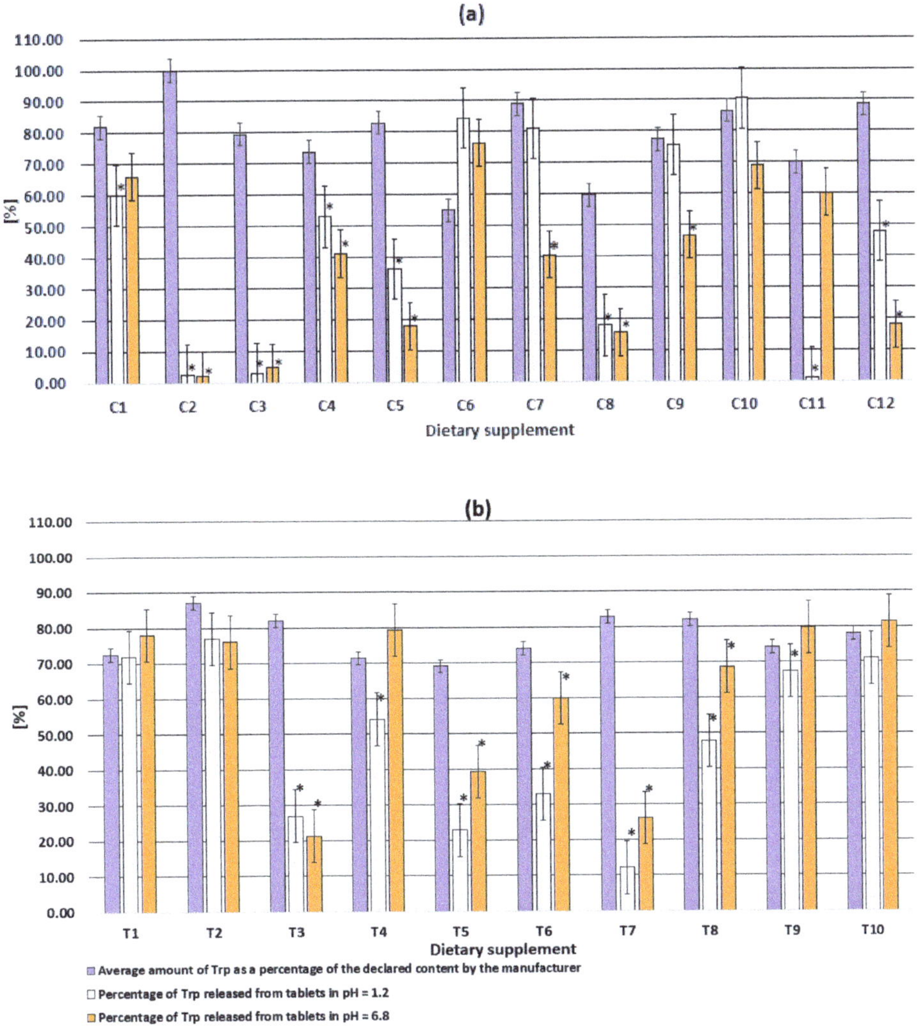


**Figure 4.** Comparison of the amount of Trp released at pH 1.2 (simulated gastric conditions), pH 6.8 (simulated intestinal conditions) with the amount determined in dietary supplements in capsules (**a**) and tablets (**b**); * significant differences (results not equal within the uncertainty) of the amount released with the amount detected.

In summary, none of the analysed dietary supplements contained 80% or more Trp (for each tablet/capsule), which means that these formulations do not meet the release requirements for medicinal products. Comparative data on Trp release from other dietary supplements are not available. Applying our criteria to dietary supplements containing lutein [8], it also the case that none of these dietary supplements would meet these requirements. However, a comparison between the two studies is not easy because the release of the lutein supplements was performed using unconventional parameters. The dissolution test fluid for tablets was 2% P80 ($w/v$) and for capsules 2% P80 ($w/v$) with 25%

ethanol. In the case of Trp supplements, the release test fluid was 0.1 mol/L hydrochloric acid (simulated gastric conditions) and 0.05 mol/L phosphate buffer (pH 6.8, simulated intestinal conditions) regardless of the form of the dietary supplement. Low release of active ingredients such as calcium carbonate [13], melatonin [14], folic acid [15,16], iron, zinc, manganese [17] and Grape seed extract [19], have also been observed in other dietary supplements in solid form (Table A2). However, in these cases, the reasons for the low release are not known due to the lack of data on the content of the main ingredient in these supplements. Low release may be due to insufficient content of the main ingredient, improperly selected process parameters, and/or improperly selected excipients. Only for food supplements containing triiodothyronine ($n$ = 3) or prehormone thyroxine ($n$ = 1) was the main component release above 93% [18].

In our study, Trp release was higher from tablets (12.1–81.4%) than from capsules (1.22–90.4%). Similar results were previously obtained for lutein [8] and folic acid [16]. For lutein supplements, release from capsules (made in the USA), despite containing adequate amounts of lutein, showed alarming results due to poor dissolution properties (less than 20% after 180 min of testing). These results may contribute to the lack of bioavailability of lutein. Unlike the capsules, the lutein tablets (made in Brazil) released more than 80% of the lutein within 180 min.

Analysis of the content, identity, and release of active ingredients from products is important to assess their quality [66]. The dissolution test determines the amount of active substance released and is mandatory for solid drug forms, but not for the same forms of dietary supplements [67]. The in vivo absorption of the active ingredient from solid formulations can be predicted to some extent using this assay [68]. A low release rate means low absorption and no intended effect. Thus, even the substance is in labelled amounts in the supplement but is not released, the consumer will not be able to achieve the effect.

Ease of marketing the supplement and low level of control combined with high popularity and high market value make dietary supplements a group of products particularly vulnerable to negligence or intentional manipulation, which poses a threat to consumers' interests and sometimes even their health [21,69,70]. Determining the quality of dietary supplements is challenging and can be more difficult than for pharmaceutical products because such products often contain multiple vitamins, minerals [71], many of which are derived from plants [72] or other biological sources [73]. However, quality control of supplements should meet the same standards as pharmaceutical products because in both cases they are intended for consumer use [74]. The results of our studies developed with the use of a gold standard in analytics—mass spectrometry coupled with liquid chromatography, provide important data on the quality of the analysed dietary supplements. We hope that our results will encourage further research and increase public awareness about the purposefulness and safety of taking dietary supplements. An informed consumer will choose tested supplements, which will encourage manufacturers to test. In the case of the Food and Drug Administration, Good Manufacturing Practice in Manufacturing, Packing, Labelling, or Holding Operations for Dietary Supplements were already established. Applying GMPs to dietary supplements would be a further step to ensure products are consistently produced and controlled to the quality standards appropriate to their intended use [74].

A limitation of our study is the inability to detect compounds present at very low concentrations. These compounds require appropriate sample preparation. In addition, the targeted analysis should be chosen over non-targeted screening in their case.

## 3. Materials and Methods

### 3.1. Samples

The study was conducted on twenty-two Trp supplements, which is 10% of all dietary supplements with Trp in tablets or capsules registered in Poland, and all available on the market. There were two types of dosage forms: capsules (C1–C12) and tablets (T1–T10). All supplements were manufactured in the EU (Poland, UK, France, Germany, Czech Republic) and the USA. Six supplements were purchased in a Polish online e-commercial platform, the rest in pharmacies or online pharmacies in Poland.

*3.2. Reagents*

L-Trp (≥99%) (standard) and doxepin hydrochloride (internal standard) (≥98%) were purchased from Merck (Darmstadt, Germany). Hydrochloric acid (35–38%) solution pure p.a., sodium hydroxide (≥98.8%) pure p.a., potassium phosphate monobasic (≥99.5%) pure p.a. were purchased from Chempur (Piekary Śląskie, Poland). HPLC-grade methanol, acetonitrile, and formic acid were purchased from Merck (Darmstadt, Germany).

*3.3. Sample Preparation*

Three tablets or capsules were randomly selected from each supplement. The total weight of three tablets, or of the contents of three capsules, were determined. For tablets, a grinding step was applied. In the next step, the tablet's mass or capsule content equivalent to 10 mg Trp was weighed and 1.00 mL of acetonitrile/methanol/water (1:1:1; *v*/*v*/*v*) mixture was added. The mixture was sonicated for 15 min and centrifuged for 5 min. The supernatant was then diluted with mobile phase to a concentration of 500 ng/mL or 100 ng/mL for qualitative and quantitative analysis, respectively. For quantitative analysis, an internal standard (doxepin) was added in the last step to a final concentration of 500 ng/mL.

*3.4. Qualitative Analysis*

Instrumental analysis was performed using a UHPLC Dionex Ultimate 3000 with a Q-Exactive hybrid quadrupole-orbitrap mass spectrometer system equipped with heat electrospray ionization (HESI), an online vacuum degasser, a quaternary pump, an autosampler, and a thermostatted column compartment. The HESI was operated in positive mode. Full MS scans were acquired over the $m/z$ 100–1400 range with a resolution of 70,000 ($m/z$ 200). Fragmentation was performed in different runs with a normalized collision energy of 20, 35, 50 eV. The ion selection threshold was $8 \times 10^3$ counts, and the maximum allowed ion accumulation times were set to auto both for full MS scans and for the tandem mass spectrum. Standard mass spectrometric conditions for all experiments were: spray voltage, 3.5 kV; sheath gas pressure: 60 arb; aux gas pressure: 20 arb; sweep gas pressure: 0 arb, heated capillary temperature: 320 °C; loop count: 3; isolation window: $m/z$ 1.0; and dynamic exclusion: 6.0 s. For all full scan measurements, lock-mass ions from ambient air ($m/z$ 445.1200 and 291.2842) were used as internal calibrants.

Chromatographic separation was achieved with an Accucore C-18 column (100 mm × 4.6 mm, 2.6 µm) supplied by Thermo Fisher Scientific (Waltham, MA, USA) equipped with a security guard. The column was maintained at 40 °C at a flow rate of 0.3 mL/min. The mobile phases consisted of HPLC grade water with 0.1% formic acid as eluent A and acetonitrile with 0.1% formic acid as eluent B. The gradient (% B) was as follows: 0 min 10%; 1 min 10%; 10 min 95%; 15 min 95%. The volume of injection was 10 µL.

The results obtained were analysed using Compound Discoverer 3.0 software supplied by Thermo Fisher Scientific (Waltham, MA, USA).

The structures of the metabolites were proposed based on:

1. The $m/z$ of the compound. The difference between experimental and theoretical molecular weight should be no higher than 5 ppm;
2. The isotopic pattern. The relative intensity tolerance to be used for the isotope search was set at 30%;
3. Fragmentation of the compound. The fragmentation spectrum was compared with experimental data found in the mass spectra library or the literature (confidence level 2), in silico fragmentation (confidence level 3) or reference standard (confidence level 1).

### 3.5. Quantitative Analysis

The instrumental analysis was performed using an Agilent 1260 Infinity (Agilent Technologies, Santa Clara, CA, USA), equipped with a degasser, autosampler, and binary pump coupled to a QTRAP 4000 hybrid triple quadrupole/linear ion trap mass spectrometer (AB Sciex, Framingham, MA, USA). The Turbo Ion Spray source was operated in positive mode. The curtain gas, ion source gas 1, ion source gas 2, and collision gas (all high-purity nitrogen) were set at 0.24 MPa, 0.41 MPa, 0.28 MPa, and "medium" instrument units, respectively. The ion spray voltage and source temperature were 4500 V and 600 °C, respectively. The target compounds were analysed in multiple reaction monitoring (MRM) mode. The compounds parameters, viz. declustering potential (DP), collision energy (CE), entrance potential (EP), and collision exit potential (CXP), were 76, 27, 12 V and 71, 25, 14 V for Trp and doxepin, respectively.

Chromatographic separation was achieved with a Kinetex C18 column (100 mm × 4.6 mm, 2.6 µm, Phenomenex, Milford, MA, USA). The injection volume was 10 µL. The flow rate was 0.75 mL/min. The mobile phases consisted of HPLC grade water with 0.2% formic acid as eluent A and acetonitrile with 0.2% formic acid as eluent B. The gradient (% B) was as follows: 0 min 5%; 1 min 5%; 2 min 95%; 3 min 95%.

The analysis of the Trp content in dietary supplements was preceded by method validation. The parameters tested were selectivity, precision, accuracy, linearity, and limit of quantification. The range of the calibration curve was selected as 0.01–10 µg/mL. Accuracy and precision were determined in triplicate at four concentration levels (0.01, 0.05, 5.0 and 10.0 µg/mL).

Calculations were made using the Analyst 1.6.3 software (AB Sciex, Framingham, MA, USA).

### 3.6. Dissolution Test for Tablets or Capsules

Trp release study was performed using a USP II Varian VK 7025 or USP I Varian VK 7025 dissolution tester (Erweka GmbH, Heusenstamm, Germany) for tablets and capsules, respectively. Six tablets or capsules were randomly selected and individually placed in the dissolution vessels. Each vessel contained 900 mL of dissolution medium. The stirring speed of 50 rpm or 100 rpm was used for tablets and capsules, respectively. The temperature was set at 37 ± 0.5 °C. Aliquots (1.5 mL) of the medium were manually collected using 5 mL syringes after 30 min of the test and filtered through a Millex-HA 0.45 µm filter. Each aliquot withdrawn was replaced with 1.5 mL of fresh medium. The experiment was performed both in hydrochloride acid pH 1.2 (simulated gastric conditions) and phosphate buffer pH 6.8 (simulated intestinal conditions). The Trp content was measured as described in Section 3.5 (Quantitative analysis).

### 3.7. Expanded Uncertainty

To assess whether the amount of Trp in the dosage unit and amount of the compound released is equal within the uncertainty range, extended uncertainty was determined using Equation (1).

$$U(x_1 - x_2) = 2\sqrt{[u(x_1)]^2 + [u(x_2)]^2} \quad (1)$$

The measurement results were equal if:

$$|x_1 - x_2| < U(x_1 - x_2)$$

$x_1$—mean [mg] Trp content determined in dosage unit using quantitative analysis ($n = 3$).

$x_2$—mean [mg] amount of Trp released from six dosage units.

$u(x_2), u(x_1)$—standard uncertainties of the measured values: $x_1$ and $x_2$ determined according to the formula:

$$u(x_1) = \frac{S}{\sqrt{n}}$$

*S*—standard deviation of the average amount of Trp in dosage unit [mg] or standard deviation of the released amount of Trp [mg].

*n*—the number of tablets or capsules analysed.

## 4. Conclusions

A new analytical approach based on liquid chromatography coupled to mass spectrometry provided the opportunity to obtain reliable results on the quality of dietary supplements. The quality of supplements is lower than that of pharmaceuticals with lower than claimed amounts of the main ingredient and a lack of uniform distribution between units. Sometimes, the release of the main ingredient is low, resulting in a lower probability of absorption and physiological effect. Contaminants were detected in all dietary supplements analysed, based on untargeted analysis. These substances, in the amounts determined, may not affect health or show significant unknown effects. The study confirms issues with the quality of dietary supplements and provides an important contribution to the discussion on the regulation of dietary supplements. We believe that the new analytical approach will have broad applicability in the assessment of supplement quality.

**Author Contributions:** Conceptualization, J.G.; methodology, J.G., K.A.S.; software, K.A.S.; validation, J.G., K.A.S.; investigation, K.A.S.; resources, K.A.S.; data curation, K.A.S.; writing—original draft preparation, K.A.S.; writing—review and editing, K.A.S., J.G.; visualization, K.A.S.; supervision, J.G.; project administration, J.G. All authors have read and agreed to the published version of the manuscript.

**Funding:** This research received no external funding.

**Institutional Review Board Statement:** Not applicable.

**Informed Consent Statement:** Not applicable.

**Data Availability Statement:** Data is contained within the article.

**Acknowledgments:** The authors express their sincere gratitude to Anna Harasimiuk and Jakub Niewiarowski for their help in performing the Trp dissolution test in pH 1.2, which was part of their master thesis.

**Conflicts of Interest:** The authors declare no conflict of interest.

## Appendix A

**Table A1.** A review of research on contaminants in dietary supplements.

| Type of Dietary Supplement | Number of Supplements | Contaminants | Country of Sale | Year | Method Applied | Ref. |
|---|---|---|---|---|---|---|
| **Plant based (e.g., Ginkgo biloba, Ginseng, flower pollen), algae** | 24 | Cd, Pb, Hg | Mexico | 2007 | ASA | [75] |
| **Mainly plant-based (herbs or botanicals as major components)** | 95 | As, Cd, Pb, Hg | USA | 2003 | ICP-MS | [76] |
| **Mainly plant-based (e.g., ginger, gingko biloba, ephedra), minerals** | 40 | Hg | USA | 2005 | ASA | [77] |
| **Plant-based and algae (e.g., gingko biloba)** | 16 | As | Denmark | 2013 | ICP-MS, LC-ICP-MS | [78] |
| **Iron supplements** | 15 | As | Brazil/Spain | 2017 | LC-ICP-MS | [79] |
| **Multimineral supplements** | 168 | Pb | Poland | 2018 | MIP-OES | [80] |

**Table A1.** *Cont.*

| Type of Dietary Supplement | Number of Supplements | Contaminants | Country of Sale | Year | Method Applied | Ref. |
|---|---|---|---|---|---|---|
| **Herbal (improve hair, skin, and nails; regulate glucose levels)** | 24 | Hg | Poland | 2018 | ASA | [81] |
| **Prenatal and children supplements** | 10 | As | USA | 2014 | IC-ICP-MS | [82] |
| **Prenatal vitamin supplements** | 51 | As, Cd, Pb, Hg | Canada | 2018 | ICP-MS | [83] |
| **Health clays products** | 27 | As, Cd, Pb, Hg | Netherlands | 2013 | ICP-MS | [84] |
| **Calcium supplements** | 45 | Pb | USA | 2007 | ICP-MS | [85] |
| **Shark cartilage powder** | 16 | Cyanobacterial toxin (N-methylamino-L-alanine) and its isomers (2,4-diaminobutyric acid and N(2-aminoethy) glycine)), Hg | USA | 2014 | LC-FLD, LC–MS, CVAFS | [86] |
| **Ginkgo** | 9 | 250 toxic substances including pesticides (e.g., hymexazol, tebufenozide) and mycotoxins (e.g., aflatoxin B1, aflatoxin B2, T-2 toxin), Insecticides, Fungicides, Herbicides | Spain, Poland, USA | 2015 | LC-HRMS | [46] |
| **Grape** | 24 | Mycotoxin (Ochratoxin A) | Italy | 2015 | LC–FLD | [87] |
| **Different plants (used for liver problems, menopause, for general health improvement)** | 69 | 57 mycotoxins (e.g., zearalenone, enniatins) | Czech Republic, USA | 2015 | LC-MS | [88] |
| **Brewer's yeast** | 51 | Mycotoxin (Ochratoxin A) | Germany | 2002 | LC-FLD | [89] |
| **Blue green algae** | 17 | Microcystins | Italy | 2012 | LC–MS, ELISA | [90] |
| **Blue green algae and Chlorella** | 18 | Microcystins | Germany | 2012 | PPIA, ELISA, LC–MS | [91] |
| **Ginseng** | 23 | Insecticide, Fungicides | USA | 2016 | GC-MS | [44] |
| **Soya** | 14 | Herbicides | Spain | 2016 | LC-MS | [92] |
| **Fish, seal and vegetable** | 30 | Insecticides | Canada | 2009 | GC-MS | [93] |
| **Omega-3** | 9 | Polychlorinated dibenzo-p-dioxins | Spain | 2017 | GC-MS | [94] |
| **Plant-based (weight loss)** | 11 | Sibutramine and its analogues, phenolphthalein | China | 2008 | LC-MS | [95] |
| **Plant-based (weight loss)** | 24 | Sibutramine and its analogues, rimonabant, phenolphthalein | Netherland | 24 | LC-DAD-MS | [96] |
| **Plant-based (naturally enhance sexual performance)** | 74 | PDE-5 inhibitors and their analogues | USA | 2013 | LC-DAD-MS | [97] |

**Table A1.** *Cont.*

| Type of Dietary Supplement | Number of Supplements | Contaminants | Country of Sale | Year | Method Applied | Ref. |
|---|---|---|---|---|---|---|
| Plant-based (enhance sexual potency) | 23 | PDE-5 inhibitors and their analogues | Netherland | 2013 | LC-DAD-MS | [98] |
| Tryptophan | 22 | Untargeted screening, Trp products generated during production, storage, transport | Poland | 2022 | LC-HRMS | Current study |

ASA—atomic absorption spectrometry, CVAFS—cold vapor atomic fluorescence spectrometry, DAD—diode array detection, ELISA—enzyme linked immuno-system, GC—gas chromatography, IC—ion chromatography, ICP-MS—inductively coupled plasma—mass spectrometry, LC-liquid chromatography, LC-HRMS—liquid chromatography-high resolution mass spectrometry, LC-FLD—liquid chromatography fluorescence detector, MIP-OES—microwave-induced plasma optical emission spectrometry, MS—mass spectrometry, PPIA—phosphatase inhibition assay.

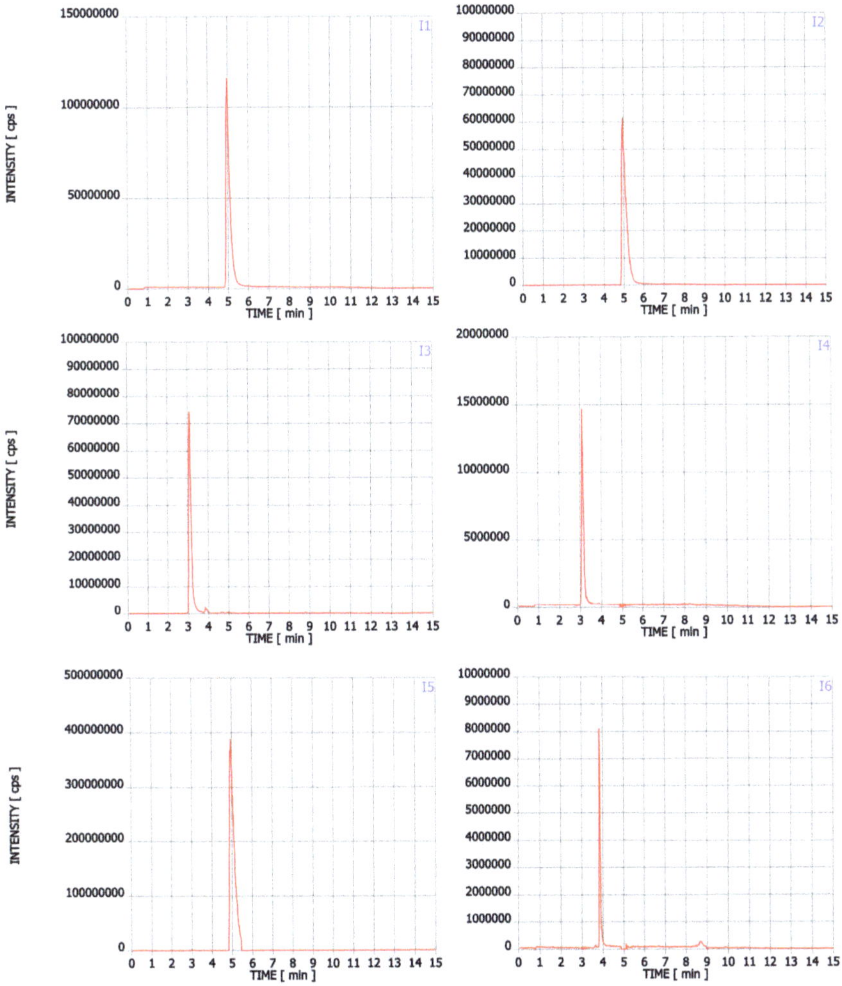


**Figure A1.** *Cont.*

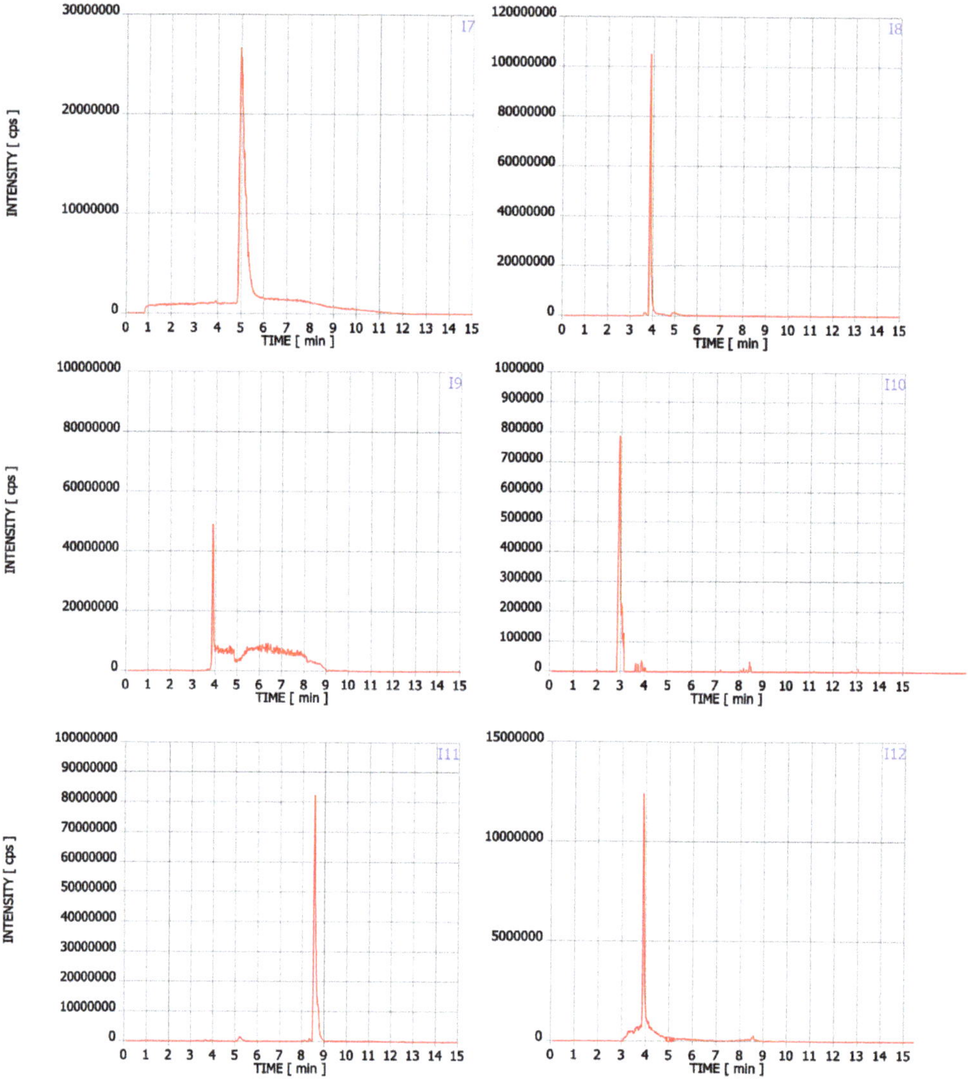


**Figure A1.** *Cont.*

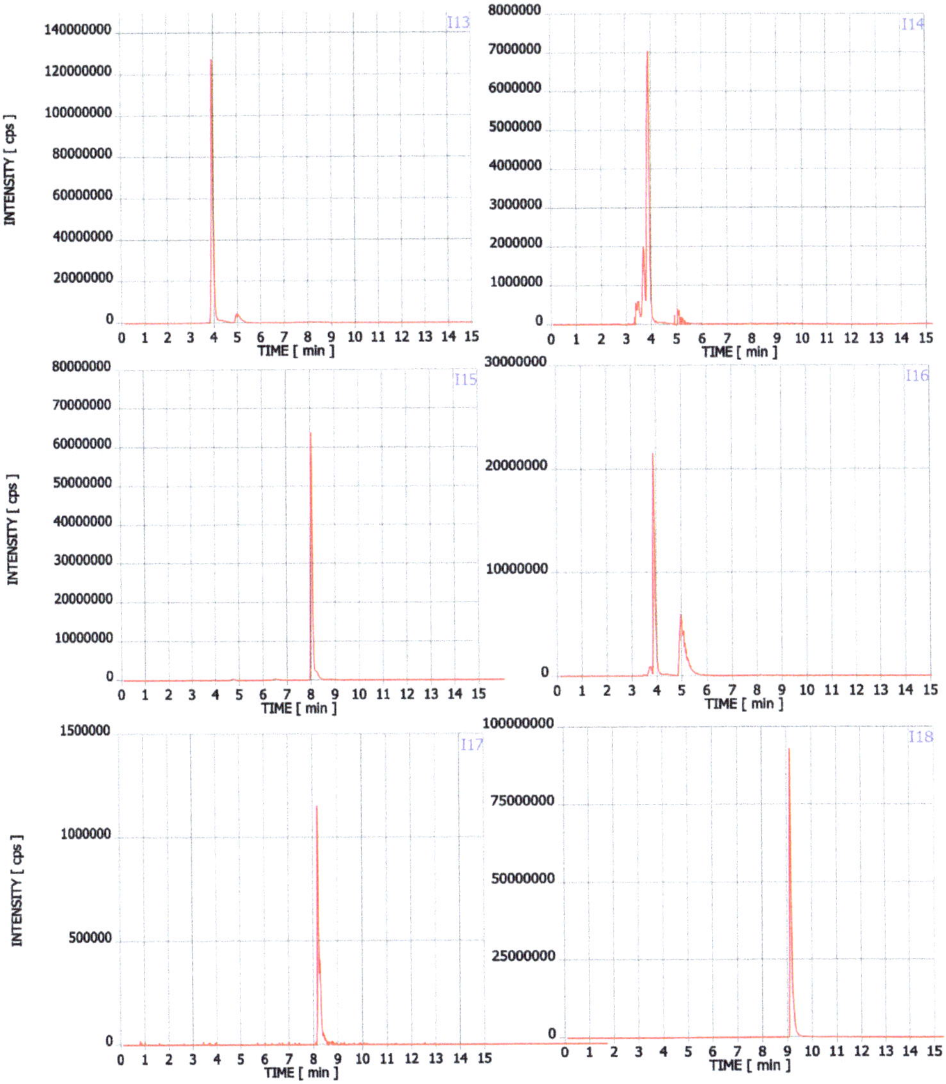


**Figure A1.** *Cont.*

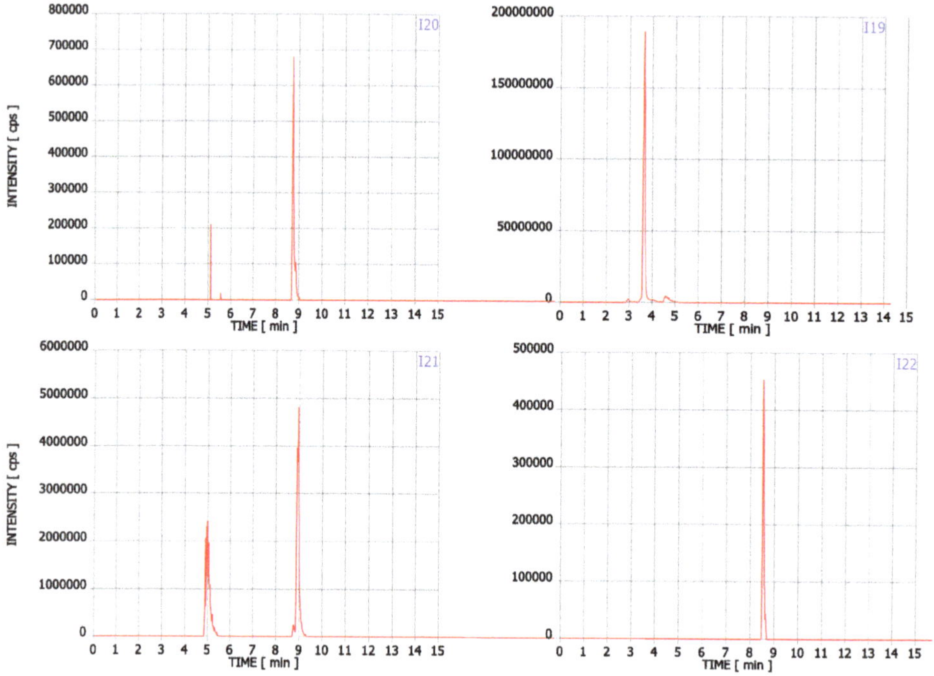


**Figure A1.** Chromatograms of Trp contaminants (I1-I22) detected in Trp dietary supplements.

**Table A2.** Review of studies on the release assay of the active substance from dietary supplements.

| Main of Ingredient Dietary Supplement | Year | Country of Sale | Dosage Form | Number of Supple-ments | Dissolution Test | The Average Percentage of Trp Amount Released from a Dosage Form (Dissolution Medium) | Reference |
|---|---|---|---|---|---|---|---|
| **Calcium Carbonate** | 1990 | USA | tablet | 27 | Yes | 5/27—below 75% (HCl pH 1.0) 4/27—between 33–75% (HCl pH 1.0) 18/27—less than 33% (HCl pH 1.0) | [13] |
| **Melatonin** | 1999 | USA | Immediate-release | 9 | Yes | 4/9 above 75% (HCl pH 1.0) | [14] |
| | | | Controlled-release | 2 | | 1/2 above 90% (HCl pH 1.0) | |
| **Folic Acid** | 2001 | United Kingdom | capsule tablet | 11 | Yes | 6/11—below 70% (0.1 M sodium hydroxide) 4/11—above 70% (0.1 M sodium hydroxide) | [15] |
| **Folic Acid** | 2009 | USA | tablet | 14 | | 45.0% (NaCl, pH 1.5) 104.5% (phosphate buffer, pH 7.5) | [16] |
| | | | capsule | 1 | | 15.2% (NaCl, pH 1.5) 47.4% (phosphate buffer, pH 7.5) | |

**Table A2.** *Cont.*

<table>
<tr><th>Main of Ingredient Dietary Supplement</th><th>Year</th><th>Country of Sale</th><th>Dosage Form</th><th>Number of Supple-ments</th><th>Dissolution Test</th><th>The Average Percentage of Trp Amount Released from a Dosage Form (Dissolution Medium)</th><th>Reference</th></tr>
<tr><td>Iron, zinc, manganase</td><td>2016</td><td>Poland</td><td>tablet</td><td>4</td><td>Yes</td><td>Iron—1/4 above 80% (HCl, pH 1.2)<br>Zinc—1/4 above 80% (HCl, pH 1.2)<br>Manganase—4/4–60% or less (HCl, pH 1.2)</td><td>[17]</td></tr>
<tr><td rowspan="2">Lutein</td><td rowspan="2">2018</td><td rowspan="2">Brazil<br>USA</td><td>tablet</td><td>4</td><td rowspan="2">Yes</td><td>41.7% (2% polysorbate 80)</td><td rowspan="2">[8]</td></tr>
<tr><td>capsule</td><td>6</td><td>122.5% (2% polysorbate 80 with 25% ethanol)</td></tr>
<tr><td>Triiodothyronine</td><td>2019</td><td>United Kingdom</td><td>tablet</td><td>3</td><td>Yes</td><td>Above 93.5% (fasted-state simulated gastric fluid)</td><td>[18]</td></tr>
<tr><td>Prehormone thyroxine</td><td>2019</td><td>United Kingdom</td><td>tablet</td><td>1</td><td>Yes</td><td>Above 97.4% (fasted-state simulated gastric fluid)</td><td>[18]</td></tr>
<tr><td>Grape seed extract</td><td>2021</td><td>USA</td><td>capsule</td><td>1</td><td>Yes</td><td>73.09, 67.9, 71.06, 59.75% of gallic acid, catechin, procyanidin B2, and epicatechin, respectively (acetate buffer pH 4.6), 96.49, 89.09, 87.65, 78.84% of gallic acid, catechin, procyanidin B2, and epicatechinin, respectively (HCl pH 1.2)</td><td>[19]</td></tr>
<tr><td>Trans-resveratrol</td><td>2021</td><td>China</td><td>capsule</td><td>1</td><td>Yes</td><td>Above 75% (acetate buffer pH 4.6) Above 75% (HCl pH 1.2)</td><td>[19]</td></tr>
<tr><td rowspan="2">Tryptophan</td><td rowspan="2">2022</td><td rowspan="2">Poland</td><td>tablet</td><td>10</td><td rowspan="2">Yes</td><td>48.5% (HCl, pH 1.2) 61.0% (phosphate buffer pH 6.8)</td><td rowspan="2">Current study</td></tr>
<tr><td>capsule</td><td>12</td><td>46.1% (HCl, pH 1.2) 38.2% (phosphate buffer pH 6.8)</td></tr>
</table>

## References


1. Cerezo, A.B.; Leal, Á.; Álvarez-Fernández, M.A.; Hornedo-Ortega, R.; Troncoso, A.M.; García-Parrilla, M.C. Quality control and determination of melatonin in food supplements. *J. Food Compos. Anal.* **2016**, *45*, 80–86. [CrossRef]
2. EC. Directive 2002/46/EC of the European parliament and of the council of 10 June 2002 on the approximation of the laws of the member states relating to food supplements. *Off. J. Eur. Communities Legis.* **2002**, *45*, 51–57.
3. FDA. *Dietary Supplement Health and Education Act of 1994*; FDA: St. Louis, MO, USA, 1995; Volume 1.
4. Moret, S.; Prevarin, A.; Tubaro, F. Levels of creatine, organic contaminants and heavy metals in creatine dietary supplements. *Food Chem.* **2011**, *126*, 1232–1238. [CrossRef]
5. Tumir, H.; Bošnir, J.; Vedrina-Dragojević, I.; Dragun, Z.; Tomić, S.; Puntarić, D.; Jurak, G. Monitoring of metal and metalloid content in dietary supplements on the Croatian market. *Food Control* **2010**, *21*, 885–889. [CrossRef]
6. Andrews, K.W.; Roseland, J.M.; Gusev, P.A.; Palachuvattil, J.; Dang, P.T.; Savarala, S.; Han, F.; Pehrsson, P.R.; Douglass, L.W.; Dwyer, J.T.; et al. Analytical ingredient content and variability of adult multivitamin/mineral products: National estimates for the dietary supplement ingredient database. *Am. J. Clin. Nutr.* **2017**, *105*, 526–539. [CrossRef]
7. De Ceglie, C.; Calvano, C.D.; Zambonin, C.G. MALDI-TOF MS for quality control of high protein content sport supplements. *Food Chem.* **2015**, *176*, 396–402. [CrossRef] [PubMed]
8. Anselmo, C.S.; Mendes, T.C.; Cabral, L.M.; Sousa, V.P. Physicochemical quality profiles of commercial oral tablets and capsules containing lutein—Impact of insufficient specific sanitary regulations. *An. Acad. Bras. Cienc.* **2018**, *90*, 3063–3073. [CrossRef] [PubMed]

9. Kowalski, A.; Frankowski, M. Levels and potential health risks of mercury in prescription, non-prescription medicines and dietary supplements in Poland. *Regul Toxicol. Pharm.* **2015**, *73*, 396–400. [CrossRef] [PubMed]
10. Martinez-Sanz, J.M.; Sospedra, I.; Ortiz, C.M.; Baladia, E.; Gil-Izquierdo, A.; Ortiz-Moncada, R. Intended or unintended doping? A review of the presence of doping substances in dietary supplements used in sports. *Nutrients* **2017**, *9*, 1093. [CrossRef] [PubMed]
11. Fernandes, A.R.; Rose, M.; White, S.; Mortimer, D.; Gem, M. Dioxins and polychlorinated biphenyls (PCBs) in fish oil dietary supplements: Occurrence and human exposure in the UK. *Food Addit. Contam.* **2006**, *23*, 939–947. [CrossRef]
12. Bengtson Nash, S.M.; Schlabach, M.; Nichols, P.D. A nutritional-toxicological assessment of Antarctic krill oil versus fish oil dietary supplements. *Nutrients* **2014**, *6*, 3382–3402. [CrossRef] [PubMed]
13. Brennan, M.J.; Duncan, W.E.; Wartofsky, L.; Butler, V.M.; Wray, H.L. In vitro dissolution of calcium carbonate preparations. *Calcif. Tissue Int.* **1991**, *49*, 308–312. [CrossRef] [PubMed]
14. Hahm, H.; Kujawa, J.; Augsburger, L. Comparison of melatonin products against USP's nutritional supplements standards and other criteria. *J. Am. Pharm. Assoc.* **1999**, *39*, 27–31. [CrossRef]
15. Sculthorpe, N.; Davies, B.; Ashton, T.; Allison, S.; McGuire, D.; Malhi, J. Commercially available folic acid supplements and their compliance with the British pharmacopoeia test for dissolution. *J. Public Health* **2001**, *23*, 195–197. [CrossRef] [PubMed]
16. Younis, I.R.; Stamatakis, M.K.; Callery, P.S.; Meyer-Stout, P.J. Influence of pH on the dissolution of folic acid supplements. *Int. J. Pharm.* **2009**, *367*, 97–102. [CrossRef]
17. Blicharska, E.; Szczęsna, B.; Kocjan, R.; Gumieniczek, A.; Komsta, Ł. Analysis of dissolution profiles of iron, zinc, and manganese from complex dietary supplements by ion chromatography and chemometrics. *J. Liq. Chromatogr. Relat. Technol.* **2016**, *39*, 30–34. [CrossRef]
18. Bowerbank, S.L.; Carlin, M.G.; Dean, J.R. Dissolution testing of single-and dual-component thyroid hormone supplements. *Separations* **2019**, *6*, 18. [CrossRef]
19. Lyu, W.; Omar, T.; Patel, H.; Rodriguez, D.; Ferruzzi, M.G.; Pasinetti, G.M.; Murrough, J.W.; Muzzio, F.J.; Simon, J.E.; Wu, Q. Dissolution study on grape polyphenol hard gelatin capsule dietary supplements. *Front. Nutr.* **2021**, *8*, 780260. [CrossRef] [PubMed]
20. Palaric, C.; Pilard, S.; Fontaine, J.-X.; Cailleu, D.; Mesnard, F.; Renaud, T.; Petit, A.; Molinié, R. Combined MS-NMR approach for the quality control of food supplements containing amino acids. *Food Control* **2018**, *88*, 217–228. [CrossRef]
21. Dwyer, J.T.; Coates, P.M.; Smith, M.J. Dietary supplements: Regulatory challenges and research resources. *Nutrients* **2018**, *10*, 41. [CrossRef] [PubMed]
22. Marozoff, S.; Veugelers, P.J.; Dabravolskaj, J.; Eurich, D.T.; Ye, M.; Maximova, K. Diet quality and health service utilization for depression: A prospective investigation of adults in alberta's tomorrow project. *Nutrients* **2020**, *12*, 2437. [CrossRef] [PubMed]
23. Park, S.J.; Lee, D.K.; Kim, B.; Na, K.S.; Lee, C.H.; Son, Y.D.; Lee, H.J. The association between omega-3 fatty acid intake and human brain connectivity in middle-aged depressed women. *Nutrients* **2020**, *12*, 2191. [CrossRef] [PubMed]
24. Galts, C.P.C.; Bettio, L.E.B.; Jewett, D.C.; Yang, C.C.; Brocardo, P.S.; Rodrigues, A.L.S.; Thacker, J.S.; Gil-Mohapel, J. Depression in neurodegenerative diseases: Common mechanisms and current treatment options. *Neurosci. Biobehav. Rev.* **2019**, *102*, 56–84. [CrossRef]
25. Lieberman, H.R.; Agarwal, S.; Fulgoni, V.L., 3rd. Tryptophan intake in the US adult population is not related to liver or kidney function but is associated with depression and sleep outcomes. *J. Nutr.* **2016**, *146*, 2609S–2615S. [CrossRef] [PubMed]
26. Palego, L.; Betti, L.; Rossi, A.; Giannaccini, G. Tryptophan biochemistry: Structural, nutritional, metabolic, and medical aspects in humans. *J. Amino Acids* **2016**, *2016*, 8952520. [CrossRef] [PubMed]
27. Fukuwatari, T. Possibility of amino acid treatment to prevent the psychiatric disorders via modulation of the production of tryptophan metabolite kynurenic acid. *Nutrients* **2020**, *12*, 1403. [CrossRef] [PubMed]
28. Soh, N.L.; Walter, G. Tryptophan and depression: Can diet alone be the answer? *Acta Neuropsychiatr.* **2014**, *23*, 3–11. [CrossRef]
29. Samad, N.; Yasmin, F.; Manzoor, N. Biomarkers in drug free subjects with depression: Correlation with tryptophan. *Psychiatry Investig.* **2019**, *16*, 948–953. [CrossRef] [PubMed]
30. Jenkins, T.A.; Nguyen, J.C.; Polglaze, K.E.; Bertrand, P.P. Influence of tryptophan and serotonin on mood and cognition with a possible role of the gut-brain axis. *Nutrients* **2016**, *8*, 56. [CrossRef]
31. Binks, H.; Vincent, G.E.; Gupta, C.; Irwin, C.; Khalesi, S. Effects of Diet on Sleep: A Narrative Review. *Nutrients* **2020**, *12*, 936. [CrossRef]
32. Strasser, B.; Fuchs, D. Diet versus Exercise in Weight Loss and Maintenance: Focus on Tryptophan. *Int. J. Tryptophan Res.* **2016**, *9*, 9–16. [CrossRef] [PubMed]
33. Gibson, E.L. Tryptophan supplementation and serotonin function: Genetic variations in behavioural effects. *Proc. Nutr. Soc.* **2018**, *77*, 174–188. [CrossRef] [PubMed]
34. Thomas, S.R. l-Tryptophan as an antidepressive agent in the management of treatment-resistant unipolar depression in borderline personality disorder: Three case reports. *Adv. Integr. Med.* **2015**, *2*, 68–71. [CrossRef]
35. European Parliament; Council of the European Union. Regulation (EU) No 609/2013 of the European parliament and of the council of 12 June 2013 on food intended for infants and young children, food for special medical purposes, and total diet replacement for weight control and repealing council directive 92/52/EEC, commission directives 96/8/EC, 1999/21/EC, 2006/125/EC and 2006/141/EC, directive 2009/39/EC of the European parliament and of the council and commission regulations (EC) No 41/2009 and (EC) No 953/2009. *Off. J. Eur. Union* **2013**, *181*, 35–56.

36. Simat, T.J.; Kleeberg, K.K.; Müller, B.; Sierts, A. Contamination of commercially available L-tryptophan by related substances. *Eur. Food Res. Technol.* **2003**, *216*, 241–252. [CrossRef]
37. Bellmaine, S.; Schnellbaecher, A.; Zimmer, A. Reactivity and degradation products of tryptophan in solution and proteins. *Free. Radic. Biol. Med.* **2020**, *160*, 696–718. [CrossRef] [PubMed]
38. Sheng, X.; Himo, F. Enzymatic pictet-spengler reaction: Computational study of the mechanism and enantioselectivity of norcoclaurine synthase. *J. Am. Chem. Soc.* **2019**, *141*, 11230–11238. [CrossRef]
39. Simat, T.; Van Wickern, B.; Eulitz, K.; Steinhart, H. Contaminants in biotechnologically manufactured L-tryptophan. *J. Chromatogr. B Biomed. Sci. Appl.* **1996**, *685*, 41–51. [CrossRef]
40. Zelentsova, E.A.; Sherin, P.S.; Snytnikova, O.A.; Kaptein, R.; Vauthey, E.; Tsentalovich, Y.P. Photochemistry of aqueous solutions of kynurenic acid and kynurenine yellow. *Photochem. Photobiol. Sci.* **2013**, *12*, 546–558. [CrossRef] [PubMed]
41. Gugliandolo, E.; Peritore, A.F.; Impellizzeri, D.; Cordaro, M.; Siracusa, R.; Fusco, R.; D'Amico, R.; Di Paola, R.; Schievano, C.; Cuzzocrea, S. Dietary supplementation with palmitoyl-glucosamine co-micronized with curcumin relieves osteoarthritis pain and benefits joint mobility. *Animals* **2020**, *10*, 1827. [CrossRef] [PubMed]
42. Amaral, F.G.d.; Cipolla-Neto, J. A brief review about melatonin, a pineal hormone. *Arch. Endocrinol. Metab.* **2018**, *62*, 472–479. [CrossRef]
43. Nieto-García, A.J.; Romero-González, R.; Frenich, A.G. Multi-pesticide residue analysis in nutraceuticals from grape seed extracts by gas chromatography coupled to triple quadrupole mass spectrometry. *Food Control.* **2015**, *47*, 369–380. [CrossRef]
44. Chen, Y.; Lopez, S.; Hayward, D.G.; Park, H.Y.; Wong, J.W.; Kim, S.S.; Wan, J.; Reddy, R.M.; Quinn, D.J.; Steiniger, D. Determination of multiresidue pesticides in botanical dietary supplements using gas chromatography–Triple-quadrupole mass spectrometry (GC-MS/MS). *J. Agric. Food Chem.* **2016**, *64*, 6125–6132. [CrossRef] [PubMed]
45. Martínez-Domínguez, G.; Romero-González, R.; Arrebola, F.J.; Frenich, A.G. Multi-class determination of pesticides and mycotoxins in isoflavones supplements obtained from soy by liquid chromatography coupled to orbitrap high resolution mass spectrometry. *Food Control* **2016**, *59*, 218–224. [CrossRef]
46. Martinez-Dominguez, G.; Romero-Gonzalez, R.; Frenich, A.G. Determination of toxic substances, pesticides and mycotoxins, in ginkgo biloba nutraceutical products by liquid chromatography orbitrap-mass spectrometry. *Microchem. J.* **2015**, *118*, 124–130. [CrossRef]
47. Yue, K.X.; Li, H.L.; Pi, Z.F.; Song, F.R.; Wang, Y.P.; Liu, Z.Y. Multi-residue screening of pesticides in panax ginseng CA meyer by ultra-high-performance liquid chromatography with quadrupole time-of-flight mass spectrometry. *J. Sep. Science.* **2022**, 1–9. [CrossRef]
48. Hou, X.; Liu, L.; Wei, L.; Feng, D.; Lv, M.; Wang, X.; Yu, X.; Lu, Z.; Hou, Z. A Novel Analysis Method for Simultaneous Determination of 31 Pesticides by High-Performance Liquid Chromatography-Tandem Mass Spectrometry in Ginseng. *J. Anal. Methods Chem.* **2022**, *2022*, 4208243. [CrossRef]
49. Nasiri, M.; Ahmadzadeh, H.; Amiri, A. Sample preparation and extraction methods for pesticides in aquatic environments: A review. *TrAC Trends Anal. Chem.* **2020**, *123*, 115772. [CrossRef]
50. Montone, C.M.; Moneta, B.G.; Aita, S.E.; Aulenta, F.; Cavaliere, C.; Cerrato, A.; Fazi, S.; Laganà, A.; Paolini, V.; Petracchini, F. Untargeted analysis of contaminants in river water samples: Comparison between two different sorbents for solid-phase extraction followed by liquid chromatography-high-resolution mass spectrometry determination. *Microchem. J.* **2022**, *172*, 106979. [CrossRef]
51. Nemati, M.; Tuzen, M.; Farazajdeh, M.A.; Kaya, S.; Mogaddam, M.R.A. Development of dispersive solid-liquid extraction method based on organic polymers followed by deep eutectic solvents elution; application in extraction of some pesticides from milk samples prior to their determination by HPLC-MS/MS. *Anal. Chim. Acta* **2022**, *1199*, 339570. [CrossRef]
52. Zohrabi, P.; Shamsipur, M.; Hashemi, M.; Hashemi, B. Liquid-phase microextraction of organophosphorus pesticides using supramolecular solvent as a carrier for ferrofluid. *Talanta* **2016**, *160*, 340–346. [CrossRef] [PubMed]
53. Alvi, T.; Asif, Z.; Khan, M.K.I. Clean label extraction of bioactive compounds from food waste through microwave-assisted extraction technique-A review. *Food Biosci.* **2022**, *46*, 101580. [CrossRef]
54. Moret, S.; Scolaro, M.; Barp, L.; Purcaro, G.; Conte, L.S. Microwave assisted saponification (MAS) followed by on-line liquid chromatography (LC)—Gas chromatography (GC) for high-throughput and high-sensitivity determination of mineral oil in different cereal-based foodstuffs. *Food Chem.* **2016**, *196*, 50–57. [CrossRef] [PubMed]
55. Zhou, Z.; Zhang, J.; Zhang, W.; Bai, Y.; Liu, H. Rapid screening for synthetic antidiabetic drug adulteration in herbal dietary supplements using direct analysis in real time mass spectrometry. *Analyst* **2011**, *136*, 2613–2618. [CrossRef]
56. Patel, D.N.; Li, L.; Kee, C.-L.; Ge, X.; Low, M.-Y.; Koh, H.-L. Screening of synthetic PDE-5 inhibitors and their analogues as adulterants: Analytical techniques and challenges. *J. Pharm. Biomed. Anal.* **2014**, *87*, 176–190. [CrossRef] [PubMed]
57. Haller, C.A.; Benowitz, N.L. Adverse cardiovascular and central nervous system events associated with dietary supplements containing ephedra alkaloids. *N. Engl. J. Med.* **2000**, *343*, 1833–1838. [CrossRef] [PubMed]
58. Schymanski, E.L.; Jeon, J.; Gulde, R.; Fenner, K.; Ruff, M.; Singer, H.P.; Hollender, J. Identifying small molecules via high resolution mass spectrometry: Communicating confidence. *Environ. Sci. Technol.* **2014**, *48*, 2097–2098. [CrossRef] [PubMed]
59. Rochat, B. Proposed confidence scale and ID score in the identification of known-unknown compounds using high resolution MS data. *J. Am. Soc. Mass Spectrom.* **2017**, *28*, 709–723. [CrossRef] [PubMed]

60. Jakubowska, E.; Ciepluch, N. Blend segregation in tablets manufacturing and its effect on drug content uniformity—A review. *Pharmaceutics* **2021**, *13*, 1909. [CrossRef]
61. Farquharson, A.; Gladding, Z.; Ritchie, G.; Shende, C.; Cosgrove, J.; Smith, W.; Brouillette, C.; Farquharson, S. Drug content uniformity: Quantifying loratadine in tablets using a created raman excipient spectrum. *Pharmaceutics* **2021**, *13*, 309. [CrossRef]
62. Khashaba, P.Y.; Abdelgaleel, M.; Derayea, S.M.; Nagi, D.M. Development of a simple validated spectrofluorometric method for the assay of midodrine in tablets dosage form; Application to content uniformity testing. *Spectrochim. Acta Part A Mol. Biomol. Spectrosc.* **2022**, *273*, 121046. [CrossRef] [PubMed]
63. Ganzetti, R.; Logrippo, S.; Sestili, M.; Caraffa, A.; Cespi, M.; Pelliccioni, G.; Blasi, P.; Bonacucina, G. Tablet splitting in elderly patients with dementia: The case of quetiapine. *Pharmaceutics* **2021**, *13*, 1523. [CrossRef] [PubMed]
64. Peeters, M.; Peeters, E.; De Leersnyder, F.; Van Hauwermeiren, D.; Cogoni, G.; Liu, Y.; De Beer, T. Determination and understanding of dynamic lead lag between in-line nir tablet press feed frame and off-line nir tablet measurements. In Proceedings of the 2019 AIChE Annual Meeting, Orlando, FL, USA, 12 November 2019.
65. FDA. *Dissolution Testing and Acceptance Criteria for Immediate-Release Solid Oral Dosage form Drug Products Containing High Solubility Drug Substances Guidance For Industry*; FDA: Rockville, MD, USA, 2018.
66. Yu, L.X.; Amidon, G.; Khan, M.A.; Hoag, S.W.; Polli, J.; Raju, G.; Woodcock, J. Understanding pharmaceutical quality by design. *AAPS J.* **2014**, *16*, 771–783. [CrossRef]
67. Al-Gousous, J.; Langguth, P. Oral solid dosage form disintegration testing—The forgotten test. *J. Pharm. Sci.* **2015**, *104*, 2664–2675. [CrossRef] [PubMed]
68. Gray, V.A. Power of the dissolution test in distinguishing a change in dosage form critical quality attributes. *AAPS PharmSciTech* **2018**, *19*, 3328–3332. [CrossRef] [PubMed]
69. Binns, C.W.; Lee, M.K.; Lee, A.H. Problems and prospects: Public health regulation of dietary supplements. *Annu. Rev. Public Health* **2018**, *39*, 403–420. [CrossRef] [PubMed]
70. Jędrejko, K.; Lazur, J.; Muszyńska, B. Risk associated with the use of selected ingredients in food supplements. *Chem. Biodivers.* **2021**, *18*, e2000686. [CrossRef] [PubMed]
71. Moses, G. The safety of commonly used vitamins and minerals. *Aust. Prescr.* **2021**, *44*, 119. [CrossRef]
72. Fibigr, J.; Šatínský, D.; Solich, P. Current trends in the analysis and quality control of food supplements based on plant extracts. *Anal. Chim. Acta* **2018**, *1036*, 1–15. [CrossRef] [PubMed]
73. Rautiainen, S.; Manson, J.E.; Lichtenstein, A.H.; Sesso, H.D. Dietary supplements and disease prevention—A global overview. *Nat. Rev. Endocrinol.* **2016**, *12*, 407–420. [CrossRef] [PubMed]
74. Costa, J.G.; Vidovic, B.; Saraiva, N.; do Ceu Costa, M.; Del Favero, G.; Marko, D.; Oliveira, N.G.; Fernandes, A.S. Contaminants: A dark side of food supplements? *Free Radic. Res.* **2019**, *53*, 1113–1135. [CrossRef] [PubMed]
75. Garcia-Rico, L.; Leyva-Perez, J.; Jara-Marini, M.E. Content and daily intake of copper, zinc, lead, cadmium, and mercury from dietary supplements in Mexico. *Food Chem. Toxicol.* **2007**, *45*, 1599–1605. [CrossRef] [PubMed]
76. Dolan, S.P.; Nortrup, D.A.; Bolger, P.M.; Capar, S.G. Analysis of dietary supplements for arsenic, cadmium, mercury, and lead using inductively coupled plasma mass spectrometry. *J. Agric. Food Chem.* **2003**, *51*, 1307–1312. [CrossRef] [PubMed]
77. Levine, K.E.; Levine, M.A.; Weber, F.X.; Hu, Y.; Perlmutter, J.; Grohse, P.M. Determination of mercury in an assortment of dietary supplements using an inexpensive combustion atomic absorption spectrometry technique. *J. Autom. Methods Manag. Chem.* **2005**, *2005*, 211–216. [CrossRef] [PubMed]
78. Hedegaard, R.V.; Rokkjær, I.; Sloth, J.J. Total and inorganic arsenic in dietary supplements based on herbs, other botanicals and algae—A possible contributor to inorganic arsenic exposure. *Anal. Bioanal. Chem.* **2013**, *405*, 4429–4435. [CrossRef] [PubMed]
79. Araujo-Barbosa, U.; Peña-Vazquez, E.; Barciela-Alonso, M.C.; Ferreira, S.L.C.; Dos Santos, A.M.P.; Bermejo-Barrera, P. Simultaneous determination and speciation analysis of arsenic and chromium in iron supplements used for iron-deficiency anemia treatment by HPLC-ICP-MS. *Talanta* **2017**, *170*, 523–529. [CrossRef]
80. Poniedziałek, B.; Niedzielski, P.; Kozak, L.; Rzymski, P.; Wachelka, M.; Rzymska, I.; Karczewski, J.; Rzymski, P. Monitoring of essential and toxic elements in multi-ingredient food supplements produced in European union. *J. Consum. Prot. Food Saf.* **2018**, *13*, 41–48. [CrossRef]
81. Brodziak-Dopierała, B.; Fischer, A.; Szczelina, W.; Stojko, J. The content of mercury in herbal dietary supplements. *Biol. Trace Elem. Res.* **2018**, *185*, 236–243. [CrossRef] [PubMed]
82. Wolle, M.M.; Rahman, G.M.; Pamuku, M. Speciation analysis of arsenic in prenatal and children's dietary supplements using microwave-enhanced extraction and ion chromatography—Inductively coupled plasma mass spectrometry. *Anal. Chim. Acta* **2014**, *818*, 23–31. [CrossRef]
83. Schwalfenberg, G.; Rodushkin, I.; Genuis, S.J. Heavy metal contamination of prenatal vitamins. *Toxicol. Rep.* **2018**, *5*, 390–395. [CrossRef] [PubMed]
84. Reeuwijk, N.M.; KLerx, W.N.; Kooijman, M.; Hoogenboom, L.; Rietjens, I.M.; Martena, M.J. Levels of lead, arsenic, mercury and cadmium in clays for oral use on the Dutch market and estimation of associated risks. *Food Addit. Contam. Part A* **2013**, *30*, 1535–1545. [CrossRef] [PubMed]
85. Kauffman, J.F.; Westenberger, B.J.; Robertson, J.D.; Guthrie, J.; Jacobs, A.; Cummins, S.K. Lead in pharmaceutical products and dietary supplements. *Regul. Toxicol. Pharmacol.* **2007**, *48*, 128–134. [CrossRef] [PubMed]

86. Mondo, K.; Glover, W.B.; Murch, S.J.; Liu, G.; Cai, Y.; Davis, D.A.; Mash, D.C. Environmental neurotoxins β-N-methylamino-l-alanine (BMAA) and mercury in shark cartilage dietary supplements. *Food Chem. Toxicol.* **2014**, *70*, 26–32. [CrossRef] [PubMed]
87. Solfrizzo, M.; Piemontese, L.; Gambacorta, L.; Zivoli, R.; Longobardi, F. Food coloring agents and plant food supplements derived from Vitis vinifera: A new source of human exposure to ochratoxin A. *J. Agric. Food Chem.* **2015**, *63*, 3609–3614. [CrossRef] [PubMed]
88. Veprikova, Z.; Zachariasova, M.; Dzuman, Z.; Zachariasova, A.; Fenclova, M.; Slavikova, P.; Vaclavikova, M.; Mastovska, K.; Hengst, D.; Hajslova, J. Mycotoxins in plant-based dietary supplements: Hidden health risk for consumers. *J. Agric. Food Chem.* **2015**, *63*, 6633–6643. [CrossRef] [PubMed]
89. Gareis, M. Ochratoxin a in brewer's yeast used as nutrient supplement. *Mycotoxin Res.* **2002**, *18*, 128–131. [CrossRef] [PubMed]
90. Vichi, S.; Lavorini, P.; Funari, E.; Scardala, S.; Testai, E. Contamination by microcystis and microcystins of blue–green algae food supplements (BGAS) on the italian market and possible risk for the exposed population. *Food Chem. Toxicol.* **2012**, *50*, 4493–4499. [CrossRef] [PubMed]
91. Heussner, A.H.; Mazija, L.; Fastner, J.; Dietrich, D.R. Toxin content and cytotoxicity of algal dietary supplements. *Toxicol. Appl. Pharmacol.* **2012**, *265*, 263–271. [CrossRef]
92. Domingos Alves, R.; Romero-González, R.; López-Ruiz, R.; Jiménez-Medina, M.; Garrido Frenich, A. Fast determination of four polar contaminants in soy nutraceutical products by liquid chromatography coupled to tandem mass spectrometry. *Anal. Bioanal. Chem.* **2016**, *408*, 8089–8098. [CrossRef] [PubMed]
93. Rawn, D.F.; Breakell, K.; Verigin, V.; Nicolidakis, H.; Sit, D.; Feeley, M. Persistent organic pollutants in fish oil supplements on the Canadian market: Polychlorinated biphenyls and organochlorine insecticides. *J. Food Sci.* **2009**, *74*, T14–T19. [CrossRef] [PubMed]
94. Garcia-Bermejo, A.; Herrero, L.; Gonzalez, M.-J.; Gomara, B. Occurrence and estimated dietary intake of PCBs and PCDD/Fs in functional foods enriched with omega-3 from Spain. *J. Agric. Food Chem.* **2017**, *65*, 3396–3405. [CrossRef] [PubMed]
95. Wang, J.; Chen, B.; Yao, S. Analysis of six synthetic adulterants in herbal weight-reducing dietary supplements by LC electrospray ionization-MS. *Food Addit. Contam.* **2008**, *25*, 822–830. [CrossRef] [PubMed]
96. Reeuwijk, N.M.; Venhuis, B.J.; de Kaste, D.; Hoogenboom, R.L.; Rietjens, I.M.; Martena, M.J. Active pharmaceutical ingredients detected in herbal food supplements for weight loss sampled on the Dutch market. *Food Addit. Contam. Part A* **2014**, *31*, 1783–1793. [CrossRef] [PubMed]
97. Campbell, N.; Clark, J.P.; Stecher, V.J.; Thomas, J.W.; Callanan, A.C.; Donnelly, B.F.; Goldstein, I.; Kaminetsky, J.C. Adulteration of purported herbal and natural sexual performance enhancement dietary supplements with synthetic phosphodiesterase type 5 inhibitors. *J. Sex. Med.* **2013**, *10*, 1842–1849. [CrossRef] [PubMed]
98. Reeuwijk, N.M.; Venhuis, B.J.; de Kaste, D.; Hoogenboom, L.; Rietjens, I.M.; Martena, M.J. Sildenafil and analogous phosphodiesterase type 5 (PDE-5) inhibitors in herbal food supplements sampled on the Dutch market. *Food Addit. Contam. Part A* **2013**, *30*, 2027–2034. [CrossRef] [PubMed]

*pharmaceuticals*

*Article*

# UHPLC-MS/MS Analysis of Antibiotics Transfer and Concentrations in Porcine Oral Fluid after Intramuscular Application

**Anna Gajda [1,*], Ewelina Nowacka-Kozak [1], Małgorzata Gbylik-Sikorska [1] and Piotr Cybulski [2]**

1 Department of Pharmacology and Toxicology, National Veterinary Research Institute, Partyzantow 57, 24-100 Pulawy, Poland; ewelina.nowacka@piwet.pulawy.pl (E.N.-K.); malgorzata.gbylik@piwet.pulawy.pl (M.G.-S.)
2 Goodvalley Agro S.A., Dworcowa 25, 77-320 Przechlewo, Poland; piotr.cybulski.dvm@gmail.com
* Correspondence: anna.gajda@piwet.pulawy.pl; Tel.: +48-81-889-3240

**Abstract:** The monitoring of antibiotic use in animals is a crucial element to ensure food safety. The main goal of this study was to analyse the distribution of selected antibiotics to porcine oral fluid, as well as to demonstrate that an oral fluid is an alternative to other biological matrices used in the control of antibacterials. Therefore, an animal study with pigs treated using seven different antibiotics was performed. Sulfadoxine (SDX) with trimethoprim (TRMP), lincomycin (LIN), tiamulin (TIAM), tylosin (TYL), amoxicillin (AMX) and penicillin G (PEN G) were injected intramuscularly to pigs, and concentrations of these analytes in the oral fluid were assessed. Ultra-high-performance liquid chromatography coupled with mass spectrometry (UHPLC-MS/MS) was used to quantify the analytes. On the first day of medication, the highest concentrations for SDX and TRMP at the level of 22,300 μg/kg and 14,100 μg/kg were found, respectively. The concentrations of LIN (10,500 μg/kg) and TIAM (7600 μg/kg) were also relatively high. The peak of TYL was recorded on the second day of drug administration. Most of the analytes were present in oral fluid for 30 days, apart from TYL, which was detected for 27 days. It was found that AMX and PEN G were quantified only for 5 and 8 days, respectively, at very low concentrations. It was found that oral fluid can be used for the verification of antibiotics on pig farms.

**Keywords:** antibiotics; food safety; oral fluid; pigs; UHPLC-MS/MS

**Citation:** Gajda, A.; Nowacka-Kozak, E.; Gbylik-Sikorska, M.; Cybulski, P. UHPLC-MS/MS Analysis of Antibiotics Transfer and Concentrations in Porcine Oral Fluid after Intramuscular Application. *Pharmaceuticals* **2022**, *15*, 225. https://doi.org/10.3390/ph15020225

Academic Editors: Jan Oszmianski and Sabina Lachowicz-Wiśniewska

Received: 24 January 2022
Accepted: 8 February 2022
Published: 14 February 2022



## 1. Introduction

Antimicrobial residues are of food safety concern. The residues of drugs may result in many biological adverse effects and allergic reactions in consumers, as well as the spread of drug-resistant bacteria and bacterial resistance acquisition [1]. Consolidation of pig production requires ensuring adequate conditions for the maintenance of animals with high health status, consistent with the guidelines for animal welfare [2,3]. At the same time, the control of antibiotics in animals is an important element providing the high quality of pig farming, as well as the protection of consumers. In the official residues monitoring programs implemented in the Member States of the European Union, antibiotics are mainly inspected in tissues, collected from animals at the slaughterhouses. However, pigs should be tested for potential antimicrobial residues on a farm because postmortem analysis does not enable the monitoring of antibiotics administration during rearing.

It is possible to determine drugs level in blood plasma; however collection of this type of material is associated with exposure of animals to stress, as well as considerable inconvenience for the person taking the samples. The pig's safety, stress reduction, time-saving, and material collection convenience, determine different sampling techniques. Even though urine and milk samples can be successfully utilised as matrices for the detection of drugs in animals, the first material in question is used solely for the analysis of banned compounds, whereas usage of the latter is limited to cows only [4]. In the antemortem drugs analysis, drinking water and feed are generally used, but it gives information about the source of contamination, not about the presence of residues in animals.

There are many benefits of using oral fluid in the control of antimicrobial residues compared to matrices used so far. Collecting oral fluid is animal-friendly and noninvasive, devoid of any needles or snout loop [5]. The technique of sampling requires the use of cotton rope, which is bitten and chewed to reduce the stress of animals. Oral fluid can be collected from a single animal, but it is usually collected from a group of animals, which offers the time and cost-effective approach [6]. A large population of pigs can be checked on the presence of antibacterials by taking one bulk sample, as one rope should be sufficient for even 40–50 pigs. The cost of analysing one bulk sample is significantly reduced compared to testing hundreds of individual samples, as in plasma analysis.

Oral fluid was demonstrated as a matrix in the evaluation of some metabolic diseases in humans by detecting a variety of analytes in the olden days [7]. After these early reports, developments in oral fluid diagnostics were generally obfuscated by improvements in the detection of some compounds in blood or serum [8]. Nevertheless, in the 1990s, intensive research on the oral fluid-based assays as a medium for infectious and non-infectious diseases, drugs, hormones and disease marker detection in humans were started and implemented again [9–11]. In veterinary medicine, significant research was carried out for the detection of many viral infections [12,13]. Literature data demonstrate a high probability of drug transmission after their administration [14]. However, very limited information about the detection of antibiotics or other compounds in oral fluid is presented. Recently, there were a few reports presenting the determination of some antibacterials such as ceftiofur and oxytetracycline in swine oral fluid [15]. In that study, both compounds were only qualitatively determined as positive or negative, using an ELISA test. In another work, described by Oruc et al., 2013, the application of a biochip array-based immunoassay in the detection of some antimicrobials in oral fluids was demonstrated, but this technique works only for clean samples collected under controlled research [16]. A suitable analytical method is an essential analytical tool in the determination of drugs in the matrix of interest. Therefore, in the paper of Gajda et al., 2017, a liquid chromatography–tandem mass spectrometry (LC-MS/MS) method for the determination of oxytetracycline and 4-epi oxytetracycline in swine oral fluid after intramuscular injection was reported [17]. The results of this study demonstrate that the tested material from pigs seems to be an effective tool for the control of proper medical treatment and residues prevention. However, to prove that oral fluid collected from medicated pigs is an appropriate matrix for the non-invasive detection of antimicrobials, more analytes should be tested by an accurate and sensitive method.

In this paper, the transfer and concentrations of seven different antibiotics in swine oral fluid, commonly used in pigs, were studied for the first time. In six separated experimental groups, animals were therapeutically treated with a veterinary product containing sulfadoxine (SDX) with trimethoprim (TRMP), lincomycin (LIN), tiamulin (TIAM), tylosin (TYL), amoxicillin (AMX), and penicillin G (PEN G), one drug per group. The aim of this study was to develop and validate a multiresidue method for the simultaneous determination of seven antibiotics as well as demonstrate an oral fluid as an alternative to other biological matrices used so far in the control of antibacterials.

## 2. Results

### 2.1. Validation of an Analytical Method

The UHPLC-MS/MS method used in this research for the determination of seven different compounds in oral fluid was fully validated. All matrix-matched calibration curves showed good linearity ($r^2 > 0.997$) for all analytes. The coefficients of variation (CVs, %) for repeatability, were lower than 10% and below 15% for within-laboratory reproducibility. The selectivity testing allowed us to verify that no peaks from endogenous compounds were detected in the retention time corresponding to each analyte or internal standard. The average recoveries were in the range of 87.6–107%, depending on the analyte.

The method was satisfactorily sensitive with limits of quantification (LOQs) established in the range of 2–10 µg/L, whereas limits of detection (LODs) were in the range of 1–5 µg/L. The results are shown in Table 1. The chromatograms of oral fluid samples spiked at validation level (VL) with all analysed compounds are presented in Figure 1.

**Table 1.** Validation parameters.

| Analyte | Repeatability (CV, %) | Reproducibility (CV, %) | Recovery (%) | LOD (µg/L) | LOQ (µg/L) |
|---|---|---|---|---|---|
| Sulfadoxine | 6.39 ± 1.3 | 12.7 ± 2.6 | 87.6 ± 6.6 | 2 | 5 |
| Trimetophrim | 8.92 ± 1.1 | 11.45 ± 2.4 | 97.3 ± 4.4 | 2 | 5 |
| Lincomycin | 7.07 ± 1.5 | 13.2 ± 2.9 | 94.2 ± 4.2 | 1 | 2 |
| Tiamulin | 4.66 ± 1.1 | 10.1 ± 2.2 | 106 ± 5.6 | 1 | 2 |
| Tylosin | 9.59 ± 1.6 | 14.2 ± 3.1 | 87.9 ± 4.9 | 2 | 5 |
| Amoxycyllin | 9.55 ± 1.8 | 14.9 ± 3.8 | 107 ± 4.6 | 5 | 10 |
| Penicillin G | 9.91 ± 1.9 | 12.0 ± 2.7 | 101 ± 4.1 | 2 | 5 |

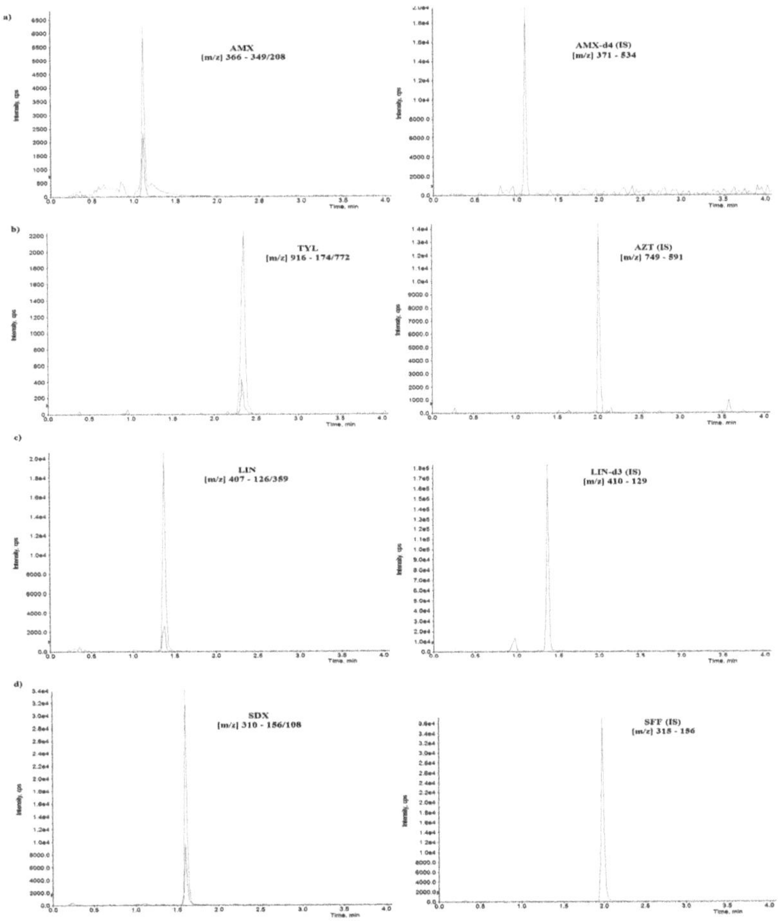


**Figure 1.** *Cont.*

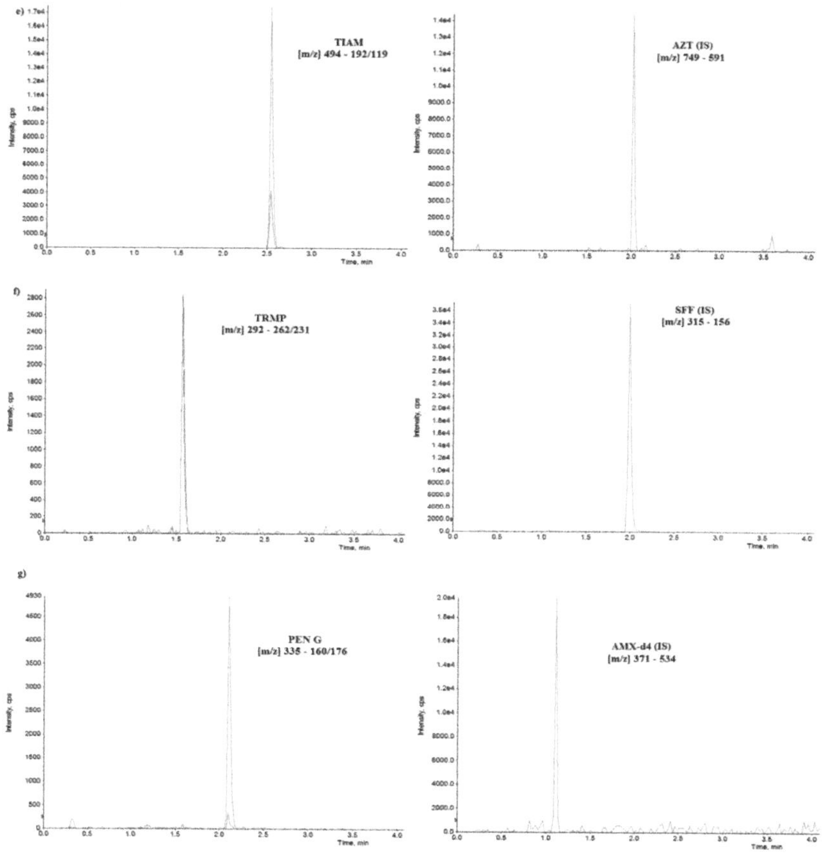


**Figure 1.** Chromatograms of oral fluid samples spiked at validation level (VL) with analysed compounds and corresponding internal standards (IS), (**a**) AMX—amoxicillin, (**b**) TYL—tylosin, AZT—azithromycin, (**c**) LIN—lincomycin, (**d**) SDX—sulfadoxine, SFF—sulfaphenazole, (**e**) TIAM—tiamulin, (**f**) TRMP—trimethoprim, (**g**) PEN G—penicillin G.

### *2.2. Detection and Quantification of SDX, TRMP, TYL, TIAM, LIN, AMX and PEN G in Oral Fluid Samples*

For the quantitative analysis of SDX, TRMP, TYL, TIAM, LIN, AMX and PEN G, an equation from the regression analysis of the matrix-matched calibration curve was used, and the corresponding internal standards were used: SFF for SDX and TRMP, AZT for TYL and TIAM, LIN-d3 for LIN, AMX-d4 for AMX and PEN G. Before the start of the experiment, all pigs were checked to be negative for all tested substances. After intramuscular injection of medicines according to therapeutic indications: two doses with 48h interval for SDX + TRMP and AMX, and three doses with 24 h interval for TYL, TIAM, LIN, PEN G, all compounds were detectable in oral fluid. However, significant differences in concentration on the first day were observed. The highest concentrations for SDX and TRMP at the level of 22,300 μg/L and 14,100 μg/L was found, respectively. The concentrations of LIN (10,500 μg/L) and TIAM (7600 μg/L) were also relatively high on the first day of medication. The peak of TYL was recorded on the second day of drug administration, but concentrations in the oral fluid were much lower from the abovelisted compounds. It was found that AMX and PEN G were poorly passed to oral fluid, and the level of these analytes was determined as 11.3 μg/L for AMX and 93 μg/L for PEN G. The concentrations of antimicrobials at each time point are listed in Table 2. The content of analytes decreased gradually over time. After three days of treatment, the highest concentration of 4010 μg/L for SDX was recorded. The levels of LIN and TIAM after the finishing of drugs administration were similar, with concentrations of 1550 μg/L for LIN and 1560 μg/L for TIAM. The sampling took 30 days, and after this time, SDX + TRMP were detected at the level of 23.0 μg/L and 22.4 μg/L, respectively. Slightly higher concentrations were recorded for LIN and TIAM (58.1 μg/L and 46.9 μg/L, respectively) on the 30th day. TYL persisted a little less and the 27th day was the last day with a quantified concentration. In contrast, PEN G was present in oral fluid for 8 days, while AMX persisted for only 5 days. Different withdrawal periods for each antimicrobial were established. The length of withdrawal periods for each drug used in the experiment are shown in Table 3, while the concentrations found at this time are presented in Table 2. At a withdrawal period of 8 days for SDX + TRMP, the concentrations were determined as 247 μg/L and 67.1 μg/L, respectively. The highest level at withdrawal time for LIN—437 μg/L was recorded. Low concentration for TYL at this time point was found (6.6 μg/L), while AMX and PEN G were below the limit of quantification of the applied analytical method.

**Table 2.** Results of oral fluid samples analysis.

| | | Avarage Concentrations (μg/L) | | | | | | |
|---|---|---|---|---|---|---|---|---|
| **Period** | **Time (d)** | **SDX + TRMP** | | **LIN** | **TIAM** | **TYL** | **AMX** | **PEN G** |
| Treatment | 1 | 22,300 ± 3555 | 14,100 ± 2556 | 10,500 ± 1889 | 7600 ± 1055 | 396 ± 68 | 11.3 ± 1.6 | 93 ± 21 |
| | 2 | 5860 ± 1155 | 2810 ± 556 | 2960 ± 446 | 2520 ± 355 | 958 ± 166 | 15.2 ± 2.3 | 50 ± 15 |
| | 3 | 4570 ± 766 | 1810 ± 311 | 4250 ± 559 | 2460 ± 311 | 832 ± 142 | 14.6 ± 1.8 | 11 ± 4.6 |
| Posttreatment | 4 | 4010 ± 601 | 1030 ± 221 | 1550 ± 211 | 1560 ± 225 | 654 ± 111 | 11.4 ± 1.4 | 16 ± 5.4 |
| | 5 | 5780 ± 788 | 1400 ± 255 | 1674 ± 199 | 1540 ± 233 | 538 ± 96 | 11.2 ± 1.6 | 5.2 ± 1.1 |
| | 6 | 4070 ± 669 | 1280 ± 198 | 1790 ± 203 | 2960 ± 299 | 475 ± 89 | <LOQ | 52 ± 13 |
| | 7 | 310 ± 98 | 264 ± 73 | 1480 ± 186 | 2570 ± 254 | 175 ± 54 | <LOQ | 32 ± 7.2 |
| | 8 | 392 ± 88 | 125 ± 41 | 437 ± 101 * | 875 ± 156 | 118 ± 41 | <LOQ | 6 ± 1.2 |
| | 9 | 386 ± 72 | 128 ± 38 | 457 ± 96 | 567 ± 111 | 158 ± 62 | <LOQ | <LOQ * |
| | 10 | 134 ± 49 | 36.4 ± 15 | 420 ± 83 | 621 ± 124 | 49.3 ± 13 | <LOQ | <LOQ |
| | 11 | 247 ± 62 * | 67.1 ± 20 * | 194 ± 55 | 279 ± 66 | 54.4 ± 16 | <LOQ * | <LOQ |
| | 12 | 74.8 ± 21 | 25.2 ± 9.6 | 257 ± 76 | 254 ± 69 | 140 ± 36 | <LOQ | <LOQ |
| | 13 | 68.0 ± 16 | 22.0 ± 8.7 | 131 ± 41 | 138 ± 43 * | 32.3 ± 11 | <LOQ | <LOQ |
| | 14 | 77.4 ± 18 | 22.1 ± 6.4 | 125 ± 33 | 269 ± 71 | 30.6 ± 8.4 | <LOQ | <LOQ |
| | 15 | 69.8 ± 14 | 31.6 ± 11 | 277 ± 61 | 156 ± 41 | 19.2 ± 5.2 | <LOQ | <LOQ |
| | 16 | 78.4 ± 12 | 24.2 ± 8.2 | 337 ± 69 | 279 ± 68 | 52.8 ± 33 | <LOQ | <LOQ |
| | 17 | 47.0 ± 18 | 18.9 ± 4.8 | 128 ± 28 | 182 ± 49 | 31.9 ± 10 | <LOQ | <LOQ |
| | 18 | 44.2 ± 16 | 13.0 ± 3.6 | 302 ± 51 | 176 ± 52 | 13.9 ± 3.6 | <LOQ | <LOQ |
| | 19 | 58.4 ± 11 | 42.0 ± 14 | 279 ± 44 | 202 ± 45 | 16.1 ± 4.6 | <LOQ | <LOQ |
| | 20 | 76.4 ± 14 | 36.4 ± 10 | 376 ± 52 | 133 ± 36 | 25.4 ± 7.7 | <LOQ | <LOQ |
| | 21 | 21.4 ± 7.3 | 10.2 ± 4.4 | 252 ± 39 | 148 ± 32 | 9.2 ± 2.1 | <LOQ | <LOQ |
| | 22 | 25.3 ± 6.9 | 46.4 ± 16 | 142 ± 24 | 220 ± 44 | 5.5 ± 1.1 | <LOQ | <LOQ |
| | 23 | 15.9 ± 7.2 | 6.9 ± 2.4 | 298 ± 33 | 146 ± 29 | 5.4 ± 0.8 | <LOQ | <LOQ |
| | 24 | 37.0 ± 5.4 | 19.4 ± 5.5 | 326 ± 47 | 310 ± 55 | 6.6 ± 0.9 * | <LOQ | <LOQ |
| | 25 | 31.5 ± 6.3 | 17.8 ± 4.1 | 180 ± 44 | 193 ± 33 | 5.0 ± 1.3 | <LOQ | <LOQ |
| | 26 | 19.1 ± 4.2 | 15.9 ± 3.6 | 160 ± 36 | 66.4 ± 12 | 3.0 ± 0.5 | <LOQ | <LOQ |
| | 27 | 19.3 ± 3.7 | 10.4 ± 2.9 | 205 ± 43 | 54.3 ± 14 | 6.8 ± 1.7 | <LOQ | <LOQ |
| | 28 | 35.0 ± 8.1 | 16.5 ± 6.2 | 103 ± 26 | 35.8 ± 17 | <LOQ | <LOQ | <LOQ |
| | 29 | 30.4 ± 7.6 | 21.2 ± 9.4 | 85.6 ± 16 | 40.6 ± 21 | <LOQ | <LOQ | <LOQ |
| | 30 | 23.0 ± 6.2 | 22.4 ± 9.1 | 58.1 ± 12 | 46.9 ± 19 | <LOQ | <LOQ | <LOQ |

LOQ—limit of quantification. *—withdrawal period for each compound.

**Table 3.** Drugs and dosage administered to animals.

| Administered Drug | Active Substance | Group of Antibacterials | Withdrawal Period for Tissues (Days) | Dosage (Active Substance Per 1 kg of Body Weight) | Practical Dosage |
|---|---|---|---|---|---|
| Borgal 24% | trimethoprim | diaminopyrimidines | 8 | 3 mg of trimethoprim and 12 mg of sulfadoxine | 1 mL for 16 kg, 2× with 48 h interval |
| | sulfadoxine | sulfonamides | | | |
| Biotyl 200 | tylosin | macrolide | 21 | 5 mg | 1 mL for 40 kg, 3× with 24 h interval |
| Lincomycin VMD | lincomycin | lincosamides | 5 | 10 mg | 1 mL for 10 kg, 3× with 24 h interval |
| Probencil | penicillin G | penicillins | 6 | 10 mg | 1 mL for 30 kg, 3× with 24 h interval |
| Tiamowet 200 | tiamuline | pleuromutilins | 10 | 6 mg | 1 mL for 20 kg, 3× with 24 h interval |
| Vetrimoxin LA | amoxicillin | penicillins | 8 | 15 mg | 1 mL for 10 kg, 2× with 48 h interval |

## 3. Discussion

The antimicrobials overused in domestic animals can lead to residues in tissues and other animal products intended for human consumption. In some cases, the lack or inadequate cleaning of water supply systems or feed dispensers after previous treatment can also give positive results. Apart from the risk to public health, the presence of antibiotics in tissues often leads to the removal and destruction of a significant quantity of meat, which contributes to a serious economic consequence. Therefore, a new method for the control of antibacterials during rearing is strongly needed. The fast and simple sampling of oral fluid with the cotton rope is an effective tool for the monitoring of reasonable treatment. In case of positive oral fluid analysis results, appropriate measures can be taken to prevent the slaughtering of positive animals intended for sale in order to avoid huge costs connected with the disposal and utilization of contaminated meat.

For the analysis of antibacterials in oral fluid, a UHPLC-MS/MS analytical method was developed. The extraction used in the presented method allows for simultaneous analyses of seven different compounds from six various chemical groups. Additionally, the simple and fast sample preparation, as well as a cleanup step by filtration, give an opportunity to test a lot of samples in a short time. The paper described a biochip array-based immunoassay method for the detection of six different antimicrobial drugs, tetracyclines were very poorly recovered in dirty samples (5%), and for other drugs (norfloxacin, ceftiofur and florfenicol), recovery values were much higher than 100% [16]. This means that the presented technology with a biochip array-based immunoassay can work only for clean oral fluid samples and is of value if it is being used under typical field conditions. Field oral fluid samples are reflective of the environment in which pigs are housed. Therefore, it was necessary to develop such a method enabling the analysis of all compounds with sufficient sensitivity, accuracy and precision with reliable recovery values. In our previous research, we tested the passage and concentrations of oxytetracycline after i.m. injection [17]. For that purpose, a sensitive LC-MS/MS method was developed with satisfactory validation results. That experiment proved that oxytetracycline is present in oral fluid for a long time after medication. At the withdrawal period (21st day), the concentration was 30.8 μg/kg in the group where all pigs were treated and 11.6 μg/kg in the group where the antibiotic was administered to half of the animals in one pen. However, in the present study, the experiment with seven other compounds was performed for the first time.

In the described work, oral fluid was tested for the detection of some antibacterials for food safety applications. To demonstrate the validity of this matrix, seven drugs from various groups of antibacterials with different chemical and therapeutic properties were administered by intramuscular injection. The compounds tested in this experiment were selected based on a review of the use of antimicrobial drugs used in swine [18]. The route of medicine application is an important element in this research. In pigs treated with medicated feed or water, the presence of drugs can be expected in saliva for most drugs. However, some reports described that β-lactam antibiotics are neutralized by different enzymes present in saliva [19]. Therefore, in our study, AMX and PEN G, which belong to this group of antibacterials, were chosen in the experiment to check the penetration of these substances and analyse the matrix after intramuscular (i.m.) injection. The value of sulphonamides as single antimicrobial agents was greatly diminished both by widespread acquired resistance and by their relatively low potency. However, when combined with diaminopirymidynes such as trimethoprim, their usefulness was enhanced thus SDX with TRMP were applied.

Oral fluid is a mixture of saliva, oral mucosal transudate and gingival crevicular fluid [20]. Antimicrobials can be transported from blood to saliva by simple diffusion and/or active transport [11,21–23]. Factors such as lipid solubility, molecular size, degree of ionization of the drug molecule, as well as the effect of salivary pH and the degree of protein binding of the drug, determine the drug availability in oral fluids [24]. Concerning antimicrobials that are weak bases, lipophilic compounds and small molecules diffuse more easily and may reach high concentrations in saliva [24]. SDX and TRMP, which are relatively small molecules (310.33 g/mol and 290.32 g/mol, respectively), reached the highest concentrations in oral fluid at 24 h post-treatment (22,300 μg/L for SDX and 14,100 μg/L for TRMP), while the level of TYL, with molecular weight at 916.1 g/mol, was quantified at 396 μg/L at the same time. The results of our study demonstrate that high lipid-soluble compounds diffuse more easily and reach higher levels in oral fluid. SDX and TRMP are lipophilic compounds, similar to TYL, TIAM and LIN, which are high lipid-soluble substances, and the result of these substances was found greater than for hydrophilic AMX or high water-soluble PEN G.

The pharmacokinetics properties determine, to some extent, the distribution of drugs in many structures of the organism. The results of our research demonstrate the strong relationship between the pharmacokinetics and penetration to oral fluid for most of the analysed substances. Concentrations of most compounds presented in the experiment were high and reached great levels in oral fluid. Most of the analysed antibiotics are basic compounds with relatively high volumes of distribution. LIN is a basic compound with pKa values of about 7.6. It has high lipid solubility and consequently a large apparent volume of distribution. LIN is widely distributed in many fluids and tissues. This antibiotic diffuses across the placenta in many species, and the apparent volume of distribution is >1 L/kg [25]. For TIAM, which is a weak base, pKa 7.6, high levels in the oral fluid were observed, similar to LIN [25]. TYL, as a weak base, pKa 7.1, with a good volume of distribution, reaches high concentrations in many distribution spaces, e.g., in kidneys, lungs, spleen, liver or milk, as well as oral fluid, where good distribution of this macrolide antibiotic was observed [26].

PEN G, for which concentrations in the oral fluid were very low, is an organic acid with pKa 2.7. The group of penicillins have relatively small apparent volumes of distribution (0.2–0.3 L/kg) and short half-lives (0.5–1.2 h) in all species of domestic animals [27]. Acid hydrolysis in the stomach limits the systematic availability of most penicillins from oral preparations. After absorption, they are widely distributed in the extracellular fluid of the body but cross biologic membranes slightly since they are ionized and poorly lipid-soluble [28]. However, for SDX, a weak organic acid with a pKa value of 6.16, the concentrations in the oral fluid were high and lasted for a long time. Most sulfonamides exist predominantly in a non-ionized form in biologic fluids with a pH lower than their pKa [29].

The transport of some drugs to the oral fluid can also be determined by the degree of protein binding. It was observed that compounds with high protein binding show a greater ability for diffusion to saliva. Sulfonamides and diaminopyrimidines, which are generally highly bound to proteins in serum (60–90%), as well as lincosamides, in which serum protein binding is 72%, reached high concentrations in oral fluid [25,29]. Substances with a low binding degree to plasma proteins, i.e., TYL < 50% and AMX ≈ 25%, penetrate to oral fluid in lesser extent [28,30].

## 4. Materials and Methods

### *4.1. Animal Experiment and Sample Collection*

All animals were born in the sow farm (8000 DanBred sows) located in Northern Poland, after crossing ♀(♂Landrace × ♀Yorskshire) × ♂Duroc. Piglets were weaned after 4 weeks of lactation with an average weight of 6 kg and then moved to the weaner stable (23,000 weaners). After the next 7 weeks, animals were transported to the 9000-places fattening farm with an average body weight of 28 kg, where the trial was performed. All farms in the aforementioned three-phase model used weekly batches and an all-in all-out system. Specialist veterinary and production management care were provided in each location.

For the purpose of the study, 240 immunologically castrated boars were divided into 8 groups, 30 animals in each pen. For the control group, 30 pigs were used in separate pens. Their initial body weight was between 40 and 60 kg. Pigs were housed at 20–25 °C. Fatteners were fed ad libitum with antimicrobial-free water and feed provided by feeders throughout the study. Before the experiment, water and feed were checked for antimicrobials, and no antimicrobial treatments were administered to these pigs before. All animals in one pen were treated due to the same health issue using disposable needles and the same drug and dosage. Pens were separated from each other with concrete walls. The drugs and dosage administered to each experimental group with an indication of withdrawal period for tissues are listed in Table 3. The oral fluid samples were collected using three cotton ropes (70 cm length with 1 cm diameter) in each pen suspended separately for 25 min on the front wall of the pen with bottoms at the pigs' shoulder height, away from water and feed.

The research material was one pooled sample of oral fluid from 10 pigs, in triplicate, per pen. Ropes were removed immediately after collection. The first oral fluid sample was collected 24 h after the first injection. The collection of each subsequent sample took place at the same time (11:30 A.M.) for 35 consecutive days. Additionally, oral fluid samples were collected from pigs not treated with any antibiotics. Oral fluid from each pen was pooled into one 50-mL Falcon plastic tube and centrifuged at 9447× rcf for 15 min. Centrifuged material was transferred to a 15 mL plastic tube and frozen at −20 °C until analysis.

### *4.2. Quantitative Analysis by UHPLC-MS/MS*

#### 4.2.1. Reagents and Chemicals

All reagents used were of analytical grade. Reference standards of SDX, TRMP, LIN, TIAM, TYL, AMX and PEN G were used for the analysis and quantification of the analytes. Lincomycin-$d_3$ (LIN-$d_3$), amoxicillin-$d_4$ (AMX-$d_4$), sulfaphenazole (SFF) and azithromycin (AZT) were used as internal standards. All standards were manufactured and obtained from LGC Standards (Teddington, Middlesex, UK). Formic acid was from Sigma Aldrich (St. Louis, MO, USA). Acetonitrile was obtained from J.T. Baker (Deventer, The Netherlands). Heptafluorobutyric acid was from Fluka (St. Louis, MO, USA). Syringe 0.22 μm Hydrophilic Polyvinylidene Fluoride (PVDF) Membrane Filters were from Restek (College, PA, USA).

#### 4.2.2. Preparation of the Standard Stock Solutions and Working Solutions

Individual stock standard solutions (1000 μg/mL) for AMX, AMX-d4, and PEN G were prepared in ultrapure water and stored in polypropylene vessels. LIN, LIN-d3, TYL, TIAM, AZT, SDX, TRMP and SFF stock standard solutions (1000 μg/mL) were dissolved in methanol. All individual stock standard solutions were stable for at least 6 months when retained in a dark place at −18 °C. A working solution for each analyte was prepared from stock solutions by dilution in ultrapure water and stored at 4–8 °C for 1 month.

4.2.3. Extraction and Cleanup

Before the analysis, the oral fluid samples were centrifuged at 9447× rcf. Next, an aliquot (1 mL) of oral fluid was placed into a 2 mL polypropylene centrifuge tube, and the 30 μL of the internal standard at the concentration of 2 μg/mL was added. For the extraction of analysed antibiotics, 600 μL of 0.5% formic acid was added, vortexed for 30 s and centrifuged for 10 min at 9447× rcf. Then, the supernatant was filtered through a 0.22 mm PVDF filter into an amber glass vial before chromatographic analysis.

4.2.4. LC-MS/MS Analysis

The analysis of SDX, TRMP, LIN, TIAM, TYL, AMX and PEN G was conducted by an UHPLC Shimadzu Nexera X2 (Shimadzu, Kyoto, Japan) system connected to the SCIEX 4500 triple quadrupole mass spectrometer (Sciex, Framingham, MA, USA) controlled by Analyst 1.6.2 software (SCIEX, Framingham, MA, USA). The mass spectrometry detection was operated in the positive ESI mode with multiple reaction monitoring (MRM). The operating parameters were set as follows: temperature—450 °C, curtain gas ($N_2$)—20; nebulizer gas ($N_2$)—60; collision gas ($N_2$)—medium; auxiliary gas—65; ion spray voltage—4500 V. Mass spectrometric conditions are shown in Table 4.

**Table 4.** MS/MS parameters and summary of the MRM monitored for presented analytes.

| Analyte | Parent Ion (*m*/*z*) | Daughter Ion(s) (*m*/*z*) | CXP (V) | DP (V) | CE (eV) | Dwell Time (msec) |
|---|---|---|---|---|---|---|
| SDX | 310.9 | 156; 108 | 13 | 60 | 25; 40 | 50 |
| TRMP | 292.1 | 262; 231 | 5 | 52 | 36; 33 | 70 |
| LIN | 407.2 | 126; 359 | 7 | 74 | 36; 28 | 50 |
| TIAM | 494.4 | 192; 119 | 18 | 128 | 30; 56 | 50 |
| TYL | 916 | 174; 772 | 20 | 110 | 52; 42 | 50 |
| AMX | 366.1 | 349; 208 | 8 | 45 | 12; 18 | 100 |
| PEN G | 335.1 | 160; 176 | 12 | 60 | 17; 48 | 100 |
| SFF (IS) | 315.1 | 156 | 15 | 90 | 26 | 50 |
| LIN $d_3$ (IS) | 410.0 | 129 | 13 | 66 | 44 | 50 |
| AMX $d_4$ (IS) | 371.0 | 354 | 13 | 35 | 15 | 50 |
| AZT (IS) | 759.0 | 591 | 13 | 89 | 40 | 50 |

The chromatographic separation assay was performed using an Agilent Zorbax SB (50 × 2.1 mm, 1.8 μm) column (Agilent, St Clara, CA, USA) with an octadecyl guard column (2 × 4 mm) maintained at 35 °C. The mobile phase consisted of 0.025% heptafluorobutyric acid (A) and acetonitrile (B) at a flow rate of 0.6 mL/min with an injection volume of 5 μL. Gradient elution was conducted as follows: 0–4 min 90% A, 4–5.3 min 20% A and finally from 5.31 to 7 min back to 90% A. The total run time was set as 7 min.

*4.3. Method Validation*

The method was validated according to the Commission Decision 2002/657/EC [31]. The following parameters were established: linearity, selectivity, precision (repeatability and within-laboratory reproducibility) and recovery. In addition, LOD and LOQ were estimated (EUR 28,099 EN) [32]. The linearity in the matrix of oral fluid was checked by preparing two calibration curves at six different spiked levels and at low concentration ranges: 0, 0.5, 1.0, 1.5, 2.0, 5.0× validation level (VL), as well as high concentrations 10, 20, 50, 80, 100, 120× VL. VLs were different depending on the analyte, and the following levels were set: VL = 100 μg/L for TIAM, LIN, SDX, TRMP, TYL and VL = 10 μg/l for PEN G and AMX. The linearity was calculated as the squares linear regression by plotting the analyte/internal standard area ratio response versus the analyte/internal standard added concentration. The selectivity was evaluated by analysing different origin oral fluid samples ($n$ = 20) and checked for potential interferences with endogenous substances. The precision for each analyte was determined by the repeated analysis ($n$ = 6) of oral fluid samples spiked with SDX, TRMP, TYL, TIAM, LIN, AMX and PEN G at three concentrations corresponding to 0.5, 1.0, 1.5× VL. The precision was calculated and expressed as the percentage coefficients of variation (CV, %). For repeatability, samples were analysed on the same day by the same operator. For within laboratory reproducibility, another two sets of fortified samples at the same concentration levels as for the repeatability were analysed on two different days with different operators. The extraction recovery experiment was carried out at the three concentration levels, in the same experiment as precision, by comparing the average area obtained for each analyte with the concentrations of analytes in spiked samples. Additionally, LOQ as the lowest point of the matrix-matched calibration curve, where the least abundant diagnostic ion for quantitative analysis can be detected was determined. The LOD was estimated as the signal to noise ratio (S/N) of 3, and LOQ was determined as the lowest validated concentration that produced a signal to noise ratio > 10.

## 5. Conclusions

The results of conducted experiment illustrate that oral fluid can be used to monitor most antibacterial substances administered in pig farms. This paper shows the usefulness of oral fluid as a biological material for the non-invasive detection of drugs administered on pig farms. An analytical UHPLC-MS/MS method for the simultaneous determination of seven antibacterials in the oral fluid has been developed and validated. The implemented method is suitable for the simultaneous rapid detection and quantitation of SDX, TRMP, TYL, TIAM, LIN, AMX and PEN G in porcine oral fluid. The concentrations of these compounds found in this work showed high distribution to oral fluid for SDX, TRMP, TYL, TIAM and LIN, except for AMX and PEN G, where low penetration was observed.

Analysis of antibiotics in oral fluid offers a cost-effective approach for the screening of an individual animal or populations of pigs. The results of this study demonstrate the oral fluid analysis as an important "security tool" for the rapid verification of the declaration of treatment on pig farms.

**Author Contributions:** Conceptualization, A.G., E.N.-K., M.G.-S. and P.C.; methodology, A.G. and E.N.-K.; validation, A.G., E.N.-K. and M.G.-S.; investigation, A.G., E.N.-K. and M.G.-S.; writing—original draft preparation, A.G., E.N.-K., M.G.-S. and P.C.; writing—review and editing, A.G., E.N.-K., M.G.-S. and P.C.; visualization, A.G., E.N.-K., M.G.-S. and P.C.; supervision, A.G., M.G.-S. and P.C. All authors have read and agreed to the published version of the manuscript.

**Funding:** This research received no external funding.

**Institutional Review Board Statement:** Ethical review and approval were waived for this study, as the analysed material originated from routine diagnostic investigations ordered by the farm owners.

**Informed Consent Statement:** Not applicable.

**Data Availability Statement:** Data is contained within the article.

**Conflicts of Interest:** The authors declare no conflict of interest.

## References

1. Brill, N.; Krasse, B.O. The passage of tissue fluid into the clinically healthy gingival pocket. *Acta Odontol. Scand.* **1958**, *16*, 233–245. [CrossRef]
2. Council Directive 2008/120/EC of 18 December 2008 laying down minimum standards for the protection of pigs. *Off. J. Eur. Union L* **2009**, *47*, 5–13.
3. Van de Veerd, H.; Ison, S. Providing Effective Environmental Enrichment to Pigs: How Far Have We Come? *Animals* **2019**, *9*, 254. [CrossRef] [PubMed]
4. Council Directive 96/23/EC on Measures to Monitor Certain Substances and Residues Thereof in Live Animals and Animal Products and Repealing Directives 85/358/EEC and 86/469/EEC and Decisions 89/187/EEC and 91/664/EEC. Available online: https://eur-lex.europa.eu/legal-content/en/ALL/?uri=CELEX:31996L0023 (accessed on 1 July 2019).
5. Prickett, J.R.; Zimmerman, J.J. The development of oral fluid-based diagnostics and applications in veterinary medicine. *Anim. Health Res. Rev.* **2010**, *11*, 207–216. [CrossRef] [PubMed]
6. White, D.; Rotolo, M.; Olsen, C.; Wang, C.H.; Prickett, J.; Kittawornrat, A.; Panyasing, Y.; Main, R.; Rademacher, C.H.; Hoogland, M.; et al. Recommendations for pen-based oral-fluid collection in growing pigs. *J. Swine Health Prod.* **2014**, *22*, 138–141.

7. Michaels, P. Saliva as an aid in the detection of diathetic diseases. *Dent. Digest.* **1901**, *7*, 105–110.
8. Drobitch, R.K.; Svensson, C.K. Therapeutic drug monitoring in saliva. *An update. Clin. Pharamcokinet.* **1992**, *23*, 365–379. [CrossRef]
9. Pfaffe, T.; Cooper-White, J.; Beyerlin, P.; Kostner, K.; Punyadeera, C.H. Diagnostic potential of saliva: Current state and future applications. *Clin. Chem.* **2011**, *57*, 675–687. [CrossRef]
10. Mandel, I.D. Salivary diagnosis: Promises, promises. *Ann. N. Y. Acad. Sci.* **1993**, *694*, 1–10. [CrossRef]
11. Ghalaut, P.; Ghalaut, V.; Yadav, S.; Lekhvani, S.; Yadav, A. Diagnostic applications of saliva. *J. Clin. Diagn. Res.* **2010**, *4*, 2330–2336.
12. Prickett, J.R.; Kim, W.; Simer, R.; Yoon, K.J.; Zimmerman, J. Oral-fluid samples for surveillance of commercial growing pigs for porcine reproductive and respiratory syndrome virus and porcine circovirus type 2 infections. *J. Swine Health Prod.* **2008**, *16*, 86–91.
13. Kittawornrat, A.; Engle, M.; Johnson, J.; Prickett, J.; Schwartz, T.; Whitney, D.; Olsen, C.; Chittick, W.; Schwartz, K.; Wang, C.; et al. Porcine reproductive and respiratory syndrome virus (PRRSV) in serum and oral fluid samples from individual boars:will oral fluid replace serum for PRRSV surveillance? *Virus Res.* **2010**, *154*, 170–176. [CrossRef] [PubMed]
14. Pink, R.; Simek, J.; Vondrakova, J.; Faber, E.; Michl, P.; Pazdera, J.; Indrak, K. Saliva as a diagnostic medium. *Biomed. Pap.* **2009**, *153*, 103–110. [CrossRef]
15. Meiszberg, A.; Karriker, L.; Zimmerman, J.; Irwin, C.H.; Coetzee, J. Detection of ceftiofur and oxytetracycline in oral fluids of swine with a pen-side competitive ELISA test after intramuscular injection. *J. Vet. Pharmacol. Ther.* **2011**, *34*, 515–517. [CrossRef] [PubMed]
16. Oruc, H.H.; Olsen, C.H.; Rumbeiha, W.K.; Schrunk, D.E.; Ensley, S. Simultaneous detection of six different groups of antimicrobial drugs on porcine oral fluids using a biochip array-based immunoassays. *Kafkas Univ. Vet. Fak. Derg.* **2013**, *19*, 407–412.
17. Gajda, A.; Jablonski, A.; Bladek, T.; Posyniak, A. Oral fluid as a biological material for antemortem detection of oxytetracycline in pigs by liquid chromatography—Tandem mass spectrometry. *J. Agric. Food Chem.* **2017**, *65*, 494–500. [CrossRef]
18. Pejsak, Z.; Truszczynski, M. Antimicrobial drugs used in swine. *Zycie Wet.* **2016**, *91*, 254–257.
19. Sommer, M.O.A.; Dantas, G.; Church, G.M. Functional characterization of the antibiotic resistance reservoir in the human microflora. *Science* **2009**, *325*, 1128–1131. [CrossRef]
20. Delima, A.J.; Van Dyke, T.E. Origin and function of the cellular components in gingival cervice fluid. *Periodontol* **2003**, *31*, 55–76. [CrossRef]
21. Kaufman, E.; Lamster, I.B. The diagnostic applications of saliva—A revive. *Crit. Rev. Oral Biol. Med.* **2002**, *13*, 197–212. [CrossRef]
22. Haeckel, R. Factor influencing the saliva/plasma ratio of drugs. *Ann. N. Y. Acad. Sci.* **1993**, *20*, 128–142. [CrossRef] [PubMed]
23. Barza, M. Challenges to antibiotic activity in tissue. *Clin. Infect. Dis.* **1994**, *19*, 910–915. [CrossRef]
24. Haeckel, R.; Hanecke, P. Application of saliva for drug monitoring. An in vivo mode for transmembrane transport. *Eur. J. Clin. Chem. Clin. Biochem.* **1996**, *34*, 171–191. [PubMed]
25. Giguere, S. Lincosamides, Pleuromutilins and Streptogramins. In *Antimicrobial Therapy in Veterinary Medicine*, 4th ed.; Chapter 11; Giguere, S., Prescott, J.F., Baggot, J.D., Walker, R.D., Dowling, P.M., Eds.; Blackwell Publishing: Ames, IA, USA, 2006; pp. 179–191.
26. Giguere, S. Macrolides, Azalides and Ketolides. In *Antimicrobial Therapy in Veterinary Medicine*, 4th ed.; Chapter 12; Giguere, S., Prescott, J.F., Baggot, J.D., Walker, R.D., Dowling, P.M., Eds.; Blackwell Publishing: Ames, IA, USA, 2006; pp. 191–207.
27. Giguere, S. Beta-lactam Antibiotics: Penam Penicillins. In *Antimicrobial Therapy in Veterinary Medicine*, 5th ed.; Chapter 8; Giguere, S., Prescott, J.F., Baggot, J.D., Walker, R.D., Dowling, P.M., Eds.; Blackwell Publishing: Ames, IA, USA, 2013; pp. 133–152.
28. Riviere, J.; Lees, P. USP Veterinary Pharmaceutical Information Monographs—Antibiotics: Aminopenicillins. *J. Vet. Pharmacol. Ther.* **2003**, *26*, 33–46.
29. Giguere, S. Sulfonamides, Diaminopyrimidines and Their Combinantions. In *Antimicrobial Therapy in Veterinary Medicine*, 5th ed.; Chapter 17; Giguere, S., Prescott, J.F., Baggot, J.D., Walker, R.D., Dowling, P.M., Eds.; Blackwell Publishing: Ames, IA, USA, 2013; pp. 279–294.
30. Riviere, J.; Lees, P. USP Veterinary Pharmaceutical Information Monographs—Antibiotics: Macrolides. *J. Vet. Pharmacol. Ther.* **2003**, *26*, 119–144.
31. Commission Decision of 12 August 2002 implementing Council Directive 96/23/EC concerning the performance of analytical methods and the interpretation of results (2002/657/EC). *Off. J. Eur. Union L* **2002**, *221*, 8–36.
32. Wenzl, T.; Haedrich, J.; Schaechtele, A.; Robouch, P.; Stroka, J. *Guidance Document on the Estimation of LOD and LOQ for Measurements in the Field of Contaminants in Feed and Food*; JRC Technical Report; European Commission: Brussels, Belgium, 2016; EUR 28099 EN; Available online: https://publications.jrc.ec.europa.eu/repository/handle/JRC102946 (accessed on 11 June 2019).

MDPI
St. Alban-Anlage 66
4052 Basel
Switzerland
www.mdpi.com

*Pharmaceuticals* Editorial Office
E-mail: pharmaceuticals@mdpi.com
www.mdpi.com/journal/pharmaceuticals

www.ingramcontent.com/pod-product-compliance
Lightning Source LLC
Chambersburg PA
CBHW061621220926
51850CB00005B/564
*9783036592299*